# Refrigeration
## and
# Air Conditioning

# Refrigeration
# and
# Air Conditioning

Billy C. Langley

Reston Publishing Company, Inc.
*A Prentice-Hall Company*
Reston, Virginia

**Library of Congress Cataloging in Publication Data**

Langley, Billy C
   Refrigeration and air conditioning.

   Includes index.
   1.  Refrigeration and refrigerating machinery.
2.  Air conditioning.  I.  Title.
TP492.L36    621.5′6    77-20218
ISBN 0-8359-6612-7

© 1978 by Reston Publishing Company, Inc.
*A Prentice-Hall Company*
Reston, Virginia 22090

10  9  8  7  6  5  4

Printed in the United States of America

# Contents

# Preface

*Refrigeration and Air Conditioning* is intended to be used as a curriculum guide, a textbook, or a course for independent study. It covers practical fundamentals as well as recommended service procedures, and serves as a comprehensive textbook for the beginning student and a valuable reference for the experienced service technician. The material in the text is organized to accommodate either short survey courses or complete two-year technical programs.

The text consists of 19 chapters, each of which covers a specific area of this exciting industry, including solar energy. Each chapter begins with an introduction to that particular phase of study and advances through recommended service operations where applicable. Safety procedures are integrated into the text as well as summarized separately at the end of each chapter. A summary and questions that cover the minimum material with which a reader should be familiar conclude each chapter. English-metric equivalents are presented throughout to aid the reader in converting to the metric system.

*Refrigeration and Air Conditioning* provides not only the necessary theory for the student and service technician alike, but examples to help reinforce that theory as well. Operating instructions from various manufacturers who provide service and/or maintenance are incorporated wherever practical.

Upon completion of this text, the reader will have the knowledge and confidence necessary to properly install and service refrigeration, air conditioning, and heating equipment.

# Fundamentals
# of
# Refrigeration

The modern way of life depends on refrigeration equipment for the preservation of food and for human comfort. Many industrial and commercial processes, such as printing, textile manufacturing, food storage etc., are heavily dependent upon the proper conditioning of air for efficient, economical operation of the necessary equipment.

The air conditioning and refrigeration field affords many opportunities to those interested in designing, selling, manufacturing, installing, and servicing its equipment. There is a heavy demand for people who are willing to become proficient in any of these areas. But in order to become proficient, the basic principles must be thoroughly understood.

The air conditioning and refrigeration industry encompasses many fields, such as chemistry and physics. Mechanical ability is also of great importance in order to design, install, or service these systems. Customer relations is another important area of competence for success in this field.

## BRIEF HISTORY OF REFRIGERATION

Mold, yeast, and bacteria require warmth in order to grow and multiply. Their growth and multiplication can, therefore, be retarded by cooling the food to a sufficiently low temperature. This reduction in temperature may be done by modern refrigeration equipment. Refrigeration is actually the most valuable method of protecting foods against spoilage. It is the only method which preserves food in its original state and does not appreciably affect its flavor, appearance, or nutritive value.

Before the existence of refrigeration, most food had to be eaten shortly after being obtained. Therefore, people who lived in one area would have difficulty enjoying foods which were plentiful in another area. Also, foods which were plentiful in one season of the year could not be enjoyed in another season.

It was discovered many years ago that

several benefits are derived from cooling food. All food, except frozen foods which are properly prepared and protected, should be kept below 50°F (10°C) * and above 32°F (0°C) to prevent spoilage by either microbic growth or by freezing. The zone between 50°F (10°C) and 32°F (0°C) is recognized as the *food safety zone* and is commonly referred to as *safety zone refrigeration.*

The icebox came into existence when natural ice became available. People who lived in warm, dry sections of the country, however, could not afford natural ice because transportation was expensive. It was also discovered that some of the natural ice was contaminated and could not be used safely.

More than 150 years ago, an English scientist changed ammonia gas to a liquid by applying pressure and lowering the temperature. As the pressure was released, the ammonia liquid boiled off rapidly and changed back to a gas. As the liquid boiled off, heat was absorbed from the surroundings. This discovery proved to be of tremendous importance in the development of modern refrigeration equipment.

The first commercial ice machine was introduced in 1825. The ice produced by the machine was cleaner and purer than natural ice. The ice also could be produced regardless of weather conditions.

With the development of the modern apartment house and the increased demand for more and more cooling, the need for refrigeration equipment became more apparent.

---

* The metric equivalents of the English measuring system will be included in this text where applicable. The metric system is included to aid in the transition from the English system to the metric measuring system.

From the previous discussion we can define refrigeration as *a process of removing heat from an enclosed space or material and maintaining that space or material at a temperature lower than its surroundings.* As heat is removed, a space or material becomes colder. The more heat is removed, the colder the object becomes. Cold, therefore, is a relative term signifying a condition of lower temperature or less heat.

## MOLECULAR THEORY

All matter is made up of molecules and may exist in three states: *solids, liquids,* and *gases.* A molecule is the smallest particle to which a compound can be reduced before it breaks down into its original elements. For instance, water is made up of two elements, hydrogen and oxygen. The movement or vibration of the molecules determines the amount of heat present in a given body. This heat is generated by the friction of these moving molecules. The attraction of these molecules to each other is reduced as the temperature increases.

**SOLIDS:** In solids, the vibrating rate of the molecules is very low. Therefore, the attraction of the molecules to each other is very strong and a solid must have support or it will fall. The force on a solid is upward. (See Figure 1-1(a).)

**LIQUIDS:** In liquids, the vibrating motion of the molecules is faster than in solids. Therefore, the attraction of the molecules to each other is less and a liquid must be kept in a container of some type. A liquid will take the shape of the container (see Figure 1-1(b)). The higher the temperature of the molecules, the faster they vibrate. The warmer molecules will move upward

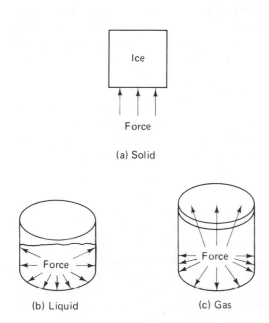

(a) Solid

(b) Liquid      (c) Gas

**Figure 1-1.** Three states of a substance

in the container toward the surface of the liquid because they are less attracted to each other and require more space. Therefore, they become lighter and rise upward.

The force exerted by a liquid is toward the sides and to the bottom of the container. The force will be greater on the bottom than on the sides because of the weight of the liquid. As the surface of the liquid is approached, the force will decrease because of the reduced weight of the liquid.

**GASES:** In gases, the vibrating motion of the molecules is even faster than in liquids. Therefore, the attraction of the molecules to each other is very small and a gas must be kept in a closed container or it will escape into the atmosphere. A gas will take the shape of the container on all sides. (See Figure 1-1(c).) The molecules of a gas have little or no attraction for any substance. Ac-

tually they bounce off each other as well as molecules of other substances.

The force exerted by a gas is equal in all directions. Most gases are lighter than air. Therefore, they tend to float upward causing a weightless condition.

With the proper regulation of temperature any substance can be made to remain in any of the three forms: solid, liquid, or gas. Also, any substance can be made to change from one form to another by the proper use of temperature and pressure. This change in form is known as the *change of state.*

**CHANGE OF STATE:** The addition of heat to a substance may cause, in addition to a rise in the temperature of that substance, a change of state of that substance. That is, an addition of heat may cause a substance to change from a solid to a liquid, or from a liquid to a gas. There are three states of any substance. The states of water are: ice, water, and steam, i.e., solid, liquid, and gas.

**HEAT**

There are two terms that should not be confused. They are heat and temperature. *Heat* is considered as the measure of quantity; *temperature,* the measure of degree or intensity. For instance, if we have a one-gallon container of water and a two-gallon container of water and both are boiling, the two gallons of water will contain twice as much heat as the one gallon, even though they both will be at 212°F (or the same temperature).

Temperature is measured in degrees with a thermometer. Heat is measured in Btu (British Thermal Unit(s)). A Btu is defined as *the amount of heat required to raise the temperature of one pound of pure water one degree Fahrenheit.*

**METHODS OF HEAT TRANSFER:** It is important to know that heat always flows from a warmer object to a colder object. The rate of heat flow depends on the temperature difference between the two objects. For example, consider two objects lying side by side in an insulated box. One of the objects weighs one pound and has a temperature of 400°F (204°C), while the second object weighs 1000 pounds (454.54 kilograms) and has a temperature of 390°F (199°C). The heat content of the larger object will be far greater than that of the smaller object. However, because the temperatures are different, heat will flow from the smaller object to the larger object until their temperatures are the same.

The three ways that heat travels are: (1) conduction, (2) convection, and (3) radiation.

**Conduction** is the flow of heat through an object (see Figure 1-2(a)). For heat transfer to take place between two objects, the objects must touch (Figure 1-2(b)). This is a very efficient method of heat transfer. If you have ever heated one end of a piece of metal and then touched the other end, you felt the heat that had been conducted from the heated end of the metal.

**Convection** is the flow of heat by use of a fluid, either gas or liquid. The fluids most commonly used with this method are air and water. The heated fluids are less dense and rise, while the cooler fluids are more dense and fall, thus creating a continuous movement of the fluid. Another example of convection is the heating furnace. Air is heated in the furnace and blown into a room to heat the objects in the room by convection.

**Radiation** is the transfer of heat by wave motion. These waves can be light waves or radio frequency waves. The form of radiation that we are most familiar with is the sun's rays. When heat is transferred by radiation, the air between the objects is not heated, as can be noticed when a person steps from the shade into the direct sunlight. The air temperature is about the same in either place. However, you feel warmer in the sunlight (Figure 1-3). This is because of the heat being conducted by the rays of the sun. There is little radiation at low temperatures, and at small temperature differences. Therefore, heat transfer by radiation is of little importance in actual refrigeration applications. However, if the refrigerated space is located in the direct rays of the sun, the cabinet will absorb heat. This heat absorption from direct sun rays can be a major factor in the calculation of the heat load of a refrigerated space.

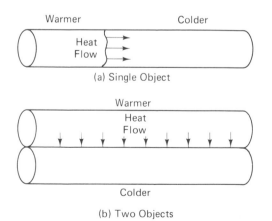

Warmer          Colder

Heat
Flow

(a) Single Object

Warmer

Heat
Flow

Colder

(b) Two Objects

**Figure 1-2.** Heat transfer by conduction

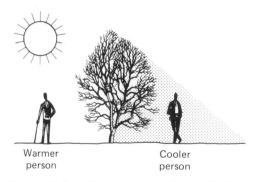

Warmer              Cooler
person               person

**Figure 1-3.** Heat transfer by radiation

Heat will travel in a combination of these processes in a normal refrigeration application. The ability of a piece of refrigeration equipment to transfer heat is known as the *overall rate* of heat transfer. As was learned earlier, heat transfer cannot take place without a temperature difference. Different materials have different abilities to conduct heat. Metal is a good conductor of heat, while asbestos is a poor conductor of heat, and can be used as an insulator.

**Sensible heat** is the heat added to or removed from a substance resulting in a change in temperature, but no change of state. The word *sensible* as applied to heat refers to that heat which can be sensed with a thermometer. An example of sensible heat is when the temperature of water is raised from 42°F to 212°F (5.6 to 100°C). There was a change in temperature of 170°F (94.4°C). This change is sensible heat. It can be measured with a thermometer (see Figure 1-4).

**Latent heat** is the heat added to or removed from a substance during a change of state, but no change in temperature. The types of latent heat are: (1) latent heat of fusion, (2) latent heat of condensation, (3) latent heat of vaporization, and (4) latent heat of sublimation.

*Latent heat of fusion* is the amount of heat which must be added to a solid to change it to a liquid at a constant tempera-

**Figure 1-4.** Sensible heat

ture. Latent heat of fusion is also equal to the heat which must be removed from a liquid to change this liquid to a solid at a constant temperature.

*Latent heat of condensation* is the amount of heat which must be removed from a gas to condense it to a liquid at a constant temperature.

*Latent heat of vaporization* is the amount of heat that must be added to a liquid to change it to a gas at a constant pressure and a constant temperature.

*Latent heat of sublimation* is the amount of heat which must be added to a substance to change it from a solid to a vapor (gas), with no evidence of going through the liquid state. This process is not possible in all substances. The most common example of this is *dry ice*. The latent heat of sublimation is equal to the sum of the latent heat of fusion and the latent heat of vaporization.

Figure 1-5 shows graphically what happens when one pound of ice is heated from 0°F to 220°F (−17.8 to 104°C) and then cooled back to 0°F. The arrows in the figure indicate that the process is reversible. That is, the water may be changed to steam by adding heat, or the steam may be condensed back to water by removing heat.

Let us follow the process starting in the lower lefthand corner. The first step in the process shows that when the ice is heated from 0°F to 32°F (−17.8 to 0°C) 16 Btus (4032 calories) of heat are absorbed by the ice. These 16 Btus are sensible heat because they did not change the state of the ice.

When 144 more Btus (36.28 kilocalories) are added to the ice, the temperature is still 32°F (°C) but the ice has all melted. Since the ice has changed to water—solid to liquid —these 144 Btus are termed latent heat of fusion.

When 180 Btus (45.36 kcal) more are added to the water, the temperature is caused to rise to 212°F (100°C), which is the boil-

**Figure 1-5.** Relationship of heat and temperature when changing ice to steam

ing point of water at sea level and atmospheric pressure. Since these 180 Btus changed the temperature but did not change the state of the water, it is termed sensible heat.

When 970 Btus (244.44 kcal) more are added to the water at 212°F (100°C), it is changed into steam at 212°F. Since there is no change in temperature, but a change of state—liquid to vapor—these 970 Btus are termed latent heat of vaporization.

When 4 Btus (1 kcal) more are added to the steam at 212°F (100°C), the temperature is raised from 212°F to 220°F (100 to 104°C). Since this heat resulted in a change in temperature but no change of state, it is termed sensible heat.

If the arrow is followed backward, it easily can be seen that this process is reversible. Exactly the same amount of heat must be given up by the substance as was absorbed by it.

**Specific heat** is the amount of heat required to raise the temperature of one pound of any substance one degree Fahrenheit. Specific heat is also the ratio between the quantity of heat required to change the tem-

perature of a substance one degree Fahrenheit and the amount of heat required to change an equal amount of water one degree Fahrenheit.

From the definition of a Btu given earlier, it can be seen that the specific heat of water must be one Btu per pound. The specific heat values of some of the more popular foods are given in Table 1-1. If you will note, after foods are frozen their specific heat values drop considerably. It may be assumed that the specific heat is a little more than one half of what it was before the foods were frozen.

**HEAT MEASUREMENT:** As was explained earlier, temperature is the measure of degree or intensity. Heat quantity is the amount of heat present. As the temperature of a body is increased, a definite amount of heat is absorbed. The greater the rise in temperature, the greater the amount of heat absorbed by the body. Heat will flow from one body to another only when there is a temperature difference. In every case the heat will flow from the warmer body to the cooler body. The greater the temperature difference, the faster the flow of heat.

## Table 1-1
### Specific Heat of Foods

| FOOD | SPECIFIC HEAT (UNFROZEN) | | SPECIFIC HEAT (FROZEN) | |
|---|---|---|---|---|
| | BTU | CAL | BTU | CAL |
| Veal | 0.70 | 176.4 | 0.39 | 98.28 |
| Beef | 0.68 | 171.36 | 0.38 | 95.76 |
| Pork | 0.57 | 143.64 | 0.30 | 75.6 |
| Fish | 0.82 | 206.64 | 0.43 | 108.36 |
| Poultry | 0.80 | 201.6 | 0.42 | 105.84 |
| Eggs | 0.76 | 191.52 | 0.40 | 100.8 |
| Butter | 0.55 | 138.6 | 0.33 | 83.16 |
| Cheese | 0.64 | 161.28 | 0.37 | 93.24 |
| Whole Milk | 0.92 | 231.84 | 0.47 | 118.44 |

Water has a greater heat capacity than almost any other substance, and, therefore, is used as a basis in the definition of the Btu. When warming one pound of pure water one degree Fahrenheit, the water absorbs one Btu. When the water is cooled one degree Fahrenheit, it will give up the same amount of heat; one Btu.

**Heat calculation** is done to determine the amount of heat in Btu which is gained or given up by a substance while it is changing temperature.

The calculation of heat gain or loss involves the use of the following formula:

Formula 1-1

$$\text{Btu (cal)} = (T_1 - T_2) \times W \times SH$$

where:

$W$ = weight of the substance in pounds

$SH$ = specific heat of the substance

$T_1$ = original temperature in degrees F (°C)

$T_2$ = final temperature in degrees F (°C)

First, let us apply this formula to a problem in which we must heat water. Notice that when this formula is used for heating the positions of $T_1$ and $T_2$ are reversed. That is, when it is used for heating the formula would read:

Formula 1-2

$$\text{Btu} = (T_2 - T_1) \times W \times SH$$

Example: If we are to heat 60 lbs (27.2169 kg) of water from 50°F (10°C) to 120°F (49°C) how many Btus are required? The specific heat of water is taken as 1.

Use Formula 1-2

$$\begin{aligned}
\text{Btu} &= (T_2 - T_1) \times W \times SH \\
&= (120 - 50) \times 60 \times 1 \\
&= 70 \times 60 \times 1 \\
&= 4200
\end{aligned}$$

$$\begin{aligned} \text{Cal} &= (T_2 - T_1) \times W \times SH \\ &= (49 - 10) \times 27.21 \text{ (kg)} \times 1 \\ &= 39 \times 27.21 \times 1 \\ &= 1.061 \text{ kcal (kilocalories)} \end{aligned}$$

Now let us cool this same 60 lbs of water from 120°F (49°C) to 50°F (10°C) and determine how many Btus (cal) are to be removed. Use Formula 1-1.

Formula 1-1

$$\begin{aligned} \text{Btu} &= (T_1 - T_2) \times W \times SH \\ &= (120 - 50) \times 60 \times 1 \\ &= 70 \times 60 \times 1 \\ &= 4200 \end{aligned}$$

$$\begin{aligned} \text{cal} &= (T_1 - T_2) \times W \times SH \\ &= (49 - 10) \times 27.21 \text{ (kg)} \times 1 \\ &= 39 \times 27.21 \text{ (kg)} \times 1 \\ &= 1.061 \text{ kcal} \end{aligned}$$

Notice particularly that in these two examples the amount of heat involved is the same. In the first case the heat is applied; in the second case the heat is removed.

In refrigeration, the total weight (W) of the material to be stored in the refrigerator must first be calculated or estimated and then multiplied by the number of degrees of difference between the original temperature and the storage temperature. This value must then be multiplied by the specific heat of the items that are being stored. This will give us the number of Btus which must be removed to lower the temperature of the stored items.

Whenever there are different materials or foods to be stored or refrigerated, each item's temperature should be calculated separately and added together. If the quantities of the different materials are approximately equal in weight, the average specific heat may be used in the calculation.

## HEAT OF COMPRESSION

Heat of compression occurs when a gas is compressed mechanically. As the gas is compressed the temperature rises. There is practically no increase in the amount of heat in the gas; only the amount of heat caused by the friction in the compressor is added to the gas. The temperature is raised because the gas molecules are forced closer together causing more friction, which in turn causes the increase in temperature.

In practice the low-temperature, low-pressure gas is drawn from the evaporator in a refrigeration system to the compressor where its volume is reduced and its temperature is increased to a point above the temperature of the air (or water) which cools the condenser. The heat will be removed from the gas by the surrounding air or water, causing a change of state to occur. The gas will be changed to a liquid.

## TEMPERATURE

As stated earlier, temperature is the measure of degree or intensity of heat. The device used to measure temperature is a thermometer. There are two types of thermometers—Fahrenheit and Centigrade. In the United States the Fahrenheit thermometer is most often used, while in Europe the Centigrade is in common usage.

**Fahrenheit thermometers** use 32°F as the freezing temperature of water and 212°F as the boiling temperature of water. There is a difference of 180 divisions between these two points. (See Figure 1-6.) The scale is divided evenly into degrees Fahrenheit. These temperature measurements were proven when taken at sea level because a different atmospheric pressure will change the boiling and freezing temperatures.

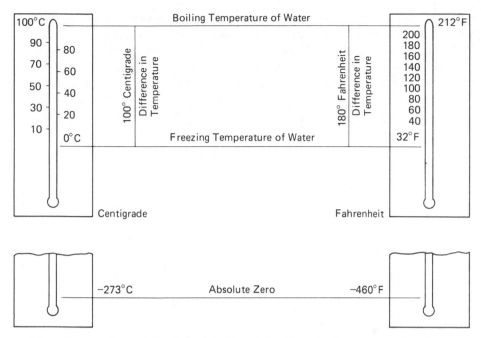

Figure 1-6. Comparison of the Fahrenheit and Centigrade thermometer scales

**Centigrade thermometers** use 0°C as the freezing temperature of water and 100°C as the boiling temperature of water. There is a difference of 100° on the Centigrade thermometer between these two points. The scale is divided into degrees Centigrade.

Measurement of temperature is obtained by placing the thermometer in the area to be tested and reading the scale directly. As Figure 1-6 illustrates, there is a great difference in temperature between any certain number of degrees on the different thermometer scales. It is often necessary when doing laboratory work to change one type of thermometer reading to the other. Since there are 180° difference between the freezing and the boiling temperatures on the Fahrenheit scale and only 100° difference on the Centigrade scale, we can see that a Centigrade degree is 1.8 times larger than the Fahrenheit degree. Because of the recent popularity of metric scales in the United States, it may be necessary to convert from one scale to the other.

Figure 1-7. Refrigeration compound gauge (*Courtesy of Marshalltown Instruments*)

To convert Centigrade degrees to Fahrenheit degrees, multiply the number of Centigrade degrees by 1.8 and then add 32. The 32 is added because the Fahrenheit zero is 32° below the freezing temperature of water.

Formula 1-3

Fahrenheit Temperature = (°C × 1.8) + 32

Example: Change 90°C to Fahrenheit

$$\text{Fahrenheit Temperature} = (°C \times 1.8) + 32$$
$$= (90 \times 1.8) + 32$$
$$= 162 + 32$$
$$= 194°F$$

To convert Fahrenheit degrees to Centigrade degrees, subtract 32 from the Fahrenheit reading and then divide by 1.8.

Formula 1-4

$$\text{Centigrade Temperature} = \frac{°F - 32}{1.8}$$

Example: Change 40°F to Centigrade

$$\text{Centigrade Temperature} = \frac{°F - 32}{1.8}$$
$$= \frac{40 - 32}{1.8}$$
$$= \frac{8}{1.8}$$
$$= 4.4°C$$

Absolute zero is the lowest possible temperature that can be achieved with any substance. This is the temperature at which all molecular movement in a substance stops. On the Fahrenheit scale absolute zero is −460°F. On the Centigrade scale absolute zero is −273°C. It is practically impossible to reach absolute zero.

**Critical temperature** is the highest temperature at which a gas may be liquified, regardless of how much pressure is applied to it. In refrigeration applications, the condensing temperature must be kept below the critical temperature. If the critical temperature is reached in the condenser, the refrigeration process will stop.

**Saturation temperature** is a condition of both pressure and temperature at which both liquid and gas can exist in the same container simultaneously. A saturated liquid or gas is at its boiling temperature. The saturation temperature increases with an increase in pressure and decreases with a decrease in pressure.

## ENTHALPY

The calculations necessary for some refrigeration and air conditioning estimating involve the use of enthalpy. Enthalpy is simply the total heat contained within a body—both sensible and latent. It is measured in Btu per pound (kcal/kg) of that substance. In theory, enthalpy is measured from absolute zero of the particular scale being used, −460°F and −273°C. However, the large figures involved when using absolute zero as a reference point have caused other reference points to be set. The zero reference point for water is 32°F and for refrigerants is −40°F. Once the zero reference point is established, any enthalpy above that point is positive and any enthalpy below that point is negative.

The following examples show how to calculate enthalpy using letters as designations:

$E$ = enthalpy

$W$ = weight of substance

$SH$ = specific heat of substance

$t_2$ = desired temperature

$t_1$ = initial temperature

LH = latent heat of substance

Formula 1-5

$E = W \times SH \, (t_1 - t_2) + LH.$

Formula 1-6

$E = W \times SH \, (t_2 - t_1).$

Example: Calculate the enthalpy of one pound (.454 g) of steam at 212°F (100°C). (Use 32°F (−0°C) as the reference point.) Use Formula 1-5 because latent heat is involved.

$E = W \times SH(t_2 - t_1) + LH$

$= 1 \times 1(212^2 - 32^1) + 970$

$= 1 \times 1(180) + 970$

$- 180 + 970$

$= 1150$ Btu per pound

$E = W \times SH(t_2 - t_1) + LH$

$= .454 \times 1000(100 - 0) + 244.44$ (kcal)

$= 45.400$ (kcal) $+ 244.44$ (kcal)

$= 289.84$ kcal/kg

Example: Calculate the enthalpy of one pound of water at 80°F (26.7°C). (Use 32°F (0°C) as the reference point.) Use Formula 1-6 because the latent heat is not included without a change of state.

$E = W \times SH(t_2 - t_1)$

$= 1 \times 1(80 - 32)$

$= 1 \times 1(48)$

$= 1 \times 48$

$= 48$ Btu per pound

$E = W \times SH(t_2 - t_1)$

$= .454 \times 1000(26.7 - 0)$

$= 454 \times 26.7$

$= 12.122$ kcal

## PRESSURE

The air around us exerts pressure which is called *atmospheric pressure*. All liquids have a definite boiling temperature at atmospheric pressure. If the pressure over a liquid is increased, the boiling temperature also will be increased. If the pressure over a liquid is lowered, the boiling temperature also will be lowered. All liquids have a definite boiling temperature for each pressure. This is one of the basic principles used in refrigeration work.

**ATMOSPHERIC PRESSURE:** The pressure exerted on the earth by the atmosphere above us is called atmospheric pressure. At any given point, this atmospheric pressure is relatively constant except for the changes caused by the weather. As a basic reference for comparison, the atmospheric pressure at sea level has been universally accepted as being 14.7 pounds per square inch (psi). This pressure is equal to that exerted by a column of mercury 29.92 in. (1.032 kg/cm²) high.

The depth of the atmosphere is less at altitudes above sea level, therefore the atmospheric pressure is less on a mountain. For example, at 5000 feet elevation the atmospheric pressure is only 12.2 psi (.857 kg/cm²).

**ABSOLUTE PRESSURE:** The pressure measured from a perfect vacuum is called absolute pressure. The atmospheric pressure and absolute pressure are the same. Atmospheric pressure is 14.7 psia (1.032 kg/cm²). Absolute pressure is normally expressed in terms of pounds per square inch absolute (psia). Absolute pressure is equal to gauge pressure plus atmospheric pressure. To find absolute pressure, add 14.7 to the pressure gauge reading.

**GAUGE PRESSURE** is zero pounds per square inch gauge (psig) when the gauge is not connected to a source of pressure. Pressures below 0 psig are negative gauge readings and are commonly referred to as *inches of vacuum*. Refrigeration compound gauges are calibrated in inches of mercury for readings below atmospheric pressure. (See Figure 1-7.) Since 14.7 psi is equal to 29.92 inches of mercury, 1 psi is equal to 2 inches of mercury:

$$\frac{29.92}{14.7} = 2.03 \text{ inches of mercury}$$

It should be remembered that gauge pressures are only relative to absolute pressures. See Table 1-2.

**Measurement** of low pressures requires a unit of measurement smaller than the pound or the inch of mercury. The *micron* is commonly used for measuring low pressures. A micron is a metric measurement of length and is used in measuring the vacuum in a refrigeration system. It is considered as being absolute pressure.

One micron is equal to 1/1000 of a millimeter. There are 25.4 millimeters in an inch. Therefore, one micron is equal to 1/25400 of an inch. A system that has been evacuated to 500 microns would have an absolute pressure of .02 inches of mercury. At standard conditions this would be equal

to a vacuum reading of 29.90 inches of mercury, which is impossible to read on a refrigeration compound gauge.

**Effects of temperature** on the pressure of gases is of great importance in the refrigeration industry. It must be thoroughly understood before a good knowledge of the refrigeration cycle can be obtained. There are several laws which deal with the effects of temperature on the pressure of a gas within a confined space. These laws are as follows:

**Charles' Law** of gases states that "With the pressure constant the volume of a gas is directly proportional to its absolute temperature." In mathematical form this is stated as:

$$\frac{V_1}{V_2} = \frac{T_1}{T_2} \qquad \text{Formula 1-7}$$

Use the following designations:

$V_1$ = Old volume

$V_2$ = New volume

$T_1$ = Old temperature

$T_2$ = New temperature

In practical applications this can be proven by use of a properly fitted piston within a cylinder (see Figure 1-8(a)). In this example the cylinder is fitted with a sliding piston. The cylinder is full of gas at atmospheric pressure. Heat is applied to the cylinder, causing the temperature to rise. Because the piston is easily moved, the volume of gas increases but the pressure remains constant at atmospheric pressure (Figure 1-8(b)). On the other hand, if the gas is cooled, the volume of gas will become smaller (Figure 1-8(c)). If it were possible

## Table 1-2
## Comparison of Atmospheric and Absolute Pressures
### at Varying Altitudes

| ALTITUDE | PSIA | PRESSURE IN INCHES HG | BOILING POINT OF WATER (°F) | REFRIGERANT BOILING POINTS (°F) R–12 | R–22 | R–502 |
|---|---|---|---|---|---|---|
| 0 ft | 14.7 | 29.92 | 212°F | −22°F | −41°F | −50°F |
| 1000 ft | 14.2 | 28.85 | 210 | −23 | −43 | −51 |
| 2000 ft | 13.7 | 27.82 | 208 | −25 | −44 | −53 |
| 3000 ft | 13.2 | 26.81 | 206 | −26 | −45 | −54 |
| 4000 ft | 12.7 | 25.84 | 205 | −28 | −47 | −56 |
| 5000 ft | 12.2 | 24.89 | 203 | −29 | −48 | −57 |
| | kg/cm$^2$ | cm/Hg | °C | | °C | |
| 0 m | 1.03 | 75.9 | 100 | −30 | −40.5 | −45.6 |
| 304 m | .998 | 73.2 | 99 | −30.5 | −41.7 | −46 |
| 608 m | .96 | 70.6 | 97.44 | −31.7 | −42.2 | −47 |
| 912 m | .93 | 68.1 | 96.3 | −32.2 | −42.8 | −47.8 |
| 1216 m | .89 | 65.6 | 95.8 | −33.3 | −43.9 | −48.9 |
| 1520 m | .857 | 63.2 | 94.7 | −33.9 | −44.4 | −49.5 |

to cool the gas to absolute zero temperature, −460°F (−273.15°C), the volume would be zero because there would be no molecular movement at this temperature.

**Boyle's Law** of gases states that "With the temperature constant, the volume of gas is inversely proportional to its absolute pressure." In mathematical form this is stated as:

$$P_1 V_1 = P_2 V_2 \qquad \text{Formula 1-8}$$

Use the following designations:

$P_1$ = Old pressure

$P_2$ = New pressure

$V_1$ = Old volume

$V_2$ = New volume

As before, this can be proven by use of a cylinder with a properly fitted piston. (See Figure 1-9(a).) By taking simultaneous pressure and volume readings as the gas is slowly compressed within the cylinder so that no temperature increase will be experienced, each side of Eq. 1-8 will always be equal. This is possible because a decrease in volume will always be accompanied by an increase in pressure (Figure 1-9(b)).

**Dalton's Law of Partial Pressures** states that "Gases occupying a common volume each fill that volume and behave as if the other gases were not present." This law, along with the combination of Charles' Law and Boyle's Law, forms the basis for deriving the psychrometric properties of air.

A practical application of Dalton's Law

(a)

(b)

(c)

**Figure 1-8.**  Application of Charles's Law

Old Pressure and Volume

(a)

New Pressure and Volume

(b)

**Figure 1-9.**  Application of Boyle's Law

is: the total pressure in a cylinder of compressed air, which is a mixture of oxygen, nitrogen, water vapor, and carbon dioxide, is found by adding together the pressures exerted by each of the individual gases.

**Pascal's Law** states that "The pressure applied upon a confined fluid is transmitted equally in all directions." A practical application of Pascal's Law is shown with a cylinder of liquid and a properly fitted piston in Figure 1-10. The piston has a cross-sectional area of one square inch. With 100 psig (7.03 kg/cm²) pressure applied to the piston, the pressure gauges show that the pressure exerted in all directions is equal.

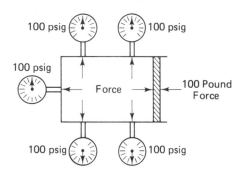

**Figure 1-10.**  Application of Pascal's Law

This is the basic principle used in most hydraulic and pneumatic systems.

**The General Gas Law** is made by combining Boyles' Law and Charles' Law of gases. The general gas law is sometimes expressed as:

Formula 1-9

$$\frac{P_1 V_1}{T_1} = \frac{P_2 V_2}{T_2}$$

A more useful form of the equation is:

Formula 1-10

$$PV = MRT$$

where:

P = absolute pressure of the gas in pounds per square foot

V = volume of the given quantity of gas in cubic feet

M = weight of the given amount of gas in pounds

R = universal gas constant of 1545.3 divided by the molecular weight of the gas.

T = absolute temperature of the gas

The general gas law may be used to study changes in the conditions of a gas as long as absolute temperature and absolute pressure are used. Gauge pressures cannot be used. This law is used in calculating the psychrometric properties of air in air conditioning work.

Example: Calculate the volume of 2 pounds (907.2 grams) of dry air at 20 pounds gauge pressure ($1.406$ kg/cm$^2$) and 80°F (26.7°C).

$$R \text{ (the gas constant for air)} = \frac{\text{universal gas constant}}{\text{molecular weight of the gas}}$$

$$= \frac{1545.3}{28.97}$$

$$= 53.4$$

$$P = 20 + 15 = 35 \times 144$$

$$= 5040$$

Use equation 1-10

$$PV = MRT$$

$$V = \frac{MRT}{P}$$

$$= \frac{2 \text{ lb} \times 53.4 \times (80 + 460)}{(20 + 15) \times 144}$$

$$= \frac{57672}{5040}$$

$$= 11.44 \text{ ft}^3$$

**Pressure-Temperature relationships** are of vital importance in refrigeration equipment design and servicing. The temperature at which a liquid will boil is dependent upon the pressure applied to it, and the pressure at which it will boil is dependent upon its temperature. From this it can be seen that for each pressure exerted on a liquid there is also a definite temperature at which it will boil, providing an uncontaminated liquid is being measured.

In practice, because all liquids react in the same manner, pressure provides us with a convenient means of regulating the temperature inside a refrigerated space. When an evaporator is part of a refrigeration system which is isolated from the atmosphere, pressure can be applied to the inside of the evaporator which is equal to the boiling pressure of the liquid at the desired cooling temperature. The liquid will boil at that temperature and, as long as heat is being absorbed by the liquid, refrigeration is being accomplished.

This process is also reversible. When the pressure over a gas is increased enough to cause the temperature of the gas to be higher than the surrounding medium, heat will be given up and condensation of the gas will occur. This is the principle used in the condenser of a refrigeration system.

**Critical pressure** of a liquid is the pressure at or above which the liquid will remain a liquid. With this pressure applied, the liquid cannot be changed to a vapor by adding heat.

## COOLING

As stated above, cooling is merely the removal of heat from a substance. This cooling can be accomplished in several ways. However, we will discuss only the evaporation and expansion methods at this time.

**EVAPORATION:** The process that causes water which is left in an open container to disappear is called evaporation. Evaporation of water depends on two things—temperature and moisture in the air. When water is left in an open container in the summertime, it dries rapidly because of the high temperature. If the temperature drops, the rate of evaporation will decrease. If the temperature goes up, the rate of evaporation will increase. Even at temperatures of $-40°F$ ($-40°C$) evaporation takes place.

If the air is nearly saturated with moisture it will absorb additional moisture very slowly. If the air is dry, it will absorb moisture very rapidly. During a hot spell when the air is muggy or humid, evaporation of the perspiration from the human body is slow because the saturated air absorbs moisture very slowly.

When water evaporates, a sufficient amount of heat will be absorbed in order to supply the latent heat of vaporization. This heat is absorbed from the water itself or from any object (or air) in contact with the water.

The cold, clammy feeling experienced by a person wearing a wet bathing suit is caused by evaporation of the water and absorption of heat from the material of the suit and from the skin of the person.

**EXPANSION:** The process which causes a cooling effect by compressing a gas, then rapidly reducing the pressure on the gas is called expansion. When a gas is compressed,

heat is generated in an amount equivalent to the amount of work done in compressing the gas. This may be demonstrated by the bicycle pump which gets hot due to the heat developed by compressing air into the tire. The greater the compression, the higher the temperature of the compressed air.

The cooling action that takes place when compressed air is allowed to expand is just the reverse of the compression effect. After a long ride over hot roads, the air in an automobile tire becomes very hot and the pressure within the tire is increased. If the valve stem were opened and the hot air allowed to escape, the air would become cool as it was released due to the expansion and resulting reduction in pressure of the air.

Some of the early refrigeration systems used this principle. The air was compressed and then cooled by passing it through a water-cooled coil. The water was used to absorb and carry away heat from the compressed air.

When the air was allowed to expand, the temperature dropped in relation to the amount of heat that was removed by the water while it was in the compressed state.

There are three steps involved in the expansion refrigeration cycle:

1. Air or gas is compressed to a high pressure.

2. The heat produced by compression is removed.

3. The air or gas is expanded causing a reduction in temperature through the absorption of heat by the air.

## HUMIDITY

The term humidity, in the general sense, refers to the moisture content of the air. The actual amount of moisture, or water vapor, that air can hold depends on the pressure and temperature of the air. As the temperature of the air is lowered it can hold less moisture. As the temperature is increased the amount of moisture it can hold at a given temperature is also increased. This amount of moisture is usually expressed as a percentage figure and is in direct relation to the amount of moisture that the air can absorb at that temperature.

**Figure 1-11.** Air at 100% relative humidity

**DEFINITION:** Humidity is the general term describing the moisture content of the air and can be expressed in two ways—absolute humidity and relative humidity. Humidity also can be defined as the water vapor within a given space.

**Absolute humidity** is all the moisture present in a given quantity of dry air. It can be expressed as grains of water per pound of dry air or as pounds of water vapor per pound of dry air.

**Relative humidity** indicates the amount of moisture, in percentage form, actually present in the air compared to the maximum amount that the air could hold under the same conditions. When air contains all the moisture it can hold at any given temperature, it is saturated. When this condition occurs, the relative humidity is 100%. If the air contains only half of the moisture it can hold at a given temperature, the relative humidity is then 50%.

When air is at the saturation point, 100% relative humidity, and the temperature is lowered further a certain amount of moisture will drop out in the form of water droplets. These droplets appear as ice on the evaporator in a refrigerator or as condensation in an air conditioner. A more common example of this is the water droplets which form on the outside of a glass containing a cold drink. (See Figure 1-11.) As the air surrounding the glass comes in contact with the cold surface, the temperature of the air in contact with the glass is reduced sufficiently to cause condensation. This condensation is in the form of water droplets.

**Effects of humidification** have a tremendous outcome on the operation of refrigeration and air conditioning installations. In refrigeration applications, when the relative humidity is too low, the food will have a tendency to dry out and become useless. If the relative humidity is too high, however, the formation of mold will occur and cause the food to spoil.

In an air conditioned building, when the relative humidity is low, the structure will dry out and cracking will appear. Also, the occupants of the building will feel cold and excessive drafts will be noticed. On the other hand, if the relative humidity is high the occupants will feel warm and clammy. Condensation may appear on the windows during the winter months causing the structure to rot and mildew.

## DEHYDRATION

There has been a great amount of money spent on researching the effects of heat, and noncondensables in a refrigeration system. Even now, many of the effects are still a mystery. We do know that their presence can result in many forms of damage in a refrigeration system such as sludging, corrosion, oil breakdown, carbon formations, and copper plating. These contaminants are usually the cause of compressor failure.

**DEFINITION:** Dehydration, also known as evacuation, is the removal of air, moisture, and noncondensables from a refrigeration system.

**Purposes** of dehydration are to protect the refrigeration system as much as possible from contaminants, and to cause it to operate as efficiently as possible with a minimum amount of equipment failure.

Methods of dehydration are many and varied. As stated above, two methods used to cause a liquid to boil are lowering the pressure exerted on it, and applying heat to the liquid. Some of the ways to eliminate moisture from a refrigeration system by the boiling process are:

1. Move the system to a higher elevation where the ambient temperature is high enough to boil water at the existing pressure.

2. Apply heat to the system, causing the moisture inside it to boil.

3. Use a vacuum pump to lower the pressure inside the refrigeration system so that the ambient temperature will boil the moisture.

In practice, the first two choices are impractical. Therefore, the vacuum pump method is the most desirable means of removing moisture and noncondensables from a system. To accomplish effective dehydration, the refrigeration system must be evacuated to at least 500 microns.

## ENERGY

Energy may be defined as the capacity to do work. Energy may be in the form of *heat* energy, *electrical* energy, or *mechanical* energy. Energy cannot be destroyed. It may, however, be converted from one form to another.

In practice, if we are to remove heat from a product in a refrigerator by mechanical means, the heat which is removed constitutes a load on the machine that is used to remove this heat. It is necessary, therefore, that we have some understanding of elementary physics.

## FORCE

Force is defined as that which tends to, or actually does, produce motion. Force may be exerted in any direction. In the simple definition of force, direction is not considered. The common unit by which force is measured is the pound.

## MOTION

Motion is the movement or constantly changing position of a body. Motion is commonly expressed in feet. The simple definition of motion does not involve the factor of time.

## VELOCITY

Velocity is the rate of change of position, or the speed of movement per unit of time. Velocity measurements are expressed in feet per minute (ft/min) or feet per second (ft/s).

## WORK

Work is the force applied multiplied by the distance through which this force acts. The unit of work is the foot-pound (ft-lb). A foot-pound is the work done by a force of one pound through a distance of one foot.

Formula 1-11

Work = Force × Distance

Example: If we are to move a 20 lb (9.07 kg) weight through a distance of 6 ft (1.8 m), how much work would be done?

$W = F \times D$ $\qquad$ $W = F \times D$

$\quad = 20 \text{ lb} \times 6 \text{ ft}$ $\qquad = 9.07 \text{ kg} \times 1.82 \text{ m}$

$\quad = 120 \text{ ft-lb}$ $\qquad = 16.34 \text{ mkg}$

## POWER

Power is the time rate of doing work. It is work done divided by time. This gives us the amount of work done in a certain amount of time. For example, a weight of 600 lbs (272.16 kg) has been moved through a distance of 10 ft (3.04 m) in 10 min.

Formula 1-12

$$Power = \frac{Work}{Time}$$

$W = F \times D$

$\quad = 272.16 \text{ kg} \times 3.04 \text{ m}$

$\quad = 872.37 \text{ kgm}$

$W = F \times D \quad = 600 \text{ lb} \times 10 \text{ ft}$

$\quad = 6000 \text{ ft-lb}$

then $\quad P = \dfrac{W}{T}$

$\quad = \dfrac{6000 \text{ ft-lb}}{10 \text{ min}}$

$\quad = 600 \text{ ft-lb/min}$

$P = \dfrac{W}{T}$

$\quad = \dfrac{872.37 \text{ kg}}{10 \text{ min}}$

$\quad = 87.237 \text{ kg/min}$

This problem is the same regardless of the direction of the applied force, whether it is applied vertically or horizontally; nor does it matter whether the resistance is caused by friction, gravity, or gas pressure as in the electric refrigeration compressor.

## HORSEPOWER (HP)

In calculation, the foot-pound is much too small a unit for ordinary work. Therefore, a larger unit called the horsepower is used. One hp is equal to 33,000 ft-lbs/min. Notice that time is involved in calculating horsepower.

Formula 1-13

$$Horsepower = \frac{Weight \times Distance}{Time \times 33000}$$

Example: A weight of 10,000 lbs is to be moved 20 ft/min. What is the horsepower required?

$Horsepower = \dfrac{Weight \times Distance}{Time \times 33000}$

$\quad = \dfrac{10000 \times 20}{1 \times 33000}$

$\quad = \dfrac{200000}{33000}$

$\quad = 6.06 \text{ hp}$

## MECHANICAL EQUIVALENT OF HEAT

The mechanical equivalent of heat is the heat produced by the expenditure of a given amount of mechanical energy. This has been determined by accurate scientific experiments. If the heat energy represented by one Btu could be changed to mechanical energy without energy loss, it would represent 778 ft-lbs of work. Thus, we have established a definite relationship between the Btu and the ft-lb. This relationship is represented as: *1 Btu of heat energy is equivalent to 778 ft-lbs.*

1 ft-lb of mechanical energy is equivalent to 1/778 Btu or 0.00128 Btu.

To obtain the heat equivalent in Btu from any number of ft-lbs, divide by 778.

Formula 1-14

$$Btu = \frac{\text{foot-pounds}}{778}$$

To obtain ft-lbs of work done from the heat in Btu, multiply by 778.

Formula 1-15

$$W = Btu \times 778$$

## BASIC REFRIGERATION SYSTEM

The basic components of a refrigeration system are compressor, condenser, flow control device, evaporator, connecting tubing, and the refrigerant. (See Figure 1-12.) The compressor is known as the heart of the system. It causes the refrigerant to circulate in the system. The refrigerant is pushed by the compressor to the condenser where both sensible and latent heat are removed. The liquified refrigerant then goes to the flow control device where the pressure is reduced, allowing the refrigerant to expand and absorb heat from within the refrigerated cabinet. This low-pressure, heat-laden refrigerant vapor is then drawn to the compressor where the cycle is repeated.

The refrigerant systems will be taken up in greater detail in a later chapter.

## SAFETY PROCEDURES

The old saying that "safety doesn't cost, it pays" is certainly true. It is most important that proper safety procedures be followed, especially by workers in the service industry. Since these people usually work alone, help may be difficult to obtain should a

Figure 1-12.   Basic refrigeration system

serious injury occur.

Safety may be divided into three general categories: safety of the worker, safety of the equipment, and safety of the contents.

**Safety of the worker** is, of course, of prime importance. Machinery and materials can always be replaced, but a human life cannot. If the equipment and tools are properly used, there is very little danger to the worker. Always use the leg muscles when lifting heavy objects; never use the back muscles. Ask for help in lifting objects weighing more than 30 to 35 pounds (13.6 kg to 15.9 kg). Keep the floor clear of water and oil. A slip while carrying heavy objects usually means serious injury.

Air conditioning and refrigeration equipment use electricity as their source of power. Be sure that all electrical circuits are disconnected from the power source before working on them. The disconnect switch or the circuit breaker will usually break the circuit. Never work a "hot" circuit.

Always wear safety goggles when working with refrigerants. If refrigerants should get in the eye, a frozen eye will result. When this happens, gently wash the eye with tap water and see a physician as soon as possible. Protect the skin from frostbite when working with refrigerants.

**Safety of the equipment** is also of great concern. Many components on refrigeration and air conditioning equipment are easily broken. Some of these components are very expensive and all of them take time to replace. Use the proper torque and sequence when tightening bolts and nuts. Before operating the equipment be sure that the fan blades and belts are clear of all objects.

**Safety of the contents** is a basic requirement, and will depend to a great extent on the care given to the equipment during installation and servicing. The contents of a given installation must be kept at the desired temperature; this is the responsibility of the service technician. He must know what the required conditions are and maintain the equipment to produce these conditions. Most manufacturers provide tables and charts showing the desired operating conditions and the required control settings.

## SUMMARY

• The growth and multiplication of mold, yeast, and bacteria can be retarded by cooling.

• Refrigeration is the only process that preserves food in its original state.

• All food, except frozen foods, should be kept between 50°F and 35°F (10 and 1.7°C).

• Refrigeration is the process of removing heat from an enclosed space or material and maintaining that space or material at a temperature lower than its surroundings.

• All matter is made up of molecules and may exist in three states—solid, liquid, and gas.

• The movement of molecules determines the amount of heat present in a body.

• In solids, the movement of the molecules is slow.

• In liquids, the molecular movement is faster than in solids.

• In gases, the molecular movement is faster than in either solids or liquids.

• The force exerted on a solid is always upward.

• The force exerted by a liquid is mainly downward.

• The force exerted by a gas is in all directions.

• The addition or removal of a sufficient quantity of heat will result in a change of state of a substance.

• The Btu is the quantity of heat.

• A Btu is the amount of heat required to raise the temperature of one pound of pure water one degree Fahrenheit.

• The three methods of heat transfer are: (1) conduction, (2) convection, and (3) radiation.

• Sensible heat is the heat added to or removed from a substance resulting in a change in temperature but no change of state.

• Latent heat is the heat added to or removed from a substance during a change of state but with no change in temperature.

• Specific heat is the amount of heat required to raise the temperature of one pound of any substance one degree Fahrenheit.

• Heat of compression occurs when a gas is compressed mechanically.

• Temperature is a measure of degree or intensity of heat.

• Critical temperature is the temperature above which a gas will never liquify.

• Fahrenheit thermometers use 32°F as

the freezing temperature of water and 212°F as the boiling temperature of water.

• Centigrade thermometers use 0°C as the freezing temperature of water and 100°C as the boiling temperature of water.

• For conversion from one type of thermometer scale to the other use these formulas:

$$F = (°C \times 1.8) + 32 \qquad C = \frac{°F - 32}{1.8}$$

• At absolute zero there is no molecular movement because no heat is present.

• Enthalpy is the total heat contained within a body.

• Atmospheric pressure is the pressure exerted on the earth by the atmosphere.

• Absolute pressure is the pressure measured from a perfect vacuum.

• To find the absolute pressure add 14.7 to the gauge pressure reading.

• Gauge pressures are relative to absolute pressures.

• One micron is equal to 1/1000 of a millimeter or 1/25400 of an inch.

• Charles' Law of gases states that, "With the pressure constant the volume of gas is directly proportional to its absolute temperature."

• Boyle's Law of gases states that, "With the temperature constant, the volume of gas is inversely proportional to its absolute pressure."

• Dalton's Law of Partial Pressures states that, "Gases occupying a common volume each fill that volume and behave as if the other gases were not present."

• Pascal's Law states that, "The pressure applied upon a confined fluid is transmitted equally in all directions."

• The temperature at which a liquid will boil is dependent upon the pressure applied to it.

• The critical pressure of a liquid is the pressure at or above which the liquid will remain a liquid.

• Evaporation of water depends on the temperature of the air and the amount of moisture in the air.

• Expansion is the process which causes a cooling effect by compressing a gas, then rapidly reducing the pressure on the gas.

• Humidity is the term generally used when referring to the moisture in the air; the actual amount of moisture the air can hold depends upon the pressure and temperature of the air.

• Absolute humidity is all the moisture present in a given quantity of dry air.

• Relative humidity indicates the amount of moisture, in percentage form, actually present in the air compared to the maximum amount that the air could hold under the same conditions.

• Dehydration, also known as evacuation, is the removal of air, moisture, and noncondensables from a refrigeration system.

• Dehydration is best accomplished by use of a vacuum pump.

• Effective dehydration requires a vacuum of at least 500 microns.

• Force is that which tends, or actually does, produce motion.

• Motion is the movement or constantly changing position of a body.

• Velocity is the rate of change of position, or the speed of movement per unit of time, measured in feet per second.

• Work is the force applied multiplied by the distance through which this force acts.

• Power is the rate of doing work. One horsepower is equal to 33,000 foot-pounds per minute.

• The mechanical equivalent of heat is the heat produced by the expenditure of a given amount of mechanical energy.

• The six components of a basic refrigeration system are: compressor, condenser, flow control device, evaporator, connecting tubing, and refrigerant.

## REVIEW EXERCISES

1. In what three states does matter exist?

2. In what direction is the force exerted by a gas?

3. What causes a change of state of a substance?

4. How is heat measured?

5. What are the three methods of heat transfer?

6. Define sensible heat.

7. Name the heat that is added to or removed from a substance during a change of state but no change of temperature.

8. Define specific heat.

9. When does the heat of compression occur?

10. How is temperature measured?

11. What is the unit of measurement used to measure heat?

12. Define critical temperature.

13. What happens to the molecular structure of a substance at absolute temperature?

14. What is the term applied to the total heat contained within a body?

15. What is atmospheric pressure at sea level?

16. Give the formula for obtaining absolute pressure.

17. If the pressure is reduced over a liquid, what will happen to its boiling temperature?

18. On what two things does the evaporation process depend?

19. Briefly define the cooling by expansion process.

20. How is relative humidity expressed?

21. What is the purpose of dehydration?

22. What is the most practical method of dehydration?

23. Convert 32°F to Centigrade.

24. Convert 25°C to Fahrenheit.

25. What are the six important parts of a basic refrigeration system?

26. Define refrigeration.

# Compression Refrigeration Systems

There are two types of refrigeration systems in use today. The *compression system,* which is the more popular of the two, and the *absorption system,* which is used in highly specialized applications. The compression system is sometimes called a mechanical system. We will deal only with the compression system in this text.

As was stated earlier, refrigeration is the process of removing heat from an enclosed space or material and keeping that space or material at a temperature lower than its surroundings.

Food is refrigerated in an ordinary icebox because the ice melts. When the ice melts it absorbs the latent heat of fusion from the foods which are stored in the refrigerator. Heat is also absorbed through the walls of the icebox. In this operation heat is taken from a warm body and space and added to the cooler body—the ice.

A liquid may be made to boil by the application of heat. If a vacuum is created above the liquid, it may be made to boil at a lower temperature. If the liquid, or refrigerant in our case, exists as a gas at atmospheric pressure and temperature, it will immediately begin to boil if it is left in an open container, absorbing heat from its surroundings and any material which is in contact with it.

Consequently any liquid, such as a refrigerant, that will boil at a temperature below the freezing temperature of water will make ice cubes and keep foods cool in a mechanical refrigerator.

## REFRIGERATION BY EVAPORATION

When we think of anything boiling, we naturally think of it as being hot. However, that is not true in all cases. Just because water boils at 212°F (100°C), does not mean that all other substances will boil at the same temperature.

The simplest way to produce refrigeration by evaporation is to place a bottle containing a refrigerant, such as R–12, inside

**Figure 2-1.** Refrigeration by evaporation

an insulated cabinet with an opening at the top. (See Figure 2-1.) The liquid R–12 immediately begins to absorb heat from its surroundings. When the liquid has absorbed enough heat to raise its temperature above −21.66°F (−30°C), it starts to boil vigorously. The temperature of the liquid remains at −21.66°F because it is absorbing the latent heat of vaporization as it changes to a gas. These things take place at a constant temperature of −21.66°F and an atmospheric pressure of 14.7 psig (1.03 kg/cm²).

This process will continue as long as the liquid R–12 in the bottle lasts, and the bottle is left open so the R–12 will evaporate. The liquid refrigerant also continues to absorb heat from the inside of the box, lowering the temperature inside the box. The arrows in Figure 2-1 indicate the circulation of the air inside the box. Notice that the air circulation is exactly the same as in an ordinary icebox. The air being cooled by the refrigerant container drops because of its own weight. The air which is warmed at the sides of the box by the heat passing through the insulation rises.

We can apply these basic principles to

the development of an elementary refrigerator cabinet as illustrated in Figure 2-2. Notice that the cabinet has no machinery connected to it. It merely contains a bottle of liquid refrigerant with a vent that allows the vapors resulting from the boiling of the liquid refrigerant to escape. As the boiling liquid absorbs heat from the inside of the refrigerator, it lowers the temperature. Even though refrigeration is accomplished, this process is not practical because refrigeration

**Figure 2-2.** Elementary refrigerator

will stop when all the liquid refrigerant in the bottle has evaporated. Furthermore, the expense and trouble of always supplying new liquid refrigerant makes this unit too costly and bothersome for practical use.

## COMPRESSION SYSTEM PRINCIPLES

To eliminate these objections, some means must be provided to capture and re-use the refrigerant vapors. This is where we can put pressure to good use in a compression refrigeration system. With pressure, we can compress these refrigerant vapors, thereby concentrating the heat they contain. When heat is concentrated in a vapor, we increase the intensity, or the degree, of the heat. In short, we increase the temperature because temperature is only a measurement of heat intensity. The most important part is that we have made the vapors hotter without actually adding any additional quantity of heat. (See Figure 2-3.)

**Figure 2-3.** Compression principle (*Courtesy of Frigidaire Division, General Motors Corp.*)

Now we have covered the scientific ground rules which apply to mechanical refrigeration. They are probably still a little hazy;

however, it is easy enough to remember these main points:

1.   All liquids absorb a lot of heat without getting any warmer when they boil into a vapor.

2.   Pressure can be used to make vapors condense back into liquid so they can be used again. With that amount of scientific knowledge, we can build a refrigerator.

## ELEMENTARY COMPRESSION REFRIGERATION SYSTEM

A bottle of liquid refrigerant can be placed in an insulated box. We know that it will boil and absorb heat from the box and its contents because it boils at such a low temperature at atmospheric pressure.

The vapors which result because of the boiling liquid can be piped to the outside of the cabinet, providing a way to carry the heat outside the cabinet (see Figure 2-4). When the heat-laden vapor reaches the outside, we can compress it with a pump. When enough pressure is applied, we can squeeze the heat out of the cold vapor even in a warm room. An ordinary radiator will help to get rid of the heat.

By removing the heat and making the refrigerant vapor into a liquid, it is the same as it was in the bottle. We can now run another pipe to the inside of the box and return the refrigerant to the bottle to be used over again.

## THE BASIC REFRIGERATION CYCLE

Mechanical refrigeration is achieved by constantly circulating, evaporating, and condensing a fixed supply of refrigerant inside a closed system. Evaporation occurs at a low pressure and low temperature. Condensation

Refrigerant as a Liquid

Refrigerant as a Vapor

Pump

Refrigerant laden with heat from the milk bottle.

Heat traveling from a warm object to a colder object.

Heat from milk bottle being discharged into surrounding air by condenser.

Condenser

Refrigerant on its way to pick up more heat from milk bottle.

**Figure 2-4.** Elementary compression refrigeration system (*Courtesy of Frigidaire Division, General Motors Corp.*)

occurs at a high temperature and high pressure. Thus it is possible to transfer heat from an area of low temperature (refrigerator cabinet) to an area of high temperature (the kitchen).

If we begin the refrigeration cycle at the evaporator inlet, (1) in Figure 2-5, the low pressure liquid expands, absorbs heat, and evaporates, changing to a low-pressure gas at the outlet of the evaporator (2).

The compressor (3) pumps this gas from the evaporator and increases its pressure. The high-pressure gas is then discharged to the condenser (4). In the condenser the heat is removed from the gas causing it to condense and become a warm, high-pressure liquid.

The next device encountered by the refrigerant as it travels through the system is the liquid line strainer/drier (5) which prevents plugging of the flow control device by

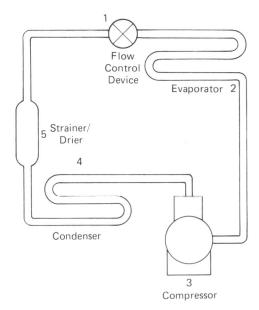

Flow Control Device

Evaporator 2

Strainer/ Drier

Condenser

Compressor

**Figure 2-5.** Compression refrigeration system

trapping scale, dirt, and moisture. The flow of refrigerant into the evaporator is controlled by a pressure differential across the flow control device.

As the warm, high-pressure liquid refrigerant enters the evaporator it is subjected to a much lower pressure due to the combined suction of the compressor and the pressure drop across the flow control device. Thus, the refrigerant tends to expand and evaporate. In order to evaporate, the liquid must absorb heat from the air passing over the evaporator.

Eventually the desired temperature is reached and a temperature control will stop the compressor. As the temperature of the air passing over the evaporator rises, the temperature control starts the compressor again. After the compressor starts, the cycle is continued.

All refrigeration systems operate at two definite pressure levels. The dividing line passes through the compressor discharge valve on one end and the orifice of the flow control device on the other end. (See Figure 2-6.) The compressor discharge valve and the orifice of the flow control device are the dividing points between the high side and the low side of the system.

The high side of the system includes all the components operating at or above the condensing pressure. In the diagram, the high side includes the discharge side of the compressor, the discharge service valve, condenser, liquid receiver, liquid line shut-off valve, dryer, liquid line strainer, and the liquid line. In practice, the complete compressor is considered as being on the high side of the system.

The low side of the system includes all components operating at or below the evaporating pressure. In the diagram, the low pressure side includes the low pressure side of the flow control device, evaporator, suc-

**Figure 2-6.** High and low side diagram

tion line, suction service shut-off valve, compressor crankcase, and the piston and suction valves of the compressor. In practice, the flow control device is considered as being on the low side of the system.

Although there are many types of compression refrigeration systems, the cycle in all of them is the same. We can summarize the cycle of a compression refrigeration system as follows:

1.   The refrigerant gas is compressed by the compressor and discharged into the condenser.

2.   The compressed gas is cooled and condensed to a liquid in the condenser.

3.   The liquid refrigerant is passed through the flow control device to the evaporator or cooling coil.

4.   Due to the reduced pressure in the evaporator, vaporization takes place. Refrigeration results from the absorption of heat during vaporization in the cooling coils.

5. The gas is drawn into the compressor and the cycle is repeated.

It is extremely important to analyze completely every system and understand the intended function of each component before attempting to determine the cause of a malfunction or failure.

## TYPES OF REFRIGERATION SYSTEMS

While there are five general types of refrigeration systems which are based on the type of flow control device, there are two more classifications based on the condition of the refrigerant in the evaporator. These two latter classifications are the *flooded* and *dry* systems.

**FLOODED SYSTEMS:** Flooded systems operate with a definite liquid refrigerant level in the evaporator. This liquid refrigerant level is maintained in the evaporator through the action of the refrigerant flow control device as shown in Figure 2-7. There are several advantages of the flooded system over the dry system. A few of these advantages are: higher efficiency (system operates at higher average suction pressures with resultant shorter operating time), lower operating costs, less cycling (starts and stops), higher rate of heat transfer, and closer control of temperature.

**Figure 2-7.** Flooded system evaporator

More liquid on the low pressure side of the system, as in the flooded system, provides a greater area of wetted surface and allows a higher rate of heat transfer through the evaporator walls and tubing.

The refrigerant flow control devices which provide a flooded evaporator are the *low-side float, high-side float* and the *capillary tube* or *restrictor*.

**DRY SYSTEMS:** Dry systems have an almost all refrigerant vapor condition in the evaporator. The flow control is accomplished by either an automatic expansion valve or a thermostatic expansion valve. The refrigerant passing through the valve is partially evaporated immediately after passing the orifice. The fine, suspended droplets of liquid refrigerant are completely evaporated as they flow through the balance of the evaporator. (See Figure 2-8.) Dry type evaporators are generally made from one continuous length of tubing. The continuous length of tubing insures better control of the evaporation of the liquid refrigerant when the flow control device is properly adjusted.

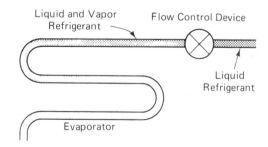

**Figure 2-8.** Dry system evaporator

**LOW SIDE FLOAT SYSTEMS:** Low side float systems have a pool of liquid refrigerant in the evaporator as shown in Figure 2-9. In operation, the refrigerant gas is compressed by the compressor. From the compressor the high-temperature, high-pressure

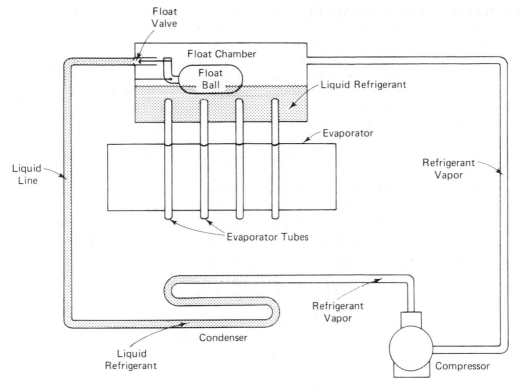

Float
Valve

Float Chamber

Float
Ball

Liquid Refrigerant

Evaporator

Liquid
Line

Refrigerant
Vapor

Evaporator Tubes

Refrigerant
Vapor

Condenser

Liquid
Refrigerant

Compressor

**Figure 2-9.** Low side float system

gas passes directly to the condenser where it is cooled and liquified, changing to a warm, high-pressure liquid. The warm, high-pressure liquid then flows into the liquid receiver. The warm, high-pressure gas fills the upper portion of the liquid receiver. The warm, high-pressure liquid sinks to the bottom. The pressure of the gas forces the warm, high-pressure liquid through the liquid line to the seat of the low side float.

As the level of the low-pressure, low-temperature liquid in the float chamber falls, the float also lowers and pulls the needle from the seat of the valve. This permits the warm, high-pressure liquid to flow into the float chamber where the pressure and temperature are reduced. The purpose of the float is to maintain a constant level of liquid

refrigerant in the evaporator. The liquid refrigerant fills the tubes of the evaporator and a part of the float chamber. As heat is absorbed by the liquid refrigerant, evaporation occurs and the resulting vapor collects in the upper portion of the float chamber. The low-pressure gas is then drawn through the suction line to the compressor and the cycle is repeated.

The refrigerant charge used on the low-side float flooded system is not critical because any excess or overcharge of refrigerant will remain in the liquid receiver. This is an important point to remember because there are other systems in which the refrigerant charge must be accurate to within a few ounces of the recommended charge or operating troubles may result.

**HIGH SIDE FLOAT SYSTEMS:** High side float systems have a majority of the liquid refrigerant on the low-pressure side of the system. The running cycle begins with the refrigerant being compressed in the compressor cylinder. It then passes through the compressor discharge valve, located above the cylinder, to the compressor discharge chamber. The high-pressure, high-temperature gas then flows to the condenser where it is cooled and liquified, changing to a warm, high-pressure liquid. The warm, high-pressure liquid then flows to the liquid receiver where it falls to the bottom and any warm, high-pressure gas present rises to the top. As the liquid level in the bottom of the receiver rises, it causes the float to rise also as illustrated in Figure 2-10. As the float rises, it pulls the needle from its seat, opening the valve which allows the warm, high-pressure refrigerant to flow into the liquid line. In this system the high side float is located inside and is a part of the receiver. There are other designs of the high-side float system that have the float located in a small, separate chamber of its own away from the receiver.

At the inlet of the evaporator the warm, high-pressure liquid flows through a pressure-reducing valve which is called a *liquid*

**Figure 2-10.**   High side float system

*temperature valve.* This valve is made of a weighted needle valve. When the pressure of the warm, high-pressure liquid is great enough to overcome the weight, it will force the valve open. The liquid temperature valve will remain open until the pressure of the warm, high-pressure liquid in the liquid line drops below the pressure required to keep the valve open. The weight will then close the valve, stopping the flow of refrigerant, and preventing moisture condensation on the suction line. The refrigerant pressure is reduced further inside the evaporator. As the liquid refrigerant flows down into the evaporator coils, evaporation occurs. The low-pressure liquid absorbs heat from its surroundings and changes to a low-pressure gas. The low-pressure gas is then drawn through the suction line to the compressor where the cycle is repeated.

In the high side float system, the flow of liquid refrigerant is controlled by both the high side float in the liquid receiver and the liquid temperature valve at the evaporator. The liquid temperature valve automatically regulates the flow of the refrigerant from the liquid line into the evaporator. The valve is caused to open when the difference in pressure between the refrigerant in the liquid line and the refrigerant in the evaporator reaches a certain point. The valve immediately closes again as soon as the difference in pressure is reduced sufficiently.

The refrigerant charge in the high-side float system is very critical because the high side float in the receiver opens and liquid refrigerant is pushed into the liquid line when the liquid refrigerant level in the receiver reaches a certain point. This causes any excess or overcharge of refrigerant in the system to pass over to the evaporator. Any excess of refrigerant would then raise the liquid refrigerant level in the evaporator, which could possibly result in *floodback*. Floodback occurs when liquid refrigerant

flows into the suction line causing frost to appear and possible damage to the compressor. If the refrigerant charge is insufficient, the liquid refrigerant level in the receiver may never rise enough to cause the float valve to open and allow refrigerant to pass into the liquid line. This condition would result in a starved evaporator and a loss of refrigeration. Therefore, the refrigerant charge in the high side float system must be kept within a few ounces of the manufacturer's recommended charge.

**CAPILLARY TUBE OR RESTRICTOR SYSTEMS:** Capillary tube systems continue to be more widely adopted by manufacturers for both refrigeration and air conditioning applications because of their low cost, simplicity, and trouble-free operation. They also lend themselves more readily to hermetically sealed type systems.

The evaporator in the system using this type of refrigerant flow control is also a flooded type. The major difference is the way in which the refrigerant is fed into the evaporator. (See Figure 2-11.) In this type of system, the capillary tube is between the filter and the evaporator. The liquid refrigerant is fed into the evaporator through a small tube. The amount of refrigerant allowed to flow is dependent mainly upon the length and inside diameter of the tube.

Capillary tube systems require an accurate charge of refrigerant for several reasons. First, any overcharge of refrigerant will be in the low side of the system resulting in floodback. Second, when the compressor stops, all the liquid refrigerant on the high side of the system tends to run over into the evaporator until the pressures balance throughout the entire system. This indicates that if there is an overcharge of refrigerant in the system, the liquid will flood the evaporator on the off cycle and cause floodback when the compressor starts the next cycle.

**Figure 2-11.** Capillary tube system

**AUTOMATIC EXPANSION VALVE SYSTEMS:** These systems use a pressure-operated refrigerant flow control device. The purpose of the expansion valve is to control the flow of refrigerant into the evaporator. When this control is properly adjusted it will keep the evaporator fully refrigerated.

In operation, the refrigerant gas is compressed in the compressor where the temperature and pressure are both raised. From the compressor the high-temperature, high-pressure gas passes to the condenser where the gas is cooled and liquified, changing to

a warm, high-pressure liquid. The warm, high-pressure liquid then flows to the liquid receiver. The warm, high-pressure gas fills the top part of the receiver and the warm, high-pressure liquid falls to the bottom. The pressure of the gas forces the warm, high-pressure liquid through the liquid line to the expansion valve. (See Figure 2-12.) The warm, high-pressure liquid passes through the automatic expansion valve into the evaporator. As the refrigerant passes through the expansion valve, both the pressure and temperature are reduced and the liquid re-

**Figure 2-12.** Automatic expansion valve system

frigerant is *atomized,* broken into fine drop-lets of liquid. The expansion valve regulates the flow of refrigerant to the evaporator in response to the refrigerant pressure in the evaporator. In the evaporator, the low-temperature, low-pressure liquid refrigerant changes to low-temperature, low-pressure gas as heat from the surroundings is absorbed. Notice that the liquid refrigerant does not collect in a pool inside the evaporator on a dry type system. Instead, it passes through the evaporator as a mist and vapor combination. As compared to the flooded type systems, there is relatively little liquid refrigerant in the evaporator. The low-pressure gas is drawn through the suction line to the compressor and the cycle is repeated.

**THERMOSTATIC EXPANSION VALVE SYSTEMS:** These systems also are included in the dry type system classification. Thermostatic control is used in combination with evaporator pressure, as in the automatic expansion valve systems, to provide more accurate control of the flow of refrigerant into the evaporator. (See Figure 2-13.) While

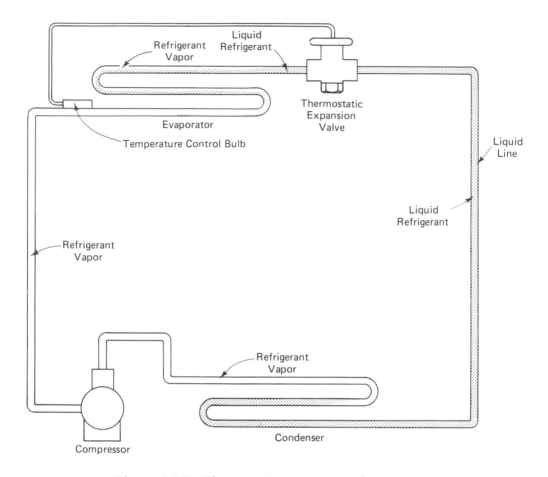

**Figure 2-13.**   Thermostatic expansion valve system

the evaporator operates basically as a dry system with the thermostatic expansion valve, control of refrigerant flow is much more positive and the operating efficiency is improved, closely paralleling that of the flooded system.

Operation of the thermostatic expansion valve is the same as the automatic expansion valve system, with one exception. The evaporator outlet temperature is used along with evaporator pressure. This additional control is accomplished by placing a temperature-sensitive bulb at or near the evaporator outlet.

**SAFETY PROCEDURES**

It should be remembered that when working with refrigeration equipment some of the components are very hot and will cause burns, while other components are very cold and can cause frostbite. Be sure to exercise caution when working near these components. Never put the tongue to an evaporator to determine if it is cold. Always use caution when working around rotating parts to prevent clothing and limbs from being caught by them. Be sure to prevent moisture from entering the system.

## SUMMARY

• The two types of systems in use today are the compression system and the absorption system.

• Food is cooled in an ordinary icebox because the ice absorbs heat while melting.

• Any liquid which will boil at a temperature below the freezing temperature of water will keep food cool and make ice cubes.

• The simplest way to produce refrigeration is to place an open container of refrigerant inside an insulated cabinet and allow the refrigerant to escape.

• The boiling temperature of R–12 at 14.7 psig (1.03 kg/cm²) is −21.66°F (−30°C). The temperature of the evaporator will cause air to circulate inside an insulated cabinet.

• A compressor increases the pressure of a vapor, which increases the temperature of the gas.

• Only the heat of compression is added to the gas during the compression process.

• The basic points to remember about a refrigeration system are: (1) all liquids absorb a lot of heat without getting any warmer when they boil into a vapor, and (2) pressure can be used to make vapors condense back into a liquid.

• Mechanical refrigeration is achieved by constantly circulating, evaporating, and condensing a fixed supply of refrigerant inside a closed system.

• It is possible to transfer heat from a cooler area to a warmer area because heat absorption takes place at a low pressure and low temperature, while condensation occurs at a high pressure and high temperature.

• In a basic refrigeration system, the compressor draws the refrigerant from the evaporator, increases its pressure and temperature, and discharges it to the condenser. The gas is cooled and changed to a liquid in the condenser. From the condenser, the refrigerant flows to the flow control device where the refrigerant flow to the evaporator is controlled. On passing through the flow control device, the pressure of the refrigerant is lowered. The refrigerant absorbs heat in the evaporator and boils into a gas which is removed from the evaporator by the compressor and the cycle is repeated.

• The dividing points between the high side and the low side of the system are the compressor discharge valve and the orifice of the flow control device.

• The high side of the system includes all the components operating at or above condensing pressure.

• The low side of the system includes all components operating at or below the evaporating pressure.

• The flooded and dry systems are classified based on the condition of the refrigerant in the evaporator.

• Flooded systems have a pool of liquid refrigerant in the evaporator.

• Dry systems have only droplets of liquid in the evaporator.

• Flooded systems use the low side float, high side float, and capillary tube or restrictor as their flow control devices.

• Dry type systems use the automatic expansion valve and the thermostatic expansion valve as their flow control devices.

## REVIEW EXERCISES

1. What method, other than the application of heat, can be used to cause a liquid to boil?

2. Why does the air circulate inside a refrigerated cabinet which does not have a fan in it?

3. How is the heat concentrated in a refrigerant gas?

4. Is any heat actually added to a refrigerant vapor during the compression process?

5. What are the two main points to be remembered as far as basic ground rules are concerned about the refrigerant process?

6. Why do vapors form inside the evaporator of a refrigeration system?

7. What two things make it possible for us to transfer heat from a cooler area to a hotter area?

8. What does the compressor in a refrigeration system do?

9. What happens to the refrigerant in the condenser?

10. What happens to the refrigerant in the evaporator?

11. What are the dividing points between the high side and the low side of a refrigeration system?

12. What components are located in the high side of the refrigeration system?

13. Summarize the cycle of a compression system.

14. Name the two classifications of refrigeration systems which consider the condition of the refrigerant in the evaporator.

15. Name the five general types of flow control devices used on compression refrigeration systems.

16. What does the term *critical charge* mean?

17. What are the types of flow control devices that require a critical charge?

18. What is the difference between the automatic expansion valve and the thermostatic expansion valve?

19. Does the automatic expansion valve or the thermostatic expansion valve provide better control of the refrigerant?

20. Are the expansion valve systems considered flooded or dry type systems?

# Refrigeration Materials
# and
# Hand Tools

## PIPING AND TUBING

A complete refrigeration unit consists of several parts which function to vaporize and condense the refrigerant as it circulates through the system. When these parts are located at a distance from each other where there can be no direct connection from one to the other, the refrigerant must have some means of travel. This is accomplished by the use of piping and tubing which is only a hollow bar used to transfer liquids and gases from one part of the system to another. There is no definite ruling which defines the difference between piping and tubing. However, piping is ordinarily understood to have walls of considerable thickness, while tubing has comparatively thin walls.

**WELDED AND BRAZED PIPE AND TUBING** is made from several different metals. Copper, brass, iron, and steel are the most popular for use in refrigeration and air conditioning installations. Pipe which is made from iron or steel is formed into a cylindrical shape from a flat strip so that the two edges meet and are butt-welded together. Sometimes, the edges are beveled and lap-welded together. The lap-welded tubing is stronger and better adapted for bending. Copper and brass pipe may be made in a similar manner. However, instead of welding, a process called brazing is used to join the edges. The brazed type of pipe or tubing is not suitable for use when bending is required.

**SEAMLESS PIPE AND TUBING** is a variety of pipe and tubing which is better for refrigeration use than the welded or brazed types. The piping and tubing which are used in refrigeration installations are generally of the seamless type made by forcing a solid bar of hot metal over a mandrel. This is known as the extrusion process and the tubing has no seam. This type of pipe and tubing is much stronger and better adapted to bending and forming than the welded or brazed types.

## SEAMLESS COPPER TUBING

This type of tubing is more commonly used on refrigeration systems using refrigerants other than ammonia. Copper is never used when ammonia is the desired refrigerant because it will quickly corrode and disintegrate if left in contact with the gas any length of time. It is used successfully with all of the freon group of refrigerants.* The heat conductivity of copper is the greatest of all the available commercial metals. It is quite difficult to thread copper tubing because of its softness. Therefore, soldered connections which make use of the maximum strength of the tube in construction of the joint are used. In addition to its good conductivity of heat, soft copper tubing is easy to bend which eliminates many elbows and joints.

Copper tubing is available in either soft-drawn or hard-drawn types. Soft-drawn copper tubing is available in 25-, 50-, and 100-ft (7.6, 15.2, and 30.4 m) rolls of ⅛ inch to 1⅜ inch (3.175 mm to 34.93 mm) outside diameter (OD). The most popular length is the 50-ft (15.2 m) roll as shown in Figure 3-1. Soft-drawn copper tubing also is available in straight lengths in larger sizes. Soft-drawn tubing of ⅛ inch and smaller is easy to work with because it is fairly flexible. It can be formed by hand. However, care must be taken to avoid kinks in the tubing which will restrict the flow of refrigerant. Soft-drawn copper may be joined by use of soft soldering, silver soldering, flaring techniques, and epoxy procedures.

Hard-drawn tubing is available in 20-ft (6.08 m) lengths and from ¼ inch to 6⅛

**Figure 3-1.** Roll of soft-drawn copper tubing

inches (6.5 mm to 155.57 mm) in OD. This tubing cannot be easily formed by hand. It may kink or collapse resulting in a restriction which must be removed from the system. Minor bends can be achieved with this tubing, however, if the area to be bent is heated to a cherry red and then allowed to cool in the air before bending is attempted. This process is known as *annealing*. Hard-drawn tubing is especially useful when long lengths, which cannot be supported must be used. It is also used when the appearance of the installation is important. Annealing also is required if flare type fittings are to be used. Hard-drawn copper may be joined by use of soft soldering, silver soldering, and epoxy methods.

**SAFE WORKING PRESSURES** of copper tubing will vary with the size of tubing and the temperature of the material inside the

---

* Freon is a trade name for a family of synthetic chemical refrigerants manufactured by DuPont.

## Table 3-1
## Safe Working Internal Pressures of Soft-Drawn ACR Tubing

| TUBE OD | WALL THICKNESS | WT PER FOOT | 150°F PSI | 250°F PSI | 350°F PSI | 400°F PSI |
|---------|---------------|-------------|-----------|-----------|-----------|-----------|
| ¼ | .030 | .0804 | 1230 | 1130 | 970 | 720 |
| ⅜ | .032 | .134 | 860 | 790 | 670 | 500 |
| ½ | .032 | .182 | 630 | 580 | 490 | 370 |
| ⅝ | .035 | .251 | 540 | 500 | 430 | 320 |
| ¾ | .035 | .305 | 440 | 400 | 350 | 260 |
| ⅞ | .045 | .455 | 500 | 460 | 390 | 300 |
| 1⅛ | .050 | .655 | 430 | 400 | 340 | 250 |
| 1⅜ | .055 | .884 | 390 | 360 | 300 | 230 |
| 1⅝ | .060 | 1.140 | 370 | 340 | 280 | 220 |

tube. (See Table 3-1.) When the temperature of the material (refrigerant in this case) increases, the safe working pressure of the tube decreases.

The safe working pressures for hard-drawn copper tubing are different from those for soft drawn tubing as shown in Table 3-2. Care must be used to ensure that these pressures are not exceeded.

**Classification** of seamless copper tubing is made in four general groups as follows:

1. S.A.E. soft tubing, thin wall.

2. Type K soft or hard tubing, extra thick wall.

3. Type L soft or hard tubing, thick wall.

4. Type M soft or hard tubing, standard wall.

The S.A.E. and type L are the most popular for use in refrigeration work and are usually designated as air conditioning and refrigeration (ACR) tubing. Type M tubing is used for condensate drain lines and uses where the operating pressures do not exceed 150 psig (10.55 kg/cm$^2$).

ACR tubing is cleaned, dried, and capped to prevent contamination by moisture and other foreign particles in accordance with ASTM 3280 and ANSI-B9.1-1971 refrigeration industry standards. When a piece of tubing is not used it should be sealed to keep it clean and dry for later use on refrigeration systems. In addition to cleaning and dehydrating, hard-drawn tubing is pressurized with dry nitrogen and reusable plugs are inserted in the ends. (See Figure 3-2).

Table 3-2

Safe Working Internal Pressures of Hard-Drawn ACR Tubing

| TUBE OD | WALL THICKNESS | WT PER FOOT | 150°F PSI | 250°F PSI | 350°F PSI | 400°F PSI |
|---|---|---|---|---|---|---|
| ⅜ | .030 | .126 | 900 | 870 | 570 | 380 |
| ½ | .035 | .198 | 800 | 770 | 500 | 330 |
| ⅝ | .040 | .285 | 740 | 720 | 470 | 310 |
| ¾ | .042 | .362 | 650 | 630 | 410 | 270 |
| ⅞ | .045 | .455 | 590 | 570 | 370 | 250 |
| 1⅛ | .050 | .655 | 510 | 490 | 320 | 210 |
| 1⅜ | .055 | .884 | 460 | 440 | 290 | 190 |
| 1⅝ | .060 | 1.14 | 430 | 420 | 270 | 180 |
| 2⅛ | .070 | 1.75 | 370 | 360 | 230 | 150 |
| 3⅛ | .090 | 3.33 | 330 | 320 | 210 | 140 |
| 3⅝ | .100 | 4.29 | 320 | 310 | 200 | 130 |
| 4⅛ | .110 | 5.38 | 300 | 290 | 190 | 120 |
| 5⅛ | .123 | 7.61 | 280 | 270 | 180 | 120 |

**Figure 3-2.** Dehydrated, pressurized ACR tubing

## SEAMLESS STEEL TUBING

This type of tubing is manufactured by the extrusion process. It may be either hard drawn or soft drawn. The hard-drawn tubes are the hardest and may be softened and toughened by the annealing process. The soft drawn is best for use where bending is to be done. Steel tubing is sized according to the outside diameter. The wall thickness is made to standard gauge numbers. The seamless tube is more expensive than the welded type but it will withstand a great deal more pressure. It is usually obtained in straight lengths of 20-ft (6.08 m). The most common method of joining steel tubing is soft soldering and silver soldering.

Steel tubing is being used almost exclusively in the manufacture of condensers for domestic refrigerators. The heat conducting capabilities of steel are fewer than those of copper. However, the steel is closely fastened to the shell of the refrigerator which in effect increases the area of the condenser. This increased area increases the heat transfer to closely parallel the heat transfer capabilities of an equal length of copper tubing. The difference in cost between copper and steel tubing will more than offset the cost of fastening the steel tubing to the refrigerator shell.

## ALUMINUM TUBING

This type of tubing is being used extensively in the manufacture of evaporators for domestic refrigerators. The heat transfer capabilities of aluminum are not as good as those of copper. Again, however, the cost of aluminum as compared to copper will more than offset any added cost in manufacturing. Aluminum tubing is easily formed by hand, eliminating almost entirely the need for fittings. Soft soldering and special welding alloys are used to join aluminum tubing. Sometimes epoxy is used in the joining process.

## STAINLESS STEEL TUBING

This tubing, like the others, is available in the usual ACR copper tube sizes. Stainless steel is strong and has a high resistance to corrosion. Stainless steel tubing number 304 is the most popular type for refrigeration work. This tubing has a low carbon, nickel, and chromium content. Some food processing requires this type of tubing. Joining is commonly accomplished by use of soft soldering techniques.

## FLEXIBLE TUBING

Automotive air conditioning systems and some transport refrigeration systems make use of flexible rubber tubing. (See Figure 3-3.) The vibration caused by the constant movement of the transport vehicle would work harden metal tubing and allow the system to leak. Flexible tubing usually has

**Figure 3-3.** Flexible refrigeration tubing (hose) (*Courtesy of Refrigeration and Air Conditioning Div., Parker-Hannifin*)

neoprene liners which are covered by rayon braids and bonded to provide leak-free, long-lasting service. It is usually purchased in bulk lengths and cut to meet specific needs. The fittings are easily installed on the tubing to provide flexibility in use and design.

## FITTINGS

There are several types of fittings in refrigeration work today. The type used depends largely on the tubing material used and the degree of permanency of connections that is desired. The most popular are the wrot (sweat), flare, compression, and hose fittings.

**EQUIVALENT LENGTH OF PIPE** is very important in piping procedures. The number of fittings in a refrigeration system should be kept to a minimum. Every turn encountered by the refrigerant flowing through the tubing is an added point of resistance. Excessive resistance will reduce the capacity of the equipment. Every elbow gives an added resistance (pressure drop) equal to a given length of that size tubing. (See Table 3-3.) Pressure drop and line sizing tables are usually designed on the basis of a given pressure drop for each 100 ft (30.4 m) of straight pipe. Therefore, the use of an equivalent length of pipe allows the data to be used directly.

When accurate calculations of pressure drop are required, the equivalent length for each fitting should be calculated. In practice,

Table 3-3

Equivalent Length in Feet of Straight Pipe for Valves and Fittings

| OD LINE SIZE (IN.) | GLOBE VALVE | ANGLE VALVE | 90° ELBOW | 45° ELBOW | TEE LINE | TEE BRANCH |
|---|---|---|---|---|---|---|
| ½ | 9 | 5 | .9 | .4 | .6 | 2.0 |
| ⅝ | 12 | 6 | 1.0 | .5 | .8 | 2.5 |
| ⅞ | 15 | 8 | 1.5 | .7 | 1.0 | 3.5 |
| 1⅛ | 22 | 12 | 1.8 | .9 | 1.5 | 4.5 |
| 1⅜ | 28 | 15 | 2.4 | 1.2 | 1.8 | 6.0 |
| 1⅝ | 35 | 17 | 2.8 | 1.4 | 2.0 | 7.0 |
| 2⅛ | 45 | 22 | 3.9 | 1.8 | 3.0 | 10.0 |
| 2⅝ | 51 | 26 | 4.6 | 2.2 | 3.5 | 12.0 |
| 3⅛ | 65 | 34 | 5.5 | 2.7 | 4.5 | 15.0 |
| 3⅝ | 80 | 40 | 6.5 | 3.0 | 5.0 | 17.0 |

an experienced system designer may be capable of making an accurate overall percentage allowance unless the piping arrangement is very complicated. For long runs of piping (100 ft or more) a 20% to 30% allowance of the actual length may be adequate. For short runs, an allowance as high as 50% to 75% of the lineal length may be required. Judgment and experience are necessary in making a good estimate. These estimates should occasionally be checked with actual calculations to ensure reasonable accuracy.

**WROT FITTINGS,** more commonly referred to as *sweat fittings,* are made to be used with either soft solder or silver solder. They are designed to make the necessary turns and connections with a minimum amount of pressure drop in the pipe. (See Figure 3-4.) In use, the tubing must be cut squarely so that a proper fit will be possible at the stop in the fitting as illustrated in Figure 3-5. If this joint is not properly made, excess solder may flow through and get inside the refrigeration system. Inside the system the solder will cause extra friction to the refrigerant and possible damage to other components in the system.

There is a wrot fitting to fill almost any need. The service technician should be familiar with them so that he will be able to make repairs to the system or accomplish the efficient replacement of parts.

**FLARE FITTINGS** are also available in refrigeration tube sizes. (See Figure 3-6.) The 45° flare is used in refrigeration piping work. Flare fittings are usually more easily installed because of the type of tools needed in making the connection. In practice, the flare nut is placed over the tube, which is then flared and clamped to another fitting. The flare on the tube should be the same size as the chamfer (beveled edge) on the fitting. The nut is then screwed onto the mating part, squeezing the tubing between the parts (see Figure 3-7). A drop of refrig-

**Figure 3-4.** Wrot copper fittings

eration oil should be put on each of the mating parts to prevent twisting of the tubing as the connection is tightened. The seal between the flare fittings is accomplished by the tubing between the parts. Therefore, if the fitting is tightened too much, the flare may

**Figure 3-6.** Flare tube fittings

either be cut off or the tubing squeezed too much to allow a proper seal to be made. Usually hand tight plus one turn is sufficient. If flare fittings are used where the temperature varies between the on and off cycles of the unit, a leak will sometimes occur. This leak is the result of the expansion and contraction caused by the warming and cooling of the metal. The expansion and contraction

**Figure 3-5.** Proper joint between tube and fitting

**Figure 3-7.** Flare connection

will cause the flare nut to loosen and require retightening before adding refrigerant to the system.

**HOSE (BARBED) FITTINGS** are commonly used on automotive and transport refrigeration. They are designed to allow connections to be made between a neoprene hose and the screwed fittings of system components. In practice, the threaded connection is made, then the hose is slipped over the barbed end and a clamp is tightened on the hose. Usually, once the clamp has been tightened, the hose will have to be cut off the fitting. These fittings provide a gas-tight seal and are satisfactory for refrigeration work.

**"O"-RING FITTINGS** are used on some service valves and special connections. These fittings use mating parts with an "O" ring placed in a groove to provide a seal. (See Figure 3-8.) In practice, caution must be taken to ensure that the "O" ring is inside the groove provided for it. Usually hand tight plus ½ turn will ensure a leak-tight fit. Care must be taken when making soldered connections close to these fittings because heat will destroy the "O" ring, rendering it useless. If the tubing must be heated close to these fittings, a damp cloth can be wrapped around the "O" ring fitting to keep it cool. Care must be taken to keep the cloth damp.

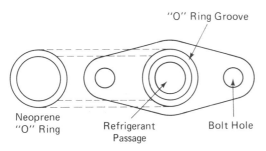

**Figure 3-8.** "O"-ring fitting

**COMPRESSION FITTINGS** are becoming popular for making quick connections to condensing units with the refrigerant lines. (See Figure 3-9.) In practice, hard-drawn tubing is preferred for use with these fittings. Soft-drawn tubing may be used if it is perfectly round and no kinks are close to the fitting. These fittings will usually provide a leak-tight joint when tightened hand tight plus 1½ turns. To prevent damage to the tubing, care should be taken not to tighten these fittings excessively. Should the tubing be dented enough to allow leakage, the tubing must be cut off past that point and the fitting reassembled and tightened.

**Figure 3-9.** Compression fitting (*Courtesy of Dearborn Division, Addison Products Co.*)

Many times the service technician will be required to do work on refrigeration piping for which he will not have all the necessary fittings. This problem can be overcome very easily in most cases, especially when changing the size of tubing. The tubing used in refrigeration work is sized in ⅛ inches (3.175 mm). This fact allows one size tube to be placed inside the next larger size tube. A joint which is made this way is not within the acceptable tightness tolerance to be soft soldered. However, silver solder will work very satisfactorily. This type of joint is commonly used in service and installation procedures.

## HAND TOOLS

In addition to the commonly used mechanic's hand tools and the refrigeration test instruments, there are some specialized hand

(a)                                                          (b)

**Figure 3-10.** Tubing cutter application

tools which are used by service and installation technicians. The most popular of these tools will be covered at this time, although all of Chapter 11 is devoted to refrigeration test instruments.

**TUBING CUTTERS** are used to cut tubing with a square end, enabling a service technician to make leak-tight connections. In practice, as shown in Figure 3-10(a), the cutter is placed around the tube and the cutter wheel is then tightened on the copper tubing (Figure 3-10(b)). Care should be taken not to tighten the cutter wheel too much because the tubing will be flattened and a leak-tight joint will be almost impossible to make. After the tubing is touched with the cutter wheel, the cutter handle should not be turned more than ½ turn. (See Figure 3-11.) Make a complete revolution of the tube with the cutter, then tighten the cutter handle no more than ½ turn again. Repeat this procedure until the tubing has been completely separated. To obtain a square end on the tubing, the tubing must be placed on the rollers so that the cutter wheel will be centered on the tube as shown in Figure 3-12.

After the tubing has been completely separated, there will be a burr on the inside of

**Figure 3-11.** Tubing cutter adjustment

the tubing. This burr must be removed or it will cause a restriction to the refrigerant flow. Most tubing cutters have an attached reaming blade. (See Figure 3-13.) This blade is rotated outward and then inserted into the tubing end and rotated to remove the burr. Care should be taken to prevent the loose metal chips from falling into the tub-

Roller type
with flare
cut-off
groove

**Figure 3-12.** Tubing cutter rollers

Retractable
locking
reamer

**Figure 3-13.** Tubing cutter reaming blade

ing or they will clog the strainers and filters which will need to be removed for cleaning or replacement.

Tubing cutters are also equipped with a flare cut-off groove in the rollers so that bad flares can be easily cut from the tubing. (See Figure 3-14.) This permits easy removal of the flare with a minimum amount of wasted tube and threading.

**Figure 3-14.** Tubing cutter with flare-removing rollers (*Courtesy of Gould Inc., Valve and Fittings Division*)

**FLARING TOOLS** (Figure 3-15) are used to make a flare on the end of tubing so that flare fittings can be used to make leak-tight connections. When a flare is to be made, insert the tubing into the flare block so that approximately ¼ inch (6.5 mm) extends above the flare block. (See Figure 3-16(a).) Swing the clamp into place against the end die and tighten it sufficiently to prevent slippage of the tube in the block. (See Figure 3-16(b).) Slide the yoke over the tubing and

**Figure 3-15.** Flaring tool (*Courtesy of Gould Inc., Valve and Fittings Division*)

turn the feed screw 5 or 6 turns after the cone touches the tubing (Figure 3-16(c)). When the flare is completed, reverse the flaring process to remove the tube from the flaring block. Backing off of the cone automatically burnishes the flare to a highly polished finish. Flares which are made in this type of flaring block are stronger because the flare is made above the flaring block die, not against it, and the original wall thickness is maintained at the base of the flare. There is no chance of "washing out" the flare. Flaring tools can be used for re-rounding and sizing bent tubing.

**SWAGING TOOLS** are used to enlarge one end of a piece of tubing so that two pieces the same size can be soldered together, eliminating the use of a coupling. Each size tubing requires a separate swaging tool (Figure 3-17). Swaging tools are used with the flaring block part of the flaring tool. The tubing is inserted into the proper sized die with enough extending out to allow the swaging process to be completed with the block tightened securely. Do not extend the tubing too much past the block because the tubing may

TUBE HEIGHT GAUGE
Slot in yoke is used for
tube height gauge.

SECURE CLAMPING
Sliding dies with lever
clamping action.

SLIP-ON YOKE
Yoke slips over top of bar,
then locks into position
with slight turn.

(a)               (b)               (c)

**Figure 3-16.** Use of flaring tool (*Courtesy of Gould Inc., Valve and Fittings Division*)

**Figure 3-17.** Swaging tool (*Courtesy of Robinair Manufacturing Company*)

bend and ruin the swage. If, however, too little tube is extended it may be cut when the swaging tool reaches the flaring block. After the tubing is properly placed in the block, the swage is placed in the tube and struck with a hammer. Between each blow with the hammer, rotate the swaging tool to prevent its sticking in the tubing. Should it stick, a few light sideways licks with the hammer will loosen the swaging tool and allow its removal from the tube.

**TUBING REAMERS** are used to remove burrs from refrigeration tubing (Figure 3-18). After the tubing has been cut the reamer is placed over the tube and rotated, removing the burrs from either the inside or the outside of the tube. Then the reamer is turned around and placed over the tube again and rotated to remove the burrs. Since these are multisized tools, one tool will fit many sizes of tubing.

**Figure 3-18.** Tubing reamer (*Courtesy of Gould Inc., Valve and Fittings Division*)

**PINCH-OFF TOOLS** are used to block the escape of refrigerant from a sealed system through the process tube after service operations have been performed. After the service operations are finished, the pinch-off tool (Figure 3-19) is placed on the process tube

**Figure 3-19.** Pinch-off tool (*Courtesy of Robinair Manufacturing Company*)

of the system and tightened. The tool closes on the tube and causes a positive seal. The service tools are then removed and the process tube is sealed with silver solder. (See Figure 3-20.) The pinch-off tool is then removed.

**Figure 3-20.** Positive seal made by pinch-off tool (*Courtesy of Robinair Manufacturing Company*)

**TUBING BENDERS** are used to make bends in tubing. Any angle of bend (turn) can be made with these tools. Their use eliminates elbows and soldered or flared connections. The tubing is measured and marked where the bend is desired. It is then placed in a bender of the proper size and the handles are pulled together to form the desired angle (see Figure 3-21). The bend should exceed the desired angle by about 3° to 5° because when the tubing is removed from the bender it will straighten

**Figure 3-21.** Tubing bender (*Courtesy of Robinair Manufacturing Company*)

some. It is almost impossible to completely straighten a piece of tubing that has been bent with a bender.

Another type of bender sometimes used is the spring bender (Figure 3-22). The spring is placed over the tubing and the bend is formed by hand. The bend should exceed the desired angle by about 5° and be re-straightened to the desired angle so the bender can be slipped from the tube. Care should be taken when using this type of bender because if too sharp a bend is attempted the tubing may crimp and make removal of the spring difficult.

**Figure 3-22.** Spring type tubing bender (*Courtesy of Gould Inc., Valve and Fittings Division*)

**HERMETIC SERVICE VALVE KITS** are used to perform service operations involving the refrigerant on sealed systems (see Figure 3-23). In use, the proper adapter is installed on the unit and the gauge manifold is then attached to the adapter. This allows checking of the refrigerant pressures and charging procedures to be completed without cutting the tubing or making sweat connections.

**REVERSIBLE RATCHETS** are used to turn the stem on service valves to allow service operations to be completed. The most common size is ¼ inch (6.5 mm). A ³⁄₁₆ inch (4.7625 mm) square is provided on the end with a ½ inch (12.7 mm) six-point socket. (See Figure 3-24.) There are a variety of sockets available to fit most any size valve stem which will enable the service technician to connect the refrigeration gauges to

**Figure 3-23.** Service valve adapter kit

**Figure 3-24.** Reversible ratchet (*Courtesy of Bonney*)

**Figure 3-26.** Packing gland nut socket (*Courtesy of Bonney*)

the system (see Figure 3-25). Also available are a variety of packing gland nut sockets which can be used with the reversible ratchet (see Figure 3-26). These are used to tighten the packing glands on certain valves to prevent leakage of refrigerant.

**GAUGE MANIFOLDS** are probably the most often used tool in the service technician's tool chest. A gauge manifold is comprised of a compound gauge, a pressure

**Figure 3-25.** Valve stem socket (*Courtesy of Bonney*)

gauge, and the valve manifold. (See Figure 3-27.) All the service operations performed involving the refrigerant, lubricating oil, and evacuation can be accomplished by use of this tool.

**COMPOUND GAUGES** (Figure 3-28) are used to read pressures both above atmospheric and below atmospheric (vacuum). In practice, they are used to determine pressures in the low side of the system. The outside scale is in pressure (psig). The inside scales are the corresponding temperatures for different refrigerants.

Gauges operate from the action of a *Bourdon tube* (see Figure 3-29). When pressure inside the Bourdon tube is increased, the element tends to straighten. As the pressure is decreased, the element tends to curve again. A Bourdon tube is a flattened metal

Blue Case
Low Side

Red Case
High Side

**Figure 3-27.** Refrigeration gauge manifold (*Courtesy of Rabinair Manufacturing Company*)

Pointer
Shaft
Gear

Gauge Case

Dashed Line Indicates
Bourdon Tube Under
Pressure

Enlarged
Cross-Section
of Bourdon
Tube

Bourdon
Tube

Operating
Link

Adapter
Fitting

**Figure 3-29.** Bourdon tube principle

tube sealed at one end, curved, and soldered to the pressure fitting on the other end. The movement of the element will pull a link which is attached to the pointer through a series of gears. This movement will be shown by the gauge hand, or pointer.

Compound retard gauges have a retarder which permits accurate readings within a

**Figure 3-28.** Compound gauge (*Courtesy of Marshalltown Instruments*)

**Figure 3-30.** Pressure gauge (*Courtesy of Marshalltown Instruments*)

Special washer allows
stem to close without turning
. . . no twisting or tearing seat

Glass-impregnated
Teflon* seat

Stem moves in and out
without turning

Flow

Chrome-plated
investment casting

Hot forged
brass body

Special "O" rings provide
positive seal

Two full turns allow
maximum flow

**Figure 3-31.** Valve manifold (*Courtesy of Robinair Manufacturing Company*)

given range. In refrigeration work this range would be between 0 psig and 100 psig. These gauges can be recognized by the change in graduations at pressures higher than those usually encountered.

**PRESSURE GAUGES** are used to determine pressures on the high side of the system (Figure 3-30). The outside scale is calibrated in psig and the inside scales indicate the corresponding temperatures of the different types of refrigerants. Some pressure gauges are not designed to operate at pressures below atmospheric pressure. Therefore, caution should be used during evacuation procedures to prevent damage to these gauges.

**VALVE MANIFOLDS** provide openings through which the various service operations are performed on the refrigeration system. (See Figure 3-31.) The proper manipulation of the hand valves will permit

almost any function. When the valves are screwed all the way in, the gauges will indicate the pressure on the corresponding line. The center line is usually connected to a vacuum pump, a refrigerant cylinder, or an oil container.

**CHARGING HOSES** are flexible (see Figure 3-32) and are used to connect the gauge manifold to the system. Charging hoses are

**Figure 3-32.** Charging hose (*Courtesy of Robinair Manufacturing Company*)

equipped with ¼ inch (6.5 mm) flare connections on each end. One end usually has a valve core depressing attachment for attaching the gauges to *schrader valves.** Charging hoses may be purchased in a variety of colors which facilitate making the connections to the unit and are designed for a working pressure of 500 psi (35.15 kg/cm²) and an average bursting pressure of 2000 psi (140.6 kg/cm²).

**POCKET THERMOMETERS** are an asset to the service technician. Through their use much time can be saved. Operating temperatures of a unit can be checked upon completion of the service work to determine whether or not the unit is operating properly. The two types curently used are the mercury and the bimetal thermometer. The bimetal type is the most popular because of its durability. (See Figure 3-33.) Pocket

---

* A schrader valve is a type of service valve which is used on systems not provided with a means for servicing.

thermometers can be purchased with a convenient shock-resistant carrying tube which can be clipped into the shirt pocket for ready use. They are accurate to within one percent.

**CAPILLARY TUBE CLEANERS** are extremely useful when servicing domestic refrigerators and freezers. The capillary tube flow control device used on these units has such a small inside diameter that it can become clogged easily. The capillary tube cleaner is connected to the plugged cap tube and up to 15,000 psi (1051.5 kg/cm²) of hydraulic pressure may be applied to force chips, filings, flux, oil, etc. from the cap tube into the evaporator where it either remains harmlessly or circulates to the strainer or filter. (See Figure 3-34.) A high grade of refrigeration oil is used to provide the hydraulic pressure.

**WELDING UNITS** are a must for the service technician. The most popular type is the portable model shown in Figure 3-35.

(b)

(c)

(a)

**Figure 3-33.** Various types of thermometers (*Courtesy of Marshalltown Instruments*) (a) mercury; (b) bimetal; (c) enlarged bimetal thermometer face.

**Figure 3-34.** Capillary tube cleaner (*Courtesy of Watsco, Inc.*)

**Figure 3-35.** Portable welding unit (*Courtesy of Union Carbide Corp.*)

These ultraportable units are convenient, economical, and easy to use. Each one includes a complete apparatus for handling a wide range of repair and maintenance work. A flame temperature near 6000°F (3298°C) can be obtained with these units.

There are two types of acetylene tanks and one oxygen tank which are used with these units (see Figure 3-36). Types MC and B acetylene tanks are commonly used with the R oxygen tank. The disposable propane (DP) tank is used for lower temperature requirements. The specifications for these tanks are shown in Table 3-4.

**Figure 3-36.** Acetylene and oxygen tanks (*Courtesy of Union Carbide Corp.*)

**CHARGEFASTER UNITS** are a tremendous time saver for the service technician. Tests have shown that when charging refrigeration systems the service technician does not need to know how much refrigerant is needed because the chargefaster method provides this information. (See Figure 3-37.) In practice, the chargefaster is connected to the low side of the gauge manifold. Liquid refrigerant can then be charged

**Figure 3-37.** Chargefaster (*Courtesy of Watsco, Inc.*)

Table 3-4

Tank Specifications *

| STYLE | CAPACITY | HEIGHT | DIAMETER | WEIGHT, FULL |
|---|---|---|---|---|
| Disposable Propane (DP) | 1 pt., 10.7 fl. oz. | 10-1/2 in. | 2-3/4 in. | 2 lb., 2 oz. |
| Type MC Acetylene | 10. cu. ft. | 14 in. | 4 in. | 8 lb. |
| Type B Acetylene | 40 cu. ft. | 23 in. | 6-1/4 in. | 26 lb. |
| Type R Oxygen | 20 cu. ft. | 14 in. | 5-3/16 in. | 13-1/2 lb. |

* Courtesy Union Carbide Corp.

directly into the low side of any refrigeration system safely and quickly. As the liquid goes through the chargefaster, it is converted to a saturated vapor at a rate of flow less than the compressor capacity.

Since the refrigerant leaves the cylinder in liquid form and expansion takes place at the outlet of the chargefaster unit, there is no pressure drop in the charging cylinder. Therefore, heating the cylinder is neither necessary nor desirable.

Because chargefaster contains a Magni-Chek valve assembly which automatically permits the flow to bypass when drawing a vacuum through the gauge manifold, there is no restriction or slowdown of evacuation of the system.*

**CHARGING CYLINDERS** are used by the service technician to charge the correct amount of refrigerant into a system in a short period of time. In practice, the charging cylinder is loaded with the recommended manufacturer's amount of refrigerant for a particular unit. (See Figure 3-38.) While

* Magni-Chek is the brand name of a valve used to prevent the backwards flow of fluids.

Standard          Heated

**Figure 3-38.** Charging cylinder (*Courtesy of Robinair Manufacturing Company*) the system is evacuated, the complete re-frigerant charge is dumped into it. Some of these units have a built-in electric heater to speed up the dumping process. These cylinders should never have a direct flame applied to them, nor should they be stored in direct sunlight or with liquid refrigerant in them.

**FITTING BRUSHES** are used to clean the inside of fittings that are to be soldered. These brushes (see Figure 3-39) can be obtained in a size for each fitting size. In practice, the brush fits tightly into the size fitting for which it has been designed. The brush is then rotated by hand to clean the fitting. If too large a brush is used or if it is rotated in more than one direction, it will be ruined.

**Figure 3-39.** Fitting brushes (*Courtesy of Schaefer Corp.*)

## MATERIALS

Various materials are peculiar to the refrigeration and air conditioning industry. Because there are so many in use, we will cover those that are most commonly used.

**SOFT SOLDER** is a low-temperature method of joining metals together. It is an adhesion process. That is, the materials be-

ing joined are not melted; only the solder is melted. Soft solder usually is a mixture of tin and lead. This includes tin and antimony; 95% tin and 5% antimony (95/5), which has a melting point of 460°F (238°C); and 50/50, which has a melting point of 421°F. Both temperatures are well below the melting point of copper, which is 1984°F (585°C). Soft solder can be purchased in one-pound spools of either $\frac{1}{16}$ inch (1.5875 mm) or $\frac{1}{8}$ inch (3.175 mm) diameter at a refrigeration supply house. (See Figure 3-40.) Soft solder's low melting temperature, low tensile strength, and inability to withstand high vibration make it unsuitable for most modern refrigeration applications. This is especially so with systems using R–22 as the refrigerant. The compressor discharge temperature is sometimes very high, and when combined with the vibration of the compressor, leaks develop quite easily. Therefore, a stronger solder is needed.

**Figure 3-40.** Soft solder (*Courtesy of J. W. Harris Co., Inc.*)

**SILVER SOLDER (STAY-SILV 15)** is the most common agent used in silver brazing joints in modern refrigeration equipment. It is sometimes referred to as silver brazing. It is available in rods $\frac{1}{8}$ inch (3.175 mm) $\times$ $\frac{1}{8}$ inch (3.175 mm) $\times$ 20 inches (50.8 cm); .050 inch (12.7 mm) $\times$ $\frac{1}{8}$ inch (3.175 mm) $\times$ 20 inches (50.8 cm); and 1, 3, 5,

**Figure 3-41.** Silver solder (*Courtesy of J. W. Harris Co., Inc.*)

and 25 troy ounces and $\frac{1}{16}$ inch (1.5875 mm) in diameter.

To silver braze copper to steel or to braze dissimilar metals, the recommended product is Stay-Silv 45 alloy, which has a melting temperature of 1145°F. Also available are the Safety-Silv cadmium-free silver solders 1200 (56% silver), Safety-Silv 1370 (45% silver), and Safety-Silv 1350 (40% silver). Their melting temperatures are according to their designated alloy number as listed. (See Figure 3-41.) These solders have strong tensile strength and are not affected by vibration. Some types of solder contain their own flux, eliminating the need for extra flux on the joint. This is especially true on copper tubing.

CAUTION: Some silver solders contain cadmium, a product that gives off poisonous fumes when heated; therefore, be sure that proper ventilation is provided when using silver solder.

**FLUX** is used when making sweat joints to aid in making leak-tight connections. It is available in both paste and liquid forms. (See Figure 3-42.)

CAUTION: Flux has an acid content and should not be allowed on clothing, near the eyes, or on open cuts. To remove, wash with soap and water or see a physician.

*There is a flux designed specifically for each type of solder which cannot be used for any other type of solder.* Flux should be put on the joint sparingly. Do not overflux a joint or allow flux to enter the refrigeration system because the acid will be harmful to the motor windings of the compressor and other internal components.

**Figure 3-42.** Soldering flux (*Courtesy of J. W. Harris Co., Inc.*)

**SAND CLOTH** is used mainly to clean the surfaces to be soldered. It is available in rolls at the refrigeration supply house (see Figure 3-43). Short strips are torn from the roll and wrapped around the tube and each end is pulled alternately until the tube is shiny bright. Caution should be used when cleaning tubing to prevent the particles from entering the tube. Perhaps the best way to accomplish this is to turn the end of the tube being cleaned downward so that the particles will fall away from it.

**Figure 3-43.**   Sand cloth

## MAKING JOINTS

The joining together of tubing is a very important operation to the installation and service technician. Because it is a common job, many people tend to be lax about using proper procedures.

**SOFT SOLDERED JOINTS** require that both pieces be thoroughly cleaned and fluxed. A good method used to prevent the entrance of flux into the system is to insert the tube into the fitting only about ¼ inch (6.5 mm). Then apply flux to the joint (see Figure 3-44). After flux has been applied, insert the tube all the way into the fitting and rotate the tube inside the fitting to spread the flux evenly.

Next apply heat to the fitting only (see Figure 3-45). Rarely will heat need to be applied to the tube. Never apply heat to the flux. Flux will burn away leaving a residue that must be removed. The joint would have to be recleaned and refluxed before the job

**Figure 3-44.**   Fluxing a joint

could be finished. Do not overheat the joint; apply only enough heat to cause the metal to melt the solder. Do not heat the solder with the flame. Test the temperature of the metal by touching the solder to the joint until it melts by only touching it to the metal.

Apply only enough solder to fill the joint. If too much solder is used it will either be wasted or will flow into the system and possibly cause damage. A piece of solder the length of the tube diameter is sufficient to make a leak-tight joint. See Figure 3-46 for a comparison of soldered joints. The space between the tube and fitting is exaggerated to provide a clear picture.

**Figure 3-45.**   Application of heat to a solder joint

**Figure 3-46.** Comparison of soldered joints

**Figure 3-47.** Silver solder line repair

**SILVER SOLDERED JOINTS** are made basically the same as soft soldered joints. The major difference is the amount of heat required. The tubing and fitting are both heated to a dull cherry red. At this temperature the silver solder will flow into the joint freely. A silver solder joint which has been cut open will appear much the same as the soft soldered joint. However, the silver solder will appear dark rather than shiny. If too much heat is applied the silver solder will boil, leaving small pin-holes which could possibly leak. Apply only enough heat to melt the silver solder.

*CAUTION:* Be sure that proper ventilation is provided to prevent cadmium oxide poisoning.

*Flare* connections were covered in the section on *flaring tools*. Refer to that section if further study is needed.

**LINE REPAIR** is sometimes needed when a cracked line occurs and the desired fittings are not available. The service technician may repair the line satisfactorily and get the system operating again by using this method. First, be certain that no pressure exists in the system because injury may result if the pressure builds enough to blow hot oil and

metal from the line. Next, clean the joint with sand cloth and apply silver solder flux.

*NOTE:* This procedure should not be attempted with soft solder.

Heat the line to a dull cherry red and apply silver solder until the area around the crack is slightly built up. (See Figure 3-47.) Be sure not to overheat the tube. This type of repair will last indefinitely if properly done.

## SAFETY PROCEDURES

Always practice safety and know the proper steps required for completing a job. The following are only a few of the safety points to be remembered.

1. Never use a screwdriver for a pry bar. The blade is hardened to retard wear and it is, therefore, brittle to sideways pressure. It may break and cause injury to the user and equipment.

2. Always grind the "mushroom" from the head of chisels and punches. When struck with a hammer, this mushroom may break and the particles may cause injury to the user or to a bystander.

3. Never use a file without a handle. The tang may cause injury to the hands.

4. Always wear goggles when drilling. Chips may break away and enter the eye, causing serious injury and possibly loss of sight.

5. Never use oxygen when pressure testing a refrigeration system. Oxygen and oil together form an explosive mixture.

6.   Never heat a refrigeration pipe containing a refrigerant to a cherry red condition. The pressure which is built up inside the pipe could cause a "blowout" and serious injury could result.

7.   Never use carbon-tetrachloride for any purpose because it is toxic and harmful to the skin.

8.   Always provide plenty of ventilation when working with silver brazing materials. These materials contain cadmium, the fumes of which are poisonous.

9.   Never use emory cloth for cleaning refrigeration tubing. The grit is extremely hard and could cause damage to the system if allowed to enter.

10.   Practice care when using epoxy bonding materials because they may irritate the skin.

11.   Never fill refrigerant cylinders to more than 85% of their rated capacity. Hydrostatic pressure may cause them to burst when they become warm.

12.   Always pull on a wrench, never push.

13.   Always use the proper size wrench to prevent injury to the user and equipment.

14.   Always "crack" cylinder service valves before opening all the way. This will allow for control of the gas if a dangerous situation occurs.

15.   Always protect an open refrigeration system from moisture. Keep all refrigeration components clean and dry.

## SUMMARY

•   In a refrigeration system, the piping is used to convey the refrigerant from one system component to another.

•   The types of tubing used are welded or brazed, and seamless tubing. Seamless tubing is the stronger of the two.

•   Copper tubing is never used when ammonia is the desired refrigerant. It is used for condensers and evaporators because of its availability and heat-conducting capabilities.

•   In refrigeration work, copper tubing is measured by its outside diameter (OD). It is rated as ACR for refrigeration applications.

•   Two types of copper tubing used in refrigeration work are soft- and hard-drawn.

•   Copper tubing is manufactured according to ASTM B88 and ANSI-B9.1–1971 refrigeration industry specifications.

•   For refrigeration work, tubing is cleaned, dehydrated, and capped.

•   Steel tubing is being used most extensively in modern domestic refrigerators as condensers.

•   Aluminum tubing is being used in domestic refrigerator evaporators.

•   Flexible tubing is used on automotive air conditioning and transport refrigeration.

•   The fittings most commonly used in refrigeration work are wrot (sweat), flare, compression and hose. Each fitting develops a resistance to refrigerant flow which is rated by an equivalent length of pipe.

•   Flare fittings are susceptible to leaking when used in places of varying temperatures.

•   Compression fittings are becoming popular for making quick connections to condensing units with refrigerant lines.

•   Hard-drawn tubing is preferred for use with compression fittings.

• One size tubing can satisfactorily be inserted into another piece of tubing the next size larger and silver soldered.

• Tubing cutters are used to cut a square end on tubing to enable the technician to make a leak-tight joint. To prevent damage to the tubing, the cutter should not be overtightened.

• Flaring tools are used to make flares on tubing so that threaded flare fittings can be used.

• The flaring tool should not be overtightened, ruining the flare.

• A drop of refrigeration oil should be put on both faces of the flare to prevent twisting of the tube.

• The flare fitting should be tightened hand tight plus one turn.

• Swaging tools are used to enlarge one end of a piece of tubing so that two pieces the same size can be soldered together. Care should be exercised to prevent enlarging the swage or cutting it with the swaging tool.

• Tubing reamers are used to remove burrs from tubing to reduce friction to the refrigerant. Care should be taken to prevent metal chips from entering the tubing. One size reamer fits several sizes of pipe.

• Pinch-off tools are used to aid in removing gauges from sealed systems after repairs have been made. The process tube is silver soldered before the pinch-off tool is removed.

• Hermetic service valve kits are used to gain entrance into hermetically sealed systems without cutting and soldering lines. An adapter is available for almost any hermetic system.

• Reversible ratchets are used to turn the stem of service valves. They are available in practically any size.

• Adapters are available for use with the ¼ inch (6.5 mm) size to make it more flexible in use.

• Gauge manifolds are the most used tools in the service technician's tool chest. They are used for checking pressures, adding refrigerant, or any process involving refrigerants, oil, and evacuation.

• Compound gauges are used to read pressures above and below atmospheric.

• Pressure gauges are used to read pressures above atmospheric only.

• Gauges use the Bourdon tube principle.

• Charging hoses are used to connect the gauge manifold to the system.

• Pocket thermometers are an asset to the technician because the system can be checked after completion of repairs to determine whether or not it is operating properly. The most popular type is the bimetal thermometer because of its durability.

• Capillary tube cleaners are used to clean foreign material from small diameter tubes with hydraulic pressure. A high grade of refrigeration oil is used to produce the hydraulic pressure.

• Portable welding units are very popular for work on tubing.

• A temperature of 6000°F (3298°C) can be obtained through proper adjustment of the oxygen and acetylene.

• There are two acetylene tanks and one oxygen tank available for portable use.

• Chargefaster units enable the technician to charge liquid refrigerant quickly and accurately into any size system without damage to the unit. The unit can also be left in the line during evacuation procedures.

• Charging cylinders are used to aid the technician in charging the correct amount of refrigerant into a system. These cylinders should never be heated with an open flame or stored in direct sunlight or with refrigerant left in them.

• Fitting brushes are used to clean the insides of fittings. There is a size for each size tubing. Care should be taken not to damage the brush.

• Soft solder is a low temperature method of joining metals together by adhesion.

• Fittings to be soft soldered must be cleaned to a shiny bright and flux applied. Excessive temperature must be avoided.

• A piece of soft solder the length of the diameter of the tube is sufficient quantity to make a leak-tight joint.

• Apply only enough heat to melt the solder. It has a melting point of 460°F (238°C) for 95/5 and 421°F (216.5°C) for 50/50.

• Silver solder is the most popular method of soldering joints in modern refrigeration systems. It has a high melting point and tensile strength. It is not affected by vibration. When heated it gives off a poisonous vapor from the cadmium in it.

• Flux is used in making soldered joints.

• Each type of solder requires a different flux. Care should be taken to prevent its entering the refrigeration system. It should not be allowed on clothing or skin because of its acid content.

• Repair of cracked refrigerant lines can be accomplished by applying silver solder to the area of the crack until a slight buildup is apparent. Do not use soft solder for this procedure. Be sure the refrigerant is purged from the system to prevent a refrigerant-oil blowout and possible human injury.

## REVIEW EXERCISES

1. What are the two types of tubing and pipe mentioned in this text?

2. Why is seamless tubing used in refrigeration work?

3. Why is copper tubing not used with ammonia refrigerant?

4. What are the two types of tubing used as refrigerant lines?

5. How is tubing which is used for refrigeration work measured (OD or ID)?

6. What should be done to hard-drawn tubing before attempting to bend it?

7. What two factors vary the safe working pressures of copper tubing?

8. What are the four classifications of copper tubing?

9. What two things are done to ACR tubing during the manufacturing process?

10. What should be done to copper tubing that is not used?

11. What part of the refrigeration system is more commonly made of steel?

12. What is done to increase the heat transfer capabilities of steel?

13. Where is aluminum tubing being used in refrigeration systems?

14. What is the most popular type of stain-

less steel tubing used in refrigeration work?

15. What type of tubing is more popular in automotive and transport refrigeration work?

16. What problem does flexible tubing eliminate?

17. What are the most popular types of tube fittings used in refrigeration work?

18. To what does the term equivalent length of pipe refer?

19. What percentage of the actual length of pipe is allowable on 100 ft (30.4 m) of pipe or more?

20. What is another name for wrot fittings?

21. What is the angle in degrees of the flare fittings used in refrigeration work?

22. What should be done to prevent twisting of the tubing when tightening a flare fitting?

23. What causes flare fittings to leak after being used a while?

24. What caution should be taken when assembling an "O"-ring fitting?

25. Where are compression fittings becoming popular in refrigeration work?

26. Why should tubing cutters be used to cut tubing?

27. How much should the handle on the

tubing cutter be tightened after each complete revolution of the tubing?

28. For what is the flaring tool used?

29. What tool is used to enlarge the end of tubing so that two pieces of the same size may be soldered together?

30. What is the purpose of the pinch-off tool?

31. What does a tubing bender eliminate?

32. On what type of systems are hermetic service valve kits used?

33. For what purpose are reversible ratchets used?

34. What is the most popular tool in the service technician's tool chest?

35. On what principle do refrigeration gauges operate?

36. For what purpose is a gauge manifold used?

37. What does the term 95/5 indicate when referring to soft solder?

38. What is the main criterion used when selecting the type of silver solder to be used?

39. Can any soldering flux be used on all types of soldered joints?

40. What is the length of solder required when making a sweat joint?

41. What precaution should be taken when silver soldering a joint?

# Compressors

All compression refrigeration systems use a compressor of some type to circulate the refrigerant through the system. The compressor has two functions. First, it draws the refrigerant vapor from the evaporator and lowers the pressure of the refrigerant in the evaporator to the desired evaporating temperature. Second, the compressor raises the pressure of the refrigerant vapor in the condenser high enough so that the saturation temperature is higher than the temperature of the cooling medium used to cool the condenser and condense the refrigerant.

## COMPRESSOR TYPES

Several different types of refrigeration compressors are in use today. The most popular types are the reciprocating, rotary, and centrifugal. The most popular of these is the reciprocating compressor.

**RECIPROCATING COMPRESSORS** are used in the smaller horsepower sizes for commercial refrigeration, domestic refriger-

ation, and for industrial applications.

Reciprocating compressors are quite similar in design to the automobile engine. A piston is driven from a crankshaft and makes alternating suction and compression strokes inside a cylinder, which is equipped with suction and discharge valves. (See Figure 4-1.)

As the piston is pulled down inside the cylinder, refrigerant vapor is drawn from the evaporator into the cylinder of the compressor. A reedlike flapper valve in the cylinder

**Figure 4-1.** Reciprocating compressor (*Courtesy of Frigidaire Division, General Motors Corp.*)

Heat-Laden Refrigerant
Vapor from Freezer

Flapper
Valve
Open

Compressed Refrigerant Vapor
on way to Condenser

Flapper
Valve
Open

**Figure 4-2.** Suction stroke of reciprocating compressors (*Courtesy of Frigidaire Division, General Motors Corp.*)

**Figure 4-3.** Compressor discharge stroke (*Courtesy of Frigidaire Division, General Motors Corp.*)

head acts as a check valve. It allows the refrigerant vapor to be drawn into the cylinder (See Figure 4-2). As soon as the refrigerant vapor pressure inside the cylinder is equal to the refrigerant vapor pressure in the suction line, the spring action of the reed closes the suction valve.

When the piston has reached the bottom of its stroke, it has drawn in all the refrigerant vapor possible. The piston then starts on its upstroke, pushing the entrapped refrigerant vapor ahead of it. The vapor cannot return to the evaporator because the suction valve is closed, blocking its path.

However, there is another reedlike flapper valve in the cylinder head. This one is arranged so that vapor can get out of the cylinder. All the vapor that is forced through this valve is directed along a path that leads to the condenser. The discharge valve, like the suction valve, will allow the refrigerant to flow in only one direction. It lets the vapor out of the compressor cylinder but blocks the way for it to return to the compressor. (See Figure 4-3.) When the piston has reached the top of its stroke it is ready to start down again and repeat the cycle by

drawing more refrigerant vapor into the cylinder.

Because the reciprocating compressor is a positive displacement pump, it is suitable for small displacement volumes, and it is quite efficient at high condensing pressures and high compression ratios. Other advantages of the reciprocating compressor are its adaptability to a number of different refrigerants (liquid refrigerant may easily be run through connecting piping because of the high pressure created by the compressor), its durability, basic simplicity of design, and relatively low cost.

The reciprocating motion of the compressor can be accomplished in several ways. The most popular means of producing this motion in modern refrigeration equipment is by the *crank throw* type crankshaft to which the piston is joined by a connecting rod, again, much like the automobile engine. (See Figure 4-4.)

The other most popular type of reciprocating compressor is the *eccentric disc* type. The eccentric disc uses a straight shaft which can be very inexpensively machined and requires no special forging like the crank throw type. (See Figure 4-5.) The piston is

**Figure 4-4.** Crank type reciprocating compressor

| | |
|---|---|
| 1. Base | 30. Lock Nut |
| 2. Base Gasket | 31. Lock Washer |
| 9. Crankshaft | 32. Oil Plug |
| 12. Crankshaft Thrust Ball | 33. Oil Plug Gasket |
| 14. Crankshaft Thrust Plug | 34. Piston |
| 15. Crankshaft Thrust Plug Gasket | 35. Piston Pin |
| 16. Cylinder Block | 36. Piston Pin Button |
| 17. Cylinder Head | 37. Piston Ring |
| 19. Discharge Valve | 39. Seal |
| 20. Discharge Valve Guide | 40. Seal Bellows |
| 21. Discharge Valve Plate | 41. Seal Cover |
| 22. Discharge Valve Plate Gasket, Lower | 42. Seal Cover Gasket |
| 23. Discharge Valve Plate Gasket, Upper | 43. Seal Guide |
| 24. Discharge Valve Safety Spring | 44. Seal Spring |
| 25. Discharge Valve Safety Spring Retainer | 45. Set Screw |
| 27. Discharge Valve Spring | 46. Suction Valve |
| 28. Eccentric Disc | 47. Suction Valve Guide |
| 29. Eccentric Strap and Rod | 48. Woodruff Key |

**Figure 4-5.** **Eccentric disc type reciprocating compressor**

connected to the eccentric disc on the crank-shaft by an eccentric strap instead of a con-necting rod. The eccentric disc is fastened to the shaft by an eccentric lock bolt or set screw. The strap fits over the eccentric with-

out being bolted.

The difference in shaft construction of the crank throw and the eccentric disc type of compressors is shown in Figures 4-6 and 4-7.

**Figure 4-6.** Piston and crankshaft details

**Figure 4-7.** Eccentric disc details

**ROTARY COMPRESSORS** are much more simple in construction than the reciprocating types. In some respects, the very essence of their simplicity makes it a little more difficult to understand how they work. (See Figure 4-8.) Actually, the only moving parts in a rotary compressor consist of a steel ring, an eccentric or cam, and a sliding barrier. (See Figure 4-9.)

Both the ring and the cam are housed in a steel cylinder. The steel ring is a little smaller in diameter than the cylinder. It is situated off-center so that one point on the outer circumference of the ring is always in contact with the wall of the cylinder (see Figure 4-10). This, of course, leaves an open, crescent-shaped space on the opposite side between the ring and the cylinder wall.

**Figure 4-8.** Rotary compressor (*Courtesy of Frigidaire Division, General Motors Corp.*)

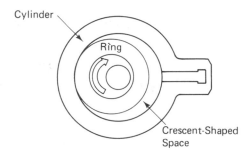

**Figure 4-10.** Internal view of a rotary compressor (*Courtesy of Frigidaire Division, General Motors Corp.*)

An electric motor rotates the cam. As the cam rotates, it carries the ring around with it, imparting a peculiar rolling motion to the ring. (See Figure 4-11.) The ring literally rolls on its outer rim around the wall of the cylinder.

**Figure 4-9.** Rotary compressor parts (*Courtesy of Frigidaire Division, General Motors Corp.*)

**Figure 4-11.** Rotary compressor motion (*Courtesy of Frigidaire Division, General Motors Corp.*)

If a port is installed in the cylinder wall and used as an entry to permit refrigerant vapor from the evaporator to flow into the crescent-shaped space (see Figure 4-12), and if the cam is rotated just a fraction of a complete turn, the ring will almost immediately cover up the port. As soon as this happens, the refrigerant vapor is trapped in the crescent-shaped space. There is no place for it to escape.

**Figure 4-12.** Rotary compressor compression phase (*Courtesy of Frigidaire Division, General Motors Corp.*)

However, if another port is installed near the other end of the crescent-shaped space, the entrapped vapor will have a way to escape. A piece of pipe is connected from the discharge port to the condenser. As the cam continues to rotate and roll the ring around in the cylinder, it pushes the crescent-shaped wedge of refrigerant ahead of it, compressing the vapor as it is forced out through the discharge port to the condenser.

As can be seen in Figure 4-12, the ring cannot roll very far before it uncovers both the suction and discharge parts. Obviously, there must be some means of directing the refrigerant vapor out through the discharge port while at the same time blocking its passage to the suction port.

The most simple and effective way to block the passage is to place a barrier between the two ports. This barrier must be flexible or sliding because one end of it is constantly moving back and forth as the ring rolls around in the cylinder. (See Figure 4-13.) To install a barrier, a slot must be cut in the cylinder wall deep enough so that the barrier can slide all the way in. A spring is placed behind the barrier to push it out and hold it snugly against the ring no matter what position the ring is in (see Figure 4-14). As the ring rolls around in the cylinder, the movable barrier will follow every movement.

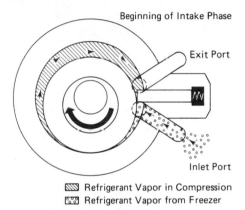

**Figure 4-13.** Sliding barrier (*Courtesy of Frigidaire Division, General Motors Corp.*)

Cam rotating as a part of the motor shaft causes the ring to roll in a circle larger than itself. In doing so, it is always in contact with both the cylinder wall and the barrier.

**Figure 4-14.** Barrier in cylinder (*Courtesy of Frigidaire Division, General Motors Corp.*)

With the movable barrier in place, the refrigerant vapor which is entrapped in the crescent-shaped space can go only one way as the ring rolls around and pushes the vapor ahead of it. (See Figure 4-15.) The only way it can go out is through the discharge port because the movable barrier blocks its passage to the inlet port.

Finish of Compression Phase

To Condenser

▨ Refrigerant Vapor in Compression
⚬⚬⚬ Refrigerant Vapor from Freezer

**Figure 4-16.** Finish of compression phase of a rotary compressor (*Courtesy of Frigidaire Division, General Motors Corp.*)

Compression and Intake
Phases Half Completed

To Condenser

▨ Refrigerant Vapor in Compression
⚬⚬⚬ Refrigerant Vapor from Freezer

**Figure 4-15.** Rotary compressor compression cycle (*Courtesy of Frigidaire Division, General Motors Corp.*)

The foregoing discussion describes the simple cycle of operation of a rotary type compressor. As the ring rolls around in the cylinder, its point of contact literally *runs in a circle* around the cylinder wall. All refrigerant ahead of that point of contact is pushed toward the movable barrier which deflects it out through the discharge port. (See Figure 4-16.) Meanwhile, once the point of contact has passed the suction port, a fresh charge of refrigerant vapor is drawn from the evaporator to the compressor.

As with the reciprocating type compressor, there are numerous variations of design to be found in different makes of rotary compressors. None of them, though, is more simple or has fewer moving parts than the rotary compressor just described.

Rotary type compressors are popular in domestic refrigeration and have recently been used in small air conditioning units. These compressors must be manufactured in large volumes to be economically produced because of the precision machining which is required to provide the desired performance. A check valve is usually needed in either the suction line or the discharge line to prevent refrigerant backup from the condenser to the evaporator during the off cycle. Liquid floodback of the refrigerant to the compressor must be eliminated because the suction line enters the compression chamber directly.

Rotary compressors are suited for applications where large volumes of vapor are to be circulated and where a low compression ratio is desired. They are positive displacement pumps.

**CENTRIFUGAL COMPRESSORS** are not positive displacement pumps like those previously discussed. These compressors produce a pressure through centrifugal effect alone. Compression is caused by whirling the mass of gaseous refrigerant at a high rate of speed, causing it to be thrown outward by centrifugal force into a channel where it is caught.

This action can be related to a ball being whirled at the end of a string. The higher the speed of rotation, the greater the centrifugal pull as the ball tries to get away. If the string were cut during rotation, the ball would fly off into space. In the centrifugal compressor, the molecules of gas are treated much the same way. By whirling them at a high rate of speed, they are thrown off the outer edge of the rotating wheel. Instead of flying off into space, however, they are caught in a channel where they are compressed by the following stream of molecules. In the reciprocating and rotary compressors, the refrigerant molecules are actually squeezed together inside the cylinder by the positive action of the piston. Compression is produced and maintained by the action of the suction and discharge valves. In the centrifugal compressor there are no valves, in the ordinary sense of the word, and the only wearing parts are the bearings at the ends of the shaft.

A centrifugal compressor is similar to a water pump, but it operates at a much higher speed than a water pump. (See Figure 4-17.) A centrifugal compressor consists of a wheel made of rotors or impellers. Each rotor is in a separate compartment or stage (see Figure 4-18). In this particular compressor there are five stages and the compressor is called a five-stage compressor. Each rotor has a series of vanes which are an integral part of the rotor. In operation, the refrigerant gas enters through the suction inlet and flows through the suction passages into the first rotor, through openings near the shaft. The centrifugal force developed by the high speed of the rotor discharges the gas from the periphery of the rotor at a higher pressure than it had at the suction passages. The refrigerant vapor com-

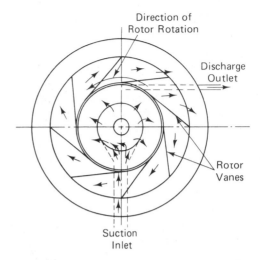

**Figure 4-17.** Cross-sectional view of a centrifugal compressor

pressed in the first rotor discharges into the space labeled C in Figure 4-18 between the first and second rotors. It passes back to the center of the compressor and enters the second rotor through the openings near

**Figure 4-18.** Longitudinal section of a centrifugal compressor

the shaft. The rotating motion of the second rotor again forces the refrigerant outward and it passes into passage D. In this manner the refrigerant follows a zigzag path through the centrifugal compressor until it is finally discharged through the discharge port marked G. From there it enters the discharge line and flows to the condenser.

Each rotor, or impeller, is progressively smaller in diameter and smaller in thickness, since the same amount of refrigerant vapor passes through each rotor, but in a more highly compressed state from stage to stage. The passages labeled B, C, D, E, F, and G are also progressively smaller. If the rotors and the passages were not made progressively smaller, the compressed gas would re-expand to fill the space and the compression would be nullified.

The refrigerant passages C, D, E, F, and G are formed of metal labyrinths, which prevent leakage of the refrigerant between the various stages of compression. The labyrinths are constructed so that they come comparatively close to the rotors without actually touching them. The clearance between these two parts is not at all the close clearance that is required in the rotary compressor. Lubrication is not needed at any place on the centrifugal compressor except at the end bearings of the shaft.

Since these end bearings are the only internal friction surfaces, the refrigerant gas which is compressed by a centrifugal compressor is free from oil, giving it the advantage of preventing an accumulation of oil on the inside surfaces of the condenser and evaporator. This situation greatly improves heat transfer.

Centrifugal compressors are designed to operate at relatively high speeds. They are quite efficient and are well suited for large capacity refrigerating plants. They are best used on systems of 250 tons up to 3000 tons. They are efficient at a wide range of operating temperatures from 50°F to 120°F (10° to 49°C). Because they are not positive displacement type compressors, they are quite flexible under varying load conditions and operate at good efficiences even when the demand is less than 40 percent of their designed capacity.

Reciprocating and centrifugal compressors are also typed according to their compressor body style. There are three styles of compressors: open, semi-hermetic, and hermetic. These may be further divided into belt-driven and direct-driven.

**Open type compressors** were used extensively in the early phases of the refrigeration industry. A separate motor was used and the power was transmitted to the compressor by a belt and a set of pulleys (see Figure 4-19). The pistons and cylinders were enclosed within a case with the crankshaft extending through the body with the attached pulley. In this type, a crankshaft seal prevented the loss of refrigerant and oil around the shaft.

**Figure 4-19.** Open type belt-driven compressor (*Courtesy of Tecumseh Products Company*)

This type of compressor is very flexible in that the speed can be changed to accommodate different refrigerants as well as to provide different capacities. Repairs are fairly easy to make and the replacement of worn parts is easily accomplished.

Even though open type compressors were widely used at one time, they have many inherent disadvantages. They are bulky and heavy because of the cast iron body. They are expensive to manufacture. The shaft seals are vulnerable to leaks, and shaft alignment is difficult on direct-drive models. They are more noisy because of the belts and other external components. The belts wear out and need to be replaced. Because of these disadvantages, the open type compressor is being replaced with the semi-hermetic or the hermetic type compressors. The open type compressor is only used in a few specialized applications, such as automotive air conditioning, transport refrigeration, large commercial and industrial units, and ammonia systems.

**Semi-hermetic compressors** are driven by an electric motor which is mounted directly on the compressor crankshaft (see Figure 4-20).

**Figure 4-20.** Semi-hermetic compressor (*Courtesy of Copeland Corp.*)

The working parts of both the motor and compressor are sealed within a common housing. The troublesome belts and shaft seal are eliminated. The motors can be sized specifically for the load which is to be handled resulting in a compact, economical, efficient, and basically maintenance-free design.

The removable heads, stator covers, bottom plates, and housing covers allow access for easy field repairs in the event of compressor damage. (See Figure 4-21.) The cast iron used in the manufacture of open and semi-hermetic type compressors must be of the close grained type to prevent refrigerant leakage through the pores of the metal.

**Hermetic compressors,** sometimes referred to as welded hermetic motor-compressors, are produced in an effort to further decrease the size and the cost of manufacturing refrigeration units. They are widely used in smaller horsepower self-contained equipment. They are popular in up to about the five-ton size in air conditioning applications. The motor and compressor are mounted on a common shaft as in the semi-hermetic units. However, the body is a formed metal shell hermetically sealed by welding (see Figure 4-22). No internal field repairs can be performed on this type of compressor since the only means of access is by cutting open the compressor shell.

The major internal parts of a hermetic compressor are listed in the same sequence as that of the refrigerant gas flow through the compressor. (See Figure 4-23.) First the suction is drawn into the compressor shell, then to and through the electric motor which provides power to the crankshaft. The crankshaft revolves in its bearings, driving the piston in the cylinder. The crankshaft is designed to carry oil from the oil pump in the bottom of the compressor housing to all bearing surfaces. Refrigerant gas

**MOTOR PROTECTOR—** Internal inherent "Center of the Wye" line break

**SUCTION VALVES—** Maximum life due to Swedish steel and finish

**PISTON PIN—**Hardened and superfinished

**VENTURI CHECK VALVE—** Insures positive oil return

**PISTON—**Close grain cast iron with same thermal expansion as cylinders, allowing small controlled clearance for low "blow-by" and oil pumping

**CONNECTING RODS—** Aluminum for light weight—precision bored for good bearing surfaces

**OIL PUMP—**Developed and manufactured by Copeland

**CRANKSHAFT—**High strength iron alloy

**SIGHT GLASS—** "Sealed in Glass"

**OIL CHECK VALVE—** Allows oil to flow into crankcase during running—prevents oil from flowing out of crankcase on start-up

**OIL SCREEN—**Fine mesh monel screen

**DISCHARGE VALVES—** Swedish steel, especially finished for long life

**TERMINALS —** "Sealed in glass"

**CASTING—**High Nickel content alloy cast iron

**STATOR—**Made to Copeland's exacting specifications with the latest insulating materials

**ROTOR KEY—**To insure no rotational slippage of rotor and crankshaft

**SUCTION SCREEN—**Full size, fine-mesh monel wire cloth

**ROTOR LOCK—**For absolute locking of rotor to crankshaft

**BEARINGS—**Solid cast bronze

**ROTOR—**Die cast aluminum end rings to assure proper electrical performance

**2-CYLINDER COPELAMETIC**

**Figure 4-21.** Cross-sectional view of semi-hermetic compressor (Courtesy of Copeland Corp.)

**Figure 4-22.** Hermetic compressor (*Courtesy of Copeland Corp.*)

surrounds the compressor crankcase and motor as it is drawn through the compressor shell and into the cylinder through the suction muffler and suction valves. As the gas is compressed by the moving piston, it is released through the discharge valves, discharge muffler, and compressor discharge tube.

## COMPRESSOR COMPONENT CONSTRUCTION

The various components which are assembled together to build a compressor must be manufactured to certain specifications. If these specifications are not met, the quality of the compressor will be lowered. Most of these specifications are met at the factory; however, the service technician must be acquainted with them so that repairs to a compressor can be efficiently and economically completed.

**Figure 4-23.** Internal view of a hermetic compressor (*Courtesy of Tecumseh Products Company*)

Compressor
Cylinder

**Figure 4-24.** Compressor cylinder (Courtesy of Copeland Corp.)

**CYLINDERS** must be manufactured of a good grade of close-grained hard cast iron. (See Figure 4-24.) In the factory, the cylinder is usually subjected to a slow running in process for a considerable period of time. This process enables the inside of the cylinder to attain a smooth, hardened, polished surface; in other words, the cylinder is very carefully broken in. Careful breaking in of the compressor will enable it to wear a long time. To prevent gas leakage past the piston, it is necessary that the cylinder bore be very carefully honed to a perfect circular shape.

**PISTONS** are made from a hollow cylinder casting which is closed on one end and provided with a piston pin (see Figure 4-25). The piston pin is used to connect the piston to the connecting rod. The diameter of the

piston is a few thousandths of an inch smaller than the bore of the cylinder. This allows space for expansion and allows the lubricating oil to enter between the cylinder wall and piston. Rings are provided on the piston to prevent excessive oil from passing into the cylinder. The lubricating oil serves a dual purpose of lubricating the movement of the piston and sealing the pressure of the compressed refrigerant to prevent its leaking between the piston and the cylinder wall.

**VALVES** used in most modern refrigeration compressors are termed *flapper valves*. They are used to control the flow of refrigerant through a compressor. The proper operation of a compressor depends greatly on their condition and operation. There are two such valves for each cylinder in a compressor. They are termed *suction* and *discharge* valves. The suction valve controls the flow of refrigerant vapor into the compressor, while the discharge valve controls the flow of refrigerant from the compressor.

The flapper valve consists of a thin flat strip of special valve steel. This type of valve operates successfully at extremely high compressor speeds and does not hammer down the valve seat like heavier type valves. It has very little valve drag due to its extremely light weight and as a result, the volumetric efficiency of the compressor is greater than that of a compressor using a heavier type valve with a coil spring. At high compressor speeds, heavy valves prove extremely inefficient.

The purpose of all compressor suction valves is to open and admit the refrigerant vapor to the cylinder chamber on the suction stroke of the piston. On the compression stroke, the suction valve closes and traps the refrigerant vapor within the cylinder chamber so that it can be compressed to a higher pressure and temperature. Suction valves are operated automatically by the

Piston Ring
Grooves

Piston Head

Wrist Pin Bearing

Piston

**Figure 4-25.** Compressor piston

differences in pressures within the compressor. In most types a spring is not used. The valve consists of a thin piece of steel which may take the shape of a ring, disc, or reed (see Figure 4-26). In each case, whatever the shape may be, the thin metal covers a port when it is closed and is lifted from the port when open.

Suction valves may be placed in the head of the piston or located on the valve plate of the compressor. When the suction valve is located in the piston it is commonly known as a piston valve (see Figure 4-27).

Every type of valve must have some means of limiting its motion. If the valve movement is too great, it may not close with

**Figure 4-27.** Suction valve locations (*Courtesy of Copeland Corp.*)

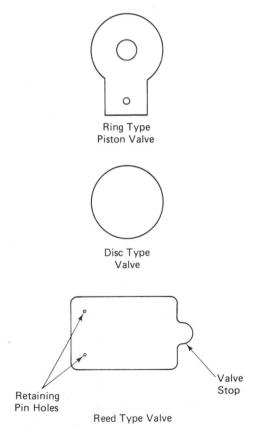

Ring Type
Piston Valve

Disc Type
Valve

Retaining
Pin Holes

Valve
Stop

Reed Type Valve

**Figure 4-26.** Suction valve types

sufficient rapidity on the compression stroke to seal properly. In addition to the different types of valves, the means whereby valve motion is limited also vary. In the piston valve of the ring type the piston head plate, or cage, is screwed tightly to the piston and keeps the various parts of the valve in place (see Figure 4-28). The valve seat consists of a circular raised edge on top of the piston. This edge is machined to a smooth surface and lapped. There is also another lapped circular edge or seat concentrically located within the valve seat. Between these two lapped circular seats is the opening through which the refrigerant vapor passes when the valve ring is raised on the down stroke of

Top View of Assembled
Piston

Valve Cover

Valve Retainer ⎫ Retainer
Valve Spring   ⎰ Assembly

Valve Disc

Piston

**Figure 4-28.** Suction valve stops

the piston. The valve ring has an anchoring tongue extension with a hole which slips over the valve retainer pin. The valve ring has a rather limited up and down motion since it can only move between the edge of the valve seat and the underside of the piston head plate. Its motion is opposed by a small spiral spring fastened to the piston head. The primary purpose of this spring is to prevent the valve from opening and closing abruptly, as this would introduce a clicking noise during operation of the compressor. In addition to its noise-eliminating action, the spring also helps keep the valve closed when the compressor is idle. This reduces the possibility that any refrigerant that has

leaked past the discharge valve will enter a crankcase.

Not all suction valves are located in the head of the piston. Many compressors have the suction valves located in the compressor cylinder head. The valves are mounted on a head plate which also carries the discharge valves. Suction valves and discharge valves move in opposite directions. The plate is known as a valve plate and is installed between the compressor cylinder proper and the cylinder head. (See Figure 4-29.) The advantage of such construction is that the cool refrigerant vapor returning from the evaporator helps to cool the cylinder head. Because the gas does not pass through the compressor crankcase, the possibility of pumping oil is reduced. Suction valves located in the valve plates have much the same construction as those located in the piston.

The reed type suction valve is made of very thin spring steel and is a spring in itself. It is flexed, that is bent downward, by the pressure of the incoming gas (see Figure 4-30). The reed constantly tends to hold itself closed except when its spring tension is

**Figure 4-29.** Valve plate location (*Courtesy of Copeland Corp.*)

**Figure 4-30.** Reed type suction valve

**Figure 4-31.** Bridged flapper type discharge valve

overcome by the pressure of the suction gas.

The purpose of discharge valves is to allow the compressed refrigerant gas to be discharged to the condenser and to maintain a gas-tight seal between the high and low sides of the system when the compressor is at rest. This prevents the high side condensed refrigerant from vaporizing and passing into the low side. The discharge valve in effect is a check valve since it checks the compressed refrigerant from re-expanding back into the cylinder chamber.

Discharge valves are almost always located in the cylinder head on the valve plate. There are several designs and types: the ring type; the disc type; and the reed type, or flapper valve as it is sometimes called. Most discharge valves are designed with an auxiliary safety discharge which comes into play if a quantity of oil or liquid refrigerant enters the cylinder chamber. Oil and liquid refrigerant are incompressible and precautions must be taken to allow them to pass safely out of the cylinder chamber.

There are many designs of the flapper type discharge valve. The bridged flapper valve is the most common and this discussion will be limited to it. The bridged flapper type discharge valve is so called because the valve bridges across the opening between the two valve guides (see Figure 4-31). The valve guides consist of specially shaped ma-

chine screws which are screwed into the valve plate. The discharge valve proper consists of a steel reed secured at both ends by the valve guides. Directly above the reed is a valve spring stop. The valve spring stop has a slight upward bend or curvature in the middle so that the thin reed can be forced sufficiently upward by gas pressure from the cylinder chamber to allow the compressed gas to pass. Should a quantity of oil be trapped in the cylinder chamber it will force the valve up against the valve spring stop. The valve spring stop will be forced upward compressing the valve springs. This action allows the oil to escape without permanently deforming or breaking the reed, since the amount of flexing of the reed is limited by the valve spring stop. The valve springs constitute a safety discharge feature and are often known as safety springs. The valve guides are wired together at the top by a locking wire to prevent possible loosening

A thin film of oil between the valve and its seat maintains the actual seal in all cases regardless of the type of valve. When the oil charge in the crankcase is correct, a sufficient amount of oil flows through the system with the refrigerant to lubricate the valves and effect a gas-tight seal. It is therefore important that there be sufficient oil in the crankcase of the compressor. A compressor loses its efficiency when the discharge or suction valves leak.

To check the condition of the valves, first stop off the flow of refrigerant to the suction side of the compressor. This is usually done by *front-seating* (turning all the way in) the suction service valve with the gauge manifold attached. Run the compressor for a few minutes. The compressor should pump a vacuum of approximately 20 inches of mercury. If it does not, stop the compressor for a few minutes, then start the compressor and see if it will pump the desired vacuum. If not, the suction valves are leaking and must be replaced.

To check the discharge valves, attach the gauge manifold, close off the flow of refrigerant to the suction side of the compressor, and run the compressor until it pumps as deep a vacuum as possible. Stop operating the compressor and observe the reading on the compound gauge. The pressure should not increase over 3 or 4 inches of mercury. If it does, try pumping another vacuum and again observe the pressure reading on the compound gauge. If the pressure still increases, the discharge valve is leaking. One step more will relieve any doubt about the condition of the discharge valve. While the vacuum remains on the compressor, front seat the discharge service valve. If the pressure reading on the compound gauge stops rising, the discharge flapper valve is positively bad and must be replaced.

The discharge valve on a rotary compressor prevents the compressed refrigerant from re-entering the cylinder chamber after it has been discharged from the compressor. The valve seat is ground and polished or lapped to a mirror finish. The valve is a reed or flapper type made of spring steel and provided with a stop spring which returns the valve to its seat with each discharge (see Figure 4-32). The valve cover and the muffler reduce the noise resulting from the discharge pulsations. The valve reed

**Figure 4-32.** Rotary compressor discharge flapper valve

is assembled to the plate by means of a screw and lock washer. Not all types of rotary compressor discharge valves use a valve stop spring, since on some rotary compressors the valve is not a positive type and may allow some refrigerant to leak back into the compressor housing. A check valve is used to prevent the leakage of refrigerant from the compressor housing into the evaporator.

**THE CONNECTING ROD** is the link which connects the piston to the crankshaft. It is provided with bearings on each end to take the wear and thrust at the piston pin and at the crankshaft (see Figure 4-33). The

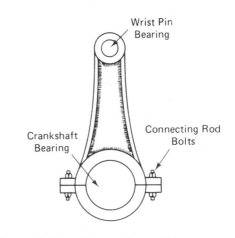

**Figure 4-33.** Connecting rod

wearing surfaces are provided with close-fitting bushings or bearings which are made of a bronze alloy or babbit.

**CRANKSHAFTS** are made of forged steel. The crankshaft must be sufficiently heavy to resist deflection. A crankshaft which deflects interferes with the smooth up and down motion of the piston which results in excessive vibration (whipping). In a two-cylinder compressor the crank throws are usually located at opposite sides of the shaft to balance each other. This enables the movement of the pistons also to balance and reduces vibration to a great extent. Crankshafts are usually counterbalanced to prevent vibration.

The amount of vibration in a compressor is directly related to the number of cylinders it contains. A single-cylinder compressor will vibrate more than a compressor of the same capacity with two cylinders. Likewise, a two-cylinder compressor will vibrate more than a compressor of the same capacity with four cylinders. This is because the total volume is divided equally among the cylinders, reducing the quantity of refrigerant pumped by each cylinder and allowing a more constant flow of the discharged refrigerant.

## COMPRESSOR SPEED

Early refrigeration compressors were designed to operate at relatively low speeds, usually less than 1000 rpm. Semi-hermetic and hermetic motor-compressors were designed to operate at 1750 rpm and made use of the standard four-pole electric motors. The increasing demand for lighter weight and more compact air conditioning equipment has been instrumental in the development of hermetic motor-compressors equipped with two-pole electric motors which operate at 3450 rpm.

Certain special applications such as automotive air conditioning, aircraft, or military air conditioning equipment make use of even higher speed compressors. For the normal commercial and domestic applications, however, the existing sixty-cycle electric power supply will generally limit compressor speeds to the currently available 1750 and 3450 rpm.

Compressors operating at higher speeds tend to have lubrication problems and a shorter life. These factors, together with cost, size, and weight must be considered in compressor design and application.

## LUBRICATION

An adequate supply of oil must be maintained in the crankcase at all times to ensure continuous compressor lubrication. The normal oil level should be maintained at or slightly above the center of the *oil sight glass,* sometimes called the bull's eye. (See Figure 4-34.)

Only a good grade of refrigeration oil should be used to lubricate refrigeration compressors. Refrigeration oil is made of a high grade of mineral oil. It is wax free and has additives to reduce foaming. At the factory, special precautions are taken to remove any moisture and foreign particles from it. After a container of oil has been opened, it should be tightly closed and precautions taken to prevent moisture and dirt from getting into the oil.

**Figure 4-34.** Oil sight glass

Oil of the specific viscosity should be used for each temperature application. Each manufacturer has a recommended viscosity for each type of equipment application and, for best results, their suggestions should be followed.

There are two basic methods used to lubricate the compressor parts: *splash* and *forced*. Almost all compressors use a combination of these two.

**SPLASH LUBRICATION** is the simplest method of lubricating the compressor parts. The crankcase oil is splashed onto the moving parts by the rotating crankshaft. This method supplies oil to the cylinder walls and bearing surfaces. The splash method of compressor lubrication was satisfactory for the slower speed compressors. It became inadequate when the large modern high-speed compressors were designed because of their higher bearing and friction surface temperatures. Lubricating oil is not only used to reduce the friction between moving parts, but to carry away some of the heat produced by this friction.

**FORCED LUBRICATION** is provided on almost all modern refrigeration compressors. There are several ways to force lubricate a compressor. On the smaller size compressors, up to about 3 hp, the oil is forced to the desired points by providing rifled passageways. When this type of system is used, the oil is provided to the passageway by splashing. As the compressor parts move, the oil is forced to the bearing surface in the same manner that a projectile is forced to spin in a gun barrel.

In the larger horsepower models, over 3 hp, a more positive type lubrication system is used. Compressor lubrication is provided by means of a positive displacement oil pump. The pump is mounted on the rear

**Figure 4-35.** Oil pump location

bearing housing and is driven from a slot in the crankshaft into which the flat end of the oil pump drive shaft is fitted. (See Figure 4-35.)

Oil is forced through a hole in the crankshaft to the compressor bearings and connecting rods. A spring-loaded ball check valve serves as a pressure relief device, allowing oil to bypass directly to the compressor crankcase if the oil pressure rises above its setting (see Figure 4-36).

Because the pump intake is directly connected to the compressor crankcase, the oil pump inlet pressure will always be equal to the crankcase pressure, and the oil pump outlet pressure will be the sum of the crankcase pressure plus the oil pump pressure.

**Figure 4-36.** Location of spring-loaded bypass assembly (*Courtesy of Copeland Corp.*)

Therefore, the net oil pressure is always equal to the pump outlet pressure minus the crankcase pressure. When the compressor is operating with the suction pressure in a vacuum, the crankcase pressure is negative and must be added to the pump outlet pressure to determine the net oil pump pressure. A typical compound gauge is calibrated in inches of mercury for vacuum readings, and 2 inches of mercury is approximately equal to 1 psi. For example:

| CRANKCASE PRESSURE | PUMP OUTLET PRESSURE | NET OIL PUMP PRESSURE |
|---|---|---|
| 50 psi (3.51 kg/cm²) | 90 psi (6.3 kg/cm²) | 40 psi (2.81 kg/cm²) |
| 8″ vacuum (equivalent to a pressure reading of minus 4 psig) (.276 kg/cm²) | 36 psi (2.53 kg/cm²) | 40 psi (2.81 kg/cm²) |

In normal operation, the net oil pressure will vary depending on the size of the compressor, the temperature and viscosity of the oil, and the amount of clearance in the compressor bearings. Net oil pressures of 30 to 40 psi (2.19 to 2.81 kg/cm²) are normal, but adequate lubrication will be maintained at pressures down to 10 psi (.703 kg/cm²). The bypass valve is set at the factory to prevent the net oil pressure from exceeding 60 psi (4.22 kg/cm²).

The oil pump may be operated in either direction; the reversing action is accomplished by a friction plate which shifts the inlet and outlet ports. After prolonged operation in one direction, however, wear, corrosion, varnish formation, or burrs may develop on the reversing plate which can prevent the pump from reversing. Therefore, on installations where the compressors have been in service for some time, care must be taken to maintain the original rotation of the motor if for any reason the electrical connections are disturbed.

The presence of liquid refrigerant in the compressor crankcase can materially affect the operation of the oil pump. Violent foaming on start-up can result in a loss of oil from the crankcase, and a resulting loss of oil pressure until some of the oil returns to the crankcase. If liquid refrigerant or a refrigerant-rich mixture of oil and refrigerant is drawn into the oil pump, the resulting flash gas may result in large variations and possibly a loss of oil pressure. The crankcase pressure may vary from the suction pressure because liquid refrigerant in the crankcase can pressurize the crankcase for short intervals. Therefore, the oil pressure safety control low-pressure connection should always be connected to the crankcase.

## COMPRESSOR SHAFT SEALS

On the conventional open type compressor, where the motor and compressor are two separate components, the shaft of the compressor extends out through the compressor crankcase. Some method of sealing around the rotating shaft must be used. This seal must positively prevent leaks both when the compressor is operating and when it is at rest. The seal must not allow gas or oil to escape out of the compressor, nor must it allow air and moisture to leak into the compressor.

When the compressor is operating, the oil is splashed around the interior of the crankcase, and is conducted to and lubricates the seal faces. It is the oil film between the seal faces that does the actual sealing. Therefore, it is important that this oil film never be permitted to dry out as this would immediately cause the seal to leak. A dry seal is also apt to squeal when the compressor is operating and wear out quickly.

Some refrigerants dissolve considerable oil. When an open type compressor with a shaft seal is allowed to stand idle for a considerable length of time, the refrigerant may dissolve the oil from the seal faces. This type of compressor should not be allowed to stand idle for long periods of time unless it has been *pumped down;* that is, unless the refrigerant has been pumped into and stored in the liquid receiver and condenser with the compressor suction and discharge service valves closed.

*CAUTION:* Never attempt to operate a compressor with the discharge service valve closed because considerable damage could be done to the compressor.

**PACKING GLAND SEALS** were used in early model compressors. This type of shaft seal is still used on large capacity ammonia compressors and compressors that operate at slow speeds. The packing gland seal guards the drive shaft against loss of refrigerant on compressors operating above atmospheric pressure. On compressors operating below atmospheric pressure, it prevents air from being drawn into the system. The early type of packing consisted of a composition made of asbestos, fiber, and graphite. This material was placed in the housing surrounding the shaft and held in tightly by means of a metal packing gland (see Figure 4-37). The compressor shaft extends through the compressor crankcase and is sealed by rings of semimetalic graphite impregnated packing. The packing gland is drawn down tightly by means of the take-up bolts. When the bolts are tightened, the packing is compressed, forming a seal between the shaft and the housing.

Packing glands have the disadvantage that they must be frequently tightened to prevent leakage; even so, there is a tendency

**Figure 4-37.** Packing gland type shaft seal

for loss of refrigerant and oil. Another difficulty is that too tight a packing gland might interfere with proper starting of the compressor. Also, the pressure required to effect a tight seal causes considerable wear on the packing which is softer than the shaft and makes frequent replacement of the packing necessary.

**STATIONARY UNBALANCED BELLOWS SEALS** use one end of the bellows as a flange while the other end has a seal nose. (See Figure 4-38.) The spring is of a smaller diameter than the bellows, which makes this seal the internal spring type. The seal nose and the flange are solidly soldered to the bellows. When the bellows seal is assembled on the compressor, the seal nose makes contact and rubs against the shoulder on the crankshaft (see Figure 4-39). The bellows flange is clamped between the com-

pressor body and the removable seal cover plate. This forms the second gas-tight seal by means of a gasket. The seal cover plate is secured to the compressor body by six cap screws. The shoulder of the shaft is case hardened, ground, and lapped to a very smooth surface. The seal nose also must be ground and lapped and free from any surface marks or defects. Only if both the face of the shoulder on the shaft and the nose of the seal are absolutely smooth will the oil be able to maintain a perfect seal.

**DIAPHRAGM TYPE SEALS** are also known as the *balanseal*. The diaphragm is soldered to the seal nose and clamped between the compressor body and the cover plate with a gasket. The seal nose does not ride directly on the shoulder of the crankshaft. Instead, it rubs against a hardened seal ring collar which rotates with the shaft. In case of wear the collar, as well as the seal nose, are easily replaceable. (See Figure 4-40.)

**Figure 4-38.** Stationary unbalanced bellows seal

**Figure 4-39.** Unbalanced bellows seal installed on a compressor

Figure 4-40. Diaphragm type seal

The diaphragm maintains a practically constant pressure between the seal nose and the hardened seal collar regardless of the crankcase pressure. This is accomplished by a balancing effect created by a fulcrum located at the proper point outside the diaphragm. The exaggerated sections 1, 2, and 3 in Figure 4-40 indicate how the diaphragm acts and how the pressure between the two sealing surfaces is equalized when affected by varying pressures in the crankcase. Section 1 indicates the position of the diaphragm when the seal is first assembled and the crankcase is at atmospheric pressure. The fulcrum presses against the diaphragm, which pressure in turn is transmitted by the diaphragm to the seal nose causing it to press against the hardened steel collar. When the pressure in the crankcase is increased, the diaphragm is caused to bulge outward as shown in section 2. This outward bulging action of the diaphragm working across the fulcrum causes the seal nose to be pressed more tightly against the seal collar. This compensates for the additional pressure in the crankcase which, of course, would act directly on the seal nose and tend to push it away. The additional pressure exerted by the diaphragm would cause these two pressures to equalize; thus, the friction between the two sealing surfaces would be practically unchanged.

In section 3, the appearance of the diaphragm shows that the pressure within the crankcase has decreased; that is, it has fallen below atmospheric pressure. The decreased pressure within the crankcase has caused the diaphragm to bulge inward. This action releases the pressure on the fulcrum and reduces the diaphragm pressure on the seal nose. Since the gas pressure against the seal nose has also decreased, it can readily be seen that the pressures are equalized; thus, the friction between the two sealing faces has not been materially changed.

**ROTARY SEALS** are widely used today. A rotary seal is available for replacing the bellows type seals. In these seals the bellows rotates with the shaft.

In the simplest adaptation of the rotary seal principle, the seal assembly consists of: seal face, friction ring, retaining shell, friction ring band, spring, and spring holder (see Figure 4-41).

In operation, the entire seal assembly rotates with the shaft (see Figure 4-42). A running seal joint is formed by contact between the surface of the seal face (1) within a mating stationary seal seat (A). These surfaces are finished to an extremely high degree of flatness and smoothness. The seal face (loosely encircling the shaft) is held tightly against the stationary seat by the pressure of the spring (5) bolstered by any internal gas or liquid pressure; the mating is so perfect that no liquid or gas can pass.

The back of the seal face rests against the front surface of the resilient friction ring (2),

Figure 4-41.   Rotary seal components

**Figure 4-42.**   Rotary seal installed

which fits frictionally tight on the shaft surface and prevents leakage at that point. The friction ring band (4) fits around the hub end of the friction ring and restricts its expansion, thus controlling and maintaining the tightness of the rings fit on the shaft.

A retaining shell (3), fitting snugly over the flanged portion of the friction ring and loosely encircling the seal face is provided with keys which engage in corresponding lateral grooves in the outer periphery of the seal face. This keying arrangement combines with the resiliency and snug fit of the friction ring both on the shaft and in the retaining shell to effect a positive, yet simple and flexible driving connection between the shaft and the seal assembly unit.

The spring, encircling the shaft, is compressed to a predetermined length between the back wall of the retaining shell and the spring holder (6), which may rest against any part of the shaft or any member fastened to the shaft. The pressure developed holds the seal face in perfect sealing contact with the stationary seal seat, and also provides the force to form a gasket joint between the friction ring and the back of the seal face.

It should be noted that just one part, the relatively soft and elastic friction ring, is tightly fastened to the shaft. Consequently, the seal face can freely adjust its running lane to compensate for any shaft vibration, deflection, lateral play, or slight misalignment which might tend to break the running seal joint.

The end plate is stationary and forms the stationary surface of the seal (see Figure 4-43). Thus, a seal is created between the rotating surfaces on the seal nose and the stationary seal surface of the end plate. A second seal is created by the friction ring, preventing leakage along the shaft. A third seal is formed by the gasket, which is placed under the end plate. The friction ring is made of a synthetic rubber substance, neoprene, which is not affected by refrigerant gas or oil.

## COMPRESSOR DISPLACEMENT

The displacement of a compressor is the volume displaced by the pistons. Some manufacturers rate the displacement of their compressors in terms of cubic feet per hour, while others rate their compressors in terms of cubic inch displacement per revolution, or in cubic feet per minute. For comparative purposes, compressor displacement may be calculated by the following formulas:

**Figure 4-43.**   Rotary seal

$$CFM = \frac{\pi \times D^2 \times RPM \times N}{4 \times 1728}$$

## Conversion Factors

|                | 1750 RPM | 3500 RPM |
|----------------|----------|----------|
| CFH =          | 60 × CFM | 60 × CFM |
| CFH =          | 60.78 × cu.in./rev. | 121.5 × cu.in./rev. |
| CFM =          | 1.013 × cu.in./rev. | 2.025 × cu.in./rev. |
| Cu.in./rev. =  | .01645 × CFH | .00823 × CFH |
| CFM =          | Cubic feet per minute | |
| CFH =          | Cubic feet per hour | |
| Cu.in./rev. =  | Cubic inch displacement per revolution | |
| $\pi$ =        | 3.1416 | |
| D =            | Cylinder bore in inches | |
| L =            | Length of stroke in inches | |
| N =            | Number of cylinders | |
| RPM =          | Revolutions per minute | |
| 1728 =         | Cubic inches per cubic foot | |
| $\dfrac{\pi D^2}{4}$ = | Area of a circle | |

$$CFH = \frac{\pi \times D^2 \times RPM \times N \times 60}{4 \times 1728}$$

$$\text{Cu. In./Rev.} = \frac{\pi \times D^2 \times L \times N}{4}$$

## COMPRESSION RATIO

Lack of knowledge is a tremendous factor in equipment failures because few service technicians really know how to calculate compression ratio. Many have no idea of the dangers involved in designing a system with a suction pressure which is too low. Surveys show that fewer than ten percent of service technicians are able to calculate the compression ratio of a given system. A system operating with a high compression ratio may have a discharge temperature as much as 150°F (66°C) higher than it should be.

The rate of chemical reaction caused by reactive materials such as refrigerant oil, cellulose, oxygen, moisture, acid, heat, and pressure approximately doubles for each 18°F (−7.8°C) rise in discharge temperature. It is only natural that a system running at abnormally high head temperatures will develop more problems in a shorter period of time than one operating with normal discharge temperatures. Sound refrigeration design and common sense seem to indicate that the relationship between the head pressure and the suction pressure should be well within the accepted industry bounds of a 10:1 compression ratio.

Compression ratio is defined as *the absolute head pressure divided by the absolute suction pressure.* Most service technicians realize that an ordinary compound gauge does not indicate atmospheric pressure but reads zero when it is not connected to a system under pressure. Most service technicians know that to find the absolute head pressure or absolute suction pressure at zero pounds gauge pressure or above, 15 psi (1.054 kg/cm²) must be added to the gauge reading.

The difficulty in computing the absolute suction pressure is when the suction pressure is below atmospheric pressure. This is done by subtracting the vacuum reading in inches from 30 inches (1.054 kg/cm²) and dividing the result by 2.

$$\text{Compression Ratio} = \frac{\text{Absolute Head Pressure}}{\text{Absolute Suction Pressure}}$$

Case 1—for 0 psig pressure
    or above:

Absolute Head Pressure = Gauge Reading + 15 psi (1.054 kg/cm²)
Absolute Suction Pressure = Gauge Reading + 15 psi (1.054 kg/cm²)

Case 2—when the low side reading is in a vacuum:

Absolute Head Pressure = Gauge Reading + 15 psi (1.054 kg/cm²)

$$\text{Absolute Suction Pressure} = \frac{30 - \text{Compound gauge reading in inches}}{2}$$

The calculation of compression ratio can be illustrated by the following examples:

Example 1:

$$\text{Head Pressure} = 160 \text{ psi } (11.25 \text{ kg/cm}^2)$$

$$\text{Suction Pressure} = 10 \text{ psi } (.703 \text{ kg/cm}^2)$$

$$\text{Compression Ratio} = \frac{\text{Absolute Head Pressure}}{\text{Absolute Suction Pressure}}$$

$$= \frac{160 \,(11.25 \text{ kg/cm}^2)}{10 \,(.703 \text{ kg/cm}^2)} + 15 \,(1.054 \text{ kg/cm}^2)$$

$$= \frac{175 \,(12.3 \text{ kg/cm}^2)}{25 \,(1.757 \text{ kg/cm}^2)} = 7{:}1$$

Example 2:

$$\text{Head Pressure} = 160 \text{ psi } (11.25 \text{ kg/cm}^2)$$

$$\text{Suction Pressure} = 10'' \text{ vacuum } (-351 \text{ kg/cm}^2)$$

$$\text{Absolute Head Pressure} = 160 \,(11.25 \text{ kg/cm}^2) + 15 \,(1.054 \text{ kg/cm}^2) =$$

$$= 175 \text{ psia } (12.3 \text{ kg/cm}^2)$$

$$\text{Absolute Suction Pressure} = \frac{30 - 10 \,(1.054 \text{ kg/cm}^2) - .351 \text{ kg/cm}^2}{2} = \frac{20}{2}$$

$$= 10 \text{ psia } (.703 \text{ kg/cm}^2)$$

$$\text{Compression Ratio} = \frac{\text{Absolute Head Pressure}}{\text{Absolute Suction Pressure}}$$

$$= \frac{175 \text{ psi } (12.3 \text{ kg/cm}^2)}{10 \text{ psi } (.703 \text{ kg/cm}^2)} = 17.5{:}1$$

In the two preceding examples, we can see what a great influence suction pressure has on the compression ratio of a system. A change in the head pressure does not produce such a dramatic effect. If the head pressure in both examples was 185 lbs. (13 kg/cm²), the compression ratio in Example 1 would become 8:1, and in Example 2, it would be 20:1.

Example 2 falls somewhat in the class of one of our more serious troublemakers, which is the R-12 system operating on a coil of approximately −35°F (−37.2°C), with a suction pressure of 8 inches of vacuum (.276 kg/cm²) or lower. It is interesting to compare two systems using different types of refrigerants, one system operating with R-12 and the other with R-22. At the −35°F (−37.2°C) evaporation, the R-22 system would have a 10.9:1 compression ratio, while the R-12 system would have a 17.4:1 compression ratio. The R-22 system is a borderline case, but the R-12 system is far out of the safe range and would run very hot with all the accompanying problems.

## CLEARANCE VOLUME

As previously mentioned, the volumetric efficiency of a compressor will vary with the compressor design. If the valves seat properly, the most important factor affecting compressor efficiency is the *clearance volume.*

At the end of the compression stroke, there still remains some clearance space which is essential if the piston is not to hit the valve plate (see Figure 4-44). There is also a great deal more space in the discharge valve ports which are located in the valve plate, since the discharge valves are on top of the valve plate. This residual space, which is unswept by the piston at the end of its stroke, is termed clearance volume. It remains filled with hot, compressed gas at the end of the compression stroke.

**Figure 4-44.** Compressor clearance volume

When the piston starts down on the suction stroke, the residual high pressure gas expands as its pressure is reduced. No vapor from the suction line can enter the cylinder chamber until the pressure in the cylinder has been reduced below the suction line pressure. (See Figure 4-45.) Thus, the first

**Figure 4-45.** Pressure in a compressor cylinder

part of the suction stroke is actually lost from a capacity standpoint. As the compression ratio increases, a greater percentage of the suction stroke is occupied by the residual gas.

With high suction pressures, the compression ratio is low and the clearance volume is not critical from a capacity standpoint. An additional clearance volume is also helpful in reducing the compressor noise level. Lower gas velocities through the discharge ports reduce both wear and operating power requirements. Most air conditioning compressor valve plates are designed with a greater clearance volume by increasing the diameter of the discharge ports.

On low temperature applications, it is often necessary to reduce the clearance volume to obtain the desired capacity. Low temperature valve plates, which have smaller discharge port sizes to reduce the clearance volume, are used on most low temperature compressors.

## COMPRESSOR COOLING

Air-cooled compressors require an adequate flow of cooling air over the compressor body to prevent the compressor from overheating. The air flow from the fan must be discharged directly on the motor-compressor. Air drawn through the compressor compartment usually will not cool the compressor adequately.

Water-cooled compressors are provided with a water jacket or wrapped with a copper water coil. The water must be circulated through the cooling circuit when the compressor is in operation.

Refrigerant-cooled motor-compressors are designed so that suction gas flows around and through the motor to help cool it. At evaporating temperatures below 0°F ($-17.8$°C), additional motor cooling by means of air flow is necessary because the decreasing density of the refrigerant gas reduces its cooling ability.

Another method of compressor-cooling is to cool the oil. Oil coolers are found only on low temperature refrigeration models. External tubes to the oil cooler connect to a coil or hairpin bend within the compressor oil sump. This coil or hairpin is not open inside the compressor. (See Figure 4-46.)

Oil Cooler Tubes

**Figure 4-46.** Oil cooler location (*Courtesy of Tecumseh Products Company*)

Its only function is to cool the compressor oil sump. The oil cooler tubes are most generally connected to an individually separated tubing circuit in the air-cooled condenser.

## FACTORS CONTROLLING COMPRESSOR OUTPUT

There are many factors which control the pressure which a compressor is able to pump. Briefly, these factors are: the design and condition of the compressor, the type of refrigerant used, and the conditions under which the compressor must operate. The following list includes the main factors which control the pressures attained in the compressor and, as a result, control the output of the compressor:

1. Type of refrigerant: freon, ammonia, etc.

2. Compression Ratio: The compression ratio is the ratio between absolute head pressure and absolute suction pressure.

3. Volumetric Efficiency: the ratio of the volume of refrigerant gas actually drawn into the cylinder to the volume of the piston displacement. Piston displacement is the volume of space in a cylinder swept through by the piston.

$$P.D. = \frac{\pi \times D^2 \times L}{4}$$

$$V.E. = \frac{\text{Actual Volume}}{\text{Calculated Volume}}$$

4. Cooling System Efficiency: An efficient cooling system increases the volumetric efficiency by lowering the refrigerant temperature, thus reducing expansion of the refrigerant. This allows a greater amount of gas by weight to enter the cylinder.

5. Cylinder Cooling System: The amount of compression heat removed from the refrigerant by the cylinder cooling system will also increase the volumetric efficiency.

6. Pressure of Refrigerant in the Suction Line: The lower the pressure of the refrigerant in the suction line, the smaller will be the volume of refrigerant drawn into the cylinder on the suction stroke.

7. Compressor Speed: At a high compressor speed the valves have a tendency to "drag" due to their weight and inertia, reducing the quantity of gas taken in on each stroke.

8. Type and Size of Valves: Size of the valve openings and the positiveness or speed of the valve action affect the volume of gas compressed.

9. Friction of the Refrigerant Gas: The friction of the gas against the turns in the tubing and the openings to the compressor decreases the speed of the gas entering the cylinder which reduces the quantity per stroke.

10. Mechanical Condition of the Compressor: Leaky piston rings, leaky valves, and worn bearings decrease the compressor output.

11. Lubricants: Proper lubrication decreases friction and seals the piston and valves against leakage, thus increasing the compressor output.

One of the basic difficulties in preventing compressor failure is the inability to determine what actually causes the failure. The compressor is the functioning heart of the refrigeration system, and regardless of the nature of a system malfunction, the compressor must ultimately suffer the consequences. Since the compressor is the component that fails, there often is a tendency to blame any failure on the compressor without determining the origin of the malfunction. In far too many cases, the actual cause of failure is not discovered and corrected, resulting in recurring failures that could have been prevented.

If the service technician is to help in eliminating the causes of compressor failure, he must thoroughly understand both the operation of the system and the possible causes of failure that might occur. He must be on the alert for any signs of system malfunction.

## SAFETY PROCEDURES

Refrigeration and air conditioning compressors are sometimes very heavy. Caution should therefore be exercised when removing or installing these components.

1. Never lift heavy objects with the back muscles; use the leg muscles.

2. Get help when lifting heavy objects.

3. Always protect open compressors and open refrigerant lines from dirt and moisture.

4. Clean up any oil that spills from the compressor. Slipping on oily surfaces can cause serious personal injury.

5. Use caution to prevent oil from a burnt compressor from contacting the skin. This oil is usually acidic and can cause irritation.

6. Be sure to tighten all bolts and nuts with the proper torque.

7. Always be sure that electrical power is disconnected before touching the terminals with hands or metal tools.

8. Be sure to tighten all bolts in the proper sequence to prevent ruining the parts.

9. Always wear goggles when charging or discharging the system refrigerant.

10. Exercise care when handling compressors that have been operating. There may be hot components which can cause burns.

11. Never close the discharge service valve without a pressure relief port available.

## SUMMARY

• The two functions of a compressor are (1) to draw the refrigerant from the evaporator lowering the pressure to the desired evaporating temperature, and (2) to raise the refrigerant vapor pressure high enough so that the saturation temperature is higher than the temperature of the condenser cooling medium.

• The most popular types of compressor are (1) reciprocating, (2) rotary, and (3) centrifugal.

• The most popular ways of producing reciprocating motion are: (1) crank-throw crankshaft, and (2) eccentric disc crankshaft.

• Connecting rods link the crankshaft and the piston together.

• Vibration in a compressor is in direct relation to the number of its cylinders.

• Only a good grade of refrigeration oil should be used to lubricate a compressor.

• The methods used to lubricate compressors are: (1) splash, (2) forced, and (3) a combination of the two methods. Oil pumps are used on compressors of 3 hp and up.

• Liquid refrigerant in the oil will cause erratic operation of the oil pump.

• Shaft seals are used on open type compressors to prevent leaking between the compressor shaft and body.

• The types of shaft seals are (1) packing gland, (2) stationary unbalanced bellows, (3) diaphragm, and (4) rotary.

• The displacement of a compressor is the volume displaced multiplied by the number of pistons.

• Compression ratio is the absolute head pressure divided by the absolute suction pressure.

• Accepted industry bounds of compression ratio are 10:1 or less.
Air-cooled compressors must be provided with an adequate flow of air.

• Air-cooled compressors must be provided with an adequate flow of air.

• Oil coolers use a small portion of the condenser to cool the compressor oil.

• The major factors controlling compressor output are: (1) type of refrigerant, (2) compression ratio, (3) volumetric efficiency, (4) cooling system efficiency, (5) cylinder cooling system, (6) pressure of the refrigerant in the suction line, (7) compressor speed, (8) type and size of valves, (9) friction of the refrigerant gas, (10) mechanical condition of compressor, and (11) the lubricant.

## REVIEW EXERCISES

1. What are the two functions of a compressor?

2. What are the most popular types of reciprocating compressors?

3. What are the purposes of suction valves in a compressor?

4. What are the most popular ways of producing the reciprocating motion in a compressor?

5. What is the difference between a connecting rod and an eccentric strap?

6. What are the moving parts in a rotary compressor?

7. What is the purpose of the barrier in a rotary compressor?

8. Where are most rotary compressors used?

9. Are centrifugal compressors positive displacement type pumps?

10. How is the number of stages determined in a centrifugal compressor?

11. Where is lubrication needed in a centrifugal compressor?

12. Where are centrifugal compressors best suited for use?

13. What is the difference between open, semi-hermetic, and hermetic compressors?

14. What are the different types of compressor suction and discharge valves?

15. What actually does the sealing on the shaft seals and compressor valves?

16. What is the name of the compressor component where the valves are located?

17. Describe the procedure used to check operation of the compressor suction valves.

18. Describe the procedure used to check operation of the compressor discharge valves.

19. Where should the oil level be maintained in a compressor?

20. What should the range of the net oil pump pressure be on a compressor?

21. Why is oil foaming undesirable in a compressor?

22. Name the different types of compressor shaft seals.

23. What is meant by compressor displacement?

24. Define compression ratio.

25. Why is a clearance volume necessary?

26. How does the gas confined in the clearance pocket affect the compressor efficiency?

27. Why do compressors require cooling?

28. How will a lower than normal suction pressure affect the compressor efficiency?

29. What type of compressors require oil coolers?

30. How can a service technician help to eliminate compressor failures?

# Condensers and Receivers

As the refrigerant leaves the compressor, it is still a vapor. It is now quite hot and ready to give up the heat it picked up in the evaporator along with the heat of compression. One of the easiest ways to remove this heat from the refrigerant vapor is to send it through a radiator-like device known as a condenser. As heat is given off by the high-temperature, high-pressure refrigerant vapor its temperature falls to the saturation point and the refrigerant vapor condenses to a liquid refrigerant; hence, the name condenser.

## PURPOSE

The purpose of the condenser on a refrigeration unit is to remove heat from the compressed refrigerant vapor changing it to liquid form. First, the sensible heat must be removed from the compressed refrigerant vapor until its temperature has been lowered to the condensing temperature. Then, as the

latent heat of condensation is removed from the refrigerant, it gradually changes (condenses) to a liquid.

From the foregoing discussion, it can be said that a condenser is a device used for removing heat from the refrigeration system.

## TYPES OF CONDENSERS

There are three types of condensers in use on modern refrigeration systems. They are: (1) air-cooled, (2) water-cooled, and (3) evaporative condenser. The air-cooled condenser uses air as the condensing medium, the water-cooled condenser uses water, and the evaporative condenser uses a combination of air and water.

**AIR-COOLED CONDENSERS** are the most popular type used on small commercial and domestic refrigeration and air conditioning units. They are constructed of the tube and external fin design, which dissipates

**Figure 5-1.** Air-cooled condensing unit (*Courtesy of Tecumseh Products Company*)

heat to the ambient air. Except for very small domestic units, which depend on gravity air circulation, heat transfer is efficiently accomplished by forcing large quantities of air through a compact condenser assembly. (See Figure 5-1.)

When natural draft, static type condensers are used, air circulates over the condenser by convection. When the air contacts the warm condenser surface it absorbs heat and rises. As the warm air rises, cooler air replaces it and more heat is absorbed (see Figure 5-2).

There are two types of natural draft air-cooled condensers: the tube and fin type

shown in Figure 5-2, and the plate type illustrated in Figure 5-3. In the plate type condenser, the plates are pressed together to form a condenser coil and seam-welded together.

Natural draft air-cooled condensers have very limited use. They are not capable of dissipating large quantities of heat because of the slow air movement over their surfaces. Because of this, relatively large surfaces are required, but they are economically manufactured, require very little maintenance, and are commonly used in domestic refrigerators.

Condenser capacity can be increased by

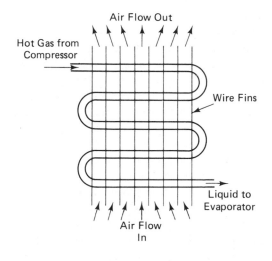

Air Flow Out

Hot Gas from Compressor

Wire Fins

Liquid to Evaporator

Air Flow In

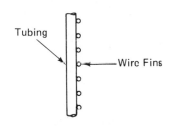

Tubing

Wire Fins

**Figure 5-2.** Natural draft condenser

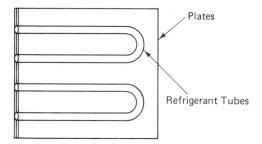

Plates

Refrigerant Tubes

**Figure 5-3.** Plate type condenser

**Figure 5-4.** Forced draft air-cooled condenser

forcing air over the surfaces. This is normally done by adding a fan to increase the air flow (see Figure 5-4). The propeller type fan or the centrifugal type may be used to move the air. The type of fan used depends on design factors such as air resistance, noise level, space requirements, etc.

Some of the earlier air-cooled condensers were of the bare tube type construction. However, the bare tube type was inefficient and gave way to the tube and fin type of condenser. The forced air type of air-cooled condenser is more practical for larger capacities than the natural draft type condenser. The major limiting factors for forced draft air-cooled condensers are economics

and available space.

Air-cooled condensers are easy to install, inexpensive to maintain, require no water, and present no danger of freezing in cold weather. A sufficient quantity of fresh air is necessary, however, and the fan may create noise problems in large installations. In very hot regions, the relatively high ambient air temperature may result in higher than normal condensing pressures. Nevertheless, if the condenser surface is properly sized, air-cooled condensers may be used satisfactorily in all climatic regions.

When space permits, condensers may be manufactured with a single row of tubing. In an attempt to obtain a compact size, however, condensers are usually constructed with a comparatively small face area and several rows of tubing in depth. When the air is forced through the condenser, it absorbs heat and the air temperature rises. Therefore, the efficiency of each succeeding row in the coil is decreased. Coils eight rows in depth are frequently used.

Draw-through fans, the type which pull the air through the condenser, produce a more uniform air flow through the condenser than the blow-through type (see Figure 5-5). Because even air distribution will increase the condenser efficiency the draw-through type fans are preferred.

Most air-cooled refrigeration systems which are operated with the condenser in low ambient temperature conditions are susceptible to compressor damage due to abnormally low head pressures unless some means of maintaining normal head pressures is provided. This is true of roof-mounted refrigeration or air conditioning

Blow-through
Condenser

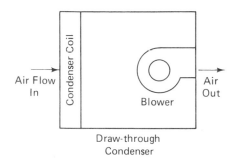

Draw-through
Condenser

**Figure 5-5.** Draw-through and blow-through condensers

systems which are exposed to low outside ambient temperatures. The capacity of refrigerant flow control devices is dependent upon the pressure difference across the device. Since these devices are selected for the desired capacity with normal operating pressures, abnormally low head pressure which reduces the pressure difference across the flow control device may result in insufficient refrigerant flow. This reduced refrigerant flow may cause erratic refrigerant feeding to the evaporator which may result in frosting of the evaporator coil on air conditioning units. The lower refrigerant velocity, and possibly lower evaporator pressure, will permit oil to settle out and be trapped in the evaporator, sometimes causing a shortage of oil in the compressor crankcase.

The low limit of the head pressure depends on the required pressure drop across the flow control device. For normal air conditioning applications the head pressure should be maintained above a condensing temperature of 90°F (32.2°C). This, in effect, corresponds to a normal lower limit of about 60°F (15.6°C) ambient air. Since air conditioning is not normally required at these lower temperatures, condenser head pressure control may not always be necessary. However, for those applications for which operation is required below 60°F ambient air temperature, three methods of condenser head pressure control are available to meet specific job requirements. They are (1) fan control, (2) damper control, and (3) refrigerant flooding the condenser.

Fan control is an automatic winter control method which will maintain a condensing pressure within reasonable limits by cycling one fan on a two-fan unit, two fans in sequence on a three-fan unit, or three fans on a four-fan unit in response to the outside air temperature entering the condensing coil.

Table 5-1

Fantrol Head Pressure Control Data *

| MODELS | DESIGN TD | MINIMUM OUTSIDE AIR TEMP. (F) † PERCENT COMPRESSOR CAPACITY | | | |
|---|---|---|---|---|---|
| | | 100 | 75 | 50 | 25 |
| Two-Fan Units—APD-008B and 010B Four-Fan Units—APD-050B thru 065B | 30 | 28 | 32 | 39 | 55 |
| | 25 | 39 | 42 | 45 | 57 |
| | 20 | 49 | 49 | 50 | 60 |
| | 15 | 61 | 58 | 55 | 63 |
| | 10 | 71 | 66 | 60 | 65 |
| Three-Fan Units—APD-015B thru 045B Six-Fan Units—APD-070B thru 100B | 30 | 5 | 16 | 28 | 49 |
| | 25 | 20 | 27 | 35 | 53 |
| | 20 | 34 | 38 | 42 | 56 |
| | 15 | 50 | 50 | 50 | 60 |
| | 10 | 64 | 60 | 56 | 63 |
| Four-Fan Units—APB-110B thru 140B | 30 | 0 | 15 | 18 | 44 |
| | 25 | 4 | 16 | 27 | 48 |
| | 20 | 21 | 28 | 36 | 53 |
| | 15 | 41 | 43 | 50 | 58 |
| | 10 | 58 | 56 | 54 | 62 |

* Courtesy of McQuay Division, McQuay-Perfex Inc.

† Minimum outside air temperature is based on maintaining 90°F condensing temperature at 100% compressor capacity, 80°F condensing temperature at 75% compressor capacity and 70°F condensing temperature at 50% and 25% compressor capacity.

Fan control of the head pressure is suitable for outside temperatures above those shown in Table 5-1. The thermostat should be set to stop the first fan when the condensing temperature has reached approximately 90°F. Table 5-2 lists approximate thermostat settings for several system temperature drops (TD). These settings are approximate because they do not take into account variations in the load.

When operation of the equipment during ambient temperatures below the range shown in Table 5-1 is required, either damper control or liquid flooding of the condenser must be used. Table 5-3 lists the minimum ambient temperatures for using the damper control method.

The damper control method of controlling head pressure utilizes the principle of modulating the air volume flow through a section of the condenser coil by use of a face and bypass damper section located on the leaving air side of the coil (see Figure 5-6).

The damper control linkage is arranged

### Table 5-2
### Fantrol Thermostat Settings (F) *

| MODELS | DESIGN TD | FAN 2(2A) CUT IN | CUT OUT | FN 3(3A) CUT IN | CUT OUT | FAN 4 CUT IN | CUT OUT |
|---|---|---|---|---|---|---|---|
| Two-Fan Units—APD-008B and 010B<br>Four-Fan Units—APD-050B thru 065B | † 30 | 65 | 55 | | | | |
| | 25 | 72 | 62 | | | | |
| | 20 | 80 | 70 | — | — | — | — |
| | 15 | 85 | 75 | | | | |
| | 10 | 90 | 80 | | | | |
| Six-Fan Units—APD-070B thru 100B<br>Three-Fan Units—APD-015B thru 045B | † 30 | 55 | 45 | 65 | 55 | | |
| | 25 | 59 | 54 | 72 | 62 | | |
| | 20 | 65 | 61 | 80 | 70 | — | — |
| | 15 | 75 | 70 | 85 | 75 | | |
| | 10 | 87 | 77 | 90 | 80 | | |
| Four-Fan Units—APB-110B thru 140B | † 30 | 33 | 28 | 55 | 45 | 65 | 55 |
| | 25 | 44 | 39 | 59 | 54 | 72 | 62 |
| | 20 | 54 | 49 | 65 | 61 | 80 | 70 |
| | 15 | 66 | 61 | 75 | 70 | 85 | 75 |
| | 10 | 76 | 71 | 87 | 77 | 90 | 80 |

* Courtesy of McQuay Division, McQuay-Perfex Inc.
† Factory setting.

so that when the face damper is completely closed, the bypass damper blades are in the wide open position. This is the normal position for full effective dampering. (See Figure 5-7.) In this position, the fan blows the air against the completely closed damper blades, out through the wide open bypass sections, back into the fan section, and then the air is recirculated. In this position very little, if any, air passes through the condenser coil.

The liquid flooding of the condenser type of head pressure control is based on liquid refrigerant flooding back into the condenser and thereby cutting down its effective condensing capacity.

Liquid flooding of the condenser system is an improved head pressure control incorporating two modulating valves. A check valve is also recommended to prevent migration to the condenser during the off cycle.

**Figure 5-6.**   Damper control location (*Courtesy of McQuay Group, McQuay-Perfex, Inc.*)

Table 5-3

Dampertrol Head Pressure Control Data *

| MODELS | DESIGN TD | MINIMUM OUTSIDE AIR TEMP. (F) † PERCENT COMPRESSOR CAPACITY | | | |
|---|---|---|---|---|---|
| | | 100 | 75 | 50 | 25 |
| | 30 | —20 | —20 | —20 | 0 |
| Two-Fan Units—APD-008B and 010B | 25 | —20 | —20 | —20 | 22 |
| Four-Fan Units—APD-050B thru 065B | 20 | —20 | —20 | —20 | 24 |
| | 15 | —20 | —20 | 3 | 37 |
| | 10 | 2 | 14 | 26 | 48 |
| | 30 | —20 | —20 | —20 | — 5 |
| Three-Fan Units—APD-015B thru 045B | 25 | —20 | —20 | —20 | 8 |
| Six-Fan Units—APD-070B thru 100B | 20 | —20 | —20 | —20 | 20 |
| | 15 | —20 | —20 | 0 | 35 |
| | 10 | 3 | 10 | 24 | 47 |
| | 30 | —20 | —20 | —20 | — 5 |
| | 25 | —20 | —20 | —20 | 8 |
| Four-Fan Units—APB-110B thru 140B | 20 | —20 | —20 | —20 | 20 |
| | 15 | —20 | —20 | 0 | 34 |
| | 10 | — 2 | —11 | 24 | 47 |

* Courtesy of McQuay Division, McQuay-Perfex Inc.
† Minimum outside air temperature is based on maintaining 90°F condensing temperature at 100% compressor capacity, 80°F condensing temperature at 75% compressor capacity and 70°F condensing temperature at 50% and 25% compressor capacity.

DAMPERTROL C head pressure control

DAMPERTROL C IN OPEN POSITION

DAMPERTROL C IN BY PASS POSITION

DAMPERTROL C head pressure control selections. Show air flow in open (normal position) and closed (winter start) position.

Figure 5-7. Damper control operation (Courtesy of McQuay Group, McQuay-Perfex, Inc.)

**Figure 5-8.** Valve arrangement for liquid flooding of the condenser (*Courtesy of Mc-Quay Group, McQuay-Perfex, Inc.*)

(See Figure 5-8.) Operation of this system is independent of any difference in elevation between the condenser and the receiver. It therefore permits the receiver and condenser to be more conveniently located to suit specific job conditions.

The main or liquid line valve is normally closed and opens on a pressure rise in the condenser. This valve is located in the liquid line between the condenser and receiver. (See Figure 5-9.)

The secondary gas or bypass valve is normally open and closes on a pressure rise. This valve is located in the line between the liquid valve and the receiver.

System operation is as follows: on system start-up, the bypass valve is normally open and the main or liquid valve is closed. Hot gas moves from the compressor, some going into the condenser, and some going through

**Figure 5-9.** Liquid valve location (*Courtesy of McQuay Group, McQuay-Perfex, Inc.*)

the bypass circuit and the open valve and into the receiver to maintain or build up pressure in the receiver as the liquid leaves. As the compressor continues to run, hot gas condenses in the condenser and raises the liquid level since the main valve on the leaving side of the condenser is still closed. As the liquid level rises, the condensing capacity of the condenser decreases. As a result, the head pressure rises. The bypass gas maintains or raises the pressure in the receiver. As the pressure in the condenser rises to the control point of the valve, the liquid valve starts modulating toward the open position, permitting liquid to leave the condenser and flow into the receiver. At the same time, the bypass valve starts modulating toward the closed position, limiting the hot gas flow into the receiver. The modulating action of the two valves maintains the proper liquid level in the condenser to maintain proper head pressure.

On multifan condensers the liquid flooding of the condenser system is ordinarily used to extend the range of a fan cycling system. The use of fan cycling in conjunction with liquid flooding of the condenser reduces the amount of refrigerant charge required for a specific minimum design temperature condition.

**WATER-COOLED CONDENSERS:** When adequate low-cost condensing water is available, water-cooled condensers are often desirable because lower condensing pressure and better control of the discharge pressure is possible. Water, especially when obtained from underground sources, is usually much colder than daytime ambient air temperatures. If cooling towers are used, the condensing water can be cooled to a point closely approaching the ambient wet bulb temperature. This allows the continuous recirculation of condensing water and reduces the water consumption to a minimum.

Shell and Tube

Tube in Tube

**Figure 5-10.** Water-cooled condensers

Because of the excellent heat transfer characteristics of water, water-cooled condensers are usually quite compact. There are several types of construction used for water-cooled condensers. The principal classifications of water-cooled condensers are (1) shell and tube, (2) tube in tube, and (3) evaporative. (See Figure 5-10.)

**Shell and tube condensers** (sometimes called condenser-receivers), insofar as external ap-

pearance is concerned, are similar to liquid receivers. The shell and tube condenser acts as both condenser and liquid receiver. It is constructed of a cylindrical tank having a refrigerant inlet and outlet. Installed within this tank is a water coil which also has inlet and outlet connections (see Figure 5-11). The shell is made of steel. The ends may be either welded in place or bolted in place. Cleaning the condenser is easier when the ends are bolted in place (see Figure 5-12).

**Figure 5-11.** Shell and tube condenser

The water coil is usually made of copper tubing. To increase the efficiency of the condenser, most manufacturers have added fins to the copper tubing. The cooling water flows through the coil from the inlet end to the outlet. The outlet may be piped to a waste water line or run to an open drain. To reduce waste, however, the cooling water is usually returned to a cooling tower and re-circulated (see Figure 5-13).

In operation, the compressed high-temperature refrigerant passes from the com-

**Figure 5-12.** Multipass water-cooled condenser with bolted-in-place ends

**Figure 5-13.** Cooling water circuit

pressor through the condenser refrigerant inlet to the interior of the shell, where it comes into contact with the cool walls of the water coil and the cooler walls of the shell. The majority of heat is conducted through the copper coil walls to the water which carries it away. Some of the heat is conducted through the shell walls to the ambient air which also aids in cooling the refrigerant. As heat is removed, the refrigerant condenses and the liquid collects in the bottom of the shell (see Figure 5-14). The refrigerant outlet tube extends from the refrigerant outlet almost to the bottom of the receiver interior in order to be below the liquid level

in the receiver. The liquid line to the evaporator connects to the receiver refrigerant outlet.

Shell and tube condensers may be either of horizontal or vertical construction. The horizontal type is most commonly used in the latest designs since it is installed underneath the base of the compressor and occupies space which would otherwise be wasted. Vertical condensers require space on top of the base which generally increases the floor area required for the condensing unit. (See Figure 5-15.)

**Tube in tube condensers** are also commonly used on water-cooled refrigeration equipment. They are made of two copper tubes, one inside the other. For example, a 5/8" OD (15.87 mm) copper tubing could be used for the outside tube and a 7/16" OD (11.1125 mm) copper tubing could be placed inside the larger tube. (See Figure 5-16.)

By using special fittings at the ends, connections can be made so that water may be circulated through the inner tube with the refrigerant flowing through the space between the two tubes. The heat is transferred from the compressed refrigerant through the thin walls of the internal tube into the cooling water which removes the heat. The arrows in Figure 5-16 show the main direction of heat flow which is from the compressed

**Figure 5-14.** Liquid refrigerant level in a shell and tube condenser

**Figure 5-15.** Shell and tube condenser location (*Courtesy of Copeland Corp.*)

High Pressure Vapor
from Compressor

Water

Outer Tube        Inner Tube

**Figure 5-16.** Cross section of a tube and tube condenser

vapor to the water. However, a portion of the heat from the refrigerant is dissipated into the surrounding air which increases the cooling capacity of the condenser.

Tube in tube condensers are arranged in coils so that the compressed refrigerant enters at the top and flows downward as it liquifies in much the same manner as in an air-cooled condenser. The cooling water enters at the bottom and flows upward. Thus the liquid refrigerant and the cool water flow in opposite directions or in a counterflow direction (see Figure 5-17).

A more efficient method used for tube in tube condenser makes use of a *counterswirl* design. (See Figure 5-18.) This design gives superior heat transfer performance. The tube construction provides for excellent mechanical stability. The water flow path is turbu-

105° F

REFRIGERANT

WATER

95° F

85° F

**Figure 5-17.** Typical counterflow path (*Courtesy of Packless Industries*)

REFRIGERANT

WATER

**Figure 5-18.** Counterswirl design principle (*Courtesy of Packless Industries*) lent, and provides a scrubbing action to maintain cleaner surfaces. The construction method also exhibits very high system pressure resistance.

The pressure from city water mains is sometimes used to circulate the water through the condenser. In very large systems, the water is often recirculated to effect economy in water costs, particularly where water is expensive or difficult to obtain in unlimited quantities. In such cases, an evaporative condenser of some type is used.

**EVAPORATIVE CONDENSERS** are frequently used where lower condensing temperatures are desired than those obtainable with air-cooled condensers, and where the available water supply may not be adequate for heavy water usage.

A simple evaporative condenser consists of an air-cooled condenser over which water is allowed to flow. The air blast from the fan causes the water to evaporate from the surface of the condenser fins, absorbing heat from the refrigerant in the evaporation process. The water which is exposed to the air flow evaporates rapidly. The latent heat required for the evaporating process is obtained by a reduction in sensible heat and, therefore, a reduction in the temperature of the remaining water. An evaporative condenser can reduce the water temperature to a point closely approaching the wet bulb temperature of the air (see Figure 5-19).

The hot refrigerant vapor is piped through a spray chamber where it is cooled by evaporation of the water coming in contact with the refrigerant tubing. Since the cooling is accomplished by evaporation of water, water consumption is only a fraction of that used in conventional water-cooled applications in which the water is discharged to a drain after it is used. Because of this, evaporative condensers are used in hot, arid regions.

**Figure 5-19.** Evaporative condenser

## COUNTERFLOW OF COOLING WATER

A water-cooled condenser should cool the refrigerant to nearly the same temperature as the cooling water entering the condenser. If a condenser is to do this, it must operate on the counterflow principle. (See Figure 5-20.) The hot refrigerant vapor enters the condenser at the top, and as it gradually liquifies it flows to the bottom. But the water enters at the bottom of the condenser and flows in the opposite direction to the top. When the counterflow method is used, the entering heated refrigerant is cooled by the warm water leaving the condenser, and upon leaving the condenser is further cooled by the cold entering water. Thus, the liquid refrigerant is cooled to very nearly the temperature of the cold entering water. If both the refrigerant and the cooling water were to travel in the same direction, the heated entering refrigerant would first make contact with the coldest water and the leaving liquid refrigerant would contact the warm leaving water. Therefore, maximum temperature difference between the two fluids would not be

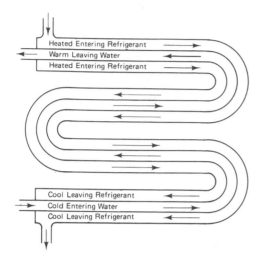

**Figure 5-20.** Counterflow principle

maintained and the flow of heat from the refrigerant to the water would be substantially reduced.

## CONDENSER CAPACITY

The heat transfer capacity of a condenser depends upon several factors:

1. Surface area of the condenser.

2. Contact between the refrigerant and the internal surface of the condenser.

3. Temperature difference between the cooling medium and the refrigerant gas.

4. The velocity of the refrigerant gas in the condenser tubes. Within the normal commercial operating range, the greater the velocity, the better the heat transfer factor, and the greater the capacity.

5. Rate of flow of the cooling medium over or through the condenser. Heat transfer increases with velocity for both air and water, and in the case of air, it also increases with density.

6. Material of which the condenser is made. Since heat transfer differs with different materials, more efficient metals will increase capacity.

7. Cleanliness of the heat transfer surface. Dirt, scale, or corrosion can reduce the heat transfer rate.

8. Rate of replacement of the condensed refrigerant by uncooled refrigerant.

For a given condenser, the physical characteristics are fixed, and the primary variable is the temperature difference between the refrigerant gas and the condensing medium.

## CONDENSING TEMPERATURE

The condensing temperature is the temperature at which the refrigerant gas becomes a liquid. This temperature should not be confused with the temperature of the cooling medium, since the condensing temperature must always be higher in order for heat transfer to occur.

In order to condense the refrigerant vapor flowing into the condenser, heat must flow from the condenser at the same rate at which heat is introduced by the refrigerant gas entering the condenser. As mentioned before, the only way in which the capacity of the condenser can be increased under a given set of conditions is by an increase in the temperature difference through the condenser walls.

Since a reciprocating compressor is a positive displacement pump, the pressure in the condenser will continue to increase until such time as the temperature difference between the cooling medium and the refrigerant condensing temperature is sufficiently great to transfer the necessary amount of heat. With a large condenser, this temperature difference may be very small. With a small condenser, in the event that the air or water flow to the condenser has been blocked, the necessary temperature difference may be very large, resulting in dangerously high discharge pressures. It is, therefore, essential that the condenser be operating properly any time a refrigeration unit is in operation.

The condensing temperature and, therefore, the condensing pressure is determined by the capacity of the condenser, the temperature of the cooling medium, and the heat content of the refrigerant gas being discharged from the compressor, which in turn is determined by the volume, density, and temperature of the discharge gas.

## CONDENSING TEMPERATURE DIFFERENCE

A condenser is normally selected for a system by sizing it to handle the compressor load at a desired temperature difference between the condensing temperature and the expected temperature of the cooling medium. Most air-cooled condensers are selected to operate on temperature differences (usually termed TD) of 20°F to 30°F (11.1°C to 16.7°C) at design conditions. Higher and lower TDs are sometimes used on special applications. Standard production air-cooled condensing units are often designed with one condenser for a wide range of applications. In order to cover as wide a range as possible, the TD at high suction pressures may be from 30°F to 40°F (16.7°C to 22.2°C), while at low evaporating temperatures the TD is often no more than 4°F to 10°F (2.2°C to 5.6°C). The design condensing temperature on water-cooled units is normally determined by the temperature of the water supply and the water flow rate available, and may vary from 90°F to 120°F (32.2°C to 49°C).

Because the condenser capacity must be greater than the evaporator capacity by an amount equal to the heat of compression and the motor efficiency loss, the condenser manufacturer may rate condensers in terms of evaporator capacity, or they may recommend a factor to allow for the heat of compression in selecting the proper size condenser.

## NONCONDENSABLE GASES

Air is composed primarily of nitrogen and oxygen. Both elements remain in a gaseous form at all temperatures and pressures encountered in commercial refrigeration and air conditioning systems. Therefore, although these gases can be liquified under extremely high pressures and extremely low temperatures, they may be considered as noncondensible in a refrigeration system.

Scientists have discovered that one of the basic laws of nature is the fact that in a combination of gases, each gas exerts its own pressure independently of others, and the total pressure existing in a system is the total of all the gases present. This is known as Dalton's Law (see Chapter 1). A second basic characteristic of a gas is that if the space in which it is enclosed remains constant, so that it cannot expand, its pressure will vary directly with the temperature. This is known as Charles' Law. Therefore, if air is sealed in a system with a refrigerant, the nitrogen and oxygen will each add their pressure to the system pressure, and this pressure will increase as the temperature rises.

Since the air is noncondensable, it will usually remain trapped in the top of the condenser and the receiver. During operation, the compressor discharge pressure will be a combination of the refrigerant condensing pressure plus the pressure exerted by the nitrogen and oxygen. The amount of pressure above normal condensing pressure that may result will depend on the amount of trapped air, but it can easily reach 40 to 50 psig (2.8 kg/cm² to 3.5 kg/cm²) or more. Any time a system is operating with abnormally high discharge pressure, air in the system is a prime suspect.

## CLEANING THE CONDENSER

Since the purpose of a condenser is to dissipate heat, it can readily be seen that a layer of dirt, lint, or scale will act as an insulator and prevent proper heat dissipation while causing the head pressure to increase. The surface of the condenser should be cleaned regularly by the service technician in order to maintain the greatest efficiency possible. An oil deposit often collects on the condenser and attracts a layer of dust on air-cooled condensers. This oil and dust must be removed. On water-cooled condenser tubes a formation of scale, much like that on the inside of a tea kettle, will appear. This scale reduces the heat transfer and must be removed. Since different parts of the country have different minerals in the water, different types of cleaners must be used. Local chemical companies should be consulted to determine the proper type and method to use.

## LOCATION OF THE CONDENSER

Since an air-cooled condenser requires a free flow of air over its surface, it is important that it be located in a well-ventilated room so that cool air will replace heated air. Insufficient quantities of cool air will result in high operating discharge pressures.

Water-cooled condensers should not be located where the temperature may drop below the freezing point of water. Freezing may crack the compressor cylinder or split the condenser. The water tubes of a shell and tube condenser may burst when frozen and allow water to get into the refrigeration system. A burst water tube requires a thorough dehydrating of the entire system including the compressor, condenser, receiver, and all the refrigerant lines.

For most satisfactory operation of the system, the temperature around a water-cooled condensing unit in use should preferably never fall below 60°F (15.6°C). If the unit is to be shut down for the winter, as a condensing unit on an air conditioning system, the cooling water should be drained out of the entire system.

The vertical type of water-cooled condenser is easily drained by opening a drain valve or removing a water drain usually provided at the bottom especially for this purpose. On the horizontal type of water-cooled condenser, however, simply disconnecting the water tubes at the ends is not sufficient to drain the coils because the bottom of each coil will not drain. The water must be removed by either blasting compressed air through the water coil to blow it out or by completely detaching the condenser from the base and raising one end to allow the water to drain out. Both the inlet and outlet must be disconnected to vent the coil; otherwise, the water will not drain out properly. When the water coil is made of continuous tubing, the easiest way to drain it is to blow it out with compressed air. This method also requires that the water connections be completely disconnected.

## WATER FLOW CONTROL VALVES

The flow of water to a condenser is controlled by means of an automatic water valve which prevents waste and maintains a constant discharge pressure on waste water type systems. On some of the larger systems, a cooling tower is used to cool the water.

The most popular type of water flow control valve is a bellows type pressure-operated valve. (See Figure 5-21.) The bellows of this type valve is connected to the discharge side of the refrigeration system, usually at the cylinder head. Consequently, the valve is actuated by the variation in the head or high side pressure, which is transmitted to the water valve through the high pressure control tube. This valve is designed to open gradually as the discharge pressure increases. A flow of water is provided in proportion to the operating requirements. The pressure of the refrigerant operates against a metallic bellows which permits more water to flow when the pressure increases. This has the effect of automatically furnishing more cooling capacity and reducing the pressure in the condenser. When the condenser or high side pressure has been reduced, the valve is automatically restricted, reducing the flow of water in accordance with the needs of the condenser. During warm weather, the pressure in a refrigeration system is higher than in cool weather. The higher pressure automatically opens the water valve wider and supplies more water.

The water valve is usually installed in the inlet water line to the condenser (see Figure 5-22). The direction of water flow should always be as indicated on the valve, usually by an arrow or "in" marked on the valve.

## LIQUID RECEIVERS

Some types of refrigeration units have sufficient space within the condenser to contain the entire refrigerant charge of the system. However, when the condenser does not have this added space, a receiver tank is usually provided. The amount of refrigerant required for proper operation of the unit will determine whether or not a receiver is used. When proper system operation requires approximately eight pounds (3.63 kg) or more of refrigerant, a receiver is generally used.

In most cases, the bottom connection from the condenser discharges into the top of the liquid receiver tank, while the liquid line connects to the bottom of the tank at

**Figure 5-21.** Water flow control valve (*Courtesy of Singer Controls Division*)

**Figure 5-22.** Location of water flow control valve

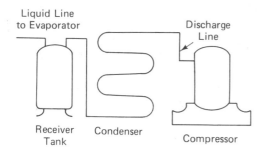

**Figure 5-23.** Liquid receiver tank location

the liquid line shut-off valve (see Figure 5-23). If the liquid line connects to the top of the receiver tank, a tube is brought to the bottom of the receiver (see Figure 5-24). To maintain normal operation, it is important that the level of the refrigerant never fall below the bottom end of the tube. On the other hand, an excess of refrigerant will not allow sufficient space for the gas above the liquid. This condition also may interfere with proper operation by causing excessive discharge pressures. The size or capacity of a liquid receiver is determined by the total quantity of refrigerant which the system requires to operate properly. These factors are determined when the unit is designed.

## LIQUID RECEIVER SAFETY DEVICES

Household units have a relatively small refrigerant charge. Therefore, city ordinances do not require safety devices for such units.

**Figure 5-24.** Extended tube in a receiver tank

In larger sized refrigeration and air conditioning units, however, local city codes and ordinances require safety protection for liquid receivers to prevent possible bursting due to excessive pressures or fire. Safety devices take the form of a fusible metal plug generally installed in the end of the receiver (see Figure 5-25). Should the liquid receiver be exposed to high temperature, as in the case of fire, the liquid receiver would not explode. Instead, the fusible plug would melt and the refrigerant would escape harmlessly. Large liquid receivers holding a considerable quantity of refrigerant are sometimes required by ordinance to provide a vent pipe connection to the fusible plug. In the event that the fusible plug should melt, the gas would be released to the outside of the building through the vent pipe.

During service operations, the service technician may have occasion to apply heat to a liquid receiver. When so doing, care should be taken that the liquid receiver is not heated to a point which will melt the fusible plug. Fusible plug melting points are commonly 165°F and 212°F (74°C and 100°C). It is very easy to accidentally heat a liquid receiver to these temperatures. If this should happen, it would be necessary to re-

**Figure 5-25.** Fusible plug location

place the fusible plug as well as the refrigerant charge.

Common practice on most installations is to charge the system until the liquid receiver is approximately one-third full. When the receiver is of the type having a liquid outlet tube extending to the bottom of the receiver, the liquid level must never fall below the lower end of the tube. In vertical receivers, the refrigerant level should extend about three inches (76.2 mm) above the lower end of the outlet tube, while in horizontal receivers one inch (25.4 mm) above is sufficient.

## SAFETY PROCEDURES

Most of the liquid refrigerant in a refrigeration system is located in condensers and receivers. The condensers are made with aluminum fins for more rapid transfer of heat. Receivers are made with safety devices incorporated. Care must be exercised when working with these devices.

1. Be careful of the aluminum fins on condensers. If the hand should slip, they may cause a serious cut, which is difficult to heal.

2. Do not bend the fins over. This prevents proper air circulation and heat transfer from the condenser.

3. Do not apply heat to a receiver tank which contains refrigerant because hydrostatic pressure may build up resulting in a possible explosion.

4. Never cap the safety relief devices on receiver tanks.

5. Never plug or use the wrong type of solder on safety devices on a receiver tank.

6. Never fill a receiver tank or condenser to more than 85% of its rated capacity.

## SUMMARY

• A condenser is a device used for removing heat from the refrigeration system.

• Types of condensers are (1) air-cooled, (2) water-cooled, and (3) evaporative.

• Air-cooled condensers are popular on small commercial and domestic refrigeration and air conditioning units.

• Types of air-cooled condensers are (1) forced draft and (2) natural draft.

• The major limiting factors for forced draft air-cooled condensers are economics and available space.

• Most air-cooled refrigeration systems which are operated with the condenser in low ambient temperature conditions are susceptible to compressor damage due to abnormally low head pressures, unless some means of maintaining normal head pressures is provided.

• Types of head pressure control for air-cooled condensers are (1) fan control, (2) damper control, and (3) refrigerant flooding of the condenser.

• When adequate low-cost condensing water is available, water-cooled condensers are often desirable because of the lower condensing pressure and possibility of better control of the discharge pressure.

• Because of the excellent heat transfer characteristics of water, water-cooled condensers are usually quite compact.

• Water-cooled condensers are classified as (1) shell and tube, (2) tube in tube, and (3) evaporative.

• The shell and tube condenser acts as both a condenser and liquid receiver.

• Shell and tube condensers are made in both horizontal and vertical models.

• Tube in tube condensers are made of two copper tubes, one inside the other.

• Tube in tube condensers transfer heat to both the surrounding air and the water.

• Evaporative condensers are used where lower condensing temperatures are desired than those obtainable with air-cooled condensers, and where the available water supply may not be adequate for heavy water usage.

• A simple evaporative condenser consists of an air-cooled condenser over which water is allowed to flow.

• A water-cooled condenser should cool the refrigerant to nearly the same temperature as the cooling water entering the condenser.

• Water-cooled condensers should operate on the counterflow principle with the refrigerant and water flowing in opposite directions.

• For a given condenser, the physical characteristics are fixed, and the primary variable is the temperature difference between the refrigerant gas and the cooling medium.

• The condensing temperature is the temperature at which the refrigerant gas is condensing to a liquid.

• The only way the capacity of a condenser can be increased under a given set of conditions is by an increase in the temperature difference through the condenser walls.

• The condensing temperature and, therefore, the condensing pressure is determined by the capacity of the condenser, the temperature of the cooling medium, and the heat content of the refrigerant gas being discharged from the compressor, which in turn is determined by the volume, density, and temperature of the discharged gas.

• A condenser is normally selected for a system by sizing it to handle the compressor load at a desired temperature difference between the condensing temperature and the expected temperature of the cooling medium.

• Since air is noncondensable, it will usually remain trapped in the top of the condenser and the receiver.

• To maintain the greatest efficiency possible, the surface of the condenser should be cleaned regularly by the service technician.

• On waste water type systems, the flow of water to a condenser is controlled by means of an automatic water valve which prevents waste and maintains a constant discharge pressure.

• The most popular type of water flow control valve is a bellows type pressure-operated valve.

• To maintain proper operation of the unit, the liquid level in a liquid receiver should never fall below the pick-up tube.

## REVIEW EXERCISES

1.  What is the purpose of a condenser in a refrigeration system?

2.  What are the three general types of condensers?

3.  What methods are used to prevent damage to a compressor when it is operated at low ambient temperatures with an air-cooled condenser?

4.  Why are water-cooled condensers

sometimes more desirable than air-cooled condensers?

5. How are water-cooled condensers classified?

6. Which water-cooled condenser serves as both condenser and receiver?

7. What two sources are used as cooling media for water-cooled condensers?

8. On what principle do water-cooled condensers operate?

9. How are condensers selected?

10. Why will air usually be trapped in the condenser or the receiver?

11. What factors determine the condensing pressure?

12. How can the capacity of a condenser be increased?

13. What are the purposes of an automatic water valve?

14. Where are evaporative condensers desirable?

15. Where is the latent heat obtained for cooling in an evaporative condenser?

16. What factors determine condenser capacity?

17. Define condensing temperature.

18. How can the capacity of a condenser be increased?

19. What is condenser TD?

20. Does the condenser or evaporator have the greater capacity?

21. Why must noncondensable gases be removed from a refrigeration system?

22. What effect will a dirty condenser have on the head pressure?

23. Where should the condenser be located?

24. What is the purpose of the water flow control valve?

25. How does a water flow control valve operate?

26. What is the purpose of the liquid receiver?

27. What is the lowest liquid refrigerant level allowed in a receiver tank?

28. What is the purpose of the fusible plug?

# Evaporators

The evaporator is an extremely important part of the refrigeration system because it is the part which performs the actual cooling. The evaporator may be defined as *a device which is used for absorbing heat into the refrigeration system*. The evaporator is placed inside the area to be cooled. The heat is absorbed because of the vaporization of the refrigerant inside the evaporator tubes.

## TYPES OF EVAPORATORS

There are two basic types of evaporators: (1) the dry or direct expansion type and (2) the flooded type. (See Figure 6-1.)

**DRY OR DIRECT EXPANSION EVAPORATORS** use a continuous piece of tubing where the refrigerant is metered into one end by the flow control device and the suction line is connected to the other end. The refrigerant flows through the tubing while it vaporizes and absorbs heat. The refrigerant does not collect in a pool as in the flooded system. Also, there is no separation line between the liquid refrigerant and the gaseous refrigerant anywhere in the evaporator.

**FLOODED EVAPORATORS** provide for the recirculation of liquid refrigerant by the addition of a *separation chamber*. The liquid refrigerant enters the separation chamber through the flow control device and flows down to the bottom of the separation chamber and the evaporator inlet. As the liquid refrigerant flows through the evaporator, it boils and absorbs heat. As the refrigerant leaves the evaporator, any liquid refrigerant present is separated from the gas in the separation chamber and recirculated.

Controlling the liquid refrigerant level and recirculating the unevaporated liquid refrigerant assure that virtually all the inside surface of the evaporator is in contact with liquid refrigerant under any load conditions in a flooded evaporator. The evaporators shown in the following illustrations are of the single pass type, but multiple pass evaporators are more commonly used for greater efficiency and economy.

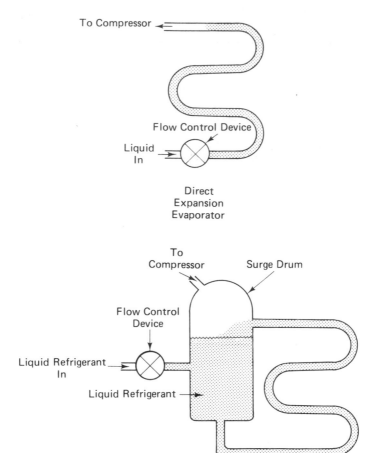

To Compressor

Flow Control Device

Liquid
In

Direct
Expansion
Evaporator

To
Compressor          Surge Drum

Flow Control
Device

Liquid Refrigerant
In

Liquid Refrigerant

Flooded Evaporator

**Figure 6-1.** Types of evaporators

## EVAPORATOR STYLES

Evaporators are made in many different shapes and styles to fill specific needs. Until the application is known, no one style can be considered superior.

The most common style is the *blower coil* or forced convection evaporator, in which the liquid refrigerant evaporates inside of finned tubes, absorbing heat from the air which is blown through the coil by a fan (see Figure 6-2).

Specific applications may, however, use *bare coils* with no fins (Figure 6-3), *gravity coils* with natural convection air flow (Figure 6-4), *flat plate surface* (Figure 6-5), or other specialized types of heat transfer surfaces.

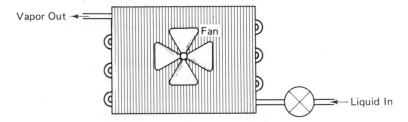

**Figure 6-2.** Blower coil evaporator

## HEAT TRANSFER IN EVAPORATORS

The transfer of heat from the space being refrigerated to the refrigerant inside of the evaporator takes place in two steps:

1. The heat from the space being refrigerated must be absorbed by the metal of the evaporator.

2. The heat must be transferred through the metal of the evaporator and absorbed by the refrigerant inside the evaporator.

The transfer of heat from the air in the refrigerated space to the metal of the evaporator depends on several factors, such as:

1. The nature of the exposed surface of the evaporator: bright or dull; rough or smooth.

**Figure 6-3.** Bare coil evaporator

2. The temperature difference between the evaporator surface and the air surrounding it.

3. The velocity of the heat-carrying air currents within the area being refrigerated.

4. The conductivity of the metal from which the evaporator is constructed.

5. The thickness of any frost coating on the evaporator.

**Figure 6-4.** Gravity evaporator

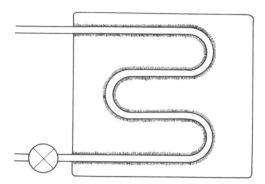

**Figure 6-5.**    Flat plate evaporator

The difference in temperature between the air and the evaporator surface is quite important. If this temperature difference is small, the rate of heat transfer will be low.

The velocity of the heat-carrying air currents within the area being cooled depends upon the design and construction of the evaporator. As the air in contact with the evaporator is cooled, it must be replaced with warm air. If the air circulation is restricted, the heat from the refrigerated space will not be properly transferred to the cooling unit, and the space will not be maintained at the desired temperature. If the air is allowed to circulate properly, the heat transferred to the air will be increased from four to six times over the rate of heat transfer to still air. In most installations, forced air circulation is induced over the surfaces of an evaporator, increasing the cooling capacity more than twenty times over natural draft installations.

The heat transfer from the evaporator surface to the refrigerant is important because the refrigerant is what really does the work. The rate of heat transfer from the evaporator surface to the refrigerant depends on such factors as:

1.    The surface area of the evaporator.

2.    The temperature difference between the evaporating refrigerant and the medium being cooled.

3.    The velocity of the refrigerant in the evaporator tubes. (In the normal commercial range, the higher the velocity, the greater the heat transfer rate.)

4.    The ratio of primary surface to secondary surface.

5.    The condition of the refrigerant (dry or flooded).

6.    Freedom from oil film.

7.    Rate of removal of the vaporized refrigerant.

8.    The type of medium being cooled. Heat flows almost five times more effectively from a liquid to the evaporator than from air to the evaporator.

9.    Dew point temperature of the entering air. If the evaporator temperature is below the dew point temperature of the entering air, latent heat as well as sensible heat will be removed.

The cooling capacity of the evaporator is proportionate to the exposed surface area. If the exposed area is small, the capacity of the evaporator also will be small. It is for this reason that fins have been added to the coils.

The ratio of primary surface to secondary surface simply means the ratio of tube area to fin area of the evaporator. Evaporators are manufactured for many different applications and the proper size evaporator for any particular installation must be determined through calculation.

If the refrigeration unit is not functioning properly and the refrigerant vapor is not removed as rapidly as it forms in the evaporator, the pressure will increase in the evaporator and reduce the evaporation process. The increase in pressure will cause the boiling point of the refrigerant to rise and the corresponding temperature of the evaporator will rise and reduce the differential between the evaporator temperature and the temperature of the circulating air. This will effectively slow down the cooling rate of the unit. In the final estimate of the capacity of a particular evaporator, it is this difference between the temperature of the refrigerant inside the evaporator and the temperature in the refrigerated space which is used. The difference in temperature is called *temperature differential*. For the ordinary household refrigerator, the refrigerant evaporating temperature is normally 5°F (−15°C). If, then, we were to assume an average refrigerator temperature of 45°F (7.2°C), we would have a temperature difference of 45°F −5°F or 40°F (22.2°C) between the temperature of the refrigerant in the evaporator and the temperature of the refrigerated space.

## EVAPORATOR CONDUCTIVITY

Metallic substances have a much higher rate of conductivity than nonmetallic substances. For this reason, metals are used as heat conductors, while the nonmetallic substances such as asbestos, cork, and glass are used as insulators since they do not readily conduct heat. The ability of metals to conduct heat is expressed by the *unit conductivity factor*. The unit conductivity factor is equal to the number of Btus (calories) which will pass through one square foot (.093 m²) of one inch (2.54 cm) thick material in one hour for each one degree Fahrenheit (1.8°C) difference in temperature between the two surfaces. The symbol used for this factor is *k*. See Table 6-1.

From Table 6-1 it can be concluded that copper has a substantially higher rate of conductivity when compared with the other commercial materials. It also may be seen that steel and iron have merely a fraction of the conductivity of copper. However, since the metal used in the evaporator is relatively thin, iron and steel can be used successfully in evaporators. The maximum heat flow is obtained by using copper, but because of the constant film resistance of the surface of the metals, the heat transfer of a copper evaporator is only 10% to 20% greater than that of a steel evaporator.

### Table 6-1
### Heat Conductivity

| SUBSTANCE | CONDUCTIVITY FACTOR K |
|---|---|
| Copper | 2680 (675.36 kcal) |
| Aluminum | 1475 (371.7 kcal) |
| Iron | 350–423 (88.2–106.6 kcal) |
| Steel | 310 (78.12 kcal) |
| Concrete (Average) | 5.8 (1.46 kcal) |
| Porcelain | 10 (2.52 kcal) |
| Water | 3.85–5 (.97–1.26 (kcal) |
| Wood (with grain) | 2.5 (.63 kcal) |
| Wood (across grain) | 1.0 (.25 kcal) |
| Lubricating Oil | 1.2 (.3 kcal) |
| Asbestos | 0.94 (.24 kcal) |
| Cork | 0.3 (.07 kcal) |
| Rock wool | 0.26 (.06 kcal) |
| Air | 0.16 (.04 kcal) |

Many of the nonmetallic substances such as asbestos, cork, and rock wool listed in Table 6-1 are used extensively as insulators in the refrigeration and air conditioning industry because of their low heat conductivity factor.

## CALCULATION OF HEAT TRANSFERRED

The total heat that may be transferred by any coil, provided a steady temperature differential exists, is determined by use of the following formula:

$$Q = U \times A \times (T_1 - T_2)$$

where:

$Q$ = Total Btu (cal) transferred or given up per hour.

$U$ = Overall heat transfer coefficient.

$A$ = Area of surface in ft² (m²) through which the heat is flowing.

$T_1$ = Space temperature in °F (°C)

$T_2$ = Refrigerant temperature in °F (°C)

Example: An evaporator is capable of absorbing 2 Btu per square foot (504 cal per .093 m²) per hour for each 1°F (.555°C) of temperature difference. Therefore, the value of $U = 2$ (504 cal). The total surface area of the evaporator: $A = 5$ ft² (.465 m²). The temperature in the refrigerated space $(T_1) = 45°F$ (7.2°C) and the refrigerant temperature $(T_2) = 5°F$ (−15°C).

$Q = 2 \times 5 \times (45 - 5)$

$\quad = 10 \times 40$

$\quad = 400$ Btu/hr.

$Q = 9.76 \times .465 \times [7.2 - (-15)]$

$\quad = 4.54 \times 22.2$

$\quad = 100.8$ kcal/hr.

From the above formula, it can be readily seen how the capacity of a coil $(Q)$ is affected by the conductivity $(U)$, the exposed surface $(A)$, and the temperature differential $(T_1 - T_2)$. If any one of these factors is increased, the capacity of the coil will be increased.

## RATE OF HEAT TRANSFER

A flooded type system will transfer approximately 50% more heat than a dry type system. The refrigerant in a dry system exists as a mist (in the gaseous state) and, therefore, does not make as much contact with the inside of the evaporator tubes as does the liquid refrigerant in a flooded system. A flooded type evaporator which uses gravity air circulation within the food compartment will absorb approximately 3 Btus (756 cal) per ft² (.093 m²) per hour for every 1°F (.555°C) difference in temperature. The dry system will not absorb as much, approximately 1½ to 2 Btu (378 to 504 cal) per ft² per hour for every 1°F difference in temperature.

## EVAPORATOR DESIGN FACTORS

Any pressure drop which occurs in an evaporator results in a reduction in system capacity due to the lower refrigerant pressure at the outlet of the evaporator coil. When a reduction in suction pressure exists, the specific volume of the refrigerant gas returning to the compressor increases, and the weight of the refrigerant gas pumped by the compressor decreases. Therefore, the length of tubing in an evaporator must be kept to a minimum. When large capacity evaporators are required, it is necessary to use several refrigerant circuits (see Figure 6-6).

There are other factors which must also be considered in evaporator design. If the evaporator tubing is too large, the refrigerant gas velocities may become so low that oil will accumulate in the tubing and will not be returned to the compressor. The only

Single Circuit

Liquid In

Liquid In

Multiple Circuit

**Figure 6-6.** Evaporator circuits

means of assuring satisfactory oil circulation is by maintaining adequate refrigerant gas velocities. The heat transfer ability of the tubing also may be decreased greatly if the refrigerant velocities are not sufficient to scrub the interior walls of the tubing and keep it clear of an oil film. Since the goals of low pressure drop and high refrigerant velocities are directly opposed by the resistance of the tubing, the final evaporator design must be a compromise.

Pressure drops of approximately 1 to 2 psi (.0703 to .1406 kg/cm²) through the evaporator are acceptable on most medium- and high-temperature applications. Low-temperature applications commonly operate with ½ to 1 psi (.0351 to .0703 kg/cm²) pressure drop in the evaporator.

## TEMPERATURE DIFFERENCE AND DEHUMIDIFICATION

In the study of evaporators, one property of air takes on a new importance. That property is the dew point temperature of the air. When air is cooled, the temperature at which moisture will start to condense is known as the dew point temperature.

Since for a given installation the physical characteristics are fixed, the primary variable is the temperature difference (TD) between the evaporating refrigerant and the medium being cooled. The colder the refrigerant is with respect to the temperature of the air entering the evaporator section, the greater the capacity of the evaporator.

Temperature differences of 5°F to 20°F (2.8°C to 11.1°C) are commonly used. Usually for best economy, the TD should be kept as low as possible because operation of the compressor will be more efficient at higher suction pressures.

The amount of moisture condensed out of the air is in direct relation to the temperature of the evaporator. An evaporator operating with too great a TD will tend to produce a low-humidity condition in the refrigerated space. In the storage of leafy vegetables, meats, fruits, and other perishable items, low humidity will result in excessive dehydration and damage to the product. For perishable commodities which require a very high relative humidity (approximately 90% Rh), a TD from 8°F to 12°F (4.4°C to 6.7°C) is recommended. For relative humidities slightly lower (approximately 80%

Rh), a TD from 12°F to 16°F (6.7°C to 8.8°C) is normally adequate.

## EVAPORATOR FROSTING

The formation of frost on the surface of the evaporator causes a considerable loss of heat. First, the evaporator must cool the air below the condensation temperature. The moisture then gives up the latent heat of condensation as it condenses. The moisture then must be cooled to the freezing point. At this temperature, the moisture gives up the latent heat of fusion as it changes to frost. The frost must then be cooled down to the temperature of the evaporator. A complete calculation will show that over 1200 Btus (302.4 kcal) must be absorbed by the evaporator to form one pound (.4536 kg) of frost on its surface. When the evaporator is defrosted, only a small amount of this heat is regained; thus, it can be readily seen that frosting on an evaporator must be kept to a minimum.

## DEFROSTING OF EVAPORATORS

Ice and frost will accumulate on evaporators which operate at or below freezing temperatures. Air flow through the evaporator will eventually be blocked unless the frost is removed. To allow continuous operation on refrigeration applications where frost accumulation can occur, periodic defrost cycles are necessary.

If the air returning to the evaporator is well above 32°F (0°C), defrosting can be accomplished by allowing the fan to continue to operate while the compressor is shut down, either for a preset time period or until the evaporator temperature rises a few degrees above 32°F (the melting temperature of frost).

For low-temperature applications, some source of heat must be supplied to melt the ice. Electric defrost systems utilize electric heater coils or rods in the evaporator. (See Figure 6-7.) Proprietary systems which use water for defrosting are available. Hot-gas defrosting is widely used, with the discharge gas from the compressor bypassing the condenser and discharging directly into the evaporator inlet. In hot-gas defrost systems, the heat of compression or some source of stored heat provides the heat for defrosting. Adequate protective devices such as re-evaporators or suction accumulators must be provided if necessary to prevent liquid refrigerant from returning to the compressor. Other systems may utilize reverse-cycle defrosting. In such a system, the flow of refrigerant is reversed to convert the evaporator into a condenser temporarily until the defrost period is complete. A drain pan heater is required on low-temperature applications to prevent refreezing of the melted condensate in the evaporator drain pan.

**Figure 6-7.** Electric defrost heaters

## SAFETY PROCEDURES

The evaporators on a refrigeration unit are on the low pressure side of the system. Many times the low pressure side is below atmospheric pressure. Therefore, when cleaning the evaporator care must be taken not to cause leaks in the refrigeration system.

1.  When defrosting an evaporator do not use sharp instruments which may puncture the metal.

2.  Never block the air flow over an evaporator since damage to the compressor might result.

3.  Care should be taken not to bend the evaporator fins.

4.  Be careful when working around tube and fin evaporators because the fins can cause serious cuts on the hands and arms.

5.  Always provide plenty of ventilation when working with coil cleaning solvents.

6.  Always provide plenty of ventilation when purging refrigerant from a system.

7.  Never heat a refrigerant line close to a flow control device without first wrapping a wet cloth around the flow control device.

8.  Never unscrew any of the fittings when the evaporator is covered with condensation.

9.  Never clean an aluminum evaporator with a caustic cleaner.

## SUMMARY

•   An evaporator is a device used for absorbing heat into the refrigeration system.

•   The two basic types of evaporators are (1) dry or direct expansion and (2) flooded.

•   The most common style of evaporator is the blower coil style.

•   The styles of evaporators are (1) tinned tube, (2) bare coil, and (3) flat plate coil.

•   Heat transfer in evaporators takes place in two steps: (1) heat from the air is absorbed by the metal, and (2) the heat is transferred through the metal to the refrigerant.

•   The rate of heat transfer will be small if the TD is small.

•   If air is allowed to circulate over an evaporator, the heat transfer will be four to six times as great as when still air flow is used.

•   When forced-air circulation is used, the heat transfer is more than twenty times as great as natural draft air flow.

•   The refrigerant inside the evaporator is what actually does the work.

•   The cooling capacity of an evaporator is proportionate to the exposed surface area.

•   The ratio of primary surface to secondary surface is the ratio of tube area to fin area of the evaporator.

•   The difference between the refrigerant temperature and the temperature of the refrigerated space is called the TD.

•   Metallic substances are used for heat conductors, while nonmetallic substances are used as insulators.

•   The ability of metals to conduct heat is expressed by the unit conductivity factor whose symbol is $k$.

•   Copper has the highest conductivity

factor $(k)$ of the commercial metals.

• The formula for determining the total heat transferred is: $Q = U \times A \times (T_1 - T_2)$.

• A flooded type system will transfer approximately 50% more heat than a dry type system.

• Any pressure drop which occurs in an evaporator results in a reduction in system capacity.

• When large capacity evaporators are required, it is necessary to use several refrigerant circuits.

• The refrigerant velocity must be kept as high as possible to increase efficiency and effect oil return to the compressor.

• Pressure drops of more than 2 psi (.1406 kg/cm²) should be avoided.

• The amount of moisture condensed out of air is in direct relation to the evaporator temperature.

• For perishable commodities which require 90% relative humidity, a TD of 8°F to 12°F (4.4°C to 6.6°C) is recommended.

• The formation of frost on an evaporator results in a great loss of heat.

• Defrost cycles are necessary on systems that require the evaporator temperature to be below 32°F (O°C).

• The types of defrost systems are: electric, water spray, and hot-gas defrost.

## REVIEW EXERCISES

1. Why is an evaporator so important?

2. Define an evaporator.

3. What are the two basic types of evaporators?

4. Which type of evaporator is the most efficient?

5. What does the actual cooling in an evaporator?

6. List four styles of evaporators.

7. What two steps take place in the transfer of heat in the evaporator?

8. Will the heat transfer be greater with a high or a low TD?

9. How much is the heat transfer increased if forced air is used rather than still air over an evaporator?

10. What is proportionate to the cooling capacity of an evaporator?

11. What does the ratio of primary surface to secondary surface mean?

12. What will be the result if the refrigerant vapor is not properly removed from the evaporator?

13. Why are metallic substances rather than nonmetallic substances used in an evaporator?

14. Write the formula for determining the total heat transferred by a refrigeration unit.

15. Which type of system will transfer the greatest amount of heat?

16. How does a pressure drop affect the performance of an evaporator?

17. Why must the refrigerant velocity be kept in an evaporator?

18. What pressure drops through a high-temperature evaporator are acceptable?

19. What air property is important when studying evaporator operation?

20. What will an evaporator operating with a high TD tend to produce?

21. What will a heavy coating of frost on an evaporator cause?

22. What can be used to reduce frost on an evaporator operating at temperatures below 32°F (0°C)?

23. What are three types of defrost systems?

24. What device should be used to prevent liquid from returning to the compressor?

25. What is used to prevent moisture from freezing in the evaporator drain pan?

# Flow Control Devices

In modern refrigeration and air conditioning practice, a wide variety of refrigerant flow control devices are used to obtain efficient economic operation. Some of the smaller systems use a manual control or the simple on–off automatic control which requires very few flow control devices. However, large systems which use more elaborate automatic controls may have a multitude of controls, and the proper operation of each one is essential to the satisfactory performance of the system.

In order to adjust a control for its most efficient performance or recognize the effect of a particular malfunction, it is necessary that the function, operation, and application of each refrigerant flow control be thoroughly understood.

## PURPOSE

Though the flow control device is one of the more difficult refrigeration system components to understand, its basic function is extremely simple. The flow control device is one which is used to control the flow of liquid refrigerant to the evaporator. The flow control device may be operated by different forces, such as temperature, pressure, or a combination of the two, but its only function is to control the flow of liquid refrigerant to the evaporator.

The proper control of refrigerant flow is also important for the following reasons:

1. The proper operation of the evaporator depends upon the correct amount of liquid refrigerant and its swirling pattern through the evaporator tubes. Too much or too little refrigerant will provide less efficiency during the heat transfer process. To obtain the best heat transfer, the inside surface of the evaporator tubes must be completely wetted by the refrigerant, except in the last section of the evaporator which is used to superheat the exiting refrigerant.

2. All the liquid refrigerant must be evaporated in the evaporator; otherwise, liquid refrigerant may return to the compressor and cause damage to the compressor valves

and bearings. This is commonly known as *compressor slugging* and must be avoided.

## TYPES OF FLOW CONTROL DEVICES

There are seven types of refrigerant flow control devices. They are: (1) hand-operated expansion valve, (2) low side float, (3) high side float, (4) automatic expansion valve, (5) thermostatic expansion valve, (6) capillary tube, and (7) thermo-electric expansion valve. The hand-operated expansion valve, low side float, and high side float have been replaced with the other more efficient types of flow control devices. Therefore, only the automatic expansion valve, thermostatic expansion valve, capillary tube, and the thermo-electric expansion valve will be discussed.

## THEORY OF OPERATION

In order to understand how the flow control device accomplishes its job, a detailed explanation of the process is required. Figure 7-1 shows what happens to the refrigerant as it passes through the flow control device.

All the area to the left of line 2 in Figure 7-1 represents the refrigerant in the liquid state. The area between lines 1 and 2 represents a gas-liquid mixture. The area to the right of line 1 represents the refrigerant in the gaseous state.

For this illustration liquid Refrigerant-12 enters the flow control device at point A under a condition of 117.1 psi (8.23 kg/cm²) and 100°F (37.8°C). As it passes through the flow control device the refrigerant pressure drops to 37 psi (2.60 kg/cm²). The temperature at that pressure is 40°F (4.4°C). This is represented by point B. At point D, a mixture of gas and liquid exists. This gas has been formed by some of the liquid being evaporated in order to remove sensible heat from the remaining liquid to 40°F (4.4°C. This evaporating gas is termed *flash gas.*

Figure 7-1 also shows how the percentage (21.5%) of flash gas is obtained. As line 2 is the saturated liquid line, at C, all the refrigerant is in the liquid state at a temperature of 40°F. At this condition, each pound (453.6 g) of Refrigerant-12 would contain 17.3 Btu (4.36 kcal). As line 1 is the saturated vapor line, at D, all the refrigerant is in a saturated vapor state at a temperature of 40°F. At this point, the refrigerant would contain 81.4 Btus per pound (44.77 kcal/kg).

**Figure 7-1.** Operation of flow control device

For all practical purposes it can be said that there is no loss or gain in total heat of the refrigerant as it passes through the flow control device. Therefore, since one pound (453.6 g) of liquid refrigerant contains 31.1 Btus (7.84 kcal) as it enters the flow control device at point A, it will also contain 31.1 Btus as it leaves the flow control device.

The total heat that one pound of refrigerant could absorb between points C and D is 64.1 Btus (16.15 kcal). As shown in Figure 7-1, 13.8 Btu (3.48 kcal), represented by the line CD, have been used in cooling the remaining liquid refrigerant from 100°F (37.8°C) to 40°F (4.4°C). The figure shows that 13.8 (3.48) divided by 64.1 (16.15) = 21.5%. Therefore, 21.5% of the liquid has been evaporated or changed to flash gas.

Actually, there is a slight loss in the heat of the refrigerant because of the heat picked up from the valve, and refrigerant lines to the evaporator, distributor, etc. But for all practical purposes the amount lost is too in-significant to be considered.

Flash gas may be caused by many factors in the refrigeration system. Figure 7-2 is the same as Figure 7-1 except that line $A^1 B^1$ has been added. This line shows the effect of an increased compression ratio on flash gas. Assuming the low side pressure will remain the same, the compression ratio has been increased by increasing the head pressure.

The line CB represents the amount of flash gas in Figure 7-1. The line $CB^1$ represents the new quantity of flash gas with a higher compression ratio. This is one very important reason why the compression ratio should be kept as low as possible.

With the theory of the metering device, the full cycle of compression, condensation, expansion, and evaporation has been explained.

Figure 7-3 shows the entire cycle imposed on one Temperature-Btu Chart. As in the previous figures, all the area to the left of line 2 represents the refrigerant in the liquid state; all the area between lines 2 and

Figure 7-2. Flash gas

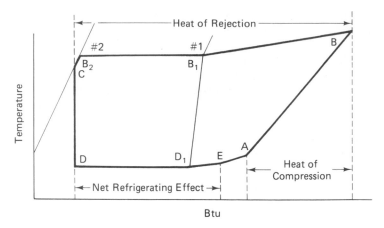

**Figure 7-3.**  Refrigeration cycle

1 represents a gas-liquid mixture; and all the area to the right of line 1 represents the refrigerant in a gaseous state.

For a brief review of the cycle, assume point A of Figure 7-3 to be the suction gas entering the compressor. From point A to point B, the gas is compressed in the compressor. Notice that not only does the temperature increase but there is an increase in the heat content of the gas. This increase in heat is the result of the actual work done to compress the gas. This increase in heat is called the *heat of compression*. From point B to point $B^1$, the superheat of the gas is removed and it is cooled to the saturated vapor line or condensing temperature. From point $B^1$ to point $B^2$, the gas is condensed From point $B^2$ to point C, the condensed liquid is subcooled in the condenser. This heat, which is removed from the system, is known as the *heat of rejection*.

From point C to point D, the liquid passes through the flow control device. Though there is a change in temperature which corresponds to the pressure, and a partial change of state, the amount of heat in the refrigerant remains the same.

The remaining part of the cycle is heat absorption. From point D to point $D^1$, the absorption is accomplished entirely by the evaporation of the refrigerant. This is latent heat since it results in a change of state of the refrigerant. From point $D^1$ to point E, the heat absorption results in a superheating of the refrigerant gas before it leaves the evaporator at point E. This quantity of heat, which is absorbed from point D to point E, is known as the *net refrigerating effect* and is the actual work done by the refrigeration system. From point E to point A, a small amount of heat is absorbed into the system through the suction line resulting in additional superheat to the refrigerant.

This cycle is the basis for all compression-refrigeration systems. If it is understood, a complete analysis can be made of any compression-refrigeration system.

## AUTOMATIC (CONSTANT PRESSURE) EXPANSION VALVES

The constant pressure or automatic expansion valve is the forerunner of the thermostatic expansion valve. The demand for energy conservation at minimum cost, however, has created a new interest in this simple, reliable control valve.

The automatic expansion valve is generally so termed because it opens and closes automatically without the aid of an external mechanical device. It maintains a nearly constant refrigerant pressure in the evaporator and on the low side of the system. Since the opening and closing of the automatic expansion valve is controlled by the refrigerant pressure in the low side of the system, this type of valve will not compensate for varying conditions in the high side or low side of the system, or for variation in load requirements.

**FUNCTION:** Automatic (constant pressure) expansion valves are basically pressure-regulating devices which respond to the refrigerant pressure of the valve outlet. They are installed at the evaporator inlet as a device to control refrigerant flow. (See Figure 7-4.) The valve meters the refrigerant to maintain a constant evaporator pressure during machine operation.

This type of valve consists of a diaphragm, a control spring, and the basic valve needle (ball) and seat. (See Figure 7-5.) The control spring above the diaphragm exerts a force to move the diaphragm downward, opening the valve (F1). On the opposite side of the diaphragm, the opposing force is developed by the low side (evaporator) pressure (F2). This force tends to close the valve.

**Figure 7-5.** Automatic expansion valve (*Courtesy of Singer Controls Division*)

During the off cycle of the machine, the valve will be closed as the evaporator pressure builds up and overcomes the control spring pressure. When the machine starts operating, the compressor quickly reduces the low side pressure. When the low side

**Figure 7-4.** Location of automatic expansion valve (*Courtesy of Singer Controls Division*)

pressure reaches the pressure of the control spring, the expansion valve is ready to open. It actually opens when the low side (evaporator) pressure is reduced just below the control spring setting.

This is the valve opening point. The valve, however, must open further to satisfy the compressor capacity at the operating pressure. This occurs as the operation of the compressor further reduces the low side pressure. The low side pressure continues to drop, and the valve continues to open until a point is reached where liquid refrigerant is entering the evaporator coil and vaporizing at a rate equal to the compressor pumping capacity. The evaporator pressure is now stabilized and the system will operate at this point for the balance of the running cycle.

**ADJUSTMENT:** Automatic expansion valves feature a manual adjustment. The adjusting screw increases or decreases the tension of the control spring above the diaphragm, changing the valve opening point (see Figure 7-6). The valve can be adjusted to open at a predetermined pressure within the range of the control spring. The valve operating point will be a few psi below the opening point. The exact differential is determined by the compressor's pumping capacity. It is the operating pressure that will be observed on a low side pressure gauge attached to a system. A change in valve outlet pressure of 1 psi ($.0703$ kg/cm$^2$) will move the valve stem approximately .001 inch (.0025cm).

The expansion valve should not be adjusted until the refrigeration unit has been operating from 24 to 48 hours. By this time, the refrigerant and the oil in the system should be properly distributed and the evaporator should be cold. When making adjustments, turn the adjusting screw one-quarter

**Figure 7-6.** Automatic expansion valve adjustment (*Courtesy of Singer Controls Division*)

turn at a time and wait 15 minutes between each adjustment. When checking the operation of an expansion valve, the refrigeration unit should be operating continuously.

**FEATURES:** Automatic expansion valves, because of their pressure-regulating performance, offer a number of operating features which are very useful on a variety of applications. Following are the performance characteristics of the automatic expansion valve:

**1. Protection against evaporator icing:** The use of automatic expansion valves eliminates the hazard of frost and ice accumulation on evaporator coils, and prevents the freezing of product loads—for example, water in a water-chilling system. Unless protected during periods of low load, air conditioners accumulate a volume of frost which retards air circulation over the evaporator coil and vastly reduces system capacity.

Automatic expansion valves maintain a constant low side pressure and, therefore, a constant evaporator temperature. Adjusting the valve for an evaporator temperature just above the freezing point of water will completely eliminate the possibility of frost formation, regardless of ambient temperatures, heat load, or the length of time the unit is operated.

Other applications which require similar protection are drinking water coolers, soda fountain water coolers, photo developing tanks, and various industrial liquid chillers and fluid coolers.

**2. Control of relative humidity:** Adjusting the value to maintain an evaporator temperature just above freezing not only prevents frost formation but maintains the low evaporator temperature required for maximum moisture removal—the means of controlling humidity in conditioned air. Humidity control is essential in conditioning air for human comfort.

**3. Motor overload protection:** Due to the close control of the system low side pressure afforded by an automatic expansion valve, the possibility of excessive current draw by the motor which is operating the unit is completely avoided. Motors require this protection at high heat load conditions. With automatic expansion valves, the low side (suction) pressure is constant and does not vary as the heat load fluctuates. Therefore, there are no variations in motor current requirements. Condensing unit current consumption is also automatically maintained within a safe maximum as dictated by the electrical wiring to the unit.

Lower cost motors and wiring can be used with this flow control device. Motor

horsepower is established on the basis of system load at normal ambient temperatures. Extreme ambients will not tax the unit because the evaporator pressure is always under control. Reserve motor capacity is not required and its cost is eliminated. Similarly, condenser surface area can be minimized.

**4. Simplification of field service:** The use of automatic expansion valves simplifies the field service of refrigeration and air conditioning units. Air conditioning unit manufacturers generally use valves with a fixed adjustment. The valves are set and sealed at the ideal operating condition for a particular unit for a refrigerant temperature above freezing. Common field difficulties such as stoppage of capillary tubes are eliminated.

The charging of expansion-valve-controlled air conditioning systems with the proper weight of refrigerant is greatly simplified. A major problem in charging capillary tube systems is adjusting the refrigerant charge to the ambient temperature at the time of charging. It is very easy to overcharge a capillary tube system which leads to operating difficulties when the ambient temperature increases.

If an expansion valve system is overcharged, the valve will automatically adjust the flow of refrigerant during the operating phase of the cycle to feed the evaporator properly. Any reserves of liquid are kept in the bottom of the condenser. Ambient temperature variations have very little effect upon the performance of an expansion-valve-equipped system.

**5. High capacity in a small size control:** The overall expansion valve size has been reduced to the point where a physically small control device provides substantial refrigerating capacity. Valve bodies now machined from bar stock are in contrast to the bulkilyforged bodies once used. Most modern valves are the bleed type—the bleed slot or bypass adding in large measure to the valve capacity. Other plus features for the reduced valve size are improved flow patterns through the valve, smaller but improved valve diaphragms, and silver solder construction.

**6. The right valve for water coolers:** On water cooler applications, automatic expansion valves are adjusted for minimum water temperatures. At the same time, the valve will provide absolute insurance against freeze-up of the water in the cooler, due to the fact that the valve maintains a constant, predetermined evaporator pressure and temperature. The valve is always adjusted for a refrigerant temperature above 32°F (0°C). It also assures no freezing at the valve orifice due to moisture in the system. A capillary tube cannot provide this assurance.

**7. An ideal valve for low-starting torque motors:** The bleed-type valve is used with split-phase and other low-starting torque motors commonly used in condensing units today. This expansion valve permits system unloading from high to low side during the system off cycle. All the advantages of automatic expansion valve operation are realized during the operating phase of the cycle.

**8. A low capacity bypass valve:** Automatic expansion valves function as high to low side pressure regulators. Installed as a bypass and responding to the outlet pressure, the valve will open when the low side pressure is reduced to the valve set point. The system's low side pressure will be maintained at the required minimum as machine capacity is effectively reduced.

**CONSTANT LOW SIDE PRESSURE:** The flow of refrigerant to the evaporator is maintained by the automatic expansion valve at a rate equivalent to the pumping capacity of the compressor. The fact that automatic expansion valves will open and remain open at a point where the refrigerant flow exactly meets compressor capacity accounts for the constant low side pressure (and, therefore, evaporator temperature) maintained during the operating phase of the refrigeration cycle.

The balance in a refrigeration system using an automatic expansion valve is, therefore, between the expansion valve and the condensing unit. Since an automatic expansion valve is a differential type control, the differential between the opening point and the operating point is automatically established by the running of the compressor. This differential provides the required valve movement to allow the flow of liquid refrigerant at the rate required by the compressor.

Automatic expansion valves are ideally suited to refrigeration applications in which control of the evaporator temperature is required.

**BLEED-TYPE VALVES FOR OFF-CYCLE UNLOADING:** Split-phase and other low-starting torque motors used in condensing units require an expansion valve that provides system unloading from the high to the low side during the off cycle.

Bleed-type valves permit pressures in the refrigeration system to reach a balance or near balance point during the off phase of the cycle. At the start of the next running phase, the motor starts under practically no load.

The bleed-type (or slotted orifice) valve is a standard automatic expansion valve with the addition of a small slot in the valve orifice (seat) to prevent complete valve close-off at the end of the machine's on cycle. Due to the bleed slot, the valve does not close completely when the unit stops operating, but permits refrigerant to continue to flow at a reduced rate. (See Figure 7-7.)

Expansion valve bleeds add to the total valve capacity. Small size, *nicked seat,* bleeds increase the valve capacity due to the flow of refrigerant through the nicks (openings) provided. The larger *drill bypass* bleeds are also open at all times to pass refrigerant when a pressure drop exists across the valve. Bleeds provide a fixed capacity depending upon (1) bleed size, (2) liquid density, and (3) the existing pressure drop.

**Figure 7-7.** Bleed slot *(Courtesy of Singer Controls Division)*

**SELECTING THE PROPER SIZE BLEED:** Some care must be taken in selecting an expansion valve with the proper size bleed for a particular application. Generally the valve selected should feature the smallest size bleed slot which will provide the re-

quired unloading (balancing of pressure) during the minimum length of machine off cycle. This is important because the bleed slots must not interfere with normal valve operation during the running phase of the cycle. A valve with a large sized bleed slot used with a relatively small condensing unit would cause difficulty at low suction temperatures, due to the bleed slot's closely matching the machine pumping capacity. Suction pressures below this point could not be reached by adjusting the automatic expansion valve.

To be certain that the bleed slot is not oversized for an application, turn the adjusting stem out to a setting lower than that at which the system will normally operate. Operate the unit and check the low side pressure with a gauge. If the suction pressure drops down to the valve setting, or at least to a point below the normal operating pressure, the user can be sure the size of the bleed will not prevent the machine from operating at the desired point.

The rate of off-cycle unloading is an especially important consideration when machines may be required to start after a comparatively short off-cycle. The larger the bleed and the greater the initial pressure drop across the valve, the faster the unloading rate.

**ELEVATION CHANGE—EFFECT UPON VALVE SETTING:** Automatic expansion valves are designed to admit atmospheric pressure to the valve diaphragm on the control spring side. The spring plus atmospheric pressure move the valve in the opening direction. The spring force can be varied by adjusting the screw, but any substantial change in altitude after a valve has been adjusted will alter the low side pressure that the valve will maintain.

**AXVS AS BYPASS VALVES:** Automatic expansion valves (AXVs) respond to valve outlet pressure. When this pressure is reduced to the valve opening point setting, the control spring above the diaphragm moves in the valve opening direction. This type of response is required for bypass valves as well as for flow control devices. Automatic expansion valves are, therefore, ideal as small capacity bypass or capacity control valves.

Installed between the system's high and low sides, the bypass valve will open at pre-set outlet (low side) pressure and bypass high pressure discharge gas to the low side of the system. Compressor capacity is modulated to prevent evaporator temperatures from dropping below a predetermined set point. In addition to capacity modulation the automatic expansion valve, when applied as a hot gas bypass valve, can provide evaporator coil freeze-up protection regardless of the type of flow control device selected to meter liquid refrigerant to the evaporator (see Figure 7-8).

1. Orifice size

2. Amount of needle movement

3. Pressure drop across the valve

4. Type of refrigerant used

5. Condensing temperature or pressure

6. Size of the bleed slot

7. Evaporating temperature or pressure

8. Liquid subcooling

## THERMOSTATIC EXPANSION VALVES

The most commonly used device for controlling the flow of liquid refrigerant into the evaporator is the thermostatic expansion valve. (See Figure 7-9.) An orifice in the valve meters the flow into the evaporator. The rate of flow is modulated as required by a needle-type plunger and seat which varies the orifice opening.

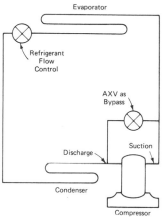

**Figure 7-8.** AXV as bypass valve

**Figure 7-9.** Thermostatic expansion valve (*Courtesy of Alco Controls Div., Emerson Electric Co.*)

**FACTORS AFFECTING VALVE CAPACITY:** The factors that affect capacity of an automatic expansion valve are:

**FUNCTION:** The thermostatic expansion valve is a precision device which is designed to meter the flow of refrigerant into an evaporator in exact proportion to the rate of evaporation of the liquid refrigerant in the evaporator, thereby preventing the return of liquid refrigerant to the compressor. By being responsive to the tempreature of the refrigerant gas leaving the evaporator and the pressure in the evaporator, the thermostatic expansion valve can control the refrigerant gas leaving the evaporator at a predetermined superheat.

A vapor is said to be superheated whenever its temperature is higher than the saturation temperature corresponding to its pressure. The amount of the superheat is, of course, the temperature increase above the saturation temperature at the existing pressure.

Consider a refrigeration evaporator operating with R–12 as the refrigerant at 37 psi (2.6 kg/cm²) suction pressure. (See Figure 7-10.) The R–12 saturation temperature at 37 psi is 40°F (4.4°C). As long as any liq-

uid refrigerant exists at this pressure, the refrigerant temperature will remain at 40°F.

As the refrigerant moves along in the evaporator, the liquid boils off into a vapor and the amount of liquid decreases. At point A in Figure 7-10 all of the liquid has evaporated due to the absorption of a quantity of heat from the surrounding atmosphere which is equal to the latent heat of vaporization of the refrigerant. The refrigerant gas continues along in the evaporator and remains at the same pressure (37 psi); however, its temperature increases due to the continued absorption of heat from the surrounding atmosphere. By the time the refrigerant gas reaches the end of the evaporator at point B, its temperature is 50°F (10°C). This refrigerant gas is now superheated and the amount of superheat is, of course, 50°F − 40°F or 10°F (5.6°C). The degree to which the refrigerant gas is superheated is a function of the amount of refrigerant being fed into the evaporator and the load to which the evaporator is exposed.

**Figure 7-10.** Basic refrigeration schematic (*Courtesy of Alco Controls Div., Emerson Electric Co.*)

**Figure 7-11.** Thermostatic expansion valve basic forces (*Courtesy of Alco Controls Div., Emerson Electric Co.*)

**OPERATION:** Three forces govern the operation of a thermostatic expansion valve (see Figure 7-11). These forces are: (1) the pressure created by the remote bulb and power assembly ($P_1$), (2) the evaporator pressure ($P_2$), and (3) the equivalent pressure of the superheat spring ($P_3$).

The remote bulb and power assembly is a closed system and in the following discussion it is assumed that the remote bulb and power assembly charge is the same refrigerant as that used in the system.

The pressure within the remote bulb and power assembly, $P_1$ in Figure 7-11, then corresponds to the saturation pressure of the refrigerant temperature leaving the evaporator and moves the valve pin in the opening direction. Opposed to this forced on the underside of the diaphragm, and acting in the closing direction, is the force exerted by the evaporator pressure ($P_2$), along with the pressure exerted by the superheat spring ($P_3$). The valve will assume a stable control position when these three forces are in equi-

librium. That is when $P_1 = P_2 + P_3$. As the temperature of the refrigerant gas at the evaporator outlet increases above the saturation temperature which coresponds to the evaporator pressure, it becomes superheated. The pressure thus generated in the remote bulb and power assembly increases above the combined pressures of the evaporator pressure and the superheat spring, causing the valve pin to move in the opening direction. Conversely, as the temperature of the refrigerant gas leaving the evaporator decreases, the pressure in the remote bulb and power assembly also decreases and the combined evaporator and spring pressures cause the valve pin to move in the closing direction.

The factory superheat setting of thermostatic expansion valves is made with the valve pin just starting to move away from the seat. These valves are so designed that an increase in superheat of the refrigerant gas leaving the evaporator, usually 4°F (−15.6°C) beyond the factory setting, is necessary for the valve pin to open to its full open position. For example, if the factory setting is 10°F (−12.2°C) superheat, the operating superheat at the rated open position (full load rating of the valve) will be 14°F (−10°C) superheat. If the system is operating at half load, with 50% compressor capacity reduction, the valve will operate at about 12°F (−11.1°C) superheat. It is important that internally adjustable type thermostatic expansion valves be ordered with the correct factory superheat setting. It is also recommended that an externally adjustable thermostatic expansion valve be used in a pilot model test to determine the correct factory superheat setting before ordering the internally adjustable type valve.

As the operating superheat setting is raised, the evaporator capacity decreases, since more of the evaporator surface is required to produce the superheat necessary to open the valve (see Figure 7-12.) It is obvious, then, that it is most important to adjust the operating superheat correctly. It is of vital importance that a minimum change in superheat is required to move the valve pin to the full open position, because this provides savings in both the initial evaporator cost and the cost of operation. Accurate and sensitive control of the liquid refrigerant flow into the evaporator is necessary to provide maximum evaporator capacity under all load conditions.

**ADJUSTMENT:** Each thermostatic expansion valve is adjusted at the factory before shipment. This factory setting will be correct and no further adjustment is required for the majority of applications. When the application or operating conditions require a different valve setting, the valve may be adjusted to obtain the required superheat.

Some expansion valves are nonadjustable for use on original equipment manufacturer's units. (See Figure 7-13.) These valves are set at a superheat which is predetermined by the manufacturer's laboratory tests and cannot be adjusted in the field.

Most nonadjustable models are modifications of standard adjustable type valves. This is done by using a solid bottom cap instead of one equipped with an adjusting stem and seal cap. (See Figure 7-14.) Adjustable bottom cap assemblies are available for converting some nonadjustable valves to the adjustable type. However, this is rarely required. If symptoms indicate that a valve adjustment is needed, carefully check the other possible causes of incorrect superheat before attempting an adjustment.

**HOW TO DETERMINE SUPERHEAT:** Thermostatic expansion valve performance cannot be analyzed properly by measuring the suction pressure or by observing the

Superheat Setting Too Low

Normal Superheat Setting

High Superheat Setting

**Figure 7-12.** Superheat setting and evaporator capacity

**Figure 7-13.** Nonadjustable thermostatic expansion valve (*Courtesy of Sporlan valve Company*)

frost formation on the suction line. The initial step in correctly determining whether or not a thermostatic expansion valve is functioning properly is to measure the superheat. The following four steps show how to calculate the superheat setting:

1.   Measure the temperature of the suction line at the point where the bulb is clamped.

2.   Obtain the suction pressure that exists in the suction line at the bulb location by either of the following methods:

a. If the valve is externally equalized, a gauge in the external equalizer line will indicate the desired pressure directly and accurately.

b. Read the gauge pressure at the suction valve of the compressor. To the pressure reading add the estimated pressure

**Figure 7-14.** Adjustable and nonadjustable thermostatic expansion valves (*Courtesy of Singer Controls Division*)

drop through the suction line between the bulb location and the compressor suction valve. The sum of the gauge reading and the estimated pressure drop will equal the approximate line pressure at the bulb.

3.   Convert the pressure obtained in step 2(a) or 2(b) above to the saturated evaporator temperature by using a Temperature-Pressure Chart as shown in Table 7-1.

## Table 7-1
## Temperature-Pressure Chart *

**BOLD FIGURES = INCHES MERCURY VACUUM   LIGHT FIGURES = PSIG**

| °F | R-12 | R-13 | R-22 | R-500 | R-502 | R-717 Ammonia |
|---|---|---|---|---|---|---|
| -100 | **27.0** | 7.5 | **25.0** | — | **23.3** | **27.4** |
| -95 | **26.4** | 10.9 | **24.1** | — | **22.1** | **26.8** |
| -90 | **25.7** | 14.2 | **23.0** | — | **20.7** | **26.1** |
| -85 | **25.0** | 18.2 | **21.7** | — | **19.0** | **25.3** |
| -80 | **24.1** | 22.2 | **20.2** | — | **17.1** | **24.3** |
| -75 | **23.0** | 27.1 | **18.5** | — | **15.0** | **23.2** |
| -70 | **21.8** | 32.0 | **16.6** | — | **12.6** | **21.9** |
| -65 | **20.5** | 37.7 | **14.4** | — | **10.0** | **20.4** |
| -60 | **19.0** | 43.5 | **12.0** | — | **7.0** | **18.6** |
| -55 | **17.3** | 50.0 | **9.2** | — | **3.6** | **16.6** |
| -50 | **15.4** | 57.0 | **6.2** | — | **0.0** | **14.3** |
| -45 | **13.3** | 64.6 | **2.7** | — | 2.1 | **11.7** |
| -40 | **11.0** | 72.7 | **0.5** | **7.9** | 4.3 | **8.7** |
| -35 | **8.4** | 81.5 | 2.6 | **4.8** | 6.7 | **5.4** |
| -30 | **5.5** | 91.0 | 4.9 | **1.4** | 9.4 | **1.6** |
| -28 | **4.3** | 94.9 | 5.9 | 0.0 | 10.6 | 0.0 |
| -26 | **3.0** | 98.9 | 6.9 | 0.7 | 11.7 | 0.8 |
| -24 | **1.6** | 103.0 | 7.9 | 1.5 | 13.0 | 1.7 |
| -22 | **0.3** | 107.3 | 9.0 | 2.3 | 14.2 | 2.6 |
| -20 | 0.6 | 111.7 | 10.1 | 3.1 | 15.5 | 3.6 |
| -18 | 1.3 | 116.2 | 11.3 | 4.0 | 16.9 | 4.6 |
| -16 | 2.1 | 120.8 | 12.5 | 4.9 | 18.3 | 5.6 |
| -14 | 2.8 | 125.7 | 13.8 | 5.8 | 19.7 | 6.7 |
| -12 | 3.7 | 130.5 | 15.1 | 6.8 | 21.3 | 7.9 |
| -10 | 4.5 | 135.4 | 16.5 | 7.8 | 22.8 | 9.0 |
| -8 | 5.4 | 140.5 | 17.9 | 8.8 | 24.4 | 10.3 |
| -6 | 6.3 | 145.7 | 19.3 | 9.9 | 26.0 | 11.6 |
| -4 | 7.2 | 151.1 | 20.8 | 11.0 | 27.7 | 12.9 |
| -2 | 8.2 | 156.5 | 22.4 | 12.1 | 29.5 | 14.3 |
| 0 | 9.1 | 162.1 | 24.0 | 13.3 | 31.2 | 15.7 |
| 2 | 10.2 | 167.9 | 25.6 | 14.5 | 33.1 | 17.2 |
| 4 | 11.2 | 173.7 | 27.3 | 15.7 | 35.0 | 18.8 |
| 6 | 12.3 | 179.8 | 29.1 | 17.0 | 37.0 | 20.4 |
| 8 | 13.5 | 185.9 | 30.9 | 18.4 | 39.1 | 22.1 |
| 10 | 14.6 | 192.1 | 32.8 | 19.8 | 41.1 | 23.8 |
| 12 | 15.8 | 198.6 | 34.7 | 21.2 | 43.3 | 25.6 |
| 14 | 17.1 | 205.2 | 36.7 | 22.7 | 45.5 | 27.5 |
| 16 | 18.4 | 211.9 | 38.7 | 24.2 | 47.8 | 29.4 |
| 18 | 19.7 | 218.8 | 40.9 | 25.7 | 50.1 | 31.4 |
| 20 | 21.0 | 225.7 | 43.0 | 27.3 | 52.5 | 33.5 |
| 22 | 22.4 | 233.0 | 45.3 | 29.0 | 55.0 | 35.7 |
| 24 | 23.9 | 240.3 | 47.6 | 30.7 | 57.5 | 37.9 |
| 26 | 25.4 | 247.8 | 49.9 | 32.5 | 60.1 | 40.2 |
| 28 | 26.9 | 255.5 | 52.4 | 34.3 | 62.8 | 42.6 |
| 30 | 28.5 | 263.2 | 54.9 | 36.1 | 65.4 | 45.0 |
| 32 | 30.1 | 271.3 | 57.5 | 38.0 | 68.3 | 47.6 |
| 34 | 31.7 | 279.5 | 60.1 | 40.0 | 71.2 | 50.2 |
| 36 | 33.4 | 287.8 | 62.8 | 42.0 | 74.1 | 52.9 |
| 38 | 35.2 | 296.3 | 65.6 | 44.1 | 77.2 | 55.7 |
| 40 | 37.0 | 304.9 | 68.5 | 46.2 | 80.2 | 58.6 |
| 45 | 41.7 | 327.5 | 76.0 | 51.9 | 88.3 | 66.3 |
| 50 | 46.7 | 351.2 | 84.0 | 57.8 | 96.9 | 74.5 |
| 55 | 52.0 | 376.1 | 92.6 | 64.2 | 106.0 | 83.4 |
| 60 | 57.7 | 402.3 | 101.6 | 71.0 | 115.6 | 92.9 |
| 65 | 63.8 | 429.8 | 111.2 | 78.2 | 125.8 | 103.1 |
| 70 | 70.2 | 458.7 | 121.4 | 85.8 | 136.6 | 114.1 |
| 75 | 77.0 | 489.0 | 132.2 | 93.9 | 148.0 | 125.8 |
| 80 | 84.2 | 520.8 | 143.6 | 102.5 | 159.9 | 138.3 |
| 85 | 91.8 | — | 155.7 | 111.5 | 172.5 | 151.7 |
| 90 | 99.8 | — | 168.4 | 121.2 | 185.8 | 165.9 |
| 95 | 108.3 | — | 181.8 | 131.2 | 199.7 | 181.1 |
| 100 | 117.2 | — | 195.9 | 141.9 | 214.4 | 197.2 |
| 105 | 126.6 | — | 210.8 | 153.1 | 229.7 | 214.2 |
| 110 | 136.4 | — | 226.4 | 164.9 | 245.8 | 232.3 |
| 115 | 146.8 | — | 242.7 | 177.3 | 262.6 | 251.5 |
| 120 | 157.7 | — | 259.9 | 190.3 | 280.3 | 271.7 |
| 125 | 169.1 | — | 277.9 | 203.9 | 298.7 | 293.1 |
| 130 | 181.0 | — | 296.8 | 218.2 | 318.0 | 315.0 |
| 135 | 193.5 | — | 316.6 | 233.2 | 338.1 | 335.0 |
| 140 | 206.6 | — | 337.3 | 248.8 | 359.1 | 365.0 |
| 145 | 220.6 | — | 358.9 | 265.2 | 381.1 | 390.0 |
| 150 | 234.6 | — | 381.5 | 282.3 | 403.9 | 420.0 |
| 155 | 249.9 | — | 405.2 | 300.1 | 427.8 | 450.0 |
| 160 | 265.12 | — | 429.8 | 318.7 | 452.6 | 490.0 |

* Courtesy of Alco Controls Division

4. Subtract the two temperatures obtained in Step 1 and 3. The difference is the superheat.

Figure 7-15 illustrates a typical example of superheat measurement on an air conditioning system using R–12 as the refrigerant. The temperature of the suction line at the bulb location is read at 51°F (10.6°C). The suction pressure at the compressor is 35 psi (2.46 kg/cm²) and the estimated pressure drop is 2 psi (.14 kg/cm²). The total suction pressure is 35 psi + 2 psi = 37 psi (2.6 kg/cm²). This is equivalent to a 40°F (4.4°C) saturation temperature. 40°F subtracted from 51°F equals 11°F (6.2°C) superheat setting.

Notice that subtracting the difference between the temperature at the inlet of the evaporator and the temperature at the outlet of the evaporator is not an accurate measure of superheat. This method is not recommended because any evaporator pressure drop will result in an erroneous superheat indication.

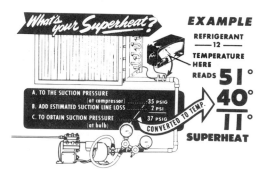

**Figure 7-15.** Determination of superheat (*Courtesy of Sporlan Valve Company*)

**HOW TO CHANGE THE SUPERHEAT SETTING:** To reduce the superheat setting, turn the adjusting stem *counterclockwise*.

To increase the superheat setting, turn the adjusting stem *clockwise*. When adjusting the superheat, make no more than one turn of the stem at a time and observe the change in superheat closely to prevent overshooting the desired setting. As much as 30 minutes may be required for the new balance to take place after an adjustment is made.

If in doubt about the correct superheat setting for a particular system, consult the equipment manufacturer. As a general rule, the proper superheat setting will depend on the amount of temperature difference (TD) between the refrigerant temperature and the temperature of the air or other substance being cooled. Where high TDs exist, such as in air conditioning applications, the superheat setting can be as high as 15°F (−9.4°C) without noticeable loss in evaporator capacity. Where low TDs exist, such as in low-temperature blower coil applications, a superheat setting of 10°F (−12.2°C) or below is usually recommended for maximum evaporator capacity.

For the correct valve setting on factory-built equipment, the manufacturer's recommendations should be followed. Some manufacturers specify the superheat directly; others may recommend valve adjustment to provide a given suction pressure at certain operating conditions, or until a certain frost line is observed. Such recommendations, however they are stated, represent the results of extensive laboratory testing to determine the best possible operation.

**BULB LOCATION AND INSTALLATION:** The location of the remote bulb is extremely important, and in some cases determines the success or failure of the refrigeration plant. For satisfactory expansion valve control, *good thermal contact* between the bulb and the suction line is essential. The bulb should be securely fastened with

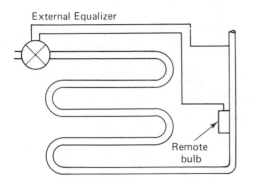

**Figure 7-16.** Remote bulb installation on vertical tubing

External Bulb on Small Suction Line

45°

External Bulb on Large Suction Line

**Figure 7-17.** Remote bulb installation on horizontal tubing (*Courtesy of Alco Controls Div., Emerson Electric Co.*)

two bulb straps to a clean, straight section of the suction line.

Installation of the bulb to a horizontal run of suction line is preferred. If a vertical installation cannot be avoided, the bulb should be mounted so that the capillary tubing comes from the top as shown in Figure 7-16.

To install, clean the suction line thoroughly before clamping the remote bulb in place. When a steel suction line is used, it is advisable to paint the line with aluminum paint to minimize future corrosion and faulty remote bulb contact with the line. On lines under 7/8 in. (22.225 mm) OD, the remote bulb may be installed on top of the line. On 7/8 in. OD and larger, the remote bulb should be installed at about the 4 or 8 o'clock position. (See Figure 7-17.)

If it is necessary to protect the remote bulb from the effects of an air stream after it is clamped to the line, use a material such as sponge rubber that will not absorb water when evaporator temperatures are above 32°F (0°C). Below 32°F cork or similar material sealed against moisture is suggested to prevent ice clogging at the remote bulb location. The use of hair felt is not recommended. When the remote bulb location is below the water or brine level of a sub-

merged coil, use a waterproofing material or pitch that does not require heating above 120°F (49°C) when applying it in order to protect the remote bulb and tubing.

Accepted principles of good suction line piping should be followed to provide a bulb location that will give the best possible valve control. Never locate the bulb in a trap or pocket in the suction line. Liquid refrigerant or a mixture of liquid refrigerant and oil boiling out of the trap will falsely influence the temperature of the bulb and result in poor valve control.

Recommended suction line piping includes a horizontal line leaving the evaporator to which the thermostatic expansion valve bulb is attached. This line is pitched slightly downward, and when a vertical riser follows, a short trap is placed immediately ahead of the vertical line. (See Figure 7-18.) The trap will collect any liquid refrigerant or oil passing through the suction line and prevent it from influencing the bulb temperature.

On multiple evaporator installations, the piping should be arranged so that the refrigerant flow from any valve cannot affect the remote bulb of another. Approved piping practices including the proper use of traps insure individual control for each valve without the influence of refrigerant and oil

**Figure 7-18.** Evaporator suction piping (*Courtesy of Sporlan Valve Company*)

flow from other evaporators. (See Figure 7-19.)

Recommended procedure for suction line piping when the evaporator is located above the compressor is to extend a vertical riser to the top of the evaporator to prevent refrigerant draining by gravity into the compressor during the off cycle. (See Figure 7-20.) When a pump-down control is used, the suction line may turn down immediately without a trap.

**Figure 7-19.** Evaporator suction piping on multiple evaporators (*Courtesy of Sporlan Valve Company*)

**Figure 7-20.** Evaporator suction piping with compressor below evaporator (*Courtesy of Sporlan Valve Company*)

**REMOTE BULB AND POWER ASSEMBLY CHARGES:** Expansion valve charges are grouped into the following general classifications:

1. Liquid Charge—usually the same as the system refrigerant. Valve control is not normally good. The charge tends to allow liquid refrigerant floodback to the compressor on system start-up.

2. Gas Charge—usually the same as the system refrigerant. Valve control is generally good. The charge will not operate in cross-ambient conitions and will condense at the coldest point. The valve will lose control if the power element or remote bulb tubing becomes colder than the remote bulb.

3. Liquid-Vapor Cross Charge—a volatile liquid refrigerant (not necessarily the same as the system refrigerant), combined with a noncondensible gas. The charge does not lose control under cross-ambient conditions. The charge shifts slightly in control point.

4. Cross-Ambient Vapor Charge—an all-purpose charge which replaces all three of the previous charge types, providing outstanding control under all conditions from $-40°F$ to $50°F$ ($-40°C$ to $10°C$).

## THERMOSTATIC EXPANSION VALVE EQUALIZER:

The operation of the thermostatic expansion valve is dependent upon the relationship of the three fundamental pressures. That is, the bulb pressure acting on top of the diaphragm must always equal the sum of the evaporator (or suction pressure) and the spring pressure applied to the evaporator side of the diaphragm.

On an internally equalized valve, the pressure at the valve outlet (or evaporator inlet) is transmitted to the evaporator side of the diaphragm via a passageway within the valve or through clearance around the push rods. (See Figure 7-21.)

On an externally equalized expansion valve the evaporator side of the diaphragm is isolated from the valve outlet pressure by packing around the push rods. The suction pressure is transmitted to the evaporator side of the diaphragm by a line usually connected between the suction line near the evaporator outlet (preferably downstream of the bulb) and an external fitting on the valve (see Figure 7-22).

While internally equalized valves may be used with evaporators that have a low pressure drop, valves with an external equalizer *must* be used when there is an appreciable drop between the valve outlet and the remote bulb location.

An internally equalized valve feeding an evaporator which, for purpose of illustration, has no pressure drop is shown in Figure 7-23. The pressure at the valve outlet and at the remote bulb location is 27 psi ($1.9$ kg/cm²). Therefore, the evaporator

**Figure 7-21.** Internally equalized expansion valve (*Courtesy of Sporlan Valve Company*)

**Figure 7-22.** Externally equalized expansion valve (*Courtesy of Sporlan Valve Company*)

Closing Pressure . . . . . . . . . . . . . . . . . . . . = 27 + 7 = 34 psi
   (Evaporator Inlet Pressure Plus Spring Pressure)

Bulb Pressure Necessary to Open Valve . . . . . . . . . . 34 psi

Bulb Temperature Equivalent to 34 psi . . . . . . . . . . . . . . . . . . .   37°F

Saturated Temperature Equivalent to Evaporator Outlet Pressure .28°F
   Superheat . . . . . . . . . . . . . . . . . . . . . . . . . . . . . . . . . . . . . . . . . . . . . 9°F
      Bulb Temperature Minus Saturated Evaporator Temperature

**Figure 7-23.**   Internally equalized valve illustration (*Courtesy of Sporlan Valve Company*)

side of the valve diaphragm senses the evaporator pressure which is also 27 psi and acts to close the valve. A spring pressure of 7 psi (.49 kg/cm²) also assists the evaporator pressure in attempting to close the valve. The valve consequently adjusts its refrigerant flow rate until the suction line vapor becomes sufficiently superheated to create a remote bulb temperature of 37°F (2.8°C) which develops a pressure of 34 psi (2.39 kg/cm²), balancing the evaporator spring. The resulting superheat is 9°F (5°C).

If this same internally equalized valve with the same spring adjustment is installed on an evaporator of equivalent nominal capacity but with a 6 psi (.42 kg/cm²) pres-

sure drop, the operating superheat will increase to 15°F (8.3°C) as shown in Figure 7-24. Now the valve senses a comparatively high pressure of 33 psi (2.32 kg/cm²) at the valve outlet. The total closing pressure is 33 (2.32) + 7 (.49 kg/cm²) = 40 psi (2.8 kg/cm²). Since the bulb pressure must equal the total closing pressure, the valve reduces the flow of refrigerant to create the necessary superheat and bulb pressure. Consequently, excessive evaporator pressure drop causes an internally equalized valve to operate at an abnormally high superheat, and a serious loss of evaporator capacity results.

The problem of improper valve control may be corrected by using a thermostatic

Closing Pressure . . . . . . . . . . . . . . . . . . . = 33 + 7 = 40 psi
   (Evaporator Inlet Pressure Plus Spring Pressure)

Bulb Pressure Necessary to Open Valve . . . . . . . . . . 40 psi

Bulb Temperature Equivalent to 40 psi . . . . . . . . . . . . . . . . . . 43°F

Saturated Temperature Equivalent to Evaporator Outlet Pressure . 28°F
   Superheat . . . . . . . . . . . . . . . . . . . . . . . . . . . . . . . . . . . . . . . . 15°F
      Bulb Temperature Minus Saturated Evaporator Temperature

**Figure 7-24.** Thermostatic expansion valve with coil pressure drop (*Courtesy of Sporlan Valve Company*)

expansion valve with an external equalizer. (See Figure 7-25.) This is the same system, but with an externally equalized expansion valve. The suction pressure at the remote bulb location is transmitted to the evaporator side of the diaphragm via the external equalizer line. The valve operation is now identical to that shown in Figure 7-23, and the superheat returns to 9°F (5°C).

**WHEN TO USE THE EXTERNAL EQUALIZER:** An internally equalized valve can tolerate less evaporator pressure drop at lower evaporator temperatures. (See Table 7-2.) There are, of course, applications which may satisfactorily employ the internal equal-

izer when a higher pressure drop is present, but this should usually be verified by laboratory tests. The general recommendations given in Table 7-2 are suitable for most field-installed systems. Use an external equalizer when the pressure drop between the valve outlet and the remote bulb location exceeds the values shown in Table 7-2.

When the expansion valve is equipped with an external equalizer connection, it must be connected—never capped—or the valve may flood, starve, or regulate erratically.

There is no operational disadvantage in using an external equalizer even if the evaporator has a low pressure drop.

Closing Pressure . . . . . . . . . . . . . . . . . . . . = 27 + 7 = 34 psi
(Suction Pressure at Bulb Plus Spring Pressure)

Bulb Pressure Necessary to Open Valve . . . . . . . . . . 34 psi

Bulb Temperature Equivalent to 34 psi . . . . . . . . . . . . . . . . . . 37° F

Saturated Temperature Equivalent to Evaporator Outlet Pressure . 28° F
Superheat . . . . . . . . . . . . . . . . . . . . . . . . . . . . . . . . . . . . . . 9° F
Bulb Temperature Minus Saturated Evaporator Temperature

**Figure 7-25.** Externally equalized expansion valve installation (*Courtesy of Sporlan Valve Company*)

Table 7-2

Maximum Pressure Drop for Which Internally
Equalized Valves May Be Used *

| REFRIGERANT | EVAPORATING TEMPERATURE | | | | |
|---|---|---|---|---|---|
| | Degrees F. | | | | |
| | 40 | 20 | 0 | —20 | —40 |
| | PRESSURE DROP—psi | | | | |
| 12 | 2 | 1.5 | 1 | 0.75 | 0.5 |
| 22 | 3 | 2 | 1.5 | 1.0 | 0.75 |
| 500 | 2 | 1.5 | 1 | 0.75 | 0.5 |
| 502 | 3 | 2.5 | 1.75 | 1.25 | 1.0 |
| 717 | 3 | 2 | 1.5 | 1.0 | 0.75 |
| (Ammonia) | | | | | |

IMPORTANT: The External Equalizer must be used on evaporators which employ a pressure drop type refrigerator distributor.
* Courtesy of Sporlan Valve Company

**PRESSURE-LIMITING THERMO-STATIC EXPANSION VALVE:** Standard thermostatic expansion valves are either pressure-limiting or nonpressure-limiting. Nonpressure-limiting valves are generally liquid-charged. Various liquid charges are used depending on the system refrigerant, application temperature, and other factors. Pressure-limiting valves provide motor overload protection at times of high heat load. The pressure-limiting feature is commonly called *maximum operating pressure* (MOP). (See Figure 7-26.)

Present day condensing units are designed to operate within a certain range of suction pressures. Operation of high side equipment above the recommended maximum

**Figure 7-26.** Maximum operating pressure expansion valve (*Courtesy of Sporlan Valve Company*)

pressure imposes an overload on condensing units which may result in eventual damage to the condensing unit motor. To prevent suction pressures from soaring during unit operation, a pressure-limiting thermostatic expansion valve should be employed.

A pressure-limiting valve will perform as a standard thermostatic expansion valve when operating at suction pressures within the range for which the condensing unit is designed. When the maximum operating pressure is reached, the valve's pressure-limiting feature takes control from the power element and prevents further increase in the flow of refrigerant to the evaporator. The system will continue to operate at this maximum suction pressure until the overload condition has passed. At this time, the suction presure will drop below the maximum point, and the valve will again function as a standard thermostatic expansion valve.

The danger of floodback to the compressor at system start-up is also reduced with a pressure-limiting valve. The power element presure above the diaphragm will not open the valve until the evaporator pressure is reduced below the pressure limit by the compressor. When the valve opens, it will operate on the valve's maximum operating pressure to minimize the initial surge of liquid refrigerant to the warm evaporator coil.

Most thermostatic expansion valves use two type of construction to provide the pressure-limiting feature: (1) limited liquid charge, and (2) liquid charge with a mechanical pressure limit.

**LIMITED LIQUID (GAS) CHARGE:** Valves with a gas charge pressure limit are charged with a limited amount of refrigerant —a limited liquid charge. When the valve pressure limit has been reached, all the charge will have vaporized. As the evaporator temperature continues to rise, the power element vapor is superheated, with no appreciable increase in power element pressure. As a result, the pressure developed above the diaphragm is limited to a fixed maximum, and the valve prevents further increase in refrigerant flow at this point. Evaporator maximum operating pressure is thus established.

Gas-charged pressure-limiting valves are common in air conditioning and other high temperature applications. However, they have one limitaton. The thermal bulb must always be colder than the capillary tubing or valve head. If not, the thermal charge will condense at the coldest point and the remote thermal bulb will lose control.

**MECHANICAL TYPE PRESSURE-LIMITING VALVES:** A thermostatic expansion valve with a mechanical pressure limit contains a standard liquid charge in the power element. The bulb always contains liquid and is aways in control.

The pressure-limiting feature is accomplished by means of a special double diaphragm. (See Figure 7-27.) At all pressures below the pressure-limit setting, the valve functions as a standard thermostatic expansion valve. The lower diaphragm operates the push pins which in turn control the valve needle.

When the valve pressure-limit setting has been reached, the increased pressure between the diaphragms offsets the spring force above the upper diaphragm. The upper diaphragm rises and the two diaphragms (now interlocked) act as one. The valve now performs as a constant pressure expansion device and meters liquid refrigerant to maintain the fixed evaporator pressure. When the overload condition has passed, the upper diaphragm moves down to a stop, and the lower diaphragm again assumes control. Normal thermostatic expansion valve operation is now resumed.

**Figure 7-27.** Mechanical pressure-limiting expansion valve (*Courtesy of Singer Controls Division*)

**DISTRIBUTORS:** When the refrigeration load is such that large evaporators are required, multiple refrigerant circuits are necessary to avoid an excessive pressure drop through the evaporator. To ensure uniform refrigerant feed from the expansion valve to each of the various circuits, a refrigerant distributor is normally used. (See Figure 7-28.) The refrigerant distributor is mounted on the evaporator with the necessary tubing as shown in Figure 7-29.

As liquid refrigerant is fed through the expansion valve, a portion of the liquid flashes into vapor in order to reduce the liquid temperature to the evaporator temperature (flash gas). This combination of liquid and flash gas is fed into the distributor from

**Figure 7-28.** Refrigerant distributor (Courtesy of Singer Controls Division)

the expansion valve, and is then distributed evenly through small feeder tubes, the number of feeders depending upon the construction of the distributor and the number of circuits required to provide proper refrigerant velocity in the evaporator.

Without the distributor, the refrigerant flow would split into separate gas and liquid layers, resulting in the starving of some evaporator circuits. To avoid variations in circuit feeding, extreme care must be taken to ensure that feeder tubes are equal in length, so that equal resistance is offered by each circuit.

There are two different approaches in the design of a distributor. A high-pressure drop distributor depends on the turbulence created by an orifice to achieve good refrigerant distribution. (See Figure 7-30.) A low-pressure drop distributor depends on a contour flow pattern to provide the proper distribution of the refrigerant flow. Both types of distributors give satisfactory performance when properly applied in accordance with the manufacturer's instructions.

Refrigerant Distributor

**Figure 7-29.** Mounted refrigerant distributor (Courtesy of Singer Controls Division)

**Figure 7-30.** Refrigerant distributor circuits (*Courtesy of Sporlan Valve Company*)

## CAPILLARY TUBES

The capillary tube is the simplest type of refrigerant flow control device used in modern refrigeration systems. However, its application is limited to single cabinets having their own condensing unit and cannot be used in multitemperature or multiple cabinet systems. A capillary tube is a small diameter tube through which the refrigerant flows to the evaporator (see Figure 7-31). The capillary tube is not a true valve since it is not adjustable and cannot be readily regulated. It is used only on flooded systems and allows the liquid refrigerant to flow into the

**Figure 7-31.** Capillary tubing

evaporator at a predetermined rate, which is determined by the size of the refrigeration machine and the load it must carry. The capillary tube acts in exactly the same manner as a small diameter water pipe which holds back water, allowing a higher pressure to be built up behind the water column with only a small rate of flow. In the same manner the small diameter capillary tube holds back the liquid refrigerant, enabling a high pressure to be built up in the condensing unit during operation of the unit, and at the same time permitting the liquid refrigerant to flow slowly into the evaporator.

Since the orifice is fixed, the rate of feed is relatively inflexible. Under conditions of constant load, and constant discharge and suction pressures, the capillary tube performs very satisfactorily. However, changes in the evaporator load or fluctuations in head pressure can result in under- or over-feeding of the evaporator with refrigerant.

When the condensing unit stops, the condenser pressure and the evaporator pressure gradually equalize as the liquid refrigerant continues to flow through the capillary tube. Eventually the compressor is able to start under balanced pressure conditions. This allows the use of low-starting torque motors, a big advantage of capillary tubes. However, it is necessary that the tube be free from obstructions because if it becomes clogged, the refrigeration system will not function.

Because of the small bore in the capillary tube, it is essential that the system be kept free from dirt and foreign matter. Usually a filter is placed before the capillary tube to prevent dirt from plugging this tube. (see Figure 7-32). If the capillary tube becomes plugged, the evaporator will defrost, the unit will run continuously, or the thermal overload may cut out. The discharge pressure will become very high unless the

**Figure 7-32.** Capillary tube and strainer (*Courtesy of Jarrow Brinda, Inc.*)

liquid receiver or condenser has capacity enough to take the entire charge of refrigerant. In many cases where the capillary tube is used, it is connected directly to the outlet end of the condenser and no liquid receiver is used. This means that most of the refrigerant is in the evaporator. In this case, it should be evident that a plugged capillary tube will result in an excessively high head pressure.

The refrigerant charge is critical in capillary tube systems because normally there is no receiver to store the excess refrigerant. Too much refrigerant will cause high head pressures, compressor motor overloading, and possible liquid floodback to the compressor during the off cycle. Too little refrigerant will allow vapor to enter the capillary tube causing a loss in system capacity.

Due to its basic simplicity, the elimination of the need for a receiver, and the low-starting torque motor requirement, a capillary tube system is the least expensive of all flow control systems.

The proper size of a capillary tube is difficult to calculate accurately, and can best be determined by actual test on the system. Once the size is determined, the proper capillary tube can be applied to identical systems, so it is well-adapted to production type units. Tables 7-3, 7-4, 7-5, 7-6, and 7-7 give tentative selection data for capillary tubes.

## THERMAL-ELECTRIC EXPANSION VALVES

Conventional type expansion valves have remained unchanged in basic design and function for more than thirty years. They have inherent limitations; their functional scope is narrow, and they do not permit a refrigeration system to operate at maximum efficiency. Manufacturers of refrigeration and air conditioning equipment are constantly improving the reliability of their products, increasing their performance, and at the same time lowering the cost. They try new ideas, develop new systems, improve performance, modernize, and revolutionize equipment in order to stay competitive.

The thermal-electric expansion valve is the only refrigerant flow control device in step with this thinking (see Figure 7-33). It is the only refrigerant control compatible with modern electronic sensing and monitoring devices. It is the only flow control device offered to the refrigeration industry in over thirty years that permits entirely new and fresh approaches in the field of refrigerant metering and system modulation.

**Figure 7-33.** Themal-electric expansion valve and thermistor (*Courtesy of Singer Controls Division*)

## Table 7-3
## Capillary Tube Selection—R–22
## High Temperature

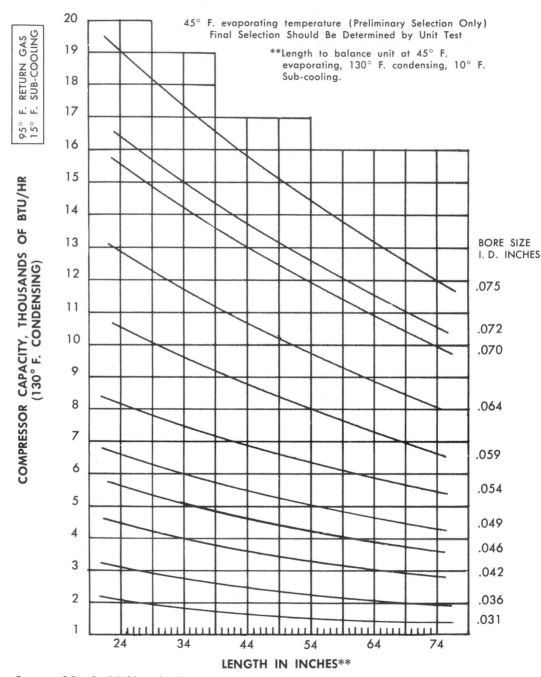

45° F. evaporating temperature (Preliminary Selection Only)
Final Selection Should Be Determined by Unit Test

**Length to balance unit at 45° F. evaporating, 130° F. condensing, 10° F. Sub-cooling.

95° F. RETURN GAS
15° F. SUB-COOLING

COMPRESSOR CAPACITY, THOUSANDS OF BTU/HR
(130° F. CONDENSING)

LENGTH IN INCHES**

BORE SIZE
I. D. INCHES

.075
.072
.070
.064
.059
.054
.049
.046
.042
.036
.031

Courtesy of Copeland Refrigeration Corp.

## Table 7-4
## Capillary Tube Selection—R–22
## Medium Temperature

95° F. RETURN GAS
15° F. SUB-COOLING

65° F. RETURN GAS
0° F. SUB-COOLING

25° F. to 10° F. Evaporating Temperature (Preliminary Selection Only)
Final Selection Should Be Determined by Unit Test

**Length to balance unit with 115° F. condensing,
5° F. sub-cooling in condenser, Heat Exchanger
to give 15° F. sub-cooling.

COMPRESSOR CAPACITY, THOUSANDS OF BTU/HR
(115° F. CONDENSING)

BORE SIZE
I. D. INCHES

.070

.064

.059

.054

.049
.046
.042

.036

.031

LENGTH IN INCHES**

# Table 7-5
## Capillary Tube Selection—R-12
## Medium Temperature

25° F. to 10° F. Evaporating Temperature (Preliminary Selection Only)
Final Selection Should Be Determined by Unit Test

**Length to balance unit with 115° F. condensing,
5° F. sub-cooling in condenser, Heat Exchanger
to give 15° F. sub-cooling.

95° F. RETURN GAS
15° F. SUB-COOLING

65° F. RETURN GAS
0° F. SUB-COOLING

COMPRESSOR CAPACITY, THOUSANDS OF BTU/HR
(115° F. CONDENSING)

BORE SIZE
I. D. INCHES

.075

.070

.064

.059

.054

.049
.046

.042

.036

.031

LENGTH IN INCHES**

**Table 7-6**
**Capillary Tube Selection—R–12**
**Low Temperature**

—15° F. to —25° F. Evaporating Temperature (Preliminary Selection Only)
Final Selection Should Be Determined by Unit Test

*Length to balance unit at 110° F. condensing
and 20° F. liquid sub-cooling
(5° F. in condenser, 15° F. in heat exchanger)

95° F. RETURN GAS
15° F. SUB-COOLING

65° F. RETURN GAS
0° F. SUB-COOLING

COMPRESSOR CAPACITY, THOUSANDS OF BTU/HR
(110° F. CONDENSING)

BORE SIZE
I. D. INCHES

.070

.064

.059

.054

.049
.046

.042

.036

.031

LENGTH IN INCHES*

Courtesy of Copeland Refrigeration Corp.

163

## Table 7-7
## Capillary Tube Selection—R-502
## Low Temperature

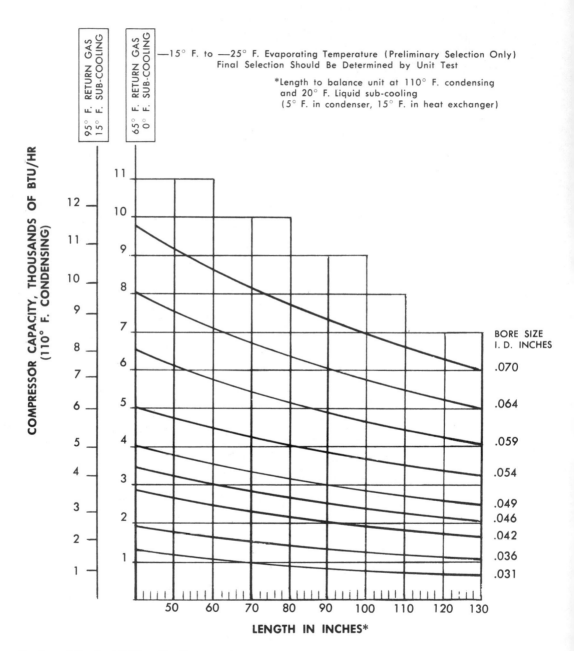

—15° F. to —25° F. Evaporating Temperature (Preliminary Selection Only)
Final Selection Should Be Determined by Unit Test

*Length to balance unit at 110° F. condensing
and 20° F. Liquid sub-cooling
(5° F. in condenser, 15° F. in heat exchanger)

95° F. RETURN GAS
15° F. SUB-COOLING

65° F. RETURN GAS
0° F. SUB-COOLING

COMPRESSOR CAPACITY, THOUSANDS OF BTU/HR
(110° F. CONDENSING)

BORE SIZE
I. D. INCHES

.070

.064

.059

.054

.049
.046

.042

.036

.031

LENGTH IN INCHES*

Courtesy of Copeland Refrigeration Corp.

164

**OPERATION:** The electric expansion valve, which is part of a valve and sensor package, is a simple heat-motor-operated needle valve infinitely positionable in response to voltage input (see Figure 7-34). An increase in voltage opens the valve and increases the flow of refrigerant. A decrease in voltage reduces the flow or closes the valve altogether.

The changes in voltage necessary to modulate this control can be established by using various temperature, pressure, or liquid refrigerant sensors (see Figure 7-35). Therefore, the function of the valve can be changed instantly by merely switching from one sensing device to another.

The valve opens to full capacity at 4.1 watts and is designed to operate at 24 volts with either AC or DC power.

**Figure 7-35.** Liquid sensing thermistor (*Courtesy of Singer Controls Division*)

The valve does not depend upon refrigerant temperature or pressure to create modulation. It responds only to voltage input. Therefore, the same valve works equally well with any common refrigerant except ammonia. It does not require external equalizer lines regardless of coil pressure drop, and does not require a charged bulb and capillary lines.

The same valve will work just as well on low-temperature freezer applications as it does on unitary air conditioners. It can be used to control coil superheat (and 0°F (−17.8°C) superheat at that) or coil pressure, or both. The capacity of this valve, as with all expansion valves, remains a function of fluid flow and refrigerant effect.

### 0°F (−17.8°C) SUPERHEAT CONTROL:
The thermal-electric expansion valve is capable of maintaining saturation conditions throughout the entire evaporator at all loads while permitting the suction line to superheat the refrigerant in order to prevent flooding the compressor. (See Figure 7-36.)

A small negative coefficient thermistor is directly exposed to the refrigerant in the suction line by mounting it in a ¼ inch (6.5 mm) male pipe fitting requiring only a short female receptacle.* The thermistor is placed

**Figure 7-34.** Thermal-electric expansion valve cross section (*Courtesy of Singer Controls Division*)

---

* A thermistor is a semiconductor whose electrical resistance varies with the temperature.

**Figure 7-36.** Refrigeration system with thermal-electric expansion valve (*Courtesy of Singer Controls Division*)

in series electrically with the expansion valve heater, making the current input to the heater a function of thermistor resistance, which in turn is a function of refrigerant conditions.

Exposure to gaseous or superheated refrigerant permits the thermistor to self-heat, lowering its resistance, and increasing the current input to the expansion valve heater. The expansion valve responds by modulating in an opening direction, increasing the flow of refrigerant to the evaporator. This process continues until saturated suction line conditions occur, permitting wet refrigerant to contact the thermistor. Liquid refrigerant or wet refrigerant gas will immediately cool

the thermistor, increasing its resistance, which causes the expansion valve to modulate in the closing direction.

The termination point of refrigerant saturation can be controlled by the location of the thermistor, and the suction line becomes a logical spot for this transition to occur. However, the ability of the thermistor to sense liquid and meter the refrigerant accordingly permits consideration of new approaches to superheat control which are difficult, if not impossible, with conventional valves. These new approaches are:

1. The $0°F$ ($-17.8°C$) superheat point can be shifted in the evaporator by using

more than one thermistor probe and the desired point of control can be switched-in on demand. This procedure could be used to size the evaporator to the compressor on systems using automatic compressor unloading.

2. The thermistor can be used to control the liquid level in a suction accumulator or riser preventing flooded or semiflooded evaporator control with dry suction gas returning to the compressor.

3. Evaporators can be split into independent sections using two or more electric valves and thermistors to control a specific series of passes or surface area. Each valve and thermistor will operate completely independently of the others. Evaporator surface can be added to, or subtracted from, the total coil by simply switching the desired

electric valves and their controlled passes off or on. Evaporator capacity, evaporator temperature, and moisture removal can be changed or regulated with this simple control arrangement.

**CONSTANT PRESSURE CONTROL ELECTRIC EXPANSION VALVE):** The electric expansion valve can be used to control evaporator pressure by placing the valve heater in electrical series with a pressure or temperature switch, sensing evaporator pressure or temperature respectively as illustrated in Figure 7-37. A slow make-and-break switch, which opens on a rise in pressure, is suitable for this function because the electrical load is light, low voltage, and noninductive. The principle of control is extremely simple.

**Figure 7-37.** Installation of electric valve as constant pressure control (*Courtesy of Singer Controls Division*)

For example: A switch set to control 70 psi (4.92 kg/cm²) closes when the evaporator drops below this pressure, energizing the expansion valve. The resultant increase in refrigerant flow will raise the evaporator pressure, opening the pressure switch and

allowing the expansion valve to modulate toward closing again. The slow, modulating nature of the heat motor expansion valve permits the system to hover at the pressure setting of the switch.

Pressure switches can be used in this

manner to limit both maximum and minimum system pressure. Switches can be placed anywhere in the system to close the expansion valve or limit its capacity when the desired levels of pressure or temperature are reached. The control of maximum head pressure is a good example of this type of superimposed function.

### 0°F (−17.8°C) SUPERHEAT AND PRESSURE-LIMIT CONTROL:

The thermal-electric expansion valve will provide both superheat control and pressure limit if the liquid-sensing thermistor is used in conjunction with pressure switches. In Figure 7-38 a high-pressure limit switch is placed in series with the expansion valve heater, preventing further increase in capacity when its pressure setting is exceeded. A low-pressure limit switch is placed in parallel with the thermistor, shunting it when the system pressure drops below its setting. (Both switches open on a rise in pressure.) This arrangement results in system operation at 0°F (−17.8°C) superheat between the high- and low-pressure limits. The pressure switches convert the expansion valve from a superheat control to a constant pressure control automatically.

**Figure 7-38.** Superheat and pressure-limit control with electric valve (*Courtesy of Singer Controls Division*)

**LIQUID LINE SHUT-OFF CONTROL:** A switch, whether automatic or manual, can be used to close the expansion valve, thus allowing the system to pump down. Being a heat motor, the valve will open slightly in ambients above room temperature, slowly bleeding off refrigerant. Below room ambient, the valve will remain closed.

**INSTALLATION AND TESTING RECOMMENDATIONS:**

**Mounting position:** The valve should be installed such that the thermal head is within 45° of an upright position to ensure maximum capacity.

**Thermistor suction line adapter:** The thermistor should be located flush, or slightly less, with the inside wall of the suction line. A 5/16 in. (7.94 mm) dimension will correctly locate the thermistor assembly with the inside wall of the suction line (see Figure 7-39). Projection of the female adapter fitting, as well as the thermistor assembly, into the suction line should be avoided. (See Figure 7-40.)

**Thermistor location and sensing positions:** The liquid-sensing thermistor assembly can

INCORRECT                    CORRECT

**Figure 7-40.** Incorrect location of thermistor on line (*Courtesy of Singer Controls Division*)

be used in any suction line with a diameter of ½ in. (12.7 mm) or larger. It will work on both vertical and horizontal suction lines, but it should never be located where liquid refrigerant is likely to accumulate or trap off; for instance, in a bottom U-bend connecting two vertical risers (see Figure 7-41).

Because suction refrigerant flow depends upon many factors including suction line size, suction gas velocity, elbow reducers, etc., it is impossible to establish firm rules regarding the best location of the liquid-sensing thermistor. The following observations have been made on a wide variety of applications and have proven to be useful guideposts for locating thermistor assemblies in suction lines:

**Figure 7-39.** Correct location of thermistor on line (*Courtesy of Singer Controls Division*)

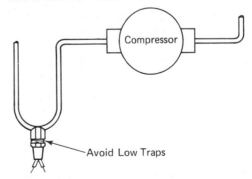

Compressor

Avoid Low Traps

**Figure 7-41.** Proper location of thermistor on line (*Courtesy of Singer Controls Division*)

1. High-velocity suction locations are preferable over low-velocity locations.

2. Smaller diameter suction locations are preferable over larger diameter locations.

3. Unless the flow pattern around an elbow or reducer is well known or specifically designed for the liquid-sensing thermistor assembly, it is best to stay at least 6 in. (152.4 mm) away when locating downstream of them.

4. Vertical suction lines make excellent locations, but trapping should be avoided (see Figure 7-41). If the line is at some angle other than vertical, the lower or gravity side is preferable. (See Figure 7-42.)

GRAVITY SIDE PREFERABLE

**Figure 7-42.** Thermistor located on slanting line (*Courtesy of Singer Controls Division*)

5. Thermistor location on horizontal suction lines is more common, and in some cases (large diameter low-velocity suction lines) permits adjustments to be made in suction gas saturation. (See Figure 7-43.) The best thermistor-sensing positions in horizontal suction lines are generally between 4 o'clock and 8 o'clock in the lower half of the suction line, although successful applications have been made with the thermistor in all axial locations. As a rule, the thermistor should be located as high axially as possible on horizontal suction lines using tolerable suction gas wetness as the limiting factor.

**Figure 7-43.** Thermistor location on horizontal lines (*Courtesy of Singer Controls Division*)

6. The most important rule regarding the location of the liquid-sensing thermistor on any application is simply this: *make sure that liquid or wet refrigerant gas can come into good contact with the thermistor at all loads.* Once a location has been established on a given application, subsequent units will show amazingly consistent repeated performance.

**EFFECTS OF AMBIENTS AND BLOWERS DURING RUNNING PERIODS:** Exposure to fan or blower air movement has almost no effect on the valve and it will operate properly in all ambients from −40°F to 150°F (−40°C to 66°C). The valve has a normal tendency to lose capacity at lower ambients and gain capacity at higher ambients. The thermistor, however, generally compensates for these changes automatically.

**EFFECT OF AMBIENT DURING OFF-CYCLE PERIODS:** During off-cycle periods, the valve will bleed refrigerant if exposed to ambients above 70°F (21.1°C) and will close off relatively tightly if exposed to ambients of less than 70°F. In short, the valve is simply calibrated to close at 70°F with no electrical energy applied.

**OFF-CYCLE SYSTEM UNLOADING:** The valve can be left energized 100% of the time, in which case it will completely and rapidly unload the system during off-cycle periods.

The valve can be de-energized during off-cycle periods with the compressor, in which case it will partially unload the system, depending upon the amount of charge and the ambient temperature.

**TESTING FOR PROPER OPERATION:** The valve and thermistor should be wired in series as shown in Figure 7-44. For pressure switch wiring circuits refer to Figures 7-37 and 7-38. All wiring should be made in accordance with U.L. low-voltage Class 2 codes.

A voltmeter placed across the terminals of the electric valve becomes a handy tool for analyzing proper valve selection and system operation as follows:

1.  A properly-sized valve will cause the voltmeter to fluctuate as it modulates refrigerant flow during normal operation. Each time wet refrigerant gas hits the thermistor, the voltmeter reading should drop; each time the suction gas dries out, it should rise. The voltmeter may read as high as 20 volts and as low as 8 volts during normal operation.

**Figure 7-44.** Wiring diagram of thermistor and valve (*Courtesy of Singer Controls Division*)

2. When the system is operating near or at peak load, the voltmeter reading should remain at higher (15–20 volts) levels longer and may even hover at some fixed point between 15 and 20 volts.

3. When the system is operating at lower loads, the voltmeter reading should remain at the lower (8–14 volt) levels longer and may even hover at some fixed point between 8 and 14 volts.

4. If the valve remains at 17–20 volts for long periods (over three minutes) without satisfying the system (suction line remains superheated), it is too small.

5. If the valve remains at 8–10 volts for long periods (over three minutes) and the system is not at minimum load conditions, the valve is too large.

6. If either the valve or thermistor is suspected of being defective, a simple ohmmeter check can be used to prove their electrical continuity.

Connect the ohmmeter to the leads of the electric expansion valve or the thermistor assembly, whichever is the case, and compare their resistances against the manufacturer's specifications. (See Table 7-8.) The valve should have 70 ohms resistance. The valve should check within 5% of the nomi-

## Table 7-8
### Liquid-Sensing Thermistor Specifications *

Fitting material . . . . . . . . . . . . . . Brass
Net weight . . . . . . . . . . . . . . . . . . 3/4 ounces
Shipping weight . . . . . . . . . . . . . . 2 ounces

| TYPICAL APPLICATIONS | THERMISTOR RESISTANCE (77°F.) | SIZE | NUMBER |
|---|---|---|---|
| Air Conditioners Heat Pumps Meat Cases Chillers Ice Makers | 50 Ohms | 1/4 MPT | 189150 |
| Low Temp. Display Cases Ice Cream Freezers Cascade Systems Low-Temp. Blower Coils | 19 Ohms | 1/4 MPT | 189485 |
| Special Applications With Extremely High Load Changes | 100 Ohms | 1/4 MPT | 189519 |

* Courtesy of Singer Controls Division

nal valve listed and the thermistor within 40%.

*NOTE:* Both the valve and thermistor, but particularly the thermistor, must be at room temperature when making this check. The reading on the thermistor must be observed within two seconds after connecting it to the ohmmeter. (The slight current input created by the ohmmeter causes the thermistor to self-heat and gradually drop in resistance.)

## SAFETY PROCEDURES

The flow control device is normally a trouble-free device. It is, however, easily damaged. Care should be exercised to prevent damage when working on or around these devices.

1.   Never apply an open flame to an expansion valve, as damage to the gaskets might result.

2.   Do not hammer an expansion valve to free it. If it is struck, it should be replaced.

3.   Never unscrew any connections while refrigerant is in the system. An explosion might occur resulting in serious personal injury.

4.   Always use two wrenches when removing a valve from the system to prevent twisting of the lines.

5.   Wear safety goggles when purging refrigerant from the system.

6.   Keep the floor clean of moisture and oil.

7.   Do not suddenly open an adjustment on an expansion valve fully because flooding of the compressor might result.

8.   Always provide plenty of ventilation when soldering or welding on a flow control device.

9.   Never try to weld or solder a line while refrigerant pressure is present.

10.   Never leave the control bulb on a thermostatic expansion valve loose because flooding of the compressor may result.

### Table 7-9
### Thermostatic Expansion Valve Service Hints *

| HIGH DISCHARGE PRESSURE | |
|---|---|
| Probable Cause | Remedy |
| Insufficient cooling water due to inadequate supply or faulty water valve. | Start pump and open water valves. Adjust, repair or replace any defective equipment. |
| Condenser or liquid receiver too small. | Replace with correct size condenser or liquid receiver. |
| Cooling water above design temperature. | Increase supply of water by adjusting water valve, replacing with a larger valve, etc. |
| Air or non-condensable gases in condenser. | Purge and recharge system. |
| Overcharge of refrigerant. | Bleed to proper charge. |
| Condenser dirty. | Clean condenser. |
| Insufficient cooling air circulation over air cooled condenser. | Properly locate condenser to freely dispel hot discharge air. Tighten or replace slipping belts or pulleys and be sure blower motor is of proper size. |

**\* Courtesy of Alco Controls Division**

## LOW SUCTION PRESSURE ... HIGH SUPERHEAT

| Probable Cause | Remedy |
|---|---|
| A. Expansion Valve Limiting Flow | |
| Inlet pressure too low from excessive vertical lift, undersize liquid line or excessive low condensing temperature. Resulting pressure difference across valve too small. | Increase head pressure. If liquid line is too small, replace with proper size. See Table 3. |
| Gas in liquid line . . . due to pressure drop in the line or insufficient refrigerant charge. If there is no sight glass in the liquid line, a characteristic whistling noise will be heard at the expansion valve. | Locate cause of liquid line flash gas and correct by use of any or all of the following methods: 1. Add charge. 2. Clean strainers, replace filter driers. 3. Check for proper line size (see Table 3). 4. Increase head pressure or decrease temperature to insure solid liquid refrigerant at valve inlet. |
| Valve restricted by pressure drop through evaporator. | Change to an expansion valve having an external equalizer. |
| External equalizer line plugged, or external equalizer connector capped without providing a new valve cage or body with internal equalizer. | If external equalizer is plugged, repair or replace. Otherwise, replace with valve having correct equalizer. |
| Moisture, wax, oil or dirt plugging valve orifice. Ice formation or wax at valve seat may be indicated by sudden rise in suction pressure after shutdown and system has warmed up. | Wax and oil indicate wrong type oil is being used. Purge and recharge system, using proper oil. Install an ALCO ADK filter-drier to prevent moisture and dirt from plugging valve orifice. |
| Valve orifice too small. | Replace with proper valve. |
| Superheat adjustment too high. | See "Measuring and Adjusting Operating Superheat". |
| Power assembly failure or partial loss of charge. | Replace power assembly (if possible) or replace valve. |
| Gas charged remote bulb of valve has lost control due to remote bulb tubing or power head being colder than the remote bulb. | Replace with a "W" cross ambient power assembly. See "Remote Bulb and Power Assembly Charges". |
| Filter screen clogged. | Clean all filter screens. |
| Wrong type oil. | Purge and recharge system and use proper oil. |
| B. Restriction in system other than expansion valve. (Usually, but not necessarily, indicated by frost or lower than normal temperatures at point of restriction.) | |
| Strainers clogged or too small. | Remove and clean strainers. Check manufacturers catalog to make sure that correct strainer was installed. Add an ALCO ADK filter-drier to system. |
| A solenoid valve not operating properly or is undersized. | Refer to ALCO "Solenoid Valve Service Hints" pamphlet. If valve is undersized, check manufacturers catalog for proper size and conditions which would cause malfunction. |
| King valve at liquid receiver too small or not fully opened. Hand valve stem failure or valve too small or not fully opened. Discharge or suction service valve on compressor restricted or not fully opened. | Repair or replace faulty valve if it cannot be fully opened. Replace any undersized valve with one of correct size. |
| Plugged lines. | Clean, repair or replace lines. |
| Liquid line too small. | Install proper size liquid line. See Table 3. |
| Suction line too small. | Install proper size suction line. See Table 2. |
| Wrong type oil in system, blocking refrigerant flow. | Purge and recharge system and use proper oil. |

## LOW SUCTION PRESSURE ... LOW SUPERHEAT

| Probable Cause | Remedy |
|---|---|
| Poor distribution in evaporator, causing liquid to short circuit through favored passes and throttling valve before all passes receive sufficient refrigerant. | Clamp power assembly remote bulb to free draining suction line. Clean suction line thoroughly before clamping bulb in place. See Figure 3. Install an ALCO Venturi-Flo distributor. Balance evaporator load distribution. |
| Compressor oversize or running too fast due to wrong size pulley. | Reduce speed of compressor by installing proper size pulley, or provide compressor capacity control. |
| Uneven or inadequate evaporator loading due to poor air distribution or brine flow. | Balance evaporator load distribution by providing correct air or brine distribution. |
| Evaporator too small . . . often indicated by excessive ice formation. | Replace with proper size evaporator. |
| Excessive accumulation of oil in evaporator. | Alter suction piping to provide proper oil return or install oil separator, if required. |

## HIGH SUCTION PRESSURE . . . HIGH SUPERHEAT

| Probable Cause | Remedy |
|---|---|
| Unbalanced system having an oversized evaporator, and undersized compressor and a high load on the evaporator. Load in excess of design conditions. | Balance system components for load requirements. |
| Compressor undersized. | Replace with proper size compressor. |
| Evaporator too large. | Replace with proper size evaporator. |
| Compressor discharge valves leaking. | Repair or replace valves. |

## HIGH SUCTION PRESSURE . . . LOW SUPERHEAT

| Probable Cause | Remedy |
|---|---|
| Compressor undersized. | Replace with proper size compressor. |
| Valve superheat setting too low. | See "Measuring and Adjusting Operating Superheat". |
| Gas in liquid line with oversized expansion valve. | Replace with proper size expansion valve. Correct cause of flash gas. |
| Compressor discharge valves leaking. | Repair or replace discharge valves. |
| Pin and seat of expansion valve wire drawn, eroded, or held open by foreign material, resulting in liquid flood back. | Clean or replace damaged parts or replace valve. Install an ALCO ADK filter-drier to remove foreign material from system. |
| Ruptured diaphragm or bellows in a constant pressure (automatic) expansion valve, resulting in liquid flood back. | Replace valve power assembly. |
| External equalizer line plugged, or external equalizer connection capped without providing a new valve cage or body with internal equalizer. | If external equalizer is plugged, repair or replace. Otherwise, replace with valve having correct equalizer. |
| Moisture freezing valve in open position. | Apply hot rags to valve to melt ice. Install an ALCO ADK filter-drier to insure a moisture-free system. |

## FLUCTUATING SUCTION PRESSURE

| Probable Cause | Remedy |
|---|---|
| Incorrect superheat adjustment. | See "Measuring and Adjusting Operating Superheat". |
| Trapped suction line. | Install "P" trap to provide a free draining suction line. |
| Improper remote bulb location or installation. | Clamp remote bulb to free draining suction line. Clean suction line thoroughly before clamping bulb in place. See Figures 3, 4 or 5. |
| "Flood back" of liquid refrigerant caused by poorly designed liquid distribution device or uneven evaporator loading. Improperly mounted evaporator. | Replace faulty distributor with an ALCO Venturi-Flo distributor. If evaporator loading is uneven, install proper load distribution devices to balance air velocity evenly over evaporator coils. Remount evaporator lines to provide proper angle. |
| External equalizer lines tapped at a common point although there is more than one expansion valve on same system. | Each valve must have its own separate equalizer line going directly to an appropriate location on evaporator outlet to insure proper operational response of each individual valve. For one example, see Figure 5. |
| Faulty condensing water regulator, causing change in pressure drop across valve. | Replace condensing water regulator. |
| Evaporative condenser cycling, causing radical change in pressure difference across expansion valve. Cycling of blowers or brine pumps. | Check spray nozzles, coil surface, control circuits, thermostat overloads, etc. Repair or replace any defective equipment. Clean clogged nozzles, coil surface, etc. |
| Restricted external equalizer line. | Repair or replace with correct size. |

## FLUCTUATING DISCHARGE PRESSURE

| Probable Cause | Remedy |
|---|---|
| Faulty condensing water regulating valve. | Replace condensing water regulating valve. |
| Insufficient charge . . . usually accompanied by corresponding fluctuation in suction pressure. | Add charge to system. |
| Cycling of evaporative condenser. | Check spray nozzles, coil surface, control circuits, thermostat overloads, etc. Repair or replace any defective equipment. Clean clogged nozzles, coil surface, etc. |
| Inadequate and fluctuating supply of cooling water to condenser. | Check water regulating valve and repair or replace if defective. Check water circuit for restrictions. |
| Cooling fan for condenser cycling. | Determine cause for cycling fan, and correct. |
| Fluctuating discharge pressure controls on low ambient air cooled condenser. | Adjust, repair, or replace controls. |

## SUMMARY

• The purpose of the flow control device is to control the flow of refrigerant to the evaporator.

• Flow control devices may be operated by forces such as: temperature, pressure, or a combination of the two.

• Proper operation of the evaporator depends on the correct amount of liquid refrigerant and its swirling pattern through the evaporator tubes.

• All the liquid refrigerant must be evaporated in the evaporator or slugging may occur.

• As liquid refrigerant passes through the flow control device flash gas occurs.

• Flash gas is equal to approximately 20% of the total liquid passing through the flow control device.

• Flash gas cools the remaining liquid refrigerant to the evaporator temperature.

• Demand for energy conservation at a minimum cost has created a new interest in the automatic expansion valve.

• The automatic expansion valve controls the flow of refrigerant without an external mechanical device.

• The automatic expansion valve maintains a nearly constant evaporator pressure.

• Opening and closing of the automatic expansion valve is controlled by the refrigerant pressure in the low side of the system.

• The automatic expansion valve will not compensate for varying conditions.

• The evaporator pressure is stabilized when the liquid refrigerant evaporating in the evaporator is equal to the compressor's pumping capacity.

• The expansion valve should not be adjusted until the system has operated 24 or 48 hours.

• Automatic expansion valves may be used for the following performance functions: (1) protection against evaporator freezing, (2) control of relative humidity, (3) motor overload protection, (4) simplification of field service, (5) high capacity in a small size control, (6) the right valve for water coolers, (7) ideal valve for low-starting torque motors, and (8) low-capacity by-pass valve.

• Bleed-type valves permit pressures in the refrigeration system to reach a balance or near balance point during the off cycle.

• The bleed-type (or slotted orifice) valve is a standard automatic valve with the addition of a small slot in the valve orifice to prevent complete valve close-off at the end of the running cycle.

• Bleeds provide a fixed-valve capacity depending upon (1) bleed size, (2) liquid density, and (3) the existing pressure drop.

• The valve must be selected with the smallest size bleed slot which will provide the required unloading during the minimum length of off cycle.

• Any substantial change in altitude after an automatic expansion valve has been adjusted will alter the low side pressure that the valve will maintain.

• Automatic expansion valves can be applied as bypass or capacity-control valves.

• Thermostatic expansion valves are the most commonly used devices for controlling the flow of liquid refrigerant into the evaporator.

• Thermostatic expansion valves are designed to meter the flow of refrigerant into an evaporator in exact proportion to the rate of evaporation of the liquid refrigerant in the evaporator.

• Thermostatic expansion valves are responsive to the temperature of the refrigerant gas leaving the evaporator and the refrigerant pressure inside the evaporator.

• The pressures that cause the thermostatic expansion valve to operate are: (1) pressure in the remote bulb and power assembly, (2) evaporator pressure, and (3) equivalent pressure of the superheat spring.

• The thermostatic expansion valve is said to be in equilibrium when $P_1 = P_2 + P_3$.

• An increase in the superheat of the refrigerant leaving the evaporator will cause the expansion valve to open.

• An increase in the evaporator pressure will cause the valve to close.

• As the superheat setting of a thermostatic expansion valve is increased, the evaporator capacity is decreased.

• The initial step in determining whether or not a thermostatic expansion valve is functioning properly is to measure the superheat.

• Determining the difference in temperature at the inlet and outlet of the evaporator is not an accurate measure of superheat.

• To reduce the superheat setting, turn the adjusting stem counterclockwise.

• As a general rule, the proper superheat setting will depend on the TD between the refrigerant temperature and the temperature of the medium being cooled.

• Good thermal contact between the remote bulb and the tube is essential for the successful operation of the unit.

• Expansion valve charges are grouped into the following general classifications: (1) liquid charge, (2) gas charge, (3) liquid-vapor cross charge, and (4) cross-ambient vapor charge.

• External equalized thermostatic expansion valves must be used on evaporators with an appreciable pressure drop between the valve outlet and the remote bulb location.

• External equalizer connections must be made—*never cap these connections*.

• Pressure-limiting (MOP) expansion valves provide motor overload protection at times of high heat load.

• When the MOP of the equipment is reached, the valve pressure-limit feature takes control from the power element and prevents further flow of refrigerant into the evaporator.

• MOP is provided by use of two features: (1) limited liquid charge and (2) liquid charge with mechanical pressure limit.

• When large evaporators are needed, distributors are required to reduce the pressure drop through the evaporator.

• The capillary tube is the simplest type of refrigerant flow control device.

• The capillary tube is limited to single cabinet and single temperature applications.

• The capillary tube system is a critical charge type system.

• The electric expansion valve is simply a heat-motor-operated needle valve.

• The required voltage change can be caused by use of various temperature, pressure, or liquid refrigerant sensors.

• The electric expansion valve opens fully at 4.1 watts of power.

• The electric expansion valve does not depend upon refrigerant temperature or pressure to create modulation.

• The thermistor is located on the suction line allowing full refrigeration of the evaporator.

• The electric expansion valve will operate with a 0°F (−17.8°C) superheat on the coil.

• A voltmeter placed across the terminals of the electric valve becomes a handy tool for analyzing proper valve selection and system operation.

## REVIEW EXERCISES

1. What is the purpose of any flow control device?

2. Why must all the liquid refrigerant be evaporated in the evaporator?

3. What is flash gas?

4. What percentage of the total refrigerant becomes flash gas?

5. What is the net refrigerating effect?

6. What forces operate an automatic expansion valve?

7. Does the automatic expansion valve open or close during the off cycle of the system?

8. What is known as the valve opening point of an automatic expansion valve?

9. Is it correct to install an expansion valve and adjust it immediately?

10. How will an altitude change affect an automatic expansion valve?

11. What are the eight performance characteristics of an automatic expansion valve?

12. At what rate equivalent is the refrigerant admitted to the evaporator?

13. Why are bleed-type valves used?

14. What are the two basic types of bleed valve seats?

15. Explain how to test whether or not the bleed-type expansion valve is the correct size.

16. How can an automatic expansion valve be used as a bypass valve?

17. What is the thermostatic expansion valve designed to do?

18. What are the forces that operate the thermostatic expansion valve?

19. When is the thermostatic expansion valve in equilibrium?

20. Explain how to check the superheat of a thermostatic expansion valve.

21. Explain how a nonadjustable thermostatic expansion valve can be adjusted.

22. What does clockwise turning of the adjusting stem of a thermostatic expansion valve do to the superheat setting?

23. Where should the remote bulb be mounted?

24. What is the purpose of the cross-ambient vapor charge on a thermostatic expansion valve?

25. Should the external equalizer connection on a thermostatic expansion valve be capped off?

26. Where is the external equalizer line connected into the suction line?

27. When is an external equalizer used?

28. What does MOP mean and how is it used?

29. What two methods are used to provide pressure limiting on a thermostatic expansion valve?

30. For what purpose are refrigerant distributors used?

31. What is a capillary tube?

32. Which flow control device discussed in this text is used as a critical charged system?

33. What opens and closes the thermal-electric expansion valve?

34. What is the lowest superheat that can be obtained with a thermal-electric expansion valve?

35. What is the purpose of the thermistor used on the thermal-electric expansion valve?

36. Where should the thermistor be located for best results?

# Accessories

There are a number of accessory items which are used in refrigeration systems for specific purposes. Their requirement in a particular system will depend upon the application of the system. An accessory will allow the basic refrigeration system to have conveniences or reach a degree of performance which could not possibly be reached by the basic system. Therefore, an accessory may be defined as an article or device that adds to the convenience or the effectiveness of the system, but is not essential.

## ACCUMULATORS

Compressors are designed to compress vapors, not liquids. Many systems, especially low-temperature systems, are subject to the return of excessive quantities of liquid refrigerant to the compressor. Liquid refrigerant returning to the compressor dilutes the oil, washes out the bearings, and in some cases causes complete loss of oil in the com-

pressor crankcase. This condition is known as oil pumping or slugging and results in broken valve reeds, pistons, rods, crankshafts, etc. The purpose of the accumulator is to act as a reservoir to temporarily hold the excess oil-refrigerant mixture and return it at a rate which the compressor can safely handle.

Proper installation of a suction accumulator in the suction line just before the compressor eliminates damage (see Figure 8-1). If the accumulator is correctly sized, relatively large quantities of liquid refrigerant may return through the suction line and the suction accumulator will prevent damage to the compressor. The liquid refrigerant is temporarily held in the suction accumulator (see Figure 8-2) and metered back to the compressor along with any oil, at a controlled rate through the metering orifice. Therefore, damage to the compressor is prevented and the compressor immediately and quietly goes to work.

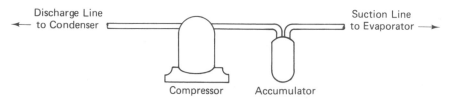

Discharge Line
← to Condenser

Suction Line
to Evaporator →

Compressor          Accumulator

**Figure 8-1.** Accumulator installation

**ACCUMULATOR SELECTION:** An accumulator must be selected for a low refrigerant pressure drop, proper oil return, and for adequate volume to insure storage of at least 50% of the total system charge. The

suction accumulator should not necessarily be selected to have the same size inlet and outlet as the compressor suction line.

## DRIERS

Most authorities agree that moisture is the single most detrimental factor in a refrigeration system. A unit can stand only a small amount of water. For this reason, the majority of refrigeration and air conditioning systems, both field- and factory-assembled, contain driers. (See Figure 8-3.) Several designs of heat pumps have two driers.

Liquid line driers (Figure 8-4) have been used in refrigeration systems for many years. Their primary function, as their name implies, is to remove moisture from the refrigerant. Moisture in a refrigeration system is a contributing factor in the formation of acids, sludge, and corrosion. Hydrochloric and hydrofluoric acids are formed by the interaction

↑ TO COMPRESSOR          ↓ FROM COIL

PATENTED INLET DEFLECTOR

**Will Not Trap Oil**

**Protects the Compressor**

LEEK PRUF FUSIBLE PLUG

METERING ORIFICE

**Figure 8-2.** Internal view of suction accumulator (*Courtesy of Refrigeration Research, Inc.*)

**Figure 8-3.** Liquid line drier (*Courtesy of Alco Controls Div., Emerson Electric Co.*)

**Figure 8-4.** Liquid line drier location

of halocarbon refrigerants and small amounts of moisture. Moisture in a refrigeration system can cause a multitude of difficulties varying from expansion valve freeze-up to possible hermetic motor compressor burn-out. In large amounts, water can directly produce mechanical malfunctions. It is important to remove quickly and completely any moisture and any acid or sludge that may have formed from a refrigeration system.

Over the years refrigeration systems have become more complex and driers are required to perform functions other than drying. Modern day liquid line driers are capable of performing the following functions:

1. Moisture removal

2. Filtering

3. Acid removal

**SELECTION OF LIQUID LINE DRIERS:** The following factors influence the selection of the correct size of drier:

1. Type and amount of refrigerant

2. Refrigeration system tonnage

3. Line size

4. Allowable pressure drop

When the refrigerant, line size, and application are known, the drier is generally selected on the basis of recommended capacities which take into account both drying and refrigerant flow capacity. The flow and moisture capacity information is published by the liquid line drier manufacturer in tabular form for all popular refrigerants. (See Table 8-1.)

**TYPES OF LIQUID LINE DRIERS:** There are two general types of liquid line driers in use. They are the straight-through sealed type shown in Figure 8-5, and the angle-replaceable core (replaceable-cartridge) type shown in Figure 8-6.

**Figure 8-5.** Straight-through liquid line drier (*Courtesy of Henry Valve Company*)

Dimension "D" is the minimum space required to remove the filter-drier core from the shell.

**Figure 8-6.** Angle-replaceable core drier (*Courtesy of Henry Valve Company*)

## Table 8-1
## Filter-Drier Selection *

| FILTER-DRIER SELECTION SERIES "H" DRI-COR | | | | | | | | RECOMMENDED TONNAGE BASED ON BOTH DRYING AND FLOW CAPACITY | |
| CATALOG NUMBERS | | SIZE CONN. | CORE FILTER AREA SQ. IN. | SHELL DIA. | OVER-ALL LENGTH | | WEIGHT POUNDS | REFRIGERANTS | |
| FLARE | O.D. SOLDER | | | | FLARE | O.D.S. | | R-12 | R-22 |
|---|---|---|---|---|---|---|---|---|---|
| H032 | H032-S | ¼" | 11 | 1⅝ | 4¼ | 3¹³⁄₁₆ | ⅜ | ¾ | ¾ |
| H052 | — | ¼" | 17 | 2⁵⁄₁₆ | 4¹¹⁄₁₆ | | ¾ | 1 | 1 |
| H053 | — | ⅜" | | | 5³⁄₁₆ | | ⅞ | 1 | 1 |
| H082 | H082-S | ¼" | 24 | 2⁵⁄₁₆ | 5⁹⁄₁₆ | 5⅛ | 1⅛ | 1 | 1 |
| H083 | H083-S | ⅜" | | | 6¹⁄₁₆ | 5¼ | 1⅛ | 2 | 2 |
| H084 | H084-S | ½" | | | 6¼ | 5⅜ | 1¼ | 2 | 2 |
| H162 | — | ¼" | 36 | 2⅞ | 6⁵⁄₁₆ | — | 1½ | 2 | 2 |
| H163 | H163-S | ⅜" | | | 6¾ | 5¹⁵⁄₁₆ | 1½ | 3 | 3 |
| H164 | H164-S | ½" | | | 7 | 6¹⁄₁₆ | 1¾ | 4 | 4 |
| H165 | H165-S | ⅝" | | | 7¼ | 6⁵⁄₁₆ | 1¾ | 5 | 5 |
| H303 | — | ⅜" | 57 | 3 | 9¹¹⁄₁₆ | — | 3⅜ | 4 | 5 |
| H304 | H304-S | ½" | | | 9⅞ | 9 | 3⅜ | 7½ | 7½ |
| H305 | H305-S | ⅝" | | | 10³⁄₁₆ | 9¼ | 3⅜ | 10 | 10 |
| — | H307-S | ⅞" | | | — | 9⅞ | 3¼ | 10 | 15 |
| H414 | — | ½" | 71 | 3½ | 9¹⁵⁄₁₆ | — | 4⅜ | 10 | 10 |
| H415 | H415-S | ⅝" | | | 10¼ | 9⁹⁄₁₆ | 4½ | 10 | 15 |
| — | H417-S | ⅞" | | | — | 9¹⁵⁄₁₆ | 4½ | 15 | 20 |
| — | H419-S | 1⅛" | | | — | 9⅞ | 4½ | 15 | 20 |
| — | H607-S | ⅞" | 106 | 3 | — | 16⁹⁄₁₆ | 6 | 20 | 25 |
| — | H609-S | 1⅛" | | | — | | | 25 | 30 |
| — | H755-S | ⅝" | 123 | 3½ | — | 15¹⁄₁₆ | 8¼ | 15 | 20 |
| — | H757-S | ⅞" | | | — | 15¹¹⁄₁₆ | | 25 | 30 |

Recommended Tonnage Rating based on both Drying and Flow Capacity is shown in table. For Refrigerants 500 and 502 use data shown for Refrigerant 12.

* Courtesy of Henry Valve Company

## SUCTION LINE FILTER

The proper suction line filter (see Figure 8-7) will protect the refrigeration compressor by collecting all foreign matter and preventing it from entering the compressor where it could damage the highly machined and polished internal working parts.

The suction line filter should be installed as close as possible to the compressor prior to system start-up. (See Figure 8-8.) When installed in this manner, it will remove foreign particles that are present even when the best installation procedures are carefully followed.

## SELECTION OF SUCTION LINE FILTERS: The following factors influence the selection of the correct size of filter.

1. Type of refrigerant

2. Suction line size

INLET

**Figure 8-7.** Suction line filter (*Courtesy of Henry Valve Company*)

**Figure 8-8.** Suction line filter installation

3. Allowable pressure drop

4. Application (low-temperature, commercial refrigeration, or air conditioning)

When the refrigerant type, line size, and application are known, the filter is selected on the basis of refrigerant flow capacity expressed in tons. Flow capacity information is published by the suction line filter manufacturer in tabular form for all popular refrigerants. (See Table 8-2.)

**Figure 8-9.** Straight-through suction line filter (*Courtesy of Henry Valve Company*)

### Table 8-2
### Suction Line Filter Selection *

| CATALOG NUMBERS | SIZE CONN. | SIZE SHELL DIA. | FILTER AREA SQ. IN. |
|---|---|---|---|
| 853A-⅝ 853A-⅞ 853A-1⅛ | ⅝″ ⅞″ 1⅛″ | 2⅞ | 25 |
| 853-1⅜ 853-1⅝ | 1⅜″ 1⅝″ | 3 | 57 |
| 853-2⅛ 853-2⅝ | 2⅛″ 2⅝″ | 4¾ | 94 |

* Courtesy of Henry Valve Company

Replaceable fluted core

**Figure 8-10.** Angle-replaceable core suction line filter (*Courtesy of Henry Valve Company*)

### SUCTION LINE FILTER-DRIER

The suction line filter-drier offers all the advantages of a suction line filter plus the ability, through a blend of activated alumina and molecular sieve desiccants (drying agents), to remove acids and moisture from the refrigerant vapor in the suction line along with any foreign matter present. (See

**TYPES OF SUCTION LINE FILTERS:** There are two general types of suction line filters in use. They are the straight-through sealed type (Figure 8-9) and the angle-replaceable core type (see Figure 8-10).

**Figure 8-11.** Suction line filter-drier (*Courtesy of Henry Valve Company*)

Figure 8-11.) A large filter area and traverse flow passageway for the refrigerant vapor permits installation in the suction line with a minimum loss of refrigeration capacity due to pressure drop. Only clean and dry refrigerant is returned to the compressor.

## STRAINERS

Strainers remove foreign matter such as dirt and metal chips from liquid or suction lines. This foreign matter could clog the small orifices of expansion and solenoid valves and also enter the compressor.

**TYPES OF STRAINERS:** There are three general types of strainers: straight-through sealed type (Figure 8-12), cleanable angle type (Figure 8-13), and the cleanable "Y" type (Figure 8-14).

**Figure 8-12.** Straight-through sealed filter (*Courtesy of Henry Valve Company*)

**Figure 8-13.** Cleanable angle filter (*Courtesy of Henry Valve Company*)

**Figure 8-14.** Cleanable "Y" filter (*Courtesy of Henry Valve Company*)

## MOISTURE-LIQUID INDICATORS

The installation of a moisture-liquid indicator is actually the most inexpensive preventive maintenance item that can be placed in the refrigeration system (see Figure 8-15). The liquid refrigerant is readily visible through the glass port of the indicator, and the presence of bubbles warns the operator or service technician of a shortage of refrigerant, a restriction in the liquid line, or improper subcooling of the refrigerant.

**Figure 8-15.** Moisture-liquid indicator (*Courtesy of Sporlan Valve Company*)

**PURPOSE OF THE MOISTURE INDICATOR:** The purpose of the moisture indicator is to reveal the presence of any moisture in the refrigerant which could be harmful to the system.

All moisture indicators have a moisture-sensing element of some type. This moisture-sensing element generally consists of a porous filter paper or cloth impregnated with an anhydrous cobalt salt. Cobalt salts are unique in that they have the capability of changing color with the addition or release of a very small amount of moisture. By changing colors, the moisture content of the refrigerant is indicated. Drier replacement is required when a "wet" system is indicated.

To ensure accuracy, the indicator must always be installed in the liquid line as shown in Figure 8-16. Only when a system is in operation with the refrigerant flowing will the sensitized element indicate the correct condition of the refrigerant.

Although the sensitized element is protected from discoloration by dirt and sludge, contaminants, excessive amounts of water, or high temperatures can permanently discolor or damage the element. If the sensitized element is damaged, it must be replaced.

**Figure 8-16.** Installation of moisture-liquid indicator

## OIL SEPARATOR

In any refrigeration system, refrigerant and oil are always present. Refrigerant is required for cooling. Oil is required for lubrication of the compressor. Refrigerant and oil are miscible in different percentages depending on the type of refrigerant, the temperature, and the pressure. Certain amounts of oil will always leave the compressor crankcase with the refrigerant.

The use of an oil separator in refrigeration systems has become quite widespread. (See Figure 8-17.) In fact, a separator is considered by most engineers to be an essential item of installation on low-temperature units and on large air conditioning units up to 150 tons. The overall efficiency of a system is much improved where an oil separator is used, particularly in open-top display cases in supermarkets where we have to deal with evaporating temperatures of $-30°F$ to $-40°F$ ($-34.4°C$ to $-40°C$). This is equally true on industrial, laboratory, and environmental equipment operating to $-100°F$ ($-73.3°C$) and lower. It is noted that most manufacturers require oil separators on their two-stage compressors.

**Figure 8-17.** Oil separator (*Courtesy of AC&R Components, Inc.*)

**PURPOSE OF AN OIL SEPARATOR:** Originally, the purpose of an oil separator was to maintain the correct level of oil in the compressor crankcase, but this has long been overshadowed by the benefits it provides in preventing free circulation of even small amounts of oil in the refrigeration or air conditioning system.

The prime function of an oil separator, however, is to separate oil from the refrigerant and return it to the compressor crankcase before it can enter the other components of

the system. Remember, oil is not a refrigerant; it belongs in the crankcase as a lubricant.

**HOW THE OIL SEPARATOR FUNCTIONS:** The oil separator is installed in the discharge line of the unit between the compressor and the condenser (see Figure 8-18).

**Figure 8-18.** Oil separator location (*Courtesy of AC&R Components, Inc.*)

The gases containing the oil enter the oil separator inlet in the form of fog and pass through the inlet baffling. In so doing, they are forced to change direction several times on the surfaces of the baffle. A good percentage of the oil is in finer particles and these can only be removed by causing them to collide with one another to form heavier particles. This can only be achieved by changing the velocity of the gas and oil fog mixture. This change in velocity occurs as

the mixture passes into the shell of the oil separator, which is designed to reduce the velocity. As this reduction occurs, the oil particles have more momentum than the refrigerant gas, causing them to collide with each other and impinge on the surfaces of the shell wall. The refrigerant gas then passes out through the outlet screens where there is a final scavaging action as the gases speed up to their original velocity and the oil-free gas then goes on to the condenser.

The oil, thus separated, drips to the bottom of the oil separator where there is an adequate reservoir (sump) to permit any foreign matter, sludge, etc., to settle out and be trapped out of circulation. A magnet is installed in the bottom of the sump to collect any metallic particles, and stop them from being circulated in the system. Oil will collect until a sufficient level is reached to raise the float and open the oil return needle valve. The oil is then returned to the crankcase of the compressor.

The float-operated oil return needle valve is located high in the sump to allow clean oil to be automatically returned to the compressor crankcase.

Only a small amount of oil is required to actuate the float mechanism, so it can be seen that only a small amount of oil is absent from the crankcase at any time when an oil separator is used in the system.

Experiments with glass oil separators have shown that oil cannot be separated completely by baffles alone. There also must be a velocity drop to determine the ratio of the volume of the shell to the cross-sectional area of the discharge line. There is relatively little pressure drop through a properly designed and sized oil separator.

**SELECTING THE SIZE OF AN OIL SEPARATOR:** Although oil separator catalogs show their capacity in tons of refrigeration or by horsepower, the actual tonnage or Btu capacity of a system may vary widely from the horsepower size of the compressor. This is due to the speed of the compressor and the density of the suction gas. It is a fact that the actual capacity of a compressor is dependent upon the suction pressure. On any given size compressor, the higher the suction pressure, the higher the actual capacity—the lower the suction pressure, the lower the actual capacity of the compressor.

The sizing of the suction line and hot gas discharge lines is based largely on obtaining the desired velocity of the gases in the system. Velocity drop in the gas and oil mixture as it passes through the oil separator is a major factor in oil separation. Therefore, the relationship of the volume of the oil separator connections to the volume of the shell and the baffling is important. The larger the capacity of the compressor, the larger the volume of the shell required, even if the connections remain the same. Since there must be minimal pressure drop, the connections must be able to carry the gases at the same pressure as the discharge line. Obviously, the connections must be the same size or larger than the discharge line.

## HEAT EXCHANGERS

Any device which brings into contact two substances of different temperatures for the purpose of heating or cooling one of these substances is called a heat exchanger. In refrigeration, the heat exchanger brings the liquid refrigerant and the suction gas into thermal contact with each other.

Refrigerant in the liquid line is ordinarily slightly above room temperature. Suction gas leaves the evaporator at a temperature near that maintained in the unit. Lowering the temperature of the liquid refrigerant before it reaches the flow control device increases the overall efficiency of the refrigeration unit by reducing the quantity of flash gas. During the heat exchange, the liquid is subcooled while any liquid present in the suction line is evaporated.

While it is sometimes possible to solder together lengths of the liquid and suction line to effect heat exchange (see Figure 8-19), a carefully designed heat exchanger allows complete and more reliable use of evaporator surface.

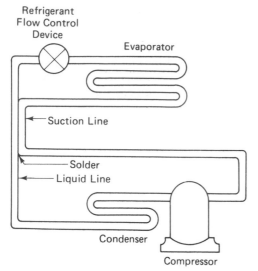

Figure 8-19. Liquid and suction lines soldered together

Figure 8-21. Coil-type heat exchanger

**TYPES OF HEAT EXCHANGERS:** There are two styles of small, highly efficient heat exchangers. One is a fountain-type heat exchanger, which is a double-tube device in which the liquid refrigerant passes through the space between the inner and outer tubes. A perforated internal tube causes the cool suction gas to impinge at high velocity on the walls of the inside tube, resulting in a high rate of heat exchange (see Figure 8-20). A second type is a coil-type heat exchanger, which has a coil made from small diameter tubing within a larger copper tube shell (Figure 8-21).

While both types give excellent heat exchange, the fountain-type is recommended if the device must be located where conden-

Figure 8-20. Fountain-type heat exchanger

sation may be a problem. Because the warm liquid refrigerant is in the outer shell, with the gas traveling through the center tube, atmospheric condensation with resultant dripping is not likely to occur.

**SELECTION OF HEAT EXCHANGERS:** Field practices vary so greatly that capacity ratings applied to heat exchangers in laboratory tests would have no practical significance. However, recommendations can be made on the basis of maximum evaporator capacity, Btu per hour, using only the liquid and suction line sizes. See Table 8-3.

**VIBRATION ELIMINATORS**

In order to prevent the transmission of noise and vibration from the compressor through the refrigeration piping, vibration eliminators are frequently installed in both the suction and discharge lines. (See Figure 8-22.)

On small units, where small diameter soft copper tubing is used for the refrigerant lines, a coil of tubing may provide adequate protection against vibration. On larger compressors, flexible metallic hose is most frequently used. (See Figure 8-23.)

**DISCHARGE MUFFLERS**

On systems where noise transmission must be reduced to a minimum, or where compressor pulsation might create vibration

### Table 8-3

### Recommendation Chart for Heat Exchangers *
MAXIMUM EVAPORATOR CAPACITIES B.T.U. PER HOUR (SEE NOTE—502)

| HEAT | | REFRIGERANT 12 SUCTION TEMP. | | REFRIGERANT 22 SUCTION TEMP. | | REFRIGERANT 502 SUCTION TEMP. | |
|---|---|---|---|---|---|---|---|
| EXCHANGER TYPE | PART NO. | 5 F | 40 F | 5 F | 40 F | 5 F | 40 F |
| FOUNTAIN | A-13730 | 4,000 | 5,000 | 5,000 | 6,000 | 4,000 | 9,000 |
| | A-14893 | 6,000 | 8,000 | 8,000 | 10,000 | 6,000 | 12,000 |
| | A-14894 | 8,000 | 11,000 | 11,000 | 13,000 | 8,000 | 18,000 |
| | A-14895 | 12,000 | 18,000 | 18,000 | 21,000 | 12,000 | 27,000 |
| | A-14896 | 18,000 | 21,000 | 21,000 | 27,000 | 18,000 | 36,000 |
| | A-14897 | 27,000 | 36,000 | 36,000 | 42,000 | 27,000 | 48,000 |
| COIL | A-14891 | 4,000 | 5,000 | 5,000 | 6,000 | 4,000 | 9,000 |
| | A-14890 | 5,000 | 6,000 | 6,000 | 7,000 | 5,000 | 11,000 |
| | A-14892 | 6,000 | 8,000 | 8,000 | 10,000 | 6,000 | 12,000 |

Note: Refrigerant 502 has a higher density than Refrigerant 22—and to minimize pressure drop, it is recommended that the evaporator capacities for Refrigerant 12 be used.

**Figure 8-22.** Vibration eliminator location

**Figure 8-23.** Metallic vibration eliminator
(*Courtesy of Packless Industries*)

problems, a discharge muffler is frequently installed to dampen and reduce compressor discharge noise. The muffler is basically a shell with baffle plates, with the internal volume required primarily dependent on the compressor displacement, although the frequency and intensity of the sound waves are also factors in muffler design (see Figure 8-24).

**Figure 8-24.** Discharge muffler

**Figure 8-25.** Discharge muffler location

### PURPOSE OF DISCHARGE MUFFLER:

The purpose of the muffler is to dampen or remove the hot gas pulsations set up by a reciprocating compressor.

Every reciprocating compressor will create some hot gas pulsations. Although great effort is made to minimize pulsations in the system and compressor design, gas pulsation can be severe enough to create two closely related problems: (1) the noise which, though irritating to users of the equipment, does not necessarily have a harmful effect on the system, and (2) vibration which can result in refrigerant line breakage. Frequently, these two problems appear simultaneously.

### DISCHARGE MUFFLER LOCATION:

The muffler is installed in the discharge line as close to the compressor as possible. In the case of hermetic compressors, the muffler is frequently installed in the compressor shell itself.

Because the normal construction of a muffler is within a shell, a natural trap is formed. Mufflers will trap oil easily and may trap liquid refrigerant. If a muffler is used, it should be installed in either a downward or a horizontal line as shown in Figure 8-25.

### DISCHARGE MUFFLER SELECTION:

The selection of a muffler can be a difficult engineering problem. Improper field selection of a muffler will sometimes lead to increased rather than decreased vibration.

### CRANKCASE HEATERS

When a compressor is installed in a location where it will be exposed to ambient temperatures colder than the evaporator, refrigerant migration to the crankcase can be aggravated by the resulting pressure difference between the evaporator and the compressor during the off cycle. To protect against the possibility of refrigerant migration, crankcase heaters are often used to keep the oil in the crankcase at a temperature high enough so that any liquid refrigerant which enters the crankcase will evaporate and create a pressure sufficient to prevent a large-scale refrigerant migration to the compressor.

If liquid refrigerant does migrate to the compressor, a reduction in suction pressure on compresor start-up will cause the liquid to boil (evaporate). When the refrigerant boils, the oil–refrigerant mixture will foam. Some of this foam will leave the crankcase and pass into the cylinder. This liquid causes a condition known as *slugging,* which will damage the compressor valves. The refrigerant also reduces the lubricating effectiveness of the oil.

### TYPES OF CRANKCASE HEATERS:

Crankcase heaters are made in two types: (1) insert type (Figure 8-26) and (2) externally mounted (see Figure 8-27).

### SELECTION OF CRANKCASE HEATERS:

The crankcase heater is a low-wattage-resistance element, normally energized continuously, and must be carefully selected to

**Figure 8-26.**   Insert type crankcase heater

**Figure 8-27.**   External type crankcase heater

avoid overheating of the compressor oil. Compressor manufacturers generally have a list of recommended crankcase heaters for any particular model.

## COMPRESSOR SERVICE VALVES

Open-type and semi-hermetic compressors are usually fitted with compressor service valves, one at the suction port and one at the discharge port of the compressor. These service valves have no operating function, but they are indispensable when service operations are to be performed on any part of the refrigeration system.

Compressor service valves are of the back-seating type, so constructed that the stem forms a seal against a seat whether the stem is full-forward (front-seated) or full-backward (back-seated). A valve packing is depended upon only when the stem is in the intermediate position. In one style of construction, the front seat, including the gauge connection, is threaded and soldered to the body after the stem has been assembled. (See Figure 8-28.)

**Figure 8-28.**   Double port compressor service valve

When the valve is full-open (the normal position when the compressor is running), the gauge and charging port plug or cap may be removed without loss of refrigerant and a charging line or pressure gauge may be attached (see Figure 8-29). It is also possible to repack the valve without interruption of service.

**Figure 8-29.**   Single port compressor service valve

Compressor valves are machined to fit standard dimensions and many special flange dimensions are used by the several compressor manufacturers. (See Figure 8-30.)

The valve bodies are fabricated of forged

**Figure 8-30.** Adjustable compressor service valve

brass or cast iron. The valve openings are designed for unrestricted refrigerant flow. Each valve must be able to withstand 300 psi (21.09 kg/cm²) without leaking refrigerant.

## SOLENOID VALVES

Solenoid valves are widely used in all types of refrigeration work. They are used as electrically-operated line stop valves and perform in the same manner as hand shut-off valves. However, being electrically operated, they may be conveniently operated in remote locations by any suitable electric control device. The automatic control of the flow of refrigerants, brine, or water frequently depends on the use of solenoid valves. These valves are very popular for use when a pumpdown cycle of the system is required.

**SOLENOID VALVE OPERATION:** A magnet in which the lines of force are produced by an electric current is called an electromagnet. This type of magnet is important to automatic control design because the magnetic field can be created or destroyed by turning an electric current on or off. A solenoid valve is a simple form of electromagnet consisting of a coil of insulated copper wire or other suitable conductor which, when energized by the flow of an electric current, produces a magnetic field that will attract magnetic materials such as iron and many of its alloys. In this way an armature (frequently called a plunger) can

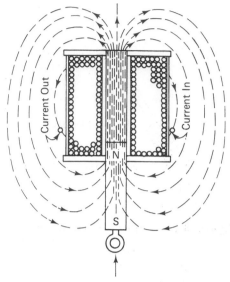

**Figure 8-31.** Solenoid coil and plunger (*Courtesy of Alco Controls Division*)

be drawn up into the core of a solenoid coil as illustrated in Figure 8-31. By attaching a stem and pin to this plunger, we are able to open or close a valve port by energizing or de-energizing the solenoid coil. This magnetic principle constitutes the basis of design of all solenoid valves.

**TYPES OF SOLENOID VALVES:** Solenoid valves can be divided into two general types: (1) direct-acting, and (2) pilot-operated. In the direct-acting type, the pull of the solenoid coil opens the valve port directly by lifting the pin out of the valve seat (see Figure 8-32).

Since this type of valve depends solely on the power of the solenoid for operation, its port size for a given operating pressure differential is limited by the practical limitations of a solenoid size. Therefore, larger solenoid valves—type 2—are usually of the pilot-operated design. (See Figure 8-33.)

In this type, the solenoid plunger does not open the main port directly, but merely

**Figure 8-32.** Direct-acting solenoid valve (*Courtesy of Alco Controls Division*)

A          B

C

**Figure 8-33.** Pilot-operated solenoid valve (*Courtesy of Alco Controls Division*)

opens the pilot port A. The pressure which is trapped on top of piston B is released through the pilot port, thus creating a pres-

sure imbalance across the piston. The pressure underneath is now greater than the pressure above and the piston moves upward, thus opening the main port C. To close, the plunger drops and closes the pilot port A. In the next step, the pressure above and the pressure below the piston equalize again, and the piston drops and closes the main valve. The pressure difference across the valve, acting upon the area of the valve seat, holds the piston in a tightly closed position.

**SOLENOID VALVE SELECTION:** The following performance characteristics are to be considered when selecting a solenoid valve for application: (1) capacity, (2) maximum operating pressure differential (MOPD), (3) electrical rating, (4) seat leak, and (5) special features.

1.   Select a valve that meets the required flow capacity. Additional costs or restricted refrigerant flow are involved when a solenoid valve is improperly selected.

2.   The MOPD should be the extreme pressure difference that the valve will operate against and not an average value. A valve failing to open due to occasional excessive pressures causes a constant "in-rush" current, and consequently coil burnouts.

3.   Choose a valve with the proper voltage and frequency rating. Installation in communities where the electrical voltage varies beyond the rated electrical input limits of the coil will result in valve failure. When the voltage is too high, the coil will eventually burn out due to excessive current draw. The valve may fail to open when the voltage is too low and again cause the coil to burn out, due to the in-rush current.

4.   It is advantageous to select a *metal-to-metal* seat valve when a seat leak is permissible. Soft seat valves are not as dependable

throughout the life span due to the resilient property of the seat material. Continuous operations eventually cause an impression (set) in the seating material conforming to the seating surface that could lead to excessive seat leak.

A metal-to-metal seat has a minimum amount of wear and is uniform. This type has the inherent property of maintaining constant or decreasing seat leak with use.

Foreign particles on the seating surface will cause excessive seat leak and sometimes are mistaken for a seat breakdown.

5. Fitting sizes, metering, materials of construction, and any other special features are also to be considered when selecting a valve for a particular application.

## CHECK VALVES

Frequently, refrigeration systems are designed in which refrigerant liquid or vapor must flow to several components but must never flow backward in a given line. In such installations, a check valve is needed (see Figure 8-34).

**Figure 8-34.** Check valve (*Courtesy of Henry Valve Company*)

**PURPOSE OF CHECK VALVES:** As its name implies, a check valve checks or prevents the flow of refrigerant in one direction, while allowing free flow in the other direction. For example, if two evaporators are controlled by a single condensing unit, a check valve should be placed in the line from the lower temperature evaporator to prevent the suction gas from the higher temperature evaporator from entering the lower temperature evaporator.

Check valves are designed to eliminate chattering and to give maximum refrigerant flow when the unit is operating with a spring tension sufficient to overcome the weight of the valve disc (see Figure 8-35). Check valves must be mounted in the proper position.

**Figure 8-35.** Spring-loaded check valve (*Courtesy of Henry Valve Company*)

## WATER-REGULATING VALVES

On water-cooled condensers, a modulating water-regulating valve is normally used to economize on water usage and to control condensing pressures within reasonable limits. (See Figure 8-36.) Water valves may be either pressure-actuated or temperature-actuated. They act to throttle water flow as required by the condenser. Proper sizing of

**Figure 8-36.** Pressure-actuated water-regulating valve (*Courtesy of Penn Controls, Inc.*)

these valves is important because oversizing will cause a fluctuating head pressure, while undersizing will cause a high head pressure.

## SAFETY PROCEDURES

Accessories are devices that are used to cause a refrigeration system to function more efficiently and economically. Care should be exercised when working on these devices to prevent upsetting their value to the system.

1. Always obtain help when lifting heavy objects.

2. Never make adjustments that are not absolutely necessary.

3. Always replace a component rather than completely removing it from the system; removal of a component may reduce the efficiency of the equipment.

4. Never touch components on the compressor discharge line while the compressor is running because they are very hot.

5. Keep hands, feet, and clothing away from belts and revolving parts.

6. Do not touch a compressor crankcase heater; severe burns may result.

7. Never force a service valve because the stem might be broken or the valve might rupture, causing a loss of refrigerant and possible personal injury.

8. Never heat a suction accumulator with an open flame; the working pressure might be exceeded, resulting in an explosion.

9. Never replace a fusible plug in a receiver with a pipe plug.

10. Never put weight on vibration eliminators; they are not designed to withstand excessive sideways force.

## SUMMARY

• Broken compressor valves, rods, pistons, crankshafts, etc. are usually the result of oil slugging.

• Accumulators can prevent slugging by trapping liquid refrigerant in the suction line.

• Accumulators are installed in the suction line of the refrigeration system.

• Accumulators should be selected for a low refrigerant pressure drop, proper oil return, and adequate volume to store at least 50% of the refrigerant charge.

• Liquid line driers are used to remove moisture from the refrigerant.

• Modern day liquid line driers perform

the following functions: (1) moisture removal, (2) acid removal, and (3) filtering.

• Selection of liquid line driers should be influenced by: (1) type and amount of refrigerant, (2) refrigeration system tonnage, (3) line size, and (4) allowable pressure drop.

• The two types of liquid line driers in use are: (1) straight-through, and (2) angle-replaceable core.

• The proper suction line filter will protect the compressor by trapping foreign matter before it enters the compressor.

• The following factors influence the selection of suction line filters: (1) type of refrigerant, (2) suction line size, (3) allowable pressure drop, and (4) application.

• The two general types of suction line filters are: (1) straight-through, and (2) angle-replaceable core.

• The suction line filter drier offers all the advantages of a suction line filter plus the ability to remove acids and moisture.

• Strainers remove foreign matter such as dirt, metal chips, etc. from the refrigerant lines.

• The three general types of strainers are: (1) straight-through, (2) cleanable angle type, and (3) cleanable "Y" type.

• Moisture-liquid indicators are installed in the liquid line and used to determine the amount of refrigerant and moisture in the refrigeration system.

• The oil separator is installed in the discharge line between the compressor and the condenser.

• The purpose of an oil separator is to maintain the correct oil level in the compressor.

• The prime function of the oil separator is to separate oil from the refrigerant and return it to the compressor before it can enter the other system components.

• The oil separator reduces the velocity of the refrigerant and oil, which allows the heavier oil to fall out and collect in the bottom of the oil separator, from where it is fed back into the compressor.

• In selecting an oil separator, one thing that should be considered is that the larger the capacity of the compressor, the larger the volume of the shell required.

• Any device which brings into contact two substances of different temperatures for the purpose of heating or cooling is called a heat exchanger.

• Heat exchangers are used to increase the capacity of refrigeration systems by bringing the liquid refrigerant into contact with the suction gas.

• Vibration eliminators are used to prevent the transmission of noise and vibration from the compressor through the refrigeration piping.

• Discharge mufflers are used to reduce noise transmissions to a minimum, or where compressor pulsations might create vibration problems.

• The discharge muffler is located in the discharge line as close to the compressor as possible.

• Improper field selection of a discharge muffler will sometimes lead to increased rather than decreased vibration.

• Crankcase heaters are used to prevent the migration of refrigerant to the compressor when installed in a colder location than the evaporator.

• Crankcase heaters are low-wattage resistance elements, normally energized continuously, and must be carefully selected to avoid overheating of the compressor oil.

• Compressor service valves are used to aid in service operations which involve the refrigerant or the oil.

• Solenoid valves are used in refrigeration work to perform the same function as hand shut-off valves.

• Two types of solenoid valves are: (1) direct-acting and (2) pilot-operated.

• Check valves prevent the flow of refrigerant in one direction while allowing free flow in the other direction.

• Water-regulating valves are used to control the head pressure and conserve water on some water-cooled condensers.

## REVIEW EXERCISES

1. What does liquid refrigerant cause in a compressor crankcase?

2. What are accumulators designed to do?

3. What would a broken valve in a compressor indicate?

4. Where should an accumulator be installed?

5. How much refrigerant should an accumulator hold?

6. How much moisture can a refrigeration system tolerate?

7. What is the purpose of a drier?

8. What three functions are modern day driers capable of performing?

9. What four factors influence the selection of a drier?

10. What are the two general types of liquid line driers?

11. How does a suction line filter protect the compressor?

12. What four factors influence the selection of suction line filters?

13. What are the two general types of suction line filters?

14. Why does the suction line filter-drier offer more advantages than the suction line filter?

15. What is the purpose of a refrigerant strainer?

16. Where is the moisture-liquid indicator installed?

17. State the purpose of a moisture indicator.

18. Can an excessive amount of moisture discolor a moisture indicator?

19. How does an oil separator function?

20. What is the purpose of an oil separator?

21. Why is an oil separator needed on some installations?

22. How are oil separators selected?

23. Define a heat exchanger.

24. How do heat exchangers work?

25. How does a heat exchanger increase the capacity of a refrigeration system?

26. How are heat exchangers selected?

27. What are vibration eliminators?

28. Where are vibration eliminators installed?

29. What is the purpose of a discharge muffler?

30. What will the improper sizing of a discharge muffler cause?

31. Where is a crankcase heater installed?

32. What two types of crankcase heaters are used in refrigeration work?

33. How do crankcase heaters operate?

34. What is meant when a compressor service valve is said to be back-seated?

35. What is the operating force in a solenoid valve?

36. Name two types of solenoid valves.

37. What performance characteristics are to be considered when selecting a solenoid valve?

38. What is the purpose of a check valve?

39. Where are check valves used?

40. What is the purpose of a water-regulating valve?

# Refrigerants

The gases or liquids used in refrigeration systems are known as refrigerants. The evaporation of the refrigerant within the evaporator extracts heat from the surrounding objects. The various parts of the refrigeration unit compress and condense the refrigerant so that it can be used over and over again. Even though there are many different types of refrigerants, only the more common fluorocarbons will be considered here.

For practical purposes, *a refrigerant is a fluid that absorbs heat by evaporating at a low temperature and pressure and gives up that heat by condensing at a higher temperature and pressure.*

## CHARACTERISTICS OF REFRIGERANTS

Most of the commonly used refrigerants exist in a gaseous state under ordinary atmospheric pressures and temperatures. To change these gases to liquid form, it is necessary to compress and cool them as is done by the condensing unit on a refrigeration system. A fluid is a liquid, gas, or vapor. The words gas and vapor are ordinarily used interchangeably, although to be perfectly technical, perhaps we should explain that a gas near its condensation point is called a vapor. All fluids have both a liquid and gaseous state. Some fluids have high boiling points, which means that they exist as a gas only when heated to a high temperature or when placed under a vacuum. Fluids which have low boiling points are in the form of gas at ordinary room temperatures and pressures. Many of the common refrigerants, such as those in the freon group are in this category. If these gases are to be liquified they must be compressed and cooled, or condensed.

Water is a fluid which exists as a liquid at atmospheric pressure and temperature. The boiling point of water under atmospheric pressure at sea level is 212°F (100°C). If the water is left in an open basin, it will evaporate very slowly. If heat is applied to the water and its temperature raised to its boiling point it will then evaporate, or boil, very rapidly. The water will change to the gaseous form of water known as steam or water vapor. If the water is boiled in an open container, its temperature will not rise above 212°F. All the heat which is supplied by the flame is used to boil off or vaporize the water.

If a liquid refrigerant is similarly placed in an open container, it will immediately begin to boil vigorously and vaporize, but at a very low temperature. Liquid Refrigerant-12, under atmospheric pressure, will boil at −21.6°F (−29.7°C). It will absorb sufficient heat from the container and the surrounding air to enable it to boil. It would not be necessary to heat it with a flame as was done with the water.

A vaporizing refrigerant will absorb heat equal to the amount of energy necessary to change the physical form of the refrigerant from the liquid state to the gaseous state. Each refrigerant will absorb an amount of heat per pound of refrigerant equal to its latent heat of vaporization.

**EFFECTS OF PRESSURE ON BOILING POINT**

The boiling point of any liquid may be raised or lowered in accordance with the amount of pressure applied to the liquid. The greater the pressure, the higher the boiling point; the lower the pressure, the lower the boiling point. Thus, a liquid can be caused to boil at a low temperature by placing it in a partial vacuum.

Certain refrigerants require high pressure

## Table 9-1

### Boiling Point of "Freon" Fluorocarbon Refrigerants*

| Product | Formula | Molecular Weight | Boiling Point | |
|---|---|---|---|---|
| | | | °F | °C |
| "FREON-14" | $CF_4$ | 88.0 | −198.4 | −128.0 |
| "FREON-23" | $CHF_3$ | 70.0 | −115.7 | −82.1 |
| "FREON-13" | $CClF_3$ | 104.5 | −114.6 | −81.4 |
| "FREON-116" | $CF_3$-$CF_3$ | 138.0 | −108.8 | −78.2 |
| "FREON-13B1" | $CBrF_3$ | 148.9 | −72.0 | −57.8 |
| "FREON-502" | $CHClF_2$/$CClF_2$-$CF_3$ (48.8/51.2% by weight) | 121.2 | −50.1 | −45.6 |
| "FREON-22" | $CHClF_2$ | 86.5 | −41.4 | −40.8 |
| "FREON-115" | $CClF_2$-$CF_3$ | 154.5 | −37.7 | −38.7 |
| "FREON-12" | $CCl_2F_2$ | 120.9 | −21.6 | −29.8 |
| "FREON-C318" | $C_4F_8$ (cyclic) | 200.0 | 21.5 | −5.8 |
| "FREON-114" | $CClF_2$-$CClF_2$ | 170.9 | 38.4 | 3.6 |
| "FREON-21" | $CHCl_2F$ | 102.9 | 48.1 | 8.9 |
| "FREON-11" | $CCl_3F$ | 137.4 | 74.8 | 23.8 |
| "FREON-114B2" | $CBrF_2$-$CBrF_2$ | 259.9 | 117.5 | 47.5 |
| "FREON-113" | $CCl_2F$-$CClF_2$ | 187.4 | 117.6 | 47.6 |
| "FREON-112" | $CCl_2F$-$CCl_2F$ | 203.9 | 199.0 | 98.2 |

* Courtesy of Freon Products Div., E.I. DuPont de Nemours & Co., Inc.

when used in a refrigeration unit. R–112 (Freon-112) is an example of such a refrigerant. (See Table 9-1.) The energy consumed by a refrigeration unit using R–112 is required chiefly on the compression side for compressing the gaseous refrigerant. R–14 and R–502 are examples of refrigerants which do not require such high pressures. R–12, R–22, R–500, and R–502 are used in most household and smaller commercial refrigeration units.

## CRITICAL TEMPERATURE

The critical temperature of a gas is the temperature above which the gas cannot be liquified regardless of the amount of pressure applied. If the gas is heated sufficiently to raise its temperature above its critical temperature, the vibrations of the molecules in the gas become so intense that pressure cannot bring them into sufficient contact with each other to exist as a liquid.

The refrigerants which are used in refrigeration units change from the liquid state to the gaseous state and back again to the liquid state while going through the refrigeration cycle. For this reason, a refrigerant must be used below its critical temperature in order to obtain the liquification phase of the refrigeration cycle. (See Table 9-2.)

Table 9-2

Critical Temperatures of
Fluorocarbon Refrigerants

| REFRIGERANT | CRITICAL TEMPERATURE | |
| | °C | °F |
| --- | --- | --- |
| R–11 | 198.0 | 388.4 |
| R–12 | 112.0 | 233.6 |
| R–22 | 96.0 | 204.8 |
| R–502 | 90.1 | 194.1 |

When the temperature of the gas is reduced, the amount of pressure required to effect liquification decreases. From this we can deduce the fact that for every temperature below the critical temperature, there is a corresponding pressure which will cause liquification.

There are graphs and tables which indicate the relationship between a temperature and the corresponding pressure which will cause liquification of the refrigerant gas (see Figure 9-1). The temperature in degrees Fahrenheit is indicated on the bottom. The temperature in degrees Centigrade is indicated on the top of the chart. Absolute pressure (psia) is on the left while the gauge pressures (psig) and inches of mercury-vacuum are on the right side of the chart. The metric equivalent values are not listed because the United States has not yet confirmed which of the SI units will be used. For example, if we have R–12 at 80°F, what will its saturation pressure be? Find 80°F on the bottom line of the chart. Follow this line vertically, until the curve for R–12 is intersected. Read to the left for absolute pressure and to the right for gauge pressure. Find R–12 at this temperature to have 99.2 psia and 84.2 psig. This is the pressure required to liquify the refrigerant at 80°F temperature.

## STANDARD CONDITIONS

The capacity of any refrigeration unit will vary as the refrigerant temperatures vary on the high and low pressure sides of the system. The latent heat of the refrigerant, its condensing pressure, and its vaporizing pressure will also vary as the temperature of the refrigerant varies. For purposes of comparing different refrigerants and refrigeration units, certain standards have been developed. The refrigeration industry has set forth conditions known as *standard conditions*. These

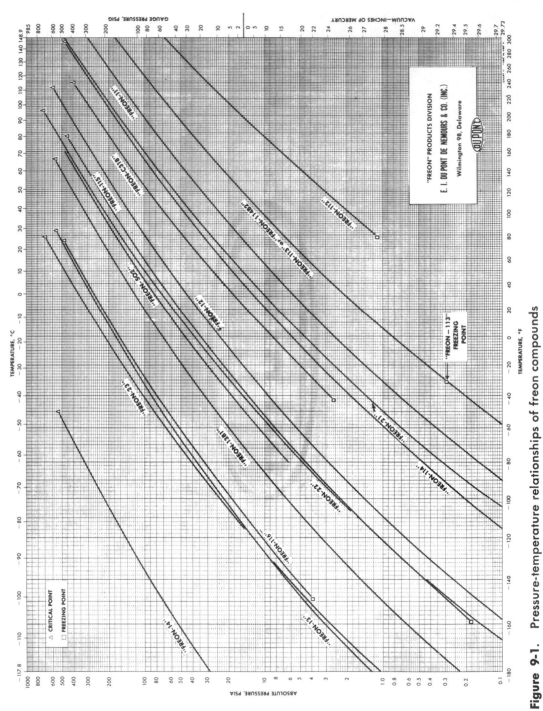

**Figure 9-1.** Pressure-temperature relationships of freon compounds

conditions are established with the following temperatures at various locations in the refrigeration cycle: a temperature in the evaporator of 5°F (−15°C); a temperature in the saturated portion of the condenser of 86°F (30°C); a liquid temperature at the flow control device of 77°F (25°C); and a suction gas temperature of 14°F (−10°C). If any two refrigerants are compared when using these standards, correct conclusions can be derived.

## CONDENSING PRESSURE

The condensing pressure will depend on the temperature at which the vapor will liquify. In refrigeration work, it is desirable to avoid high-condensing pressures if at all possible. Therefore, the condensing medium (air or water) must be as cool as possible. Ordinarily, a water-cooled condenser will operate at a lower condensing temperature and pressure than an air-cooled condenser. Because of this, there is some difference in the operating pressure for these two types of condensers. In general, it may be assumed that the condensing temperature and pressure of an air-cooled unit will be approximately 25°F to 35°F (−3.9°C to 1.7°C) higher than ambient temperatures. The actual temperature and pressure will, however, depend on the efficiency of the condenser itself, the location of the condenser, whether or not sufficient air circulation is obtained, and the cleanliness of the condenser surface. When water-cooled condensers are used, the condenser temperature is generally lower than the ambient temperature. Therefore, the condensing pressure of a water-cooled condenser is lower than the condensing pressure of an air-cooled condenser.

The saturation pressures shown in Figure 9-1 should not be considered as operating head pressures for a refrigeration unit. The operating head pressures will be higher than the saturation pressures because of the addition of the heat of compression. If condensation is to take place in the condenser, the temperature of the refrigerant inside must be higher than the temperature of the cooling medium which is used for the condenser.

## VAPORIZING PRESSURE

The vaporizing pressure of a refrigerant is important because the refrigerant must evaporate without requiring too low a suction pressure. A temperature of 5°F (−15°C) is considered the temperature of most domestic refrigerator evaporators. This is the same temperature as that set for the standard conditions used for comparison of different refrigerants and refrigeration units. In general, a refrigerant is desired that has an evaporating pressure at or near atmospheric pressure (see Table 9-3). A refrigerant which requires a vacuum to produce evaporation is not practical under ordinary conditions because of the tendency for air to leak into the system. The air will not condense and will cause the condenser pressure to become very high. This high pressure will reduce the efficiency of the refrigeration unit. Refrigerants that have a vaporizing pressure above atmospheric pressure do not allow air to be drawn into the system through a leak.

It should be noted that the pressure in the evaporator and the low side of the system will be the same. Also, that the temperature of the evaporating refrigerant will correspond to the temperature shown on the pressure-temperature chart, the pressure in the evaporator, or in the low side of the system. (See Table 9-4.)

# REFRIGERANTS

## Table 9-3

### Vaporizing Pressures of Refrigerants at 5°F

| REFRIGERANT | PRESSURE AT 5°F | PRESSURE AT —15°C |
|---|---|---|
| R—11 | —24″ Hg | .2 kg/cm² |
| R—12 | 11.8 | .83 kg/cm² |
| R—22 | 28.1 | 1.97 kg/cm² |
| R—500 | 16.4 | 1.15 kg/cm² |
| R—502 | 36 | 2.53 kg/cm² |

## Table 9-4
### Pressure-Temperature Relation Chart*

Vacuum - Inches of Mercury *Italic Figures*

| TEMPERATURE °F. | 12 | 22 | 500 | 502 | 717 |
|---|---|---|---|---|---|
| -60 | *19.0* | *12.0* | *17.0* | *7.2* | *18.6* |
| -55 | *17.3* | *9.2* | *15.0* | *3.8* | *16.6* |
| -50 | *15.4* | *6.2* | *12.8* | *0.2* | *14.3* |
| -45 | *13.3* | *2.7* | *10.4* | 1.9 | *11.7* |
| -40 | *11.0* | 0.5 | *7.6* | 4.1 | *8.7* |
| -35 | *8.4* | 2.6 | *4.6* | 6.5 | *5.4* |
| -30 | *5.5* | 4.9 | *1.2* | 9.2 | *1.6* |
| -25 | *2.3* | 7.4 | 1.2 | 12.1 | 1.3 |
| -20 | 0.6 | 10.1 | 3.2 | 15.3 | 3.6 |
| -18 | 1.3 | 11.3 | 4.1 | 16.7 | 4.6 |
| -16 | 2.0 | 12.5 | 5.0 | 18.1 | 5.6 |
| -14 | 2.8 | 13.8 | 5.9 | 19.5 | 6.7 |
| -12 | 3.6 | 15.1 | 6.8 | 21.0 | 7.9 |
| -10 | 4.5 | 16.5 | 7.8 | 22.6 | 9.0 |
| - 8 | 5.4 | 17.9 | 8.8 | 24.2 | 10.3 |
| - 6 | 6.3 | 19.3 | 9.9 | 25.8 | 11.6 |
| - 4 | 7.2 | 20.8 | 11.0 | 27.5 | 12.9 |
| - 2 | 8.2 | 22.4 | 12.1 | 29.3 | 14.3 |
| 0 | 9.2 | 24.0 | 13.3 | 31.1 | 15.7 |
| 1 | 9.7 | 24.8 | 13.9 | 32.0 | 16.5 |
| 2 | 10.2 | 25.6 | 14.5 | 32.9 | 17.2 |
| 3 | 10.7 | 26.4 | 15.1 | 33.9 | 18.0 |
| 4 | 11.2 | 27.3 | 15.7 | 34.9 | 18.8 |
| 5 | 11.8 | 28.2 | 16.4 | 35.8 | 19.6 |
| 6 | 12.3 | 29.1 | 17.0 | 36.8 | 20.4 |
| 7 | 12.9 | 30.0 | 17.7 | 37.9 | 21.2 |
| 8 | 13.5 | 30.9 | 18.4 | 38.9 | 22.1 |
| 9 | 14.0 | 31.8 | 19.0 | 39.9 | 22.9 |
| 10 | 14.6 | 32.8 | 19.7 | 41.0 | 23.8 |
| 11 | 15.2 | 33.7 | 20.4 | 42.1 | 24.7 |

| TEMPERATURE °F. | 12 | 22 | 500 | 502 | 717 |
|---|---|---|---|---|---|
| 12 | 15.8 | 34.7 | 21.2 | 43.2 | 25.6 |
| 13 | 16.4 | 35.7 | 21.9 | 44.3 | 26.5 |
| 14 | 17.1 | 36.7 | 22.6 | 45.4 | 27.5 |
| 15 | 17.7 | 37.7 | 23.4 | 46.5 | 28.4 |
| 16 | 18.4 | 38.7 | 24.1 | 47.7 | 29.4 |
| 17 | 19.0 | 39.8 | 24.9 | 48.8 | 30.4 |
| 18 | 19.7 | 40.8 | 25.7 | 50.0 | 31.4 |
| 19 | 20.4 | 41.9 | 26.5 | 51.2 | 32.5 |
| 20 | 21.0 | 43.0 | 27.3 | 52.4 | 33.5 |
| 21 | 21.7 | 44.1 | 28.1 | 53.7 | 34.6 |
| 22 | 22.4 | 45.3 | 28.9 | 54.9 | 35.7 |
| 23 | 23.2 | 46.4 | 29.8 | 56.2 | 36.8 |
| 24 | 23.9 | 47.6 | 30.6 | 57.5 | 37.9 |
| 25 | 24.6 | 48.8 | 31.5 | 58.8 | 39.0 |
| 26 | 25.4 | 49.9 | 32.4 | 60.1 | 40.2 |
| 27 | 26.1 | 51.2 | 33.2 | 61.5 | 41.4 |
| 28 | 26.9 | 52.4 | 34.2 | 62.8 | 42.6 |
| 29 | 27.7 | 53.6 | 35.1 | 64.2 | 43.8 |
| 30 | 28.4 | 54.9 | 36.0 | 65.6 | 45.0 |
| 31 | 29.2 | 56.2 | 36.9 | 67.0 | 46.3 |
| 32 | 30.1 | 57.5 | 37.9 | 68.4 | 47.6 |
| 33 | 30.9 | 58.8 | 38.9 | 69.9 | 48.9 |
| 34 | 31.7 | 60.1 | 39.9 | 71.3 | 50.2 |
| 35 | 32.6 | 61.5 | 40.9 | 72.8 | 51.6 |
| 36 | 33.4 | 62.8 | 41.9 | 74.3 | 52.9 |
| 37 | 34.3 | 64.2 | 42.9 | 75.8 | 54.3 |
| 38 | 35.2 | 65.6 | 43.9 | 77.4 | 55.7 |
| 39 | 36.1 | 67.1 | 45.0 | 79.0 | 57.2 |
| 40 | 37.0 | 68.5 | 46.1 | 80.5 | 58.6 |
| 41 | 37.9 | 70.0 | 47.1 | 82.1 | 60.1 |

Pressure - Pounds Per Square Inch Gage **Bold Figures**

| TEMPERATURE °F. | 12 | 22 | 500 | 502 | 717 |
|---|---|---|---|---|---|
| 42 | 38.8 | 71.4 | 48.2 | 83.8 | 61.6 |
| 43 | 39.8 | 73.0 | 49.4 | 85.4 | 63.1 |
| 44 | 40.7 | 74.5 | 50.5 | 87.0 | 64.7 |
| 45 | 41.7 | 76.0 | 51.6 | 88.7 | 66.3 |
| 46 | 42.6 | 77.6 | 52.8 | 90.4 | 67.9 |
| 47 | 43.6 | 79.2 | 54.0 | 92.1 | 69.5 |
| 48 | 44.6 | 80.8 | 55.1 | 93.9 | 71.1 |
| 49 | 45.7 | 82.4 | 56.3 | 95.6 | 72.8 |
| 50 | 46.7 | 84.0 | 57.6 | 97.4 | 74.5 |
| 55 | 52.0 | 92.6 | 63.9 | 106.6 | 83.4 |
| 60 | 57.7 | 101.6 | 70.6 | 116.4 | 92.9 |
| 65 | 63.8 | 111.2 | 77.8 | 126.7 | 103.1 |
| 70 | 70.2 | 121.4 | 85.4 | 137.6 | 114.1 |
| 75 | 77.0 | 132.2 | 93.5 | 149.1 | 125.8 |
| 80 | 84.2 | 143.6 | 102.0 | 161.2 | 138.8 |
| 85 | 91.8 | 155.7 | 111.0 | 174.0 | 150.7 |
| 90 | 99.8 | 168.4 | 120.6 | 187.4 | 165.9 |
| 95 | 108.2 | 181.8 | 130.6 | 201.4 | 181.1 |
| 100 | 117.2 | 195.9 | 141.2 | 216.2 | 197.2 |
| 105 | 126.6 | 210.8 | 152.4 | 231.7 | 214.2 |
| 110 | 136.4 | 226.4 | 164.1 | 247.9 | 232.3 |
| 115 | 146.8 | 242.7 | 176.5 | 264.9 | 251.5 |
| 120 | 157.6 | 259.9 | 189.4 | 282.7 | 271.7 |
| 125 | 169.1 | 277.9 | 203.0 | 301.4 | 293.1 |
| 130 | 181.0 | 296.8 | 217.2 | 320.8 | —— |
| 135 | 193.5 | 316.6 | 232.1 | 341.2 | —— |
| 140 | 206.6 | 337.2 | 247.7 | 362.6 | —— |
| 145 | 220.3 | 358.9 | 264.0 | 385.0 | —— |
| 150 | 234.6 | 381.5 | 281.1 | 408.4 | —— |
| 155 | 249.5 | 405.1 | 298.9 | 432.9 | —— |

* Courtesy of Sporlan Valve Company

## LATENT HEAT OF VAPORIZATION

The amount of heat in Btu (cal) that is required to change a liquid to a vapor, the change taking place at a constant temperature, is known as the latent heat of vaporization. This definition can now be developed so as to apply to one pound of refrigerant, the vaporization taking place at atmospheric pressure and the liquid to be at a temperature equal to its boiling point when the operation begins. Thus, the latent heat of vaporization of a liquid is the amount of heat in Btu (cal) that is required to vaporize one pound of the liquid at atmospheric pressure, the liquid to be at its boiling point when the operation begins. To convert one pound of water at $212°F$ ($100°C$) into steam at the same temperature and pressure, the water must absorb 970 Btu/lb (533.5 kcal/kg). This quantity of heat is the total latent heat of one pound of water at atmospheric pressure.

Any refrigerant, when evaporating in the evaporator, must absorb heat from within the cooled space exactly equal to its latent heat of vaporization. When a refrigerant has a high latent heat, it will absorb more heat per pound of liquid than a refrigerant with a lower latent heat of vaporization. Thus, if a refrigerant with a high latent heat of vaporization is used, a smaller compressor, condenser, and evaporator can be used. (See Table 9-5.)

Table 9-5

Latent Heat of Vaporization at $5°F$ ($-15°C$) in BTU/LB (kcal/kg)

| REFRIGERANT | LATENT HEAT | |
| | Btu/lb | kcal/kg |
| --- | --- | --- |
| R–11 | 83.459 | 45.902 |
| R–12 | 68.204 | 37.512 |
| R–22 | 93.206 | 51.263 |
| R–500 | 82.45 | 45.34 |
| R–502 at 40°F | 63.1 | 34.7 |

The latent heat of vaporization of a liquid will vary with the pressure and the corresponding temperature at which the vaporization occurs. When lower temperatures and pressures are encountered, the latent heat of vaporization increases.

## TYPES OF REFRIGERANTS

There are many types of refrigerants which are commercially available. Those used in early refrigeration units were ammonia, sulfur dioxide, methyl chloride, propane, and ethane. Ammonia is still used in large refrigeration equipment, however, all the others have been discontinued due to the fact that they are either toxic, dangerous, or have other undesirable characteristics. Spe-

cialized refrigerants have been developed for ultra-low temperature applications or for large centrifugal compressors. For normal commercial refrigeration and air conditioning systems which use reciprocating compressors, refrigerants R–12, R–22, R–500, and R–502 are presently used almost exclusively. R–11 is used in centrifugal compressors.

## R–11 TRICHLOROFLUOROMETHANE

($CCl_3F$): This refrigerant is a synthetic chemical product which is used as a refrigerant. It is considered stable, nonflammable, and has a very low toxicity rating. It operates with a low side pressure of 24 inches of vacuum (609.6 mmHg) at $5°F$ ($-15°C$), and a high side pressure of 18.3

psia (1.28 kg/cm²a) at 86°F (30°C). It has a latent heat value at 5°F of 83.459 Btu/lb (45.9 kcal/kg). (See Table 9-6.) The boiling point at atmospheric pressure is 74.87°F (23°C).

R–11 may be used as a cleaning agent to aid in cleaning systems that have been contaminated by motor compressor burnout or moisture. This process is very expensive and should be used only as a last resort. A thorough evacuation and replacement of driers in conjunction with oil sample testing is recommended to clean the contaminated systems. Leak detection is accomplished by use of a soap solution, a halide torch, or an electronic leak detector.

The cylinder color code for R–11 is orange.

### Table 9-6

### R–11 Refrigerant Standard Ton Characteristics*

| | |
|---|---|
| Evaporator Pressure at 5°F, psia | 2.9373 |
| Condenser Pressure at 86°F, psia | 18.186 |
| Compression Ratio (86°F/5°F) | 6.19 |
| Latent Heat of Vaporization at 5°F, Btu/lb | 83.459 |
| Net Refrigerating Effect, Btu/lb | 66.796 |
| Refrigerant Circulated per Ton of Refrigeration, lb/min | 2.9942 |
| Saturated Liquid Volume at 86°F, cu ft/lb | 0.010942 |
| Liquid Circulated per Ton of Refrigeration, cu in/min | 56.614 |
| Saturated Vapor Density at 5°F, lb/cu ft | 0.081933 |
| Saturated Vapor Density at 86°F, lb/cu ft | 0.44668 |
| Compressor Displacement per Ton of Refrigeration, cu ft/min | 36.544 |
| Refrigeration per Cubic Foot of Compressor Displacement, Btu | 5.473 |
| Heat of Compression, Btu/lb | 13.292 |
| Temperature of Compressor Discharge, °F | 110.9 |
| Coefficient of Performance | 5.025 |
| Horsepower per Ton | 0.9383 |

* Courtesy of Freon Products Div., E.I. DuPont De Numours & Co., Inc.

### R–12 DICHLORODIFLUOROMETHANE ($CCl_2F_2$):

R–12 is a refrigerant which is being used in nearly all facets of refrigeration and air conditioning. It has a boiling point of −21.6°F (−30°C) at atmospheric pressure. Therefore, it exists as a gas at ordinary room temperatures and atmospheric pressure. R–12 will liquify under a pressure of 76 psi (5.34 kg/cm²) at 75°F (23.9°C). (See Table 9-7.)

R–12 has very little odor, but in large concentrations a faint, sweet odor may be detected. Its critical temperature is 233.6°F (112°C). It is a colorless gas and liquid. It is nontoxic, nonflammable, and nonirritating, and is noncorrosive to any of the ordinary metals even in the presence of water. It is stable under all conditions and temperatures normally encountered in refrigeration. R–12 liquid will dissolve lubricating oil in all proportions; the oil in return will absorb refrigerant vapor. There is no

## Table 9-7

### R–12 Refrigerant Standard Ton Characteristics*

| | |
|---|---|
| Evaporator Pressure at 5°F, psia | 26.483 |
| Condenser Pressure at 86°F, psia | 108.04 |
| Compression Ratio (86°F/5°F) | 4.08 |
| Latent Heat of Vaporization at 5°F, Btu/lb | 68.204 |
| Net Refrigerating Effect, Btu/lb | 50.035 |
| Refrigerant Circulated per Ton of Refrigeration, lb/min | 3.9972 |
| Saturated Liquid Volume at 86°F, cu ft/lb | 0.012396 |
| Liquid Circulated per Ton of Refrigeration, cu in/min | 85.621 |
| Saturated Vapor Density at 5°F, lb/cu ft | 0.68588 |
| Saturated Vapor Density at 86°F, lb/cu ft | 2.6556 |
| Compressor Displacement per Ton of Refrigeration, cu ft/min | 5.8279 |
| Refrigeration per Cubic Foot of Compressor Displacement, Btu | 34.318 |
| Heat of Compression, Btu/lb | 10.636 |
| Temperature of Compressor Discharge, °F | 100.84 |
| Coefficient of Performance | 4.704 |
| Horsepower per Ton | 1.0023 |

* Courtesy of Freon Products Div., E.I. DuPont de Numours & Co., Inc.

separation or oil blanket formed in the evaporator to interfere with proper vaporization.

R–12 is only slightly soluble in water. It is this characteristic which prevents corrosion of metals in the presence of water. Leaks may be detected with a halide torch, an electronic leak detector, or a soap-bubble solution. The halide torch imparts a blue-green tinge to the flame in the presence of R–12.

The cylinder color code for R–12 is white.

## R–22     MONOCHLORODIFLUOROMETHANE (CHC1F$_2$): R–22 is a refrigerant that has been developed for refrigeration installations that have a low evaporating temperature. In most physical characteristics it is similar to R–12. However, its saturation pressures are higher than those for R–12 for the same temperatures. It has a latent heat of vaporization at 5°F (−15°C) of 93.21 Btu/lb (51.26 kcal/kg). It has a lower specific volume than R–12 and, therefore, a greater refrigeration capacity for a given volume of saturated refrigerant vapor. The boiling point is −41.36°F (−41°C) at atmospheric pressure. R–22 has an operating head pressure at 86°F (30°C) of 172.8 psia, or 158 psi (11.1 kg/cm²). (See Table 9-8.) Its evaporator pressure at 5°F is 42.88 psia, or 28 psi (1.96 kg/cm²). R–22 is nontoxic, nonflammable, noncorrosive, and nonirritating. It is a stable refrigerant under normal operating conditions. Because of its low evaporating temperatures and high compression ratios, however, the temperature of compressed R–22 vapor may become so high that damage to the compressor could result under ultralow temperature applications.

## Table 9-8

### R–22 Refrigerant Standard Ton Characteristics*

| | |
|---|---|
| Evaporator Pressure a 5°F, psia | 42.888 |
| Condenser Pressure at 86°F, psia | 172.87 |
| Compression Ratio (86°F/5°F) | 4.03 |
| Latent Heat of Vaporization at 5°F, Btu/lb | 93.206 |
| Net Refrigerating Effect, Btu/lb | 70.027 |
| Refrigerant Circulated per Ton of Refrigeration, lb/min | 2.8560 |
| Saturated Liquid Volume at 86°F, cu ft/lb | 0.013647 |
| Liquid Circulated per Ton of Refrigeration, cu in/min | 67.351 |
| Saturated Vapor Density at 5°F, lb/cu ft | 0.80422 |
| Saturated Vapor Density at 86°F, lb/cu ft | 3.1622 |
| Compressor Displacement per Ton of Refrigeration, cu ft/min | 3.5512 |
| Refrigeration per Cubic Foot of Compressor Displacement, Btu | 56.32 |
| Heat of Compression, Btu/lb | 15.022 |
| Temperature of Compressor Discharge, °F | 128.4 |
| Coefficient of Performance | 4.662 |
| Horsepower per Ton | 1.0114 |

* Courtesy of Freon Products Div., E.I. DuPont de Numours & Co., Inc.

Where size of equipment and economy are a factor, such as in packaged air conditioning units, R–22 is commonly used. If the evaporating temperature of R–22 of −40°F (−40°C) is reached, the oil will begin to separate from the refrigerant and form a film on the surface which will interfere with proper evaporation.

Leaks can be detected with a halide torch, soap-bubble solution, or with an electronic leak detector.

The cylinder color code for R–22 is green.

**R–500 REFRIGERANT ($CCl_2F_2/CH_3CHF_2$):** This refrigerant is an azeotropic mixture of 26.2 percent of R–152a and 73.8% of R–12. An azeotrope is the scientific name given to a specific mixture of different compounds in which the resulting mixture has different characteristics than either of the components, and which can evaporate and condense without a change in its composition. The vapor-pressure temperature curve is fairly constant and is different from either of the compounds used in making it. For the same size compressor, it offers approximately 20% more refrigeration capacity than R–12. R–500 has an evaporator pressure of 31.21 psia, or 19.7 psi (1.4 kg/cm²) at 5°F (−15°C). Its condensing pressure is 127.6 psia, or 113 psi (7.94 kg/cm²) at 86°F (30°C). The boiling point is −28°F (−33°C) at atmospheric pressure, and its latent heat value is 82.45 Btu/lb (45.35 kcal/kg) at 5°F.

R–500 is used in both commercial and industrial applications and only with reciprocating compressors. It also can be used when greater refrigeration capacity is required than that possible with R–12.

It is fairly soluble with oil and its solubility with water is very critical. Therefore, proper dehydration procedures and the use of driers is strongly recommended. Leak detection is accomplished with a halide leak detector, soap-bubble solution, or an electronic leak detector.

The cylinder color code for R–500 is yellow.

## R–502 REFRIGERANT (CHC1F$_2$/CC1F$_2$ CF$_3$):

R–502 is an azeotropic mixture of 48.8% R–22 and 51.2% of R–115. It is similar to R–12 and R–22 in most physical characteristics. The latent heat of vaporization is 72.5 Btu/lb (39.87 kcal/kg). Its refrigerating capacity is comparable to that of R–22, except at lower temperatures it will usually be greater. The boiling point is −49.76°F (−46°C) at atmospheric pressure.

The evaporator pressure of R–502 is 30 psia, 15 psi (1.05 kg/cm²) at −20°F (−28.9°C) evaporating temperature. The condenser pressure is 297.4 psia, or 282.4 psi (19.85 kg/cm²) at a condensing temperature of 120°F (49°C). (See Table 9-9.)

As with R–22, a compressor with a lower displacement may be used for performance equal to R–12. Because of the excellent low-temperature characteristics, R–502 is well-suited for low-temperature refrigeration applications. R–502 is recommended for all single-stage installations with an evaporating temperature of 0°F (−17.8°C) or below. It also is very satisfactory for use in two-stage systems providing they are extra low temperature applications, and is becoming very popular for use in the medium temperature range.

Table 9-9

R–502 Refrigerant Standard Ton Characteristics*

EVAPORATING TEMPERATURE = −20°F
RETURN GAS TEMPERATURE = 65°F
CONDENSING TEMPERATURE = 120°F

| | |
|---|---|
| Evaporator Pressure, psia | 30.01 |
| Condenser Pressure, psia | 297.4 |
| Compression Ratio | 9.91 |
| Net Refrigerating Effect**, Btu/lb | 45.26 |
| Refrigerant Circulated per Ton, lb/min | 4.418 |
| Saturated Liquid Volume at 120°F, cu ft/lb | 0.01472 |
| Liquid Circulated per Ton, cu in/min | 112.4 |
| Vapor Density at 65°F†, lb/cu ft | 0.6165 |
| Compressor Displacement per Ton, cu ft/min | 7.167 |
| Refrigeration Capacity, Btu/cu ft | 27.90 |
| Heat of Compression, Btu/lb | 22.52 |
| Temperature of Compressor Discharge, °F | 226.0 |
| Coefficient of Performance | 2.010 |
| Horsepower per Ton | 2.346 |

* Courtesy of Freon Products Div., E.I. DuPont de Numours & Co., Inc.

** Enthalpy of vapor at 65°F and evaporator pressure—enthalpy of liquid at 120°F.

† It is assumed that vapor enters the compressor cyclinder at a temperature of 65°F.

It is nonflammable, noncorrosive, non-toxic, and stable under all normal situations. Leak detection is accomplished by use of halide leak detectors, soap-bubble solution, or electronic leak detectors. Oil return is not a problem with this refrigerant down to −40°F (−40°C) evaporator temperatures. However, oil separators are sometimes used when lower temperatures are necessary.

The cylinder color code for R–502 is orchid.

## REFRIGERANT-OIL RELATIONSHIPS

When reciprocating compressors are used, oil and refrigerant mix continuously. Oils which are used in refrigeration are soluble in liquid refrigerants, and at room temperatures will mix completely. *Miscibility* is the term used to express the ability of liquid refrigerant to mix with oil.

Any oil which is circulating in a refrigeration system is exposed to both very high and very low temperatures. Because of the critical nature of lubrication under these extreme conditions, and the damage that can be done to the system by wax or other impurities which may be in the oil, only highly refined oils specifically prepared for use in refrigeration systems can be used.

In general, naphthenic oils are more soluble in refrigerants than paraffinic oils. Separation of the oil and refrigerant into separate layers can take place with either type of oil. However, naphthenic oils separate at somewhat lower temperatures. This separation does not necessarily affect the lubricating properties of the oil, but it could cause problems in properly supplying the oil to the working parts.

Because oil must pass through the compressor cylinders to provide lubrication of moving parts, a small amount of oil is always circulating with the refrigerant. Oil and refrigerant vapor do not mix readily. Therefore, oil can be properly circulated through the system only if the refrigerant vapor velocities are kept high enough to sweep the oil along the piping. If the velocities are not kept high enough, the oil will tend to lie on the bottom of the refrigeration tubing, which will decrease the heat transfer and, perhaps, cause a shortage of oil in the compressor. As the refrigerant evaporating temperatures become lower, the oil separation becomes more critical. The viscosity of the oil increases with a decrease in temperature. Therefore, the proper design of refrigerant piping is necessary for proper oil return.

One of the basic characteristics of a refrigerant and oil mixture inside a sealed system is that liquid refrigerant is attracted by oil. This liquid refrigerant will vaporize and migrate through the system to the compressor crankcase, even though no pressure differential exists to cause the movement. When the refrigerant reaches the compressor crankcase, it will condense back into a liquid. This refrigerant migration will continue until the compressor oil is saturated with liquid refrigerant.

The excess of liquid refrigerant in the compressor crankcase will result in violent foaming and boiling action of the oil, and may cause all of the oil to leave the compressor crankcase resulting in lubrication problems. Therefore, some provisions, such a crankcase heater, must be made to prevent the accumulation of excess liquid refrigerant in the compressor crankcase.

The refrigerants R–22 and R–502 are much less soluble in oil than is R–12. Therefore, the proper piping and system design for these two refrigerants is more critical with regard to oil return.

## REFRIGERANT TABLES

In order to accurately determine the operating performance of a refrigeration system, very precise and accurate information is required concerning the various properties of refrigerants at any pressure and temperature which may be considered. Refrigerant manufacturers have calculated and compiled this data in the form of tables listing the thermodynamic properties of each refrigerant.

An excerpt from an R–12 saturation table which lists five major saturation properties of this refrigerant is shown in Table 9-10. Pressure, volume, and density were previously discussed.

*Enthalpy* is a term used in thermodynamics to describe the heat content of a substance. When discussing refrigerants, enthalpy is expressed in terms of Btu per pound of refrigerant. An arbitrary base of saturated liquid at −40°F (−40°C) has been accepted as the standard zero value. More simply stated, the enthalpy of any liquid refrigerant is zero for liquid at −40°F. Liquid refrigerant below that temperature is considered to have a negative enthalpy, while at all temperatures above −40°F the refrigerant has a positive enthalpy value.

The difference in enthalpy values in different parts of the refrigeration system is commonly used to rate the performance of a refrigeration unit. If the heat content per pound of a refrigerant entering and leaving a cooling coil can be determined, and the flow rate of the refrigerant is known, the cooling capabilities of that coil can be calculated.

*Entropy* (see Table 9-10) is best described as a mathematical ratio which is used in thermodynamics. Thermodynamics is used in the solving of complex refrigeration engineering problems, but is seldom used in commercial refrigeration and air conditioning applications. Therefore, a discussion of it is not pertinent in this text.

## POCKET PRESSURE-TEMPERATURE CHARTS

Small pocket-sized tables which list the saturation temperatures and pressures of common refrigerants are available from expansion valve manufacturers, refrigerant manufacturers, and refrigeration supply stores. (See Table 9-11.)

These charts are a ready reference and an invaluable tool for the service technician. They are used to check the performance of a refrigeration system. The suction and discharge pressures can be readily checked by using gauges. These pressures indicate the evaporating and condensing temperatures of the refrigerant.

## HANDLING OF REFRIGERANT CYLINDERS

The pressure which is created by a liquid refrigerant in a sealed container is equal to its saturation pressure at that liquid temperature as long as there is space above the liquid for the vapor. However, if the refrigerant cylinder is overfilled, or if the cylinder is gradually and uniformly overheated, the liquid refrigerant will expand until the cylinder becomes full of liquid. When this occurs, hydrostatic pressure builds up rapidly producing pressures well above saturation pressures. Figure 9-2 illustrates the pressure-temperature relationship of a liquid refrigerant before and after the cylinder becomes full of the expanded liquid under gradual and uniform overheating. The true pressure-temperature relationship exists up to the point where expansion room is no longer available in the cylinder.

## Table 9-10
## Saturation Properties-Temperature Table of R—12*

| TEMP. °F | PRESSURE PSIA | PRESSURE PSIG | VOLUME cu ft/lb LIQUID $v_f$ | VOLUME cu ft/lb VAPOR $v_g$ | DENSITY lb/cu ft LIQUID $1/v_f$ | DENSITY lb/cu ft VAPOR $1/v_g$ | ENTHALPY Btu/lb LIQUID $h_f$ | ENTHALPY Btu/lb LATENT $h_{fg}$ | ENTHALPY Btu/lb VAPOR $h_g$ | ENTROPY Btu/(lb)(°R) LIQUID $s_f$ | ENTROPY Btu/(lb)(°R) VAPOR $s_g$ | TEMP. °F |
|---|---|---|---|---|---|---|---|---|---|---|---|---|
| −40 | 9.3076 | 10.9709* | 0.010564 | 3.8750 | 94.661 | 0.25806 | 0 | 72.913 | 72.913 | 0 | 0.17373 | −40 |
| −39 | 9.5530 | 10.4712* | 0.010575 | 3.7823 | 94.565 | 0.26439 | 0.2107 | 72.812 | 73.023 | 0.000500 | 0.17357 | −39 |
| −38 | 9.8035 | 9.9611* | 0.010586 | 3.6922 | 94.469 | 0.27084 | 0.4215 | 72.712 | 73.134 | 0.001000 | 0.17343 | −38 |
| −37 | 10.059 | 9.441* | 0.010596 | 3.6047 | 94.372 | 0.27741 | 0.6324 | 72.611 | 73.243 | 0.001498 | 0.17328 | −37 |
| −36 | 10.320 | 8.909* | 0.010607 | 3.5198 | 94.275 | 0.28411 | 0.8434 | 72.511 | 73.354 | 0.001995 | 0.17313 | −36 |
| −35 | 10.586 | 8.367* | 0.010618 | 3.4373 | 94.178 | 0.29093 | 1.0546 | 72.409 | 73.464 | 0.002492 | 0.17299 | −35 |
| −34 | 10.858 | 7.814* | 0.010629 | 3.3571 | 94.081 | 0.29788 | 1.2659 | 72.309 | 73.575 | 0.002988 | 0.17285 | −34 |
| −33 | 11.135 | 7.250* | 0.010640 | 3.2792 | 93.983 | 0.30495 | 1.4772 | 72.208 | 73.685 | 0.003482 | 0.17271 | −33 |
| −32 | 11.417 | 6.675* | 0.010651 | 3.2035 | 93.886 | 0.31216 | 1.6887 | 72.106 | 73.795 | 0.003976 | 0.17257 | −32 |
| −31 | 11.706 | 6.088* | 0.010662 | 3.1300 | 93.788 | 0.31949 | 1.9003 | 72.004 | 73.904 | 0.004469 | 0.17243 | −31 |
| −30 | 11.999 | 5.490* | 0.010674 | 3.0585 | 93.690 | 0.32696 | 2.1120 | 71.903 | 74.015 | 0.004961 | 0.17229 | −30 |
| −29 | 12.299 | 4.880* | 0.010685 | 2.9890 | 93.592 | 0.33457 | 2.3239 | 71.801 | 74.125 | 0.005452 | 0.17216 | −29 |
| −28 | 12.604 | 4.259* | 0.010696 | 2.9214 | 93.493 | 0.34231 | 2.5358 | 71.698 | 74.234 | 0.005942 | 0.17203 | −28 |
| −27 | 12.916 | 3.625* | 0.010707 | 2.8556 | 93.395 | 0.35018 | 2.7479 | 71.596 | 74.344 | 0.006431 | 0.17189 | −27 |
| −26 | 13.233 | 2.979* | 0.010719 | 2.7917 | 93.296 | 0.35820 | 2.9601 | 71.494 | 74.454 | 0.006919 | 0.17177 | −26 |
| −25 | 13.556 | 2.320* | 0.010730 | 2.7295 | 93.197 | 0.36636 | 3.1724 | 71.391 | 74.563 | 0.007407 | 0.17164 | −25 |
| −24 | 13.886 | 1.649* | 0.010741 | 2.6691 | 93.098 | 0.37466 | 3.3848 | 71.288 | 74.673 | 0.007894 | 0.17151 | −24 |
| −23 | 14.222 | 0.966* | 0.010753 | 2.6102 | 92.999 | 0.38311 | 3.5973 | 71.185 | 74.782 | 0.008379 | 0.17139 | −23 |
| −22 | 14.564 | 0.270* | 0.010764 | 2.5529 | 92.899 | 0.39171 | 3.8100 | 71.081 | 74.891 | 0.008864 | 0.17126 | −22 |
| −21 | 14.912 | 0.216 | 0.010776 | 2.4972 | 92.799 | 0.40045 | 4.0228 | 70.978 | 75.001 | 0.009348 | 0.17114 | −21 |
| −20 | 15.267 | 0.571 | 0.010788 | 2.4429 | 92.699 | 0.40934 | 4.2357 | 70.874 | 75.110 | 0.009831 | 0.17102 | −20 |
| −19 | 15.628 | 0.932 | 0.010799 | 2.3901 | 92.599 | 0.41839 | 4.4487 | 70.770 | 75.219 | 0.010314 | 0.17090 | −19 |
| −18 | 15.996 | 1.300 | 0.010811 | 2.3387 | 92.499 | 0.42758 | 4.6618 | 70.666 | 75.328 | 0.010795 | 0.17078 | −18 |
| −17 | 16.371 | 1.675 | 0.010823 | 2.2886 | 92.399 | 0.43694 | 4.8751 | 70.561 | 75.436 | 0.011276 | 0.17066 | −17 |
| −16 | 16.753 | 2.057 | 0.010834 | 2.2399 | 92.298 | 0.44645 | 5.0885 | 70.456 | 75.545 | 0.011755 | 0.17055 | −16 |
| −15 | 17.141 | 2.445 | 0.010846 | 2.1924 | 92.197 | 0.45612 | 5.3020 | 70.352 | 75.654 | 0.012234 | 0.17043 | −15 |
| −14 | 17.536 | 2.840 | 0.010858 | 2.1461 | 92.096 | 0.46595 | 5.5157 | 70.246 | 75.762 | 0.012712 | 0.17032 | −14 |
| −13 | 17.939 | 3.243 | 0.010870 | 2.1011 | 91.995 | 0.47595 | 5.7295 | 70.141 | 75.871 | 0.013190 | 0.17021 | −13 |
| −12 | 18.348 | 3.652 | 0.010882 | 2.0572 | 91.893 | 0.48611 | 5.9434 | 70.036 | 75.979 | 0.013666 | 0.17010 | −12 |
| −11 | 18.765 | 4.069 | 0.010894 | 2.0144 | 91.791 | 0.49643 | 6.1574 | 69.930 | 76.087 | 0.014142 | 0.16999 | −11 |
| −10 | 19.189 | 4.493 | 0.010906 | 1.9727 | 91.689 | 0.50693 | 6.3716 | 69.824 | 76.196 | 0.014617 | 0.16989 | −10 |
| − 9 | 19.621 | 4.925 | 0.010919 | 1.9320 | 91.587 | 0.51759 | 6.5859 | 69.718 | 76.304 | 0.015091 | 0.16978 | − 9 |
| − 8 | 20.059 | 5.363 | 0.010931 | 1.8924 | 91.485 | 0.52843 | 6.8003 | 69.611 | 76.411 | 0.015564 | 0.16967 | − 8 |
| − 7 | 20.506 | 5.810 | 0.010943 | 1.8538 | 91.382 | 0.53944 | 7.0149 | 69.505 | 76.520 | 0.016037 | 0.16957 | − 7 |
| − 6 | 20.960 | 6.264 | 0.010955 | 1.8161 | 91.280 | 0.55063 | 7.2296 | 69.397 | 76.627 | 0.016508 | 0.16947 | − 6 |
| − 5 | 21.422 | 6.726 | 0.010968 | 1.7794 | 91.177 | 0.56199 | 7.4444 | 69.291 | 76.735 | 0.016979 | 0.16937 | − 5 |
| − 4 | 21.891 | 7.195 | 0.010980 | 1.7436 | 91.074 | 0.57354 | 7.6594 | 69.183 | 76.842 | 0.017449 | 0.16927 | − 4 |
| − 3 | 22.369 | 7.673 | 0.010993 | 1.7086 | 90.970 | 0.58526 | 7.8745 | 69.075 | 76.950 | 0.017919 | 0.16917 | − 3 |
| − 2 | 22.854 | 8.158 | 0.011005 | 1.6745 | 90.867 | 0.59718 | 8.0898 | 68.967 | 77.057 | 0.018388 | 0.16907 | − 2 |
| − 1 | 23.348 | 8.652 | 0.011018 | 1.6413 | 90.763 | 0.60927 | 8.3052 | 68.859 | 77.164 | 0.018855 | 0.16897 | − 1 |
| 0 | 23.849 | 9.153 | 0.011030 | 1.6089 | 90.659 | 0.62156 | 8.5207 | 68.750 | 77.271 | 0.019323 | 0.16888 | 0 |
| 1 | 24.359 | 9.663 | 0.011043 | 1.5772 | 90.554 | 0.63404 | 8.7364 | 68.642 | 77.378 | 0.019789 | 0.16878 | 1 |
| 2 | 24.878 | 10.182 | 0.011056 | 1.5463 | 90.450 | 0.64670 | 8.9522 | 68.533 | 77.485 | 0.020255 | 0.16869 | 2 |
| 3 | 25.404 | 10.708 | 0.011069 | 1.5161 | 90.345 | 0.65957 | 9.1682 | 68.424 | 77.592 | 0.020719 | 0.16860 | 3 |
| 4 | 25.939 | 11.243 | 0.011082 | 1.4867 | 90.240 | 0.67263 | 9.3843 | 68.314 | 77.698 | 0.021184 | 0.16851 | 4 |
| 5 | 26.483 | 11.787 | 0.011094 | 1.4580 | 90.135 | 0.68588 | 9.6005 | 68.204 | 77.805 | 0.021647 | 0.16842 | 5 |
| 6 | 27.036 | 12.340 | 0.011107 | 1.4299 | 90.030 | 0.69934 | 9.8169 | 68.094 | 77.911 | 0.022110 | 0.16833 | 6 |
| 7 | 27.597 | 12.901 | 0.011121 | 1.4025 | 89.924 | 0.71300 | 10.033 | 67.984 | 78.017 | 0.022572 | 0.16824 | 7 |
| 8 | 28.167 | 13.471 | 0.011134 | 1.3758 | 89.818 | 0.72687 | 10.250 | 67.873 | 78.123 | 0.023033 | 0.16815 | 8 |
| 9 | 28.747 | 14.051 | 0.011147 | 1.3496 | 89.712 | 0.74094 | 10.467 | 67.762 | 78.229 | 0.023494 | 0.16807 | 9 |
| 10 | 29.335 | 14.639 | 0.011160 | 1.3241 | 89.606 | 0.75523 | 10.684 | 67.651 | 78.335 | 0.023954 | 0.16798 | 10 |
| 11 | 29.932 | 15.236 | 0.011173 | 1.2992 | 89.499 | 0.76972 | 10.901 | 67.539 | 78.440 | 0.024413 | 0.16790 | 11 |
| 12 | 30.539 | 15.843 | 0.011187 | 1.2748 | 89.392 | 0.78443 | 11.118 | 67.428 | 78.546 | 0.024871 | 0.16782 | 12 |
| 13 | 31.155 | 16.459 | 0.011200 | 1.2510 | 89.285 | 0.79935 | 11.336 | 67.315 | 78.651 | 0.025329 | 0.16774 | 13 |
| 14 | 31.780 | 17.084 | 0.011214 | 1.2278 | 89.178 | 0.81449 | 11.554 | 67.203 | 78.757 | 0.025786 | 0.16765 | 14 |
| 15 | 32.415 | 17.719 | 0.011227 | 1.2050 | 89.070 | 0.82986 | 11.771 | 67.090 | 78.861 | 0.026243 | 0.16758 | 15 |

* Inches of mercury below one atmosphere

* Courtesy of Freon Products Div., E.I. DuPont de Numours & Co., Inc.

## Table 9-11
## Condensed Pressure-Temperature Chart

### Vapor Pressure, Psig

| °F. | "Freon-113" | "Freon-114" | "Freon-13" | R-500** |
|---|---|---|---|---|
| -50 | | 27.2* | 57.0 | |
| -48 | | 27.0* | 60.0 | |
| -46 | | 26.8* | 63.0 | |
| -44 | | 26.6* | 66.2 | |
| -42 | | 26.3* | 69.4 | |
| -40 | | 26.1* | 72.7 | 7.9* |
| -38 | | 25.9* | 76.2 | 6.7* |
| -36 | | 25.6* | 79.7 | 5.4* |
| -34 | | 25.3* | 83.3 | 4.2* |
| -32 | | 25.0* | 87.1 | 2.8* |
| -30 | 29.3* | 24.7* | 90.9 | 1.4* |
| -28 | 29.3* | 24.4* | 94.9 | 0.0 |
| -26 | 29.2* | 24.0* | 98.9 | 0.8 |
| -24 | 29.2* | 23.7* | 103.0 | 1.5 |
| -22 | 29.1* | 23.3* | 107.3 | 2.3 |
| -20 | 29.1* | 22.9* | 111.7 | 3.1 |
| -18 | 29.0* | 22.5* | 116.2 | 4.0 |
| -16 | 28.9* | 22.1* | 120.8 | 4.9 |
| -14 | 28.9* | 21.6* | 125.7 | 5.8 |
| -12 | 28.8* | 21.1* | 130.5 | 6.8 |
| -10 | 28.7* | 20.6* | 135.4 | 7.8 |
| -8 | 28.6* | 20.1* | 140.5 | 8.8 |
| -6 | 28.5* | 19.6* | 145.7 | 9.9 |
| -4 | 28.4* | 19.0* | 151.1 | 11.0 |
| -2 | 28.3* | 18.4* | 156.5 | 12.1 |
| 0 | 28.2* | 17.8* | 162.2 | 13.3 |
| 2 | 28.1* | 17.2* | 167.9 | 14.5 |
| 4 | 28.0* | 16.5* | 173.7 | 15.7 |
| 6 | 27.9* | 15.8* | 179.8 | 17.0 |
| 8 | 27.7* | 15.1* | 185.9 | 18.4 |
| 10 | 27.6* | 14.3* | 192.2 | 19.8 |
| 12 | 27.5* | 13.5* | 198.6 | 21.2 |
| 14 | 27.3* | 12.7* | 205.2 | 22.7 |
| 16 | 27.1* | 11.9* | 211.9 | 24.2 |
| 18 | 27.0* | 11.0* | 218.8 | 25.7 |
| 20 | 26.8* | 10.1* | 225.8 | 27.3 |
| 22 | 26.6* | 9.1* | 233.0 | 29.0 |
| 24 | 26.4* | 8.1* | 240.3 | 30.7 |
| 26 | 26.2* | 7.1* | 247.8 | 32.5 |
| 28 | 26.0* | 6.1* | 255.5 | 34.3 |
| 30 | 25.8* | 5.0* | 263.3 | 36.1 |
| 32 | 25.6* | 3.9* | 271.3 | 38.0 |
| 34 | 25.3* | 2.7* | 279.5 | 40.0 |
| 36 | 25.1* | 1.5* | 287.8 | 42.0 |
| 38 | 24.8* | 0.2* | 295.3 | 44.1 |
| 40 | 24.5* | 0.5 | 305.0 | 46.2 |
| 42 | 24.2* | 1.2 | 313.9 | 48.4 |
| 44 | 23.9* | 1.9 | 322.9 | 50.7 |
| 46 | 23.6* | 2.6 | 332.2 | 53.0 |
| 48 | 23.3* | 3.3 | 341.6 | 55.4 |
| 50 | 22.9* | 4.0 | 351.2 | 57.8 |
| 52 | 22.6* | 4.8 | 361.1 | 60.3 |
| 54 | 22.2* | 5.6 | 371.1 | 62.9 |
| 56 | 21.8* | 6.4 | 381.3 | 65.5 |
| 58 | 21.4* | 7.3 | 391.7 | 68.2 |
| 60 | 21.0* | 8.1 | 402.4 | 71.0 |
| 62 | 20.6* | 9.0 | 413.3 | 73.8 |
| 64 | 20.1* | 9.9 | 424.2 | 76.7 |
| 66 | 19.7* | 10.9 | 435.6 | 79.7 |
| 68 | 19.2* | 11.9 | 447.0 | 82.8 |
| 70 | 18.7* | 12.9 | 458.8 | 85.8 |
| 72 | 18.2* | 13.9 | 470.7 | 89.0 |
| 74 | 17.6* | 15.0 | 482.9 | 92.3 |
| 76 | 17.1* | 16.1 | 495.3 | 95.6 |
| 78 | 16.5* | 17.2 | 508.1 | 99.0 |
| 80 | 15.9* | 18.3 | 521.0 | 102.5 |
| 82 | 15.3* | 19.5 | 534.1 | 106.1 |
| 84 | 14.6* | 20.7 | 547.5 | 109.7 |
| 86 | 13.9* | 22.0 | | 113.4 |
| 88 | 13.2* | 23.3 | | 117.3 |
| 90 | 12.5* | 24.6 | | 121.2 |
| 92 | 11.8* | 25.9 | | 125.1 |
| 94 | 11.0* | 27.3 | | 129.2 |
| 96 | 10.2* | 28.7 | | 133.3 |
| 98 | 9.4* | 30.2 | | 137.6 |
| 100 | 8.6* | 31.7 | | 141.9 |
| 102 | 7.7* | 33.2 | | 146.3 |
| 104 | 6.8* | 34.8 | | 150.9 |
| 106 | 5.9* | 36.4 | | 155.4 |
| 108 | 4.9* | 38.0 | | 160.1 |
| 110 | 4.0* | 39.7 | | 164.9 |
| 112 | 3.0* | 41.4 | | 169.8 |
| 114 | 1.9* | 43.2 | | 174.8 |
| 116 | 0.8* | 45.0 | | 179.9 |
| 118 | 0.1 | 46.9 | | 185.0 |
| 120 | 0.7 | 48.7 | | 190.3 |
| 122 | 1.3 | 50.7 | | 195.7 |
| 124 | 1.9 | 52.7 | | 201.2 |
| 126 | 2.5 | 54.7 | | 206.7 |
| 128 | 3.1 | 56.7 | | 212.4 |
| 130 | 3.7 | 58.8 | | 218.2 |
| 132 | 4.4 | 61.0 | | 224.1 |
| 134 | 5.1 | 63.2 | | 230.1 |
| 136 | 5.8 | 65.5 | | 236.3 |
| 138 | 6.5 | 67.7 | | 242.5 |
| 140 | 7.2 | 70.1 | | 248.8 |
| 142 | 8.0 | 72.5 | | |
| 144 | 8.8 | 74.9 | | |
| 146 | 9.6 | 77.4 | | |
| 148 | 10.4 | 80.0 | | |
| 150 | 11.2 | 82.6 | | |
| 152 | 12.1 | 85.2 | | |
| 154 | 13.0 | 87.9 | | |
| 156 | 13.9 | 90.7 | | |
| 158 | 14.8 | 93.5 | | |
| 160 | 15.7 | 96.4 | | |

*ABOVE CRITICAL TEMPERATURE* (R-500 column, below 140)

*Inches mercury below one atmosphere.
**Patented by Carrier Corporation.

### Vapor Pressure, Psig

| °F. | "Freon-11" | "Freon-12" | "Freon-502" | "Freon-22" |
|---|---|---|---|---|
| -50 | | 15.4* | 0.0 | 6.0* |
| -48 | | 14.6* | 0.8 | 4.7* |
| -46 | | 13.8* | 1.6 | 3.3* |
| -44 | | 12.9* | 2.5 | 1.8* |
| -42 | | 11.9* | 3.4 | 0.3* |
| -40 | 28.4* | 11.0* | 4.3 | 0.6 |
| -38 | 28.3* | 10.0* | 5.2 | 1.4 |
| -36 | 28.2* | 8.9* | 6.2 | 2.3 |
| -34 | 28.1* | 7.8* | 7.2 | 3.2 |
| -32 | 28.0* | 6.7* | 8.3 | 4.1 |
| -30 | 27.8* | 5.5* | 9.4 | 5.0 |
| -28 | 27.7* | 4.3* | 10.5 | 6.0 |
| -26 | 27.5* | 3.0* | 11.7 | 7.0 |
| -24 | 27.4* | 1.6* | 12.9 | 8.1 |
| -22 | 27.2* | 0.3* | 14.2 | 9.2 |
| -20 | 27.0* | 0.6 | 15.5 | 10.3 |
| -18 | 26.9* | 1.3 | 16.9 | 11.5 |
| -16 | 26.7* | 2.1 | 18.3 | 12.7 |
| -14 | 26.5* | 2.8 | 19.7 | 13.9 |
| -12 | 26.2* | 3.7 | 20.2 | 15.2 |
| -10 | 26.0* | 4.5 | 22.8 | 16.6 |
| -8 | 25.8* | 5.4 | 24.4 | 18.0 |
| -6 | 25.5* | 6.3 | 26.0 | 19.4 |
| -4 | 25.3* | 7.2 | 27.7 | 20.9 |
| -2 | 25.0* | 8.2 | 29.4 | 22.5 |
| 0 | 24.7* | 9.2 | 31.2 | 24.1 |
| 2 | 24.4* | 10.2 | 33.1 | 25.7 |
| 4 | 24.1* | 11.2 | 35.0 | 27.4 |
| 6 | 23.8* | 12.3 | 37.0 | 29.2 |
| 8 | 23.5* | 13.5 | 39.0 | 31.0 |
| 10 | 23.1* | 14.6 | 41.1 | 32.9 |
| 12 | 22.7* | 15.8 | 43.2 | 34.9 |
| 14 | 22.3* | 17.1 | 45.4 | 36.9 |
| 16 | 21.9* | 18.4 | 47.7 | 39.0 |
| 18 | 21.5* | 19.7 | 50.1 | 41.1 |
| 20 | 21.1* | 21.0 | 52.4 | 43.3 |
| 22 | 20.6* | 22.4 | 54.9 | 45.5 |
| 24 | 20.2* | 23.9 | 57.4 | 47.9 |
| 26 | 19.7* | 25.4 | 60.0 | 50.2 |
| 28 | 19.1* | 26.9 | 62.7 | 52.7 |
| 30 | 18.6* | 28.5 | 65.4 | 55.2 |
| 32 | 18.1* | 30.1 | 68.2 | 57.8 |
| 34 | 17.5* | 31.7 | 71.1 | 60.5 |
| 36 | 16.9* | 33.4 | 74.1 | 63.3 |
| 38 | 16.3* | 35.2 | 77.1 | 66.1 |
| 40 | 15.6* | 37.0 | 80.2 | 69.0 |
| 42 | 14.9* | 38.8 | 83.4 | 72.0 |
| 44 | 14.2* | 40.7 | 86.6 | 75.0 |
| 46 | 13.5* | 42.7 | 90.0 | 78.2 |
| 48 | 12.8* | 44.7 | 93.4 | 81.4 |
| 50 | 12.0* | 46.7 | 96.9 | 84.7 |
| 52 | 11.2* | 48.8 | 100.5 | 88.1 |
| 54 | 10.4* | 51.0 | 104.1 | 91.5 |
| 56 | 9.5* | 53.2 | 107.9 | 95.1 |
| 58 | 8.7* | 55.4 | 111.7 | 98.8 |
| 60 | 7.7* | 57.7 | 115.6 | 102.5 |
| 62 | 6.8* | 60.1 | 119.6 | 106.3 |
| 64 | 5.8* | 62.5 | 123.7 | 110.2 |
| 66 | 4.8* | 65.0 | 127.9 | 114.2 |
| 68 | 3.7* | 67.6 | 132.2 | 118.3 |
| 70 | 2.6* | 70.2 | 136.6 | 122.5 |
| 72 | 1.5* | 72.9 | 141.1 | 126.8 |
| 74 | 0.4* | 75.6 | 145.6 | 131.2 |
| 76 | 0.4 | 78.4 | 150.3 | 135.7 |
| 78 | 1.0 | 81.3 | 155.1 | 140.3 |
| 80 | 1.6 | 84.2 | 159.9 | 145.0 |
| 82 | 2.2 | 87.2 | 164.9 | 149.8 |
| 84 | 2.9 | 90.2 | 170.0 | 154.7 |
| 86 | 3.6 | 93.3 | 175.1 | 159.8 |
| 88 | 4.3 | 96.5 | 180.4 | 164.9 |
| 90 | 5.0 | 99.8 | 185.8 | 170.1 |
| 92 | 5.7 | 103.1 | 191.3 | 175.4 |
| 94 | 6.5 | 106.5 | 196.9 | 180.9 |
| 96 | 7.3 | 110.0 | 202.6 | 186.5 |
| 98 | 8.1 | 113.5 | 208.4 | 192.1 |
| 100 | 8.9 | 117.2 | 214.4 | 197.9 |
| 102 | 9.8 | 120.9 | 220.4 | 203.8 |
| 104 | 10.6 | 124.6 | 226.6 | 209.9 |
| 106 | 11.5 | 128.5 | 232.9 | 216.0 |
| 108 | 12.5 | 132.4 | 239.3 | 222.3 |
| 110 | 13.4 | 136.4 | 245.8 | 228.7 |
| 112 | 14.4 | 140.5 | 252.5 | 235.2 |
| 114 | 15.3 | 144.7 | 259.2 | 241.9 |
| 116 | 16.4 | 148.9 | 266.1 | 248.7 |
| 118 | 17.4 | 153.2 | 273.1 | 255.6 |
| 120 | 18.5 | 157.7 | 280.3 | 262.6 |
| 122 | 19.6 | 162.2 | 287.6 | 269.7 |
| 124 | 20.7 | 166.7 | 295.0 | 277.0 |
| 126 | 21.9 | 171.4 | 302.5 | 284.4 |
| 128 | 23.0 | 176.2 | 310.2 | 291.8 |
| 130 | 24.3 | 181.0 | 318.0 | 299.3 |
| 132 | 25.5 | 185.9 | 326.0 | 307.1 |
| 134 | 26.8 | 191.0 | 334.1 | 315.2 |
| 136 | 28.1 | 196.1 | 342.3 | 323.6 |
| 138 | 29.4 | 201.3 | 350.7 | 332.3 |
| 140 | 30.8 | 206.6 | 359.2 | 341.3 |
| 142 | 32.2 | 212.0 | 367.8 | 350.3 |
| 144 | 33.7 | 217.5 | 376.7 | 359.4 |
| 146 | 35.1 | 223.1 | 385.6 | 368.6 |
| 148 | 36.6 | 228.8 | 394.7 | 377.9 |
| 150 | 38.2 | 234.6 | 404.0 | 387.2 |
| 152 | 39.7 | 240.5 | 413.4 | 396.6 |
| 154 | 41.3 | 246.5 | 423.0 | 406.1 |
| 156 | 43.0 | 252.6 | 432.7 | 415.6 |
| 158 | 44.6 | 258.8 | 442.6 | 425.1 |
| 160 | 46.3 | 265.1 | 452.6 | 434.6 |

*Inches mercury below one atmosphere.

* Courtesy of Freon Products Div., E.I. DuPont de Numours & Co., Inc.

215

**Figure 9-2.** Hydrostatic pressure with R–12 (*Courtesy of Copeland Corp.*)

**Figure 9-3.** Ruptured refrigerant cylinder (*Courtesy of Copeland Corp.*)

The extremely dangerous pressures which can result under such circumstances can cause the rupture of the refrigerant cylinder (see Figure 9-3). Under uniform heating conditions the cylinder in Figure 9-3 ruptured at approximately 1,300 psi (91.39 kg/cm²). If, however, heat is applied with a welding torch, the area of the cylinder wall where the heat is applied may be weakened and the danger of rupture increased. In a controlled test, a cylinder such as the one shown in Figure 9-3 flew over 40 ft (12.16 m) in the air upon rupture—a dramatic demonstration showing the danger of overheating refrigerant cylinders.

The Interstate Commerce Commission's regulations prescribe that a liquified gas container shall not be full of liquid when heated to 131°F (55°C). If cylinders are filled in compliance with this regulation, liquid refrigerant may completely fill the cylinder because of the expansion of the liquid at temperatures above 131°F. Fusible metal plugs are designed to protect the refrigerant cylinder in case of fire. However, they will not protect the cylinder from a gradual uniform overheating. Fusible plugs begin to soften at 157°F (69.5°C), but the hydrostatic pressure created at this temperature far exceeds the cylinder test pressure.

The following safety rules should always be followed when handling compressed gas cylinders.

1. Never heat a cylinder above 125°F (51.5°C).

2. Never store refrigerant cylinders in the direct sunlight.

3. Never apply flame directly to a cylinder.

4. Never place an electric resistance

heater in direct contact with a refrigerant cylinder.

5. Do not drop, dent, or otherwise abuse cylinders.

6. When refilling small cylinders, never exceed the weight stamped on the refrigerant cylinder.

7. Always keep the valve cap and head cap in place when the cylinder is not in use.

8. Always open all cylinder valves slowly.

9. Secure all cylinders in an upright position to a stationary object with a strap or chain when they are not mounted in a suitable stand.

## LUBRICANTS

The common fluorocarbon refrigerants in use today are miscible with the lubricating oil in amounts depending on viscosity, temperature, suction pressure, etc. In a properly designed refrigeration system, the oil is absorbed in the refrigerant and will travel through the system with the refrigerant. The lubricating oil should be a straight-run refined mineral oil, free from moisture, sediment, acid, soap, or any substance which is not derived from petroleum. The use of such compounds as ethylene glycol, glycerine, or castor oil is not recommended because these substances are all hygroscopic —that is, they all absorb moisture and become gummy. They also may produce a sticky sludge in the compressor crankcase when the system is idle or when the oil becomes heated. The lubricating oil must be thoroughly dehydrated to avoid the freezing of moisture in the refrigerant flow control device and to prevent possible emulsification of the oil and the formation of sludge.

## SAFETY PROCEDURES

Many of the more common refrigerants do not have a disagreeable odor. It might therefore be possible to work in an area where there is a considerable amount of refrigerant vapor. Many refrigerants are heavier than air and will replace the air within the room. This is dangerous because a person must have at least 19.1% oxygen in the air breathed; otherwise, he might become unconscious.

1. Be sure that the room is thoroughly ventilated before repairing a refrigerant leak.

2. Check the type of refrigerant before charging the system with refrigerant.

3. Be sure to wear goggles when charging or purging a system.

4. Be sure to keep liquid refrigerant out of the eyes. If liquid refrigerant should enter the eyes, flush them gently with tap water to prevent freezing, and see a physician immediately.

5. Do not breathe the refrigerant vapor from a system that has a burned compressor.

6. Keep burned refrigerant off the skin to prevent acid burns.

7. Never apply an open flame to a refrigerant cylinder; an explosion and serious personal injury could result.

8. Always protect refrigerants from being contaminated with moisture, air, or other substances.

9. Do not allow oil from a burned compressor to touch the skin, as it is acidic and may burn the skin.

10. If liquid refrigerant comes in contact with the skin, flush the skin with water and treat the affected area for frostbite.

## SUMMARY

- Most of the commonly used refrigerants are in the gaseous state at atmospheric pressures and temperature.

- The words gas and vapor are ordinarily used interchangeably.

- All fluids have both a liquid and a gaseous state.

- A vaporizing refrigerant will absorb heat equal to the amount of energy necessary to change the liquid to a gas.

- Each refrigerant will absorb an amount of heat per pound of refrigerant equal to its latent heat of vaporization.

- The greater the pressure exerted on a refrigerant, the greater the boiling point.

- A liquid can be caused to boil by placing it in a partial vacuum.

- The critical temperature of a gas is the temperature above which the gas cannot be liquified, regardless of the amount of pressure applied.

- The condensing pressure will depend on the temperature at which the vapor will liquify.

- Water-cooled condensers will operate at lower condensing temperatures and pressures than air-cooled condensers.

- The vaporizing pressure of a refrigerant is important because it must evaporate without requiring a low suction pressure.

- The amount of heat in Btu (cal) that is required to change a liquid to a vapor (the change taking place at a constant temperature) is known as the latent heat of vaporization.

- Any refrigerant when evaporating in the evaporator must absorb heat from within the cooled space exactly equal to its latent heat of vaporization.

- The latent heat of vaporization of a liquid will vary with the pressure and the corresponding temperature at which the vaporization occurs.

- The fluorocarbon refrigerants may be tested for leaks with a halide torch, a soap-bubble solution, or an electronic leak detector.

- All fluorocarbon refrigerants are miscible with oil.

- Miscibility is the term used to express the ability of liquid refrigerant to mix with oil.

- In general, naphthenic oils are more soluble in refrigerants than paraffinic oils.

- Because oil must pass through the compressor cylinders to provide lubrication of moving parts, a small amount of oil is always circulating with the refrigerant.

- Excess liquid refrigerant in the compressor crankcase may result in improper lubrication of the compressor because of the foaming and boiling action on start-up.

- Proper piping and system design are required for systems using R–22 and R–502 to ensure proper oil return to the compressor.

- Enthalpy is a term used in thermodynamics to describe the heat content of a substance.

- Only the proper amount of refrigerant should be put in a refrigerant cylinder to prevent explosions resulting from hydrostatic pressure.

- Refrigeration lubricating oils should

be straight-run refined mineral oil, free from moisture, sediment, acid, soap, or any substance which is not derived from petroleum.

## REVIEW EXERCISES

1.  In what state do most refrigerants exist at atmospheric pressure and temperature?

2.  How much heat will a liquid refrigerant absorb per pound in changing from a liquid to a gas?

3.  How will an increase in pressure on a liquid refrigerant affect its boiling point?

4.  What does the term critical temperature of a refrigerant mean?

5.  What happens to the condensing pressure of a refrigerant when the temperature of the gas is reduced?

6.  List the standard conditions which were established by the refrigeration industry to compare refrigerants and refrigeration units.

7.  On what does the condensing pressure depend?

8.  Will a water-cooled condenser or an air-cooled condenser operate with a lower condensing temperature?

9.  How can the approximate condensing temperature and pressure of an air-cooled condenser be determined?

10. Why is the vaporizing pressure of a refrigerant important?

11. What will air in a refrigeration system cause?

12. Define the latent heat of vaporization of a refrigerant.

13. What methods of leak detection are used for testing fluorocarbon refrigerants?

14. What does the term miscibility mean?

15. Why is a small amount of oil always circulating in a refrigeration system?

16. What precautions should be observed on systems using R–22 and R 502?

17. What does the term enthalpy indicate?

18. What precautions should be observed when filling a refrigerant cylinder?

19. Why is it not safe to heat a refrigerant cylinder with a welding torch?

20. Why must moisture be kept out of refrigeration oil?

# Introduction to Electricity

Man has known about electrical phenomena for centuries. The word electric is derived from the Greek word *elektron* which means amber. Amber was the first substance to show identifiable electrical characteristics as far back as 600 B.C.

However, man did not learn to harness this powerful and mysterious force until the nineteenth century. Thomas A. Edison began the new "electric era" when he invented the incandescent lightbulb in 1879. After that time, practical applications for electrical energy multiplied at a rapid pace.

## THE ELECTRON PRINCIPLE

The basis of the electron principle is the fact that all known matter or substance regardless of its form—solid, liquid, or gas—is fundamentally composed of very small particles called *atoms*. These atoms, in turn, are composed of even smaller particles which are called *electrons* and *protons*. Both electrons and protons have an electrical charge. The electrons are negatively charged and the protons are positively charged. Thus, everything on earth, including human beings, is basically electrical in nature.

Further research has revealed a number of other interesting and important facts regarding the construction of an atom. For example, it was found that a positively charged proton is almost 2000 times heavier than a negatively charged electron. Also, each atom has a core or center which is called a nucleus. The nucleus contains the heavier protons along with uncharged particles called *neutrons*. A neutron is made up of one electron and one proton. The electrical charges of these particles have been neutralized. Because of the combined weight of the protons and neutrons, the nucleus is relatively stationary as compared to the electrons which are known to be rotating at high speeds in orbits or shells around the nucleus. Except for the speed, this phenomenon compares with the earth and other planets rotating in orbits around their nucleus, the sun.

This principle is the very foundation upon which electricity is based. In order to more fully understand electricity, it is necessary to learn more about the makeup of matter. It is this makeup that helps us to understand why different substances or materials are used in the construction of electrical components and devices.

## NATURE OF MATTER

Scientists tell us that everything is made up of matter. Matter is generally defined as anything which occupies space or has mass. However, some matter differs from other matter, and observable characteristics will not always allow identification of a substance. Thus, it is necessary to complete a chemical analysis to determine the nature and behavior of some substances in order to identify them. Almost all materials are made up of a combination of various kinds of matter or mixtures of matter.

It must be understood, in the study of electron theory, that some types of elementary matter may exist which are neither mixtures nor compounds and that any further effort to break this matter into parts by chemical decomposition will produce no change in any of the characteristics of the matter. When matter has been divided into its simplest components, these components are known as *elements*.

**ELEMENTS:** Even though matter is basically comprised of atoms, in a larger sense it is composed of one or more elements. An element may be defined as a substance containing only one kind of atom.

There are at the present time 102 known elements. However, only 88 of this number are considered to be natural elements. The remaining 14 are *synthetic elements,* which have been produced by man through atomic

research. As a point of interest, 11 of the 88 natural elements are gases, 2 are liquids, and the remaining 75 are solids at normal temperature and pressure.

Very few people realize the role that elements play in our daily lives. The food we eat, the clothes we wear, and even our bodies are made up basically of elements (see Figure 10-1). A human body is made up basically of 15 elements, 6 of which vary from 1% to 65% of the whole for a total of 99%. The other 9 elements combine to produce the remaining 1%.

Because each of the 102 elements is made up of only one kind of atom, it follows that there are 102 kinds of atoms. It is known, however, that the electrons and protons in one atom are identical to those in all other atoms. The difference between atoms lies in the number of electrons and protons each

**Figure 10-1.** Elements of a human body (*Courtesy of Frigidaire Division, General Motors Corp.*)

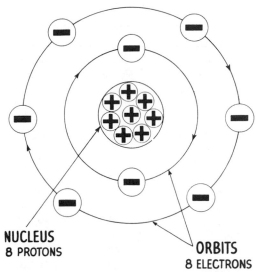

**Figure 10-2.** A hydrogen atom (*Courtesy of Frigidaire Division, General Motors Corp.*)

**Figure 10-3.** An oxygen atom (*Courtesy of Frigidaire Division, General Motors Corp.*)

contains and the arrangement of the electrons in orbit around the nucleus.

The structure of the hydrogen atom is the simplest of all atoms as shown in Figure 10-2. It has one electron rotating in orbit around one proton as the nucleus.

*NOTE:* The nucleus of the hydrogen atom also contains one neutron. In most instances, the number of neutrons of any atom is equal to the number of protons and to the number of electrons in orbit.

The hydrogen atom is relatively simple, but as the number of electrons and protons increases, the structure of the atom becomes more complex. Helium has 2 protons in the nucleus and 2 electrons in orbit around it. Carbon has 6 protons in the nucleus and 6 electrons in orbit. This establishes the pattern for all atoms; the number of protons in the nucleus is always equal to the number of electrons in orbit.

An oxygen atom has 8 protons and 8 electrons (see Figure 10-3). The electrons are divided into two groups with 2 electrons

in an orbit near the nucleus and the other 6 in an orbit at a distance farther from the nucleus. As the number of electrons in the elements increases, the number of orbits and their spacing also increases. For example, the 92 electrons in a uranium atom are divided into 7 orbits differently spaced around the 92 protons in the nucleus. All elements have an atomic number and weight. (See Table 10-1.) The number is based on the number of electrons and protons in the element.

**COMPOUNDS:** Except for the elements, all other known matter may be defined as substances which are made up of two or more elements. In a compound, two or more elements combine chemically to form a distinct substance with the same proportion of elements always present in its composition.

When an atom of one kind of element is combined with an atom of another kind, it forms a larger particle called a *molecule*. A

Table 10-1

Some Commonly Known Elements*

| ELEMENT | ATOMIC NUMBER | ATOMIC WEIGHT |
|---|---|---|
| ** Hydrogen | 1 | 1.008 |
| Carbon | 6 | 12.01 |
| ** Nitrogen | 7 | 14.008 |
| ** Oxygen | 8 | 16.00 |
| Aluminum | 13 | 26.97 |
| Iron | 26 | 55.85 |
| Nickel | 28 | 58.69 |
| Copper | 29 | 63.57 |
| Silver | 47 | 107.88 |
| Tin | 50 | 118.70 |
| Gold | 79 | 197.0 |
| † Mercury | 80 | 200.61 |
| Lead | 82 | 207.21 |
| Uranium | 92 | 238.07 |

* Courtesy Frigidaire Div., General Motors Corp.
** Gas
† Liquid

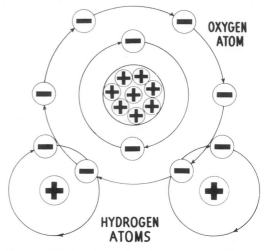

**Figure 10-4.** A molecule of water (Courtesy of Frigidaire Division, General Motors Corp.)

molecule may be defined as the smallest particle of a compound which is of the same chemical composition as the compound itself.

The chemical formula for water, $H_2O$, is known to almost everyone. The formula indicates that the two elements, hydrogen and oxygen, are chemically combined. There are 2 atoms of hydrogen for every atom of oxygen. The smallest particle of water must contain 2 hydrogen atoms and 1 oxygen atom in order to retain the same chemical properties as water itself. (See Figure 10-4.) The atoms are locked together due to the crossing orbits of the rapidly rotating electrons. Of course, it would require millions of these molecules to form a single drop of water. All other compounds are made up in the same general manner.

Many compounds are composed of more than two elements. These compounds would include all of the refrigerants which are commonly used in refrigeration and air conditioning. For example, a molecule of R–12 contains 1 carbon atom, 2 chlorine atoms, and 2 fluorine atoms which are locked together by the rotating electrons. The chemical name for R–12 is dichlorodifluoromethane.

**ELECTRON MOVEMENT:** The relationship between the electrons and protons of an atom is the basis for the well-known facts in electricity that positive charges repel each other, negative charges repel each other, and positive charges attract negative charges. This attraction of the protons in the nucleus for the surrounding electrons is the force that holds the atom together. The strength of this attraction varies widely in different atoms. As the distance from the nucleus to the electron orbit increases, the force of attraction between the nucleus and the electrons decreases proportionately.

Even though the atoms are locked together by the overlapping orbits, some of the electrons are able to move from one atom to another because of the weakness of attraction to their nucleus. These electrons are called *free electrons*. The remaining electrons are close to the nucleus and the attraction is too strong for them to leave their orbit. These electrons are called *planetary* or *bound electrons*. According to the electron principle, electricity is the movement of free electrons from one atom and molecule to another. The forces which cause these electrons to move will be explained later. Because of its free electrons (see Figure 10-5) and other desirable characteristics, carbon is used as a conductor for electricity on many applications.

Any substance with a great number of free electrons would make a good conductor of electricity. Such substances as platinum, gold, silver, and tin each have many free electrons and make excellent conductors. However, their cost and other characteristics make them impractical for commercial use. In spite of its cost, however, silver is used extensively for electrical contacts in switches and contactors to ensure good electrical performance and give long, trouble-free service. Other substances such as copper, brass, aluminum, iron, tungsten, and nichrome are widely used to either conduct electricity from one place to another or in the manufacturing of the multitude of electrical devices in use today.

On the other hand, there are many substances in which the force of attraction between the protons and the electrons is so strong that there are very few free electrons. These substances are good insulators because they have low conductance. A few examples of good insulators are glass, slate, wood, mica, and plastic. Insulation is used on wires to protect accidental short circuits and to avoid electrical contact between the conductors and any supporting devices. It is apparent that some materials resist the flow of electrons, while others offer less resistance to the flow of electrons. Because of this, resistance must always be considered as an important characteristic of an electrical circuit.

## SOURCES OF ELECTRICITY

The source of all energy is the sun and electricity is a form of energy. This means that we are continually being bombarded with untold billions of electrons and protons from the sun. This energy is absorbed and stored by the earth and everything on it. Except for the splitting of atoms, energy can neither be created nor destroyed. Therefore, it can be said that electricity cannot be made by machines or devices. Although we say that batteries and generators produce

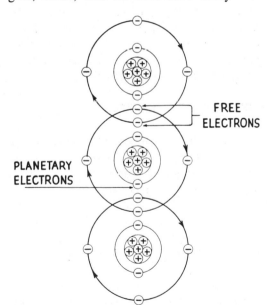

**Figure 10-5.** Combining of carbon atoms (*Courtesy of Frigidaire Division, General Motors Corp.*)

electricity, they actually are devices that make it possible for the ever-present free electrons in a substance to be moved from one place to another. There are six sources of electricity: (1) friction, (2) chemical action, (3) magnetism, (4) heat, (5) pressure, and (6) light (see Figure 10-6). Each of these sources of electricity has significance in modern everyday life.

**PRESSURE**                    **LIGHT**

**FRICTION**                    **CHEMICAL ACTION**

**Figure 10-6.** Sources of electricity (*Courtesy of Frigidaire Division, General Motors Corp.*)

**FRICTION:** Electricity which is produced by friction is commonly called *static electricity* and is produced by rubbing different types of substances together. Trucks hauling gasoline drag a chain along the ground to prevent static electricity caused by friction from igniting the gasoline vapors and producing an explosion.

**CHEMICAL ACTION:** Flashlight and automobile batteries are possible only because electricity can be produced by chemical action.

**MAGNETISM:** This is our greatest and most important source of electricity. Without this source, there would be no generators, motors, or many other devices in common use today.

**HEAT:** Although the electricity which is produced by heat may not be as significant to us as that produced by magnetism, it does make possible accurate temperature-testing devices and safety devices on modern gas furnaces.

**PRESSURE AND LIGHT:** These sources of electricity contribute to our pleasure and convenience. For example, electricity produced by pressure makes possible the recording and playing of phonograph records for homes and business establishments. Photoelectric cells and exposure meters used by photographers are possible because of electricity which is produced by light.

## ELECTRICITY BY FRICTION

As previously mentioned, electricity which is produced by friction is commonly called static electricity and is produced by rubbing certain types of materials together.

The types of materials that are used in producing static electricity are called *dielectrics*. Dielectrics are actually insulators, such as glass, rubber, paper, and plastics through which an electric current will not normally pass. They do, however, contain some free electrons which can be transferred from one substance to another to create a shortage of electrons in one and an overabundance in the other.

Due to the fact that static electricity cannot be harnessed to produce a steady flow of current, it has little practical application for the direct operation of refrigeration and air conditioning equipment. However, it is believed that a knowledge of the subject is required in order to understand the basic construction of a capacitor or condenser. A capacitor is an electrical device that is essential to the operation of numerous products such as radios, televisions, and automobiles.

Most people are familiar with many instances of static electricity. The sparks which result when a doorknob is touched after walking across a rug are static electricity developed by friction between shoes and the rug. Lightning is static electricity produced by the friction between the air and water particles. A cloud picks up electrons and then discharges them to another cloud or to the earth.

The ancient Greek philosopher Thales, who lived between 640 and 546 B.C., is said to have been the first person to have observed that amber, when rubbed, attracted straws, dried leaves, and other light objects.

There were many ancient theories as to how amber attracted objects, but none of them proved correct according to the present day knowledge of atoms. However, they made their contributions to the subject of electricity. Not much was known about electricity until Sir William Gilbert, who was a physician to Queen Elizabeth, wrote a series of books in 1600 A.D. which summarized all that was then known about the subject and added much from his own experimentation. It was he who adopted the word electric from the Greek word for amber.

In 1660, Otto Van Guericke built the first static electric generator (see Figure 10-7). He proved that electricity could be generated by a machine. His generator was a globe of sulphur which was mounted on

**Figure 10-7.** First static electric generator (*Courtesy of Frigidaire Division, General Motors Corp.*)

an axle. By rotating the globe and rubbing the dry hand on the surface, the globe became sufficiently charged to attract paper, feathers, and other light objects. Van Guericke was the first to discover that bodies were repelled as well as attracted and also that an electric charge could travel to the end of a linen thread.

For approximately the next 90 years contributions were made and theories advanced pertaining to electricity with no great significance, until Benjamin Franklin gave us the one-fluid theory of electricity. He believed that electricity existed in two states which he called positive and negative. In 1752 he performed his famous kite experiment which proved that lightning and electricity were the same thing.

To understand lightning and other manifestations of static electricity, we must accept the fact that electrons are negatively charged and protons are positively charged.

**UNCHARGED AND CHARGED ATOMS:** In an uncharged state each atom has an equal number of electrons and protons (see Figure 10-8). If one, or any number, of free electrons are removed by friction,

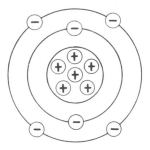

**Figure 10-8.** Uncharged atom (*Courtesy of Frigidaire Division, General Motors Corp.*)

**Figure 10-10.** Electrical charges of fur and hard rubber rod (*Courtesy of Frigidaire Division, General Motors Corp.*)

the substance becomes positively charged (see Figure 10-9). The question now arises: What happens to the free electrons that we removed? When we remove an electron from a substance it will cling to anything that is handy, thus causing that substance to become negatively charged. When this negatively charged object comes in contact with another which has fewer electrons than protons, the electro⌐ transfer from the negatively charged obj ⸝t to the positively charged object in an attempt to produce a balance or equilibrium. This electron transfer is usually accompanied by a spark or a cracking sound.

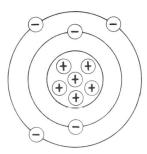

**Figure 10-9.** Charged atom (*Courtesy of Frigidaire Division, General Motors Corp.*)

**PROPERTIES:** A statically charged insulator or dielectric has a field of force and the laws of attraction and repulsion apply.

Figure 10-10 illustrates a piece of fur and a hard rubber rod, both of which are neutral or uncharged. This means that the electrons and protons in each substance are equal or in equilibrium. Figure 10-11 illustrates what happens when the rod is rubbed with the fur. The friction causes some of the free electrons in the fur to cling to the rod, leaving the fur with a shortage of electrons and the rod with an oversupply. The rod is negatively charged and the fur is positively charged.

**Figure 10-11.** Electrical charges of fur and hard rubber rod after rubbing (*Courtesy of Frigidaire Division, General Motors Corp.*)

This same thing happens to silk and a glass rod, except that rubbing the glass rod with silk causes the silk to become negatively charged, leaving the glass rod positively charged. If the positively charged glass rod is suspended by a string while the end

**Figure 10-12.** Unlike charged rods

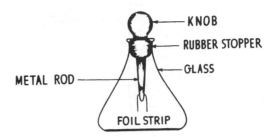

**Figure 10-14.** Metal foil electroscope (*Courtesy of Frigidaire Division, General Motors Corp.*)

of the negatively charged rubber rod is held near the free end of the glass rod, the two will be attracted, which proves that unlike charges attract. (See Figure 10-12.) Next, if a negatively charged rubber rod is suspended and another negatively charged rubber rod is held near it, the suspended rod will move away, proving that like charges repel (see Figure 10-13).

**Figure 10-13.** Like charged rods

Another characteristic of static electricity is that it has the ability to induce a charge of electricity without contact. First, a substance can be charged temporarily by induction. A device which can be used to show a temporary charge is a metal foil electroscope. (See Figure 10-14.)

When a negatively charged rubber rod is held near the knob, the free electrons, in an attempt to get away from the rod, travel to the foil strips. (See Figure 10-15.) The metal strips are shown in a repelled position. Since like charges repel, the foil strips

**Figure 10-15.** Temporarily charged metal foil electroscope (*Courtesy of Frigidaire Division, General Motors Corp.*)

move away from each other. Removing the rod permits the repelled electrons to return to their original position. Since the strips are repelled only while the rod is held close to the knob, it is temporary induction. The same results can be obtained if the rod is positively charged.

A substance also can be charged permanently with either a positive or a negative charge. Again using a metal foil electroscope, a permanent positive charge can be induced. (See Figure 10-16.) A negatively charged rod is held near the knob. While the strips are in a repelled position, the knob is touched with the finger and the free electrons on the foil strips are attracted through the hand to the ground. This leaves the foil strips short of electrons. When the rod and finger are removed, the strips are left with a shortage of electrons and they

**Figure 10-16.** Permanent charging of a metal foil electroscope (*Courtesy of Frigidaire Division, General Motors Corp.*)

**Figure 10-18.** Pith ball electroscope (*Courtesy of Frigidaire Division, General Motors Corp.*)

repel each other (see Figure 10-17). The electroscope is now permanently charged positively. The electroscope can be neutralized by holding a negatively charged rod near the knob.

**Figure 10-17.** Permanently charged electroscope (*Courtesy of Frigidaire Division, General Motors Corp.*)

**Figure 10-19.** Negatively charged rod near a pith ball electroscope (*Courtesy of Frigidaire Division, General Motors Corp.*)

A substance also can be charged by contact or conduction with either a positive or negative charge. (See Figure 10-18.) A neutral or uncharged pith will hang straight down from the stand. If the pith ball is touched with a negatively charged rod a reaction will occur (see Figure 10-19). As the rod approaches the ball, just prior to contact there is an attraction of the ball to the rod. Upon contact, the ball takes a negative

charge from the rod and is immediately repelled, again showing that like charges repel. This same thing would happen if a glass rod were rubbed with silk and then touched to another suspended pith ball, except that the ball would be positively charged.

The question might arise as to what happens when two pith balls with unlike charges are placed near each other? The answer is that they are attracted to each other, but as soon as they touch they both become neutral and separate. (See Figure 10-20.)

During this discussion of static electricity,

**Figure 10-20.** Pith ball electroscope with unlike charges (*Courtesy of Frigidaire Division, General Motors Corp.*)

it has been demonstrated that a substance can be charged either positively or negatively, temporarily or permanently, by either induction or conduction. It has also been demonstrated that positive charges repel positive charges, negative repels negative, and that unlike charges attract each other.

## ELECTRICITY BY CHEMICAL ACTION

Although electricity produced by chemical action has no association with the operation of refrigeration and air conditioning systems, it does play an important role in the operation of several types of instruments which are required for properly diagnosing service complaints. Also, due to the type of current produced by chemical action, and its relative simplicity, other phases of the subject of electricity can be more easily explained.

Any device which produces electricity by chemical action is commonly called a battery. To understand how such a device produces electricity, a knowledge of its ingredients and chemical behavior is required.

**ELECTROLYTES:** Nature stores up much energy in many chemical compounds. Due to the fact that the amount of energy stored is not the same in the various compounds,

some are more desirable than others for use in batteries.

How the stored energy of compounds can be put to a practical use is of considerable interest and is the basis of all battery operation. When certain acids or salts are dissolved in water, their molecules are broken up into positively and negatively charged particles which are called *ions*. An ion may be defined as an electrically charged atom or group of atoms. Any solution containing these ions is capable of conducting an electrical current and is called an *electrolyte*. Even a solution of common table salt (sodium chloride) and water may be considered an electrolyte: the sodium atoms become positively charged ions and the chloride atoms become negatively charged ions. Other solutions which contain water and acids such as hydrochloric and sulphuric are capable of conducting more current than those containing sodium chloride. They also have other more desirable characteristics for use in batteries.

Regardless of the type of solution, ions are capable of moving to every part of the solution.

**ELECTRODES:** In order to take advantage of the current-carrying properties of electrolytes, a battery must have electrodes to which lights or other loads can be connected by wires or other conductors.

The chemical behavior of the electrolyte and the electrodes can best be explained by use of a simple *voltaic cell*. The term voltaic is applied to the cells of batteries in honor of Alessandro Volta, an Italian professor of physics who in 1796 produced the first storage battery.

Dissimilar metals can be used as electrodes; copper and zinc are excellent for the purpose due to the fact that zinc has an abundance of negatively charged atoms, while copper contains an abundance of positively charged atoms. When plates or rods of these metals are submerged into an electrolyte, chemical action between the two metals begins. The zinc electrode accumulates a much larger negative charge due to the fact that it gradually dissolves into the electrolyte solution. The atoms, which leave the zinc during the process, are positively charged. These positively charged zinc atoms are attracted by the negatively charged ions of the electrolyte. On the other hand, they repel the positively charged ions of the electrolyte which are then forced toward the copper electrode. This causes the negative electrons to be removed from the copper, leaving it with a large excess of positive charge. If a load such as a lightbulb is connected across the terminals on the electrodes, the forces of attraction and repulsion will cause the free electrons in the negative zinc electrode, connecting wires, and lightbulb filament to move toward the positively charged copper electrode. Because of this, a flow of electric current has been established.

The life of a cell is limited because the zinc electrode dissolves. When an insufficient amount of zinc remains to provide the required chemical action, the cell is no longer capable of producing electricity and the cell is dead.

Contrary to its name, a storage battery does not store electricity; it merely converts chemical energy into electrical energy in accordance with the limitations of its ingredients.

The cell just described is known as a wet cell (see Figure 10-21) due to the aqueous solution used as the electrolyte. The instruments used for testing electrical components during diagnosis procedures are equipped with one or more dry cells.

A dry cell actually is not dry because the electrolyte is a relatively moist paste. (See

**Figure 10-21.** Wet type voltaic cell (*Courtesy of Frigidaire Division, General Motors Corp.*)

**Figure 10-22.** Dry type voltaic cell (*Courtesy of Frigidaire Division, General Motors Corp.*)

Figure 10-22.) From this, we can see that the outer shell which is made of zinc also serves as the negative electrode. The positive electrode is made of carbon. Although the electrolyte is a paste and is made of different compounds than those used in wet cells, and although carbon is used instead of copper, the chemical action within the cell is similar to the wet cell and produces the same results. Both the wet and dry type voltaic cells are called *primary cells* since they use up one of their primary ingredients.

**SECONDARY CELL:** A cell which loses its chemical energy, but which can be restored by passing an electric current through it is called a secondary cell. This is the type used in a storage battery. Most storage batteries are made up of two or more cells.

In a storage battery, the cell has electrodes or plates, one made of lead-oxide, the other of lead. Both are submerged in a solution of sulphuric acid and water. (See Figure 10-23.) The plates are kept apart by separators which are made of wood, rubber, or glass.

As in the dry type voltaic cell, the ions have picked up the electrons from the lead oxide plate, leaving it positively charged while the others give up their electrons to the lead plate, making it negatively charged. During use, the cell slowly discharges and the lead and the sulphate from the acid combine to form lead-sulphate, which is deposited on both plates (see Figure 10-24). As the deposit of lead-sulphate builds up on both plates, they eventually become lead-sulphate and the electrolyte becomes a weak solution of sulphuric acid and water. The chemical energy has ceased.

By removing the light and connecting the wires to an outside source of current, the cell can be recharged by passing current through the cell in the opposite direction to that of the discharge. The plates again become lead and lead-oxide. The solution then becomes strong with sulphuric acid. With

**Figure 10-23.** Storage battery cell (Courtesy of Frigidaire Division, General Motors Corp.)

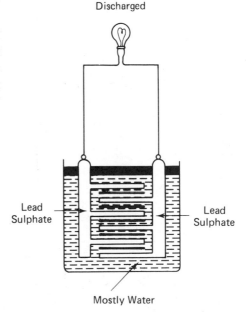

**Figure 10-24.** Discharged storage battery cell (Courtesy of Frigidaire Division, General Motors Corp.)

proper care a secondary cell can be charged and discharged many times.

**REGULAR DIRECT CURRENT:** When a load, such as a lightbulb, is connected to a battery, the flow of current is always in the same direction; from the negative to the positive terminals. Because the current is flowing in only one direction and is flowing at an unfluctuating and steady rate, it is called regular dc current. The flow of regular dc current can be compared with the flow of water produced by a centrifugal pump (see Figure 10-25).

The factors that affect the movement of water in a piped system also can be applied to the movement of electrons in an electrical circuit: pressure, quantity, rate of flow, and resistance.

**Pressure** is the force which moves water through pipes and is the force that moves electrons through conductors. Just as the

**Figure 10-25.** Regular dc current *(Courtesy of Frigidaire Division, General Motors Corp.)*

pump produces the water pressure, the battery produces the electrical pressure.

**Quantity** is the amount of water moved by the pump and is the number of electrons moved by the battery.

**Rate of flow** is the speed or strength of flow of water through the pipes and electrons through the conductors.

**Resistance** is the opposition which is produced by the pipes and fittings to the flow of water and is also the opposition produced by the conductors to the flow of electrons.

Just as the pump must produce a sufficient amount of pressure to overcome the resistance in order to move a quantity of water through a pipe, so must the battery produce a sufficient amount of pressure to move a quantity of electrons through a conductor.

Each of these four factors as they pertain to electricity requires further explanation.

### PRESSURE (ELECTROMOTIVE FORCE)

It has been established that in a battery chemical action produces negatively and positively charged particles which react with the negatively charged zinc electrode and the positively charged copper electrode. When a load is connected across the terminals, the imbalance in charges produces a pressure called an *electromotive force* (emf) which causes the free electrons *(current)* to flow through the circuit. Actually, this flow of current is many billions of free electrons repelling each other throughout the conductor. As stated before, the direction of force is always from the negative electrode to the positive electrode.

It should be remembered that any particular electron does not move from one end of a conductor to the other. This electron

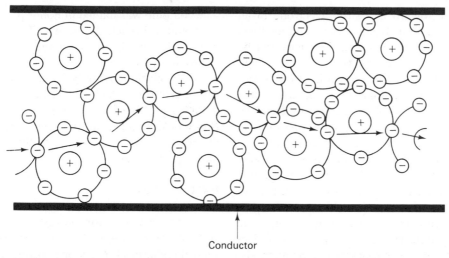

Conductor

**Figure 10-26.** Electron flow (*Courtesy of Frigidaire Division, General Motors Corp.*)

movement may be compared to a switch engine bumping a standing string of boxcars on a track. When the engine bumps one end of the string, each car progressively bumps the next car. (See Figure 10-26.)

If a free electron from a source of energy is forced into an atom at one end, it tends to upset the balance between the electrons and protons of the atom. This action forces another free electron from that atom to shift to an adjacent atom, thus upsetting its balance. This shifting or drifting of free electrons toward the source of energy at the other end of the conductor is called *dynamic electricity* or *current in motion*. All of this is possible because free electrons are repelled out of a substance having a high negative charge (the zinc electrode) and are attracted to another substance having fewer electrons or a positive charge (the copper electrode).

The greater the difference between the concentration of electrons, the greater the electromotive force will be. In everyday language, the term *voltage* is used to designate electromotive force.

**VOLT:** The term voltage is derived from the word volt, which is the unit of measure for electromotive force. This unit of measure was established by Professor Volta while developing his primary cell, and was named in his honor. How he established this unit of measure can be better explained after an explanation of the other factors pertaining to the flow of electrons in a circuit.

**POTENTIAL DIFFERENCE:** When two substances having an equal concentration of free electrons are connected by a conductor, there is no movement of electrons from one substance to the other. The two substances are said to have the same electrical potential. However, if one substance has a higher concentration of free electrons than the other, they are said to have a potential difference between them. Again, the greater the difference between the concentration of free electrons in the electrodes, the greater the potential difference will be. According to the Electron Principle, free electrons always move from a point of high potential toward a point of lower potential—from a negative electrode toward a positive electrode in an electrical circuit. Using water as an analogy, the potential difference in electricity can be compared with a tank of water on a high tower. The water in the tank has a higher pressure potential than the water in a bucket placed at the end of a pipe leading from the bottom of the tank (see Figure 10-27).

While the terms electromotive force or voltage and potential difference are closely related and are often used interchangeably, there is a technical difference between them. In brief, emf or voltage is the pressure produced throughout an electrical circuit, while potential difference indicates a ratio of electron potency or concentration between the electrodes of a battery or certain components of other current-producing devices.

**Figure 10-27.   Potential difference**

**COULOMB:** Just as units of measure have been developed to measure the quantity of liquids, so have units been established to measure the quantity of electricity. When dealing with large quantities of water, the term gallon (litre) is applied. We say that a gallon (3.785 litres) is 4 quarts, 8 pints, or 128 ounces. While the electron can be considered a quantity of electricity, it is too small to be used as a standard measure. A larger unit, comprising 6.3 billion billions of electrons, is called a coulomb and is the standard electrical unit of quantity measure.

The coulomb was named in honor of Charles Coulomb, a French physicist (1736–1806), who made investigations into magnetism and electricity.

Although the coulomb is a definite quantity of electrons which can be measured in a laboratory, it is of little concern to us from a practical standpoint. In other words, we do not measure coulombs directly in diagnostic procedures.

**RATE OF FLOW**

Coulombs can no more be used as a measure of the strength of an electric current than the gallon (litre) alone can be used as a measure of the strength of a waterfall. Both must be considered in their relation to time. If it were said that Niagara Falls spills 5 million gallons (18.93 million litres) of water, it would mean very little unless the time required for this amount to spill was also given. If it took a year to spill that much, it wouldn't be much of a waterfall. However, it is known that it takes only an hour and, from this its strength can be determined.

**AMPERE:** The ampere is the unit of measure of current strength or the rate of flow of the electrons. One ampere is equal to one coulomb per second. This means that when one coulomb passes a given point in an electric circuit each second, the rate of flow is one ampere. Coulombs are quantity; amperes are the rate of flow or the strength of the current. An ordinary lightbulb may require ½ ampere while a 36-inch searchlight may require 150 amperes. This shows that the searchlight is about 300 times stronger than the lightbulb.

AMPERE: THE UNIT OF MEASURE
OF CURRENT STRENGTH OR RATE
OF FLOW OF ELECTRONS

1 ampere = 1 coulomb per second.

This unit of measure was named in honor of André Ampere (1775-1836), a French scientist.

## RESISTANCE

The last, but very important factor in the understanding of electricity, is the resistance to the flow of electrons which is encountered in a conductor. Most metals have a low resistance and are considered good conductors because the molecules are loosely hung together and have many free electrons. In other words, the attraction between the electrons and the protons or nucleus is weak and the electrons can be readily pushed out of their orbit. Copper is one of these metals and is widely used as a conductor. Of course, all matter, including copper, presents a certain amount of resistance to the flow of current by attempting to retain its own electrons. In any circuit, the resistance of the conductor must be overcome by the force of the current. If the force is strong or the resistance small, a strong current will flow. On the other hand, if the force is weak or the resistance great, only a small amount of current will flow.

A comparison can be made between the resistance factors pertaining to water and those pertaining to electricity. There are four factors which determine the resistance to the flow of water through a pipe. These four factors are: (1) diameter of the pipe, (2) length of the pipe, (3) kind of pipe, and (4) velocity of flow.

The smaller the pipe, the longer the pipe, and the rougher the interior of the pipe, the more friction there is. Friction is resistance, so the greater the friction, the smaller the flow will be.

**ELECTRICAL CONDUCTORS:** Electrical conductors (wires) are the pipes of an electric circuit and the resistance of these wires is also dependent on four factors. These factors are: (1) diameter of wire, (2) length of wire, (3) kind of wire, and (4) temperature of the wire.

If the wire is long, or small in diameter, the flow of current will be less. Likewise, if the wire is made of iron, the flow of current will be less than it would be if copper were used. The resistance of most conductors increases as the temperature increases. If the resistance of a conductor is increased by any one of the four factors, the current flow is decreased.

From this, it can readily be seen that if one desires to install wires from a source of power to a load, such as a large condensing unit, the wires must be large enough to carry the load and should be of a low resistance material such as copper.

There is one exception to the temperature-resistance factor that is of particular interest to us from a practical standpoint. When certain substances, such as oxides of manganese, nickel, and cobalt are mixed and formed into a solid, it forms a semiconductor called a *varistor* or *thermistor*.

Thermistors are conductors whose resistance to the flow of electricity decreases as their temperature increases. In other words, at a low temperature their resistance is so high that very little current can flow through them. As the temperature of the metal increases, either by the current passing through it or by that applied externally, more current proportionately flows through it. Thermistors are being used in the control circuits of refrigeration and air conditioning equipment to provide more accurate control of the equipment.

**OHM:** Just as the volt is a unit of measure for electromotive force and the ampere is a unit of measure for the rate of flow of electrical current, the ohm is the unit of measure of resistance of a conductor to the flow of electrons. The unit of resistance was developed by George S. Ohm, a German scientist. To establish the unit of measure, he used a column of mercury 106.3 cm in height and one square millimeter in cross-section. One ohm is the resistance offered by this column of mercury to the flow of one ampere of current having an electromotive force of one volt. The temperature of the mercury column was 32°F (0°C). This column of mercury is comparable to 1000 ft (304 m) of number 10 wire which is $\frac{1}{10}$ in. (2.54 mm) in diameter or 2.4 ft (.73 m) of number 36 wire which is .005 in. (.127 mm) in diameter.

## OHM: THE UNIT OF MEASUREMENT FOR THE RESISTANCE OF A CONDUCTOR TO THE FLOW OF ELECTRONS

We can now define the relationship of volts, amperes, and ohms. A volt may be defined as the electromotive force required to push one coulomb of electrons in one second of time (one ampere) through a resistance of one ohm. Ohm did more than experiment with resistance. He combined his own discoveries with those of Volta and Ampere and, in 1827, he formulated the all-important Ohm's Law upon which all electrical measurement is based. Ohm's Law states that the current in an electrical circuit is equal to the pressure divided by the resistance.

## OHM'S LAW: THE CURRENT IN AN ELECTRICAL CIRCUIT IS EQUAL TO THE PRESSURE (EMF) DIVIDED BY THE RESISTANCE

or

$$\text{CURRENT} = \frac{\text{Pressure}}{\text{Resistance}}$$

To simplify this further, letters are used to represent the words. $E$ represents the pressure or electromotive force in volts, $I$ represents the intensity of current in amperes, and $R$ represents the resistance in ohms:

$$I = \frac{E}{R}$$

As practical examples, three simple circuits will be used. The proper arrangement of the formula is $I = E \div R$. (See Figure 10-28.) The three-cell storage battery has a voltage of 6 volts; the lightbulb has a resistance of 2 ohms. By simple arithmetic, the intensity of current is found to be 3 amperes. It is obvious that the formula $I = E \div R$ can be used to find the intensity of current only when the voltage and resistance are known. An unknown resistance can be determined if the voltage and amperage are known, however, by dividing the volts by the amperes (see Figure 10-29). The formula now is $R = E \div I$. Still the battery is supplying 6 volts and the intensity of the current is 3 amperes. Again, by simple arithmetic, the resistance is 2 ohms.

**Figure 10-28.** Example of Ohm's Law (*Courtesy of Frigidaire Division, General Motors Corp.*)

$R = E \div I$

$R = 2$ Ohms

3 Amps (I)

6 Volts (E)

**Figure 10-29.** Finding resistance using Ohm's Law (*Courtesy of Frigidaire Division, General Motors Corp.*)

$E = I \times R$

2 Ohms (R)

3 Amps (I)

$E = 6$ Volts

**Figure 10-30.** Finding voltage using Ohm's Law (*Courtesy of Frigidaire Division, General Motors Corp.*)

The formula can be arranged in still another manner. If the voltage is the only unknown factor, it can be readily determined by multiplying the amperes by the ohms. (See Figure 10-30.) The formula now reads $E = I \times R$. Here it is known that the intensity of the current is 3 amperes and the resistance is 2 ohms. By multiplying the two, it is found that the battery is supplying 6 volts of pressure.

To summarize, when using Ohm's Law, any one of the three factors current, pressure, or resistance can be computed if the other two factors have known values. Also, the current in any electrical circuit increases or decreases in proportion to an increase or decrease of either of the other two factors.

## ELECTRICAL CIRCUITS

Because it has been established that electrical circuits are the electron paths of an electrical system, and because they can be quite complex, it is important that several fundamental facts be thoroughly understood. No matter how complex any particular circuit becomes, it is one of three general types:

(1) series, (2) parallel, and (3) series-parallel.

The *series circuit* is a one-path circuit and can be recognized in two ways. It will never have more than one conductor connected to one terminal and there will be only one path from the source to the load and back to the source. (See Figure 10-31.)

In a *parallel circuit,* there are two or more paths between the terminals of the source of current. (See Figure 10-32.)

A *series-parallel circuit,* as the name implies, is a combination of both the series and parallel circuits (see Figure 10-33).

Every electrical load is designed to have a specific resistance and to operate at a certain voltage. The load resistance controls the amount of current at the rated voltage.

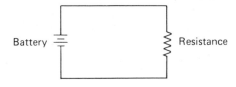

Battery          Resistance

**Figure 10-31.** Series electrical circuit

**Figure 10-32.** Parallel electrical circuit

**Figure 10-33.** Series-parallel electrical circuit

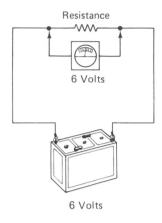

**Figure 10-34.** Voltage drop in a series circuit with total resistance (*Courtesy of Frigidaire Division, General Motors Corp.*)

**VOLTAGE IN SERIES CIRCUITS:** When a voltmeter is connected across the total circuit resistance, it will indicate the total voltage drop across the resistance. Figure 10-34 illustrates a voltage drop of 6 volts. If the voltmeter is connected across half of the resistance, it will indicate ½ the total voltage, as shown in Figure 10-35. In this case the voltage drop is 3 volts, indicating that half of the voltage is used for half of the resistance. If another voltage reading is taken across ⅓ of the resistance, the voltage for that amount of resistance would be 2 volts. This experiment gives us the law of voltage drops in a series circuit. The total voltage is the sum of all voltage drops.

A common example of this law is a string of Christmas tree lights of the series type. The rated voltage of each bulb is approximately 15 volts. When they are connected in a series of eight, the combined voltage totals approximately that of the source, 115–120 volts.

**CURRENT IN A SERIES CIRCUIT:** When two lamps and a switch are connected in series with a battery, as in Figure 10-36,

**Figure 10-35.** Voltage drop in a series circuit with only one-half total resistance (*Courtesy of Frigidaire Division, General Motors Corp.*)

the current through each part of the circuit is the same. For example, if the current strength in one bulb is 1.5 amps, the other will also be drawing 1.5 amps. If the switch is opened, the current through the switch becomes zero. Likewise, the current through each lamp becomes zero. Switches controlling electrical loads are always in series with the load.

**Figure 10-36.** Amperage in a series circuit (*Courtesy of Frigidaire Division, General Motors Corp.*)

**Figure 10-37.** Resistance in a series circuit (*Courtesy of Frigidaire Division, General Motors Corp.*)

A broken bulb is also an open circuit and would stop the current flow through the switch and the other bulbs. This is the chief reason why series connections for lighting circuits are objectionable. If a house had all the lights connected in series, turning out a light in the bedroom would cause all the lights in the house to go out.

**RESISTANCE IN SERIES CIRCUITS:** If two resistances are connected in series in an electrical circuit, the current passes through both and is, therefore, opposed by both. (See Figure 10-37.) Since each resistance is 2 ohms, the total resistance is 4 ohms because the total resistance is equal to the sum of all resistances. That is

$$R_t = R_1 + R_2$$

Again, the resistance of the wiring has been ignored.

**VOLTAGE IN PARALLEL CIRCUITS:** A parallel circuit is a multiple circuit having more than one path between the two terminals of the source of power. Figure 10-38 is a simple parallel circuit which has a conductor from each of the three lights attached to a common terminal on each side of the source of current. Three separate terminals on each side could also be used and the same result would be obtained. Regardless of the resistance offered by each lightbulb, the voltage is the same across all branches.

**CURRENT IN PARALLEL CIRCUITS:** The law for current or amperage in a parallel circuit states that the total amperage is the sum of the amperage in all branches. That is

$$I_t = I_1 + I_2 + I_3$$

**Figure 10-38.** Voltage in a parallel circuit

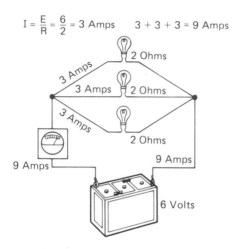

$$I = \frac{E}{R} = \frac{6}{2} = 3 \text{ Amps} \qquad 3 + 3 + 3 = 9 \text{ Amps}$$

3 Amps

3 Amps    2 Ohms

3 Amps    2 Ohms

2 Ohms

9 Amps

9 Amps

6 Volts

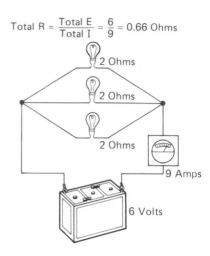

$$\text{Total R} = \frac{\text{Total E}}{\text{Total I}} = \frac{6}{9} = 0.66 \text{ Ohms}$$

2 Ohms

2 Ohms

2 Ohms

9 Amps

6 Volts

**Figure 10-39.** Current in a parallel circuit (*Courtesy of Frigidaire Division, General Motors Corp.*)

**Figure 10-40.** Resistance in parallel circuits (*Courtesy of Frigidaire Division, General Motors Corp.*)

Each of the three lights has a resistance of 2 ohms (see Figure 10-39). Applying the formula $I = E \div R$, we find that each branch has a current flow of 3 amps (6 ÷ 2). The total current then is $3 + 3 + 3 = 9$ amps.

**RESISTANCE IN PARALLEL CIRCUITS:** Because three times as much current can flow through three parallel conductors, it is plain to see that the resistance to the flow is only ⅓ as great. Therefore, it can be said that the total resistance is equal to the total volts divided by the total amps. (See Figure 10-40.) For example, a total voltage of 6 and a total amperage of 9 are given. The total resistance is 6 ÷ 9 = .66 ohms.

**SERIES-PARALLEL CIRCUITS:** Many circuits are neither simple series nor simple parallel, but a combination of both. While this may seem to present a fairly complex problem, it can be readily analyzed by distinguishing one type from the other and ap-

plying the correct laws for each type.

If a parallel-series circuit has two circuits, each with 4 lightbulbs, and the two series circuits are connected in parallel, the laws governing series and parallel circuits must be used. (See Figure 10-41.) The laws governing series and parallel circuits will now be applied and the volts, amps, and the resistance determined.

Reviewing, in a series circuit the total voltage is the sum of all voltage drops. Also, there is an emf of 6 volts in each branch, which conforms with voltages in parallel. The amperage in a series circuit is the same in all parts. The total amperage in a parallel circuit is the sum of all the amperages in each of the branches. Therefore, the rate of flow from the batteries is .3 + .3 = .6 amp. Now that the total voltage and the total amperage have been determined, the total resistance may be determined by applying Ohm's Law.

$$R_t = \frac{E_t}{I_t} = \frac{6}{.6} = 10 \text{ ohms}$$

**Figure 10-42.** Magnetite or Lodestone (*Courtesy of Frigidaire Division, General Motors Corp.*)

**Figure 10-41.** Series-parallel circuit (*Courtesy of Frigidaire Division, General Motors Corp.*)

## ELECTRICITY BY MAGNETISM

As previously stated, the production of electricity by magnetism is very important to us because it is the greatest and most common source of electrical power. By use of magnetism, electricity can be generated, transmitted, and distributed in unlimited quantities to millions of homes, business establishments, and industrial plants.

**MAGNETS:** In order to understand how magnetism can be made to work for us in many practical ways, it is necessary to have a basic knowledge of magnets.

**NATURAL MAGNETS:** The power of a magnet to attract iron was the first electrical phenomenon our ancestors observed. It was discovered that nature provided a natural magnet called magnetite or lodestone (Figure 10-42). This is a species of ore which is common throughout the earth, but especially abundant in the Province of Magnesia in Asia Minor. This type ore has a property which enables it to attract iron or steel. According to legend, a Cretian shepherd discovered while attending his flocks that the iron nails in his sandals and the iron tip of his staff were attracted to the earth. While digging in the ground, he discovered a rock which stuck to his sandals.

The earliest practical use of the magnet was a compass which was used to point out a direction. It was discovered that a suspended bar of magnetite would seek a north and south position. It was for this reason that natural magnets became known as lodestones or leading stones. Magnetic compasses were developed and widely used all over the civilized world as far back as 1492 when Columbus set sail for America.

**MAGNETIC PROPERTIES:** To understand how and why a compass can determine directions, we must know some of the properties of a magnet. The most common magnetic substances are iron, steel, nickel, and cobalt. Some substances are only slightly magnetic, while others, such as bismuth, act in the opposite direction and are repelled by a magnet. These substances are called diamagnetic substances. The most important of the nonmagnetic substances are copper, brass, and aluminum.

**ARTIFICIAL PERMANENT MAGNETS:** A magnetic substance can be artificially converted into a permanent or temporary magnet. (See Figure 10-43.) These are known as the bar type and the horseshoe type. They can be produced by stroking pieces of steel or alloys with either a natural magnet or an

artificial magnet, or by properly applying an electric current. Steel and alloys are universally used due to the fact that, although they cannot be magnetized as easily as soft iron, they retain or hold their magnetism longer after the source of magnetism has been removed.

**ARTIFICIAL TEMPORARY MAGNETS:** A temporary magnet is one which passes magnetic powers only while being subjected to the magnetic force of a permanent magnet. If a piece of unmagnetized soft iron, such as a nail, is suspended from one end of a bar magnet, or bar, as shown in Figure 10-44, a second nail may be suspended from the first one, a third nail from the second, and so forth. Each nail, in turn, becomes a magnet with less force than the preceding one. When the bar magnet is removed from the first nail, however, the others will instantly fall apart. Soft iron has

BAR TYPE

HORSE SHOE TYPE

**Figure 10-43.** Artificial magnets (*Courtesy of Frigidaire Division, General Motors Corp.*)

**Figure 10-44.** Artificial temporary magnet (*Courtesy of Frigidaire Division, General Motors Corp.*)

a high permeability, which means that it can be magnetized easily; it also has low retentivity, which means that it loses its magnetism quickly after the source has been removed.

**MAGNETIC INDUCTION:** The reason an unmagnetized piece of iron is attached to a magnet is due to *magnetic induction* (see Figure 10-45). A nail can be sufficiently magnetized to attract and hold iron filings even though it is not in direct contact with the permanent magnet. This is one of the most significant of all factors in the production and utilization of electricity by magnetism.

**Figure 10-45.** Magnetic induction (Courtesy of Frigidaire Division, General Motors Corp.)

**MAGNETIC POLARITY:** Every magnet has two poles, a north-seeking pole or *N* pole, and a south-seeking pole or *S* pole. This fact can be readily proven by stroking a common sewing needle with a magnet and floating the needle on a piece of straw in a pan of water. One end of the needle will turn toward the north. The end of the magnet which points toward the north is the *N* pole of the magnet. The fact that a magnet has poles can be readily proven by dipping a bar magnet into a quantity of iron filings. The filings will cling in tufts to both ends of the bar, but scarcely any will be seen in the center.

**REPULSION AND ATTRACTION:** Another property of magnets is their ability to repel and attract each other. Like poles repel and unlike poles attract. If the *N* poles of two bar magnets were put together, they would jump apart, but if an *N* pole of one magnet is placed to the *S* pole of the other magnet, they will be held together by magnetic attraction. This can be proven by holding one end of a bar magnet close to the end of a suspended bar magnet. If the two ends are of like polarity, the end of the suspended bar will move away from the end of the bar being held. If the polarity is different, the end of the suspended bar magnet will move toward the other magnet.

**THE EARTH AS A MAGNET:** It can also be proven that the earth itself is a magnet. This is shown by the fact that a magnetized object, such as a compass needle, always points approximately north or south. It has been discovered that there is an *S* magnetic pole near the geographic North Pole and a magnetic *N* pole near the geographic South Pole. Inasmuch as the North Pole of a compass points to the north and it is known that like poles repel each other, it is logical to assume that the magnetic pole in the north is *S*.

The magnetic *S* pole which is near the north geographic pole was discovered by

Sir James Ross in 1831. Its location was 70° 30'N latitude and 95°W longitude, which is in the north-central part of Canada about 1400 miles from the geographic pole. However, in 1905, Captain Roald Amundsen located it a little farther west at 70° 5'N and 96° 46'W. From these two observations, it is believed that the Earth's magnetic poles slowly shift their position.

Because the magnetic and geographic poles are separated, a compass needle does not always point true north and south. Its direction also changes as the compass is moved from one place to another on the surface of the earth. The number of degrees by which the needle varies from true north and south is called the *angle of declination*.

On the *agonic line,* which is an imaginary line encircling the earth from one magnetic pole to the other, a compass will point true north and south and the angle of declination will be zero. (See Figure 10-46.)

The magnetic lines of the earth are not parallel to its surface. The angle between the magnetic lines and the surface of the earth is known as the dip or inclination. The inclination can be readily determined in the following manner: Thrust an unmagnetized knitting needle through a cork and another needle through the cork at right angles to the first needle. Support the apparatus on two objects with one needle acting as an axis and pointing east and west. By means of a bent pin or some wax, balance the weight of the first needle until it is parallel with the table. Next, magnetize the first needle by stroking the north end extending from the cork with the *N* pole of a strong magnet, and in a similar manner the south end of the needle with the *S* pole of the magnet. It will be noted that the north end of the needle will dip or incline toward the table. The angle of inclination will depend on how far north of the magnetic equator the experiment is being performed. At the magnetic equator, there will be no dip. At the magnetic pole, the angle will be 90°. In Chicago, the angle of inclination is 72° 50'.

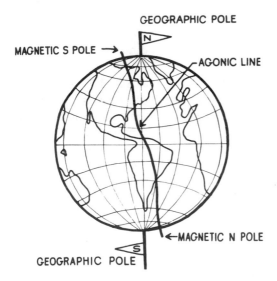

**Figure 10-46.** The earth as a magnet (*Courtesy of Frigidaire Division, General Motors Corp.*)

**MAGNETIC LINES OF FORCE:** The fact that magnets have the properties of repulsion and attraction indicates that *lines of force* are prevalent through and about them. In 1830, Michael Faraday introduced the idea that the magnetic lines of force passed in a curved line from the *N* pole of a magnet around the outside to the *S* pole and on the inside of the magnet from *S* back to *N*.

This can be proven as follows: Place a bar magnet beneath a pan of water. Place a cork with a magnetized needle through it in the water near the *N* pole. The cork will move in a curved path from *N* to *S*. This is due to the fact that the forces of repulsion and attraction are continually changing as the relative distance of the moving pole changes.

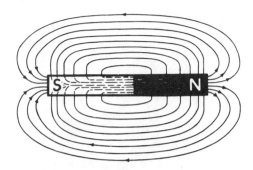

**Figure 10-47.** Magnetic field of flux (Courtesy of Frigidaire Division, General Motors Corp.)

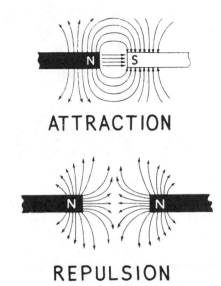

ATTRACTION

REPULSION

**Figure 10-48.** Flux pattern (Courtesy of Frigidaire Division, General Motors Corp.)

**MAGNETIC FIELD OF FLUX:** The area surrounding a magnet in which its magnetic force can be detected is known as the magnetic field of flux (see Figure 10-47). There are three important facts that should be noted: (1) none of the lines of force cross each other, (2) all lines are complete, and (3) all lines leave the magnet at right angles to the magnet. Using the fields of force as the basis of magnetism, the many characteristics of magnets can be readily understood.

Lines of force in a magnetic field may be seen by sprinkling iron filings on a piece of paper which has been placed above a bar magnet.

Earlier in our discussion of magnetism we mentioned the fact that like poles of a magnet repel each other while unlike poles attract. Now, let us discuss repulsion and attraction again briefly as they relate to the field of force or the flux pattern which is produced by a bar magnet. Figure 10-48 (a) shows the lines of force which are produced when two unlike poles are placed near each other, while (b) illustrates the lines of force when two like poles are placed near each other. They attempt to repel each other.

This can readily be seen by placing a piece of paper over the ends of two bar magnets, sprinkling iron filings over the area above the magnets, and observing the pattern made by the filings. One of the magnets should then be reversed and the change in the pattern observed.

**MAGNETIZING A SUBSTANCE:** From what we have learned thus far, we can assume that every atom and molecule of a magnetic substance has a north and south pole. Each atom and molecule is a tiny magnet.

In a piece of unmagnetized iron the molecules are jumbled together. In Figure 10-49 (a), the N poles are black and the S poles are white. They cancel out each other's force. If we stroke a magnet along the piece of iron, the strong N pole of the magnet attracts the S poles in the iron, causing some of the molecules to shift so that the S poles point to the N pole of the magnet. After each stroke, more and more of the molecules are found to have shifted. After sufficient stroking, all of the S poles are pointing

UNMAGNETIZED

MAGNETIZED

**Figure 10-49.** Molecular arrangement of iron (*Courtesy of Frigidaire Division, General Motors Corp.*)

the same direction as illustrated in Figure 10-49 (b).

When inducing magnetism, more strokes will produce more magnetism. However, there is a limit to which any substance may be magnetized. When all of the molecules are lined up as illustrated in Figure 10-49 (b) we say that the magnet is saturated.

There are other properties and characteristics of magnetism; however, we have covered the most common and important phases which have a direct bearing on the subject of electromagnetism which will be discussed later.

**INDUCED VOLTAGE AND CURRENT:** Having discussed some of the more important characteristics of magnets and their effect on metals which are capable of being magnetized, let us now discuss the effect of a magnetic field on metals that are nonmagnetic in nature. It is the effect of magnetic fields on nonmagnetic metals that makes possible the generation of electric current and its practical application to the many electrical components which are used on refrigeration and air conditioning equipment.

**Figure 10-50.** Induced voltage downward motion (*Courtesy of Frigidaire Division, General Motors Corp.*)

If a conductor is passed through a magnetic field, an electrical voltage will be induced in the conductor. (See Figure 10-50.) A horseshoe magnet, a metal conductor such as copper, and the wiring connecting a galvanometer to both ends of the conductor completes an electrical circuit. The galvanometer is an instrument which is used for measuring very small electrical currents. As the conductor is moved in a downward motion through the magnetic field of flux between the ends of the horseshoe magnet, an electric current is induced in the conductor and wires. The direction of the current flow is indicated by the arrows in Figure 10-50 and the amount of current is indicated on the galvanometer. Actually, it is not current which is induced. What really happens is that the transfer of some of the magnetic energy to the conductor forces the electrons in the conductor to flow.

When the conductor is moved in an upward direction the needle of the galvanometer moves in the opposite direction, indicating that the direction of current flow has reversed. Although a flow of current is obtained while the conductor is in motion, the flow of electrons stops when the motion stops. The direction of the induced current depends on the direction in which the flux lines are cut (see Figure 10-51).

Of course, the direction of the induced current is also determined by the direction of the field of flux. For example, if the conductor is being moved in a downward direction, the N pole is on the left side and the magnetic flux is to the right as shown in Figure 10-52 (a). The induced voltage is to the rear as indicated by the arrow. In Figure 10-52 (b), the magnetic poles and field have been reversed and, while the conductor is still cutting the field in a downward motion, the induced voltage is reversed.

An important point to remember regarding induced current in a conductor is the fact that the strength of the current is determined by the position of the conductor in the magnetic field. For example, as the conductor enters the field of flux the electron flow is weak. It increases to its maximum as the conductor reaches the center of the field. The current flow diminishes as the conductor leaves the magnetic field.

It stands to reason that the stronger the field, the stronger the current and the voltage will be. Therefore, if the poles of the

**Figure 10-51.** Induced voltage upward motion (*Courtesy of Frigidaire Division, General Motors Corp.*)

**Figure 10-52.** Induced voltage reversed field (*Courtesy of Frigidaire Division, General Motors Corp.*)

magnet were increased in magnetic strength, the wire would cut more flux lines and a stronger voltage would be induced. The strength of a magnetic pole can be increased by using electromagnetism, i.e., wrapping an electricity-carrying conductor around the magnet's poles.

Since the amount of voltage induced in a wire is determined by the number of flux lines cut per second, a higher voltage can be induced in three ways: (1) by speeding up the movement of the conductor, (2) by increasing the magnetic field strength, and (3) by coiling the conductor so that more turns of the wire will cut the magnetic field. These three factors play an important part in the construction and design of motors and generators.

The basic principle of conductors cutting a magnetic field can now be applied to generators and motors. As stated before, although generators appear to be fairly simple in construction, they are sometimes quite complex in design. Because this is an introduction to electricity, the discussion will cover only the simplest of generators.

In principle, generators and motors are identical. They are both classified as *dynamos*. If the shaft of a dynamo is connected to a prime mover, such as a turbine, and is turned, it pumps electrical power out on its lines and is called a generator. If, however, the shaft is connected to a mechanical load and takes electrical power in, it is called a motor. In other words, a generator *produces* electric current; a motor *utilizes* electric current.

**DIRECT CURRENT GENERATION:** Earlier it was shown that a battery produces a steady unfluctuating flow of electrons in one direction only. This was called regular dc current. A dc generator also produces a flow of electrons in one direction only. Because of the method by which it is produced,

however, the flow of electrons is not steady by comparison and it is called pulsating dc current.

If we use the water analogy, the dc generator may be compared to a piston type water pump connected to a closed water system. (See Figure 10-53.) On the upstroke of the piston, the intake valve closes, the check valve opens, and water flows in the direction indicated by the arrows. On the down-stroke, the check valve closes, the intake valve opens, and the flow of water momentarily stops until the piston moves upward again. Pulsating dc current can be compared also with the flow of blood in the body which gets a push every time the heart beats.

Regardless of the complexity of any generator, there are only three elements involved: (1) a magnetic field which is produced by magnets, (2) an armature for carrying a coil of wire through the magnetic

**Figure 10-53.** Pulsating dc principle (*Courtesy of Frigidaire Division, General Motors Corp.*)

field, and (3) a prime mover to turn the armature. There are two essential parts of a generator. Figure 10-54 (a) shows the stationary frame which includes the yoke and base with the two pole pieces, which in this case are permanently magnetized. The armature shown at (b) consists of a series of copper wires looped around an iron core and a *commutator*. The commutator is a series of copper segments to which the ends of the wires are attached.

Figure 10-54. Pulsating dc current generator parts (*Courtesy of Frigidaire Division, General Motors Corp.*)

**DC GENERATOR ACTION:** To better understand what happens in an armature see Figure 10-55. One turn of the coil and two segments of the commutator have been lifted from the armature and placed in the magnetic field of the pole pieces. Two views of a coil in a magnetic field are shown. At each end of the coil is a segment of the commutator and against the segments are carbon blocks called brushes. This arrangement provides a slipping contact between the ro-

Figure 10-55. Dc generator action (*Courtesy of Frigidaire Division, General Motors Corp.*)

tating armature and the stationary load. For sake of clarity, one-half of the coil and one segment are colored black and the other half is white. Note the direction of the magnetic flux and the direction of rotation.

In view (a), no voltage is produced because the coil is cutting no flux lines. In this position the brushes are contacting only the insulation between the segments. In view (b), the coil has moved at right angles. The black side of the coil is cutting downward and the white side is cutting upward. This causes the induced voltage in the coil to flow "in" the white side of the coil and "out" the black side.

In view (c), the armature has been turned another one-quarter turn. Also, note in view (c) that the same condition is apparent as in the position in view (a) except that the coil is upside down. Again the flow of current has stopped because the conductor is no longer cutting the lines of force.

In view (d), the coil is the opposite position from view (b). Now the black side of the coil is cutting upward and the current is flowing in; the white side is cutting downward and the current is flowing out. Since half of the time the current in the black portion of the coil is flowing in and the other half of the time it is flowing out, it can be said that alternating current is flowing inside the coil. In fact, a rotating coil in a magnetic field always produces alternating current. The alternating current in the loop is caused to flow in only one direction to a load [view (c) and (d)].

In view (b) and in view (d) the black brush is marked negative and the white brush is marked positive. The brushes are stationary, and when current is flowing the negative brush is always in contact with the positive brush. So, if it is desired to deliver dc current to a load, a commutator is always used. Since the commutator causes the alternating current in the armature to be delivered to the load in one direction, it is said to *rectify* the current.

A single coil rotating in a magnetic field is like an automobile hitting on one cylinder. The output is weak and fluctuating.

Dc voltage, which is produced by a single

loop rotating within a magnetic field can be shown graphically. (See Figure 10-56.) When the black portion of the loop is in the upward position, no lines of force are being cut and no voltage is produced. As the loop turns in a clockwise direction, the black portion progressively cuts more lines of force and the voltage increases accordingly to the maximum. As the black portion of the loop progresses toward the downward position, the voltage decreases to zero. The loop is again in the vertical position. At this point, the white portion of the loop begins to cut the lines of force in a downward direction as the black portion of the loop cuts them in an upward direction. This causes the voltage to increase to its maximum when the loop is horizontal and to decrease again to zero when the white portion has progressed to the downward position. Since, during this process, the black commutator has moved to the white brush and the white commutator has moved to the black brush, the current flows to and from the load in the same direction as it did before the commutator and brushes changed relationship. From this it can readily be seen that even if a one-loop coil were capable of producing sufficient current to light a lightbulb, the light would

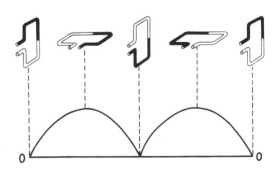

**Figure 10-56.** Graph of dc voltage and current (single loop) (*Courtesy of Frigidaire Division, General Motors Corp.*)

tend to go out every time the voltage dropped to zero.

The fluctuation of voltage can be reduced by increasing the number of loops and commutator segments. In Figure 10-57 an armature with two loops and a four-segment commutator is shown. When this armature is rotated, the black coil is going to be ¼ of a revolution behind the white coil. This means that the voltage of the black coil is at zero when the white coil is at its maximum. However, the brushes riding on the commutator contact only the coil producing the greater voltage.

The solid line portion of the graph in Figure 10-57 is the voltage picked up by the brushes. Notice that the voltage is more level with two coils than it was with one coil. It is still a pulsating current, but it doesn't get all the way down to zero.

In Figure 10-58, a four-coil armature with eight segments in the commutator is substituted for the two-coil armature and

**Figure 10-58.** Four-coil voltage (*Courtesy of Frigidaire Division, General Motors Corp.*)

the output is more level. As before, when the armature is rotated, the brushes pick up only the very peaks of each coil voltage. This is still pulsating dc, but only slightly pulsating. The voltage rise and fall is very short, as shown in the graph. Obviously, if voltage without fluctuation is desired, all that is necessary is to add sufficient coils and increase the number of segments proportionately.

**ALTERNATING CURRENT GENERATION:** While dc current, either regular or pulsating, is a one-way flow of electrons, alternating current (ac) first flows in one direction, then reverses and flows in the opposite direction.

The water analogy can be used again to illustrate ac current. In Figure 10-59 a pump with a solid piston and pipes containing water are shown. If the piston is moved upward, the water flows in a counterclockwise direction (indicated by the arrows).

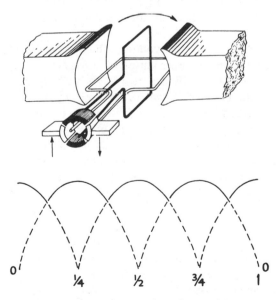

**Figure 10-57.** Two-coil voltage (*Courtesy of Frigidaire Division, General Motors Corp.*)

**Figure 10-59.** Alternating current (*Courtesy of Frigidaire Division, General Motors Corp.*)

**Figure 10-60.** Ac generator action (*Courtesy of Frigidaire Division, General Motors Corp.*)

Pulling the piston downward causes the water to flow in the opposite direction.

**AC GENERATOR ACTION:** Since we produce ac current in the rotating coil of any generator, it is just a matter of eliminating the rectifying commutator and employing slip rings (see Figure 10-60). Instead of connecting each end of the coil to a commutator segment or half a ring, each end is connected to a complete ring, and a brush is in contact with each end at all times.

In view (a) of Figure 10-60, the current is flowing "out" the black portion of the coil and flowing "in" through the white portion. The black ring and brush will be called negative and the white ring and brush will be called positive. In view (b), the coil has rotated through the neutral plane and the black portion now cuts upward, and the current flow reverses, flowing "in" through the black ring and brush and "out" through the white ring and brush. This reverses the current through the load. The white brush becomes negative and the black brush becomes positive.

A graph of ac voltage, which is produced by a single loop rotating in a magnetic field, is shown in Figure 10-61. As with dc current generation, whenever the loop is in a vertical position, no voltage or current is produced. Following the black portion of the loop through half a cycle, it can be seen that the voltage increases to its maximum positive and then decreases to zero voltage. At this point, the white portion of the loop begins its downward path through the magnetic field while the black portion begins its upward path. This causes the current to reverse its direction and, because of the slip rings, the voltage increases to the maximum negative and then back to zero. This also means that the current through the load and connecting conductors is being reversed every half turn of the loop.

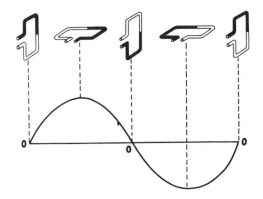

**Figure 10-61.** Graph of ac voltage and current (*Courtesy of Frigidaire Division, General Motors Corp.*)

**IRON FILINGS**

**Figure 10-62.** Magnetism by current (*Courtesy of Frigidaire Division, General Motors Corp.*)

As in the dc generator, a single loop would produce a very weak and fluctuating current and voltage. While the strength of ac current and voltage can be increased by adding more loops to the slip rings, the fluctuation cannot be changed in this manner. Fluctuation can be overcome and controlled only by increasing the speed of the armature and keeping this speed constant.

## ELECTROMAGNETISM

Until now, we have established only the basic principle of producing electricity by magnetism as applied to dc and ac generators, which have a rotating armature in a stationary magnetic field.

In order to understand the production of voltage and current in a modern ac generator, as well as the basic principles of transformers and other electrical components, a knowledge of electromagnetism is required. As the name implies, electromagnetism is the production of magnetism by an electric current.

**ELECTROMAGNETS:** An electromagnet is like a natural or artificial magnet in its

attraction, but differs in its control. Although an electromagnet can be powerful in its attraction, it can be turned on and off with the flick of a switch. Electromagnets are used to lift heavy pieces of metal from one place to another, to open and close contacts in relays, to open orifices in solenoid valves, and to turn rotors in dynamos or motors.

If a wire connected from the negative to the positive of a dry cell has been dipped into a pile of iron filings, some of the filings will cling to the wire as shown in Figure 10-62. This proves that any conductor carrying an electric current is surrounded by a field of flux and the current flow produces a magnetic field. If one end of the wire is disconnected from the dry cell, the filings drop off. This proves that the field exists only when the current is flowing.

While the filings are shown clinging to only a portion of the wire, the field surrounds the wire along its entire length.

Although a field of flux or magnetism has

been produced with a straight, single conductor, no polarity is apparent. To be able to put magnetism by electricity to a practical application, poles must be produced. To produce poles, simply form a straight conductor into a loop (see Figure 10-63).

**Figure 10-63.** Magnetic polarity of a loop (*Courtesy of Frigidaire Division, General Motors Corp.*)

When a conductor is formed into a loop, the flux bends toward the center of the loop and produces a north pole on one side and a south pole on the other. If a small number of loops of wire are combined, they form a *helix coil* which produces much stronger poles than a single loop. If a very strong magnetic coil is desired, more turns of wire are built up in layers. This produces a *solenoid coil*.

If a solenoid coil is wound around a piece of iron or other easily magnetized metal, the metal is called the core or pole piece. When the coil is energized, the core becomes temporarily magnetized and an electromagnet is

produced. The core of an electromagnet has a low retentivity and immediately becomes demagnetized when the flow of current stops. Without this characteristic, electromagnets would have no practical use.

A basic electromagnet is shown in Figure 10-64. For the sake of clarity, the illustration shows very few loops of wire as compared to those actually required. Electromagnets can be produced by using either dc or ac current.

**Figure 10-64.** Basic electromagnet (*Courtesy of Frigidaire Division, General Motors Corp.*)

Now that the basic principles of electromagnetism have been established, we can discuss the role that electromagnetism plays in the generation of electric current.

An illustration of how magnetism in the pole pieces of a dc generator can be strengthened is shown in Figure 10-65. Here we see that the pole pieces are wound with coils of wire which are connected to the

**Figure 10-65.** Electromagnetism as applied to a dc generator (*Courtesy of Frigidaire Division, General Motors Corp.*)

brushes. In operation, some of the current produced by the generator is used to strengthen its own poles. An electromagnet can always produce a stronger field than a permanent one. A permanent magnet can also be strengthened by winding an energized coil around it.

The proper polarity of the pole pieces of the generator is established and maintained by winding the poles in the opposite direction. For example, if the pole marked *N* were wound in the same direction as the *S* pole, it would also become an *S* pole and there would be no current generated by the rotating loop.

**ELECTROMAGNETISM AND AC GENERATION:** Although the simple ac generator previously described could produce voltage and current, they would both be too weak to be of practical value. By increasing the number of rotating loops and by strengthening the magnetic field through electromagnetism, stronger voltage and current can be produced. Unlike dc generators, ac generators cannot share their current to self-strengthen their pole pieces due to their inability to maintain constant polarity of the poles. They can, however, be strengthened

by electromagnetism produced by current supplied from a separate dc generator. All modern power plants use dc generators to strengthen or excite a magnetic field for ac generators. When so employed, the dc generator is called an *exciter* and the ac generator is called an *alternator*.

**ALTERNATORS:** There are basically two types of alternators. One type uses a dc exciter to strengthen the stationary field through which the rotating armature passes as shown in Figure 10-66. This type was used in the early days of ac generation and a few may still be found in operation today in small communities. It should be remembered that both the dc exciter and alternator require a prime mover. In some instances, each has its own driving turbine, but, more often, the same turbine is used for both.

**Figure 10-66.** Basic rotating armature alternator (*Courtesy of Frigidaire Division, General Motors Corp.*)

This type of alternator was the first used in the generation of ac current, but it had construction limitations which made it unsuitable for the high voltages and currents which are required for modern transmission over long distances.

The second type is the basic design of a later development called the rotating field type which uses the principles of all modern alternators. (See Figure 10-67.) Here the dc exciter is supplying current to the loop by means of brushes and slip rings. The pole pieces have coils wound from a continuous length of wire. The ends of the wire are connected to conductors which carry the ac current to the transformer and to the transmission lines. As the loop is rotated by the prime mover, the magnetic field sweeps across the pole pieces and, by magnetic induction, produces a voltage which causes the current to flow through the coils and conductors. Since the magnetic field has polarity, the induced current flows in the opposite direction every half revolution.

**Figure 10-67.** Ac single phase alternator (schematic diagram) (*Courtesy of Frigidaire Division, General Motors Corp.*)

One of the major advantages of this type of alternator over the rotating armature type pertains to the amount of current carried by the brushes and slip rings. In the rotating armature type, the slip rings are exposed and are subject to arc-overs and short circuits when they are required to carry high voltage and current.

Because the rotating field type alternator can generate exceptionally high voltage and current from the magnetism produced by a relatively low voltage from the dc exciter, arc-overs and short circuits at the slip rings are avoided.

**POWER FACTOR:** Since the introduction of fluorescent lights and the mass production of electric motor-powered refrigeration systems and appliances, the term *power factor* has come into fairly common use. Before entering into this discussion, we should not confuse the power factor with the efficiency of an electrical device.

The power factor of an electrical system is a very complex subject. For our purpose, power factor may be defined as the difference between the amount of power supplied by the power company for a device and that consumed by this device for which the company receives no payment. To understand how such a condition can exist, we must take into consideration that, except for capacitors, there are two classifications of current-consuming devices or loads: a "pure" resistance type and an induction type.

Pure resistance loads include items such as incandescent lights and all types of heater elements.

Induction type devices are those which have coils of wire wound around metal cores and in which electromagnetism is produced during its operation. Induction type devices include transformers, ac motors, relays, solenoid valves, and fluorescent light equipment. For an illustration of how electrical

LINE VOLTAGE

ALL CURRENT THAT ENTERS
MUST RETURN

INDUCTION MOTOR

FLUORESCENT
LIGHT                                    BALLAST

INCANDESCENT LIGHT

HEATER ELEMENT

⬛⬛⬛ Load Current     ⬜⬜⬜ Magnetizing Current

**Figure 10-68.** Distribution of current in a circuit (*Courtesy of Frigidaire Division, General Motors Corp.*)

current is distributed in a total parallel circuit containing both pure resistance and induction type loads or devices see Figure 10-68. The induction motor can be used to drive a refrigeration compressor, the fluorescent light illuminates the interior, the incandescent light can be used for temperature warning, and the heater element can be used to defrost the evaporator.

It should be remembered that the rate of current flow returning to the source is the same as that entering a circuit from the source. The power company is quite concerned as to how the current it provides and gets back has been utilized.

Pure resistance loads have no metal cores requiring magnetism. Therefore, the electric meter records all of the energy consumed by such devices. In these instances, the power factor is said to be 100%.

With induction type loads, however, some of the current supplied by the power company is used only to magnetize the metal core while the remaining current overcomes

resistance and performs the work. Magnetizing current does no work and, therefore, it is not recorded on the electric meter. If one-fourth of the current being supplied to an induction type device is required to magnetize the metal and only three-fourths of it is used to produce work, the device is said to have a 75% power factor. The 25% is a direct loss to the power company due to the fact that it has spent money for fuel and equipment to supply current through the resistance of distribution lines, transformers, etc. for which no payment is received.

Realizing that induction type devices are essential to our way of living and that some current losses are unavoidable, power companies have established a minimum 90% power factor as being acceptable for devices connected to their power lines. Power factor can also be expressed in decimal form but is usually expressed as a percentage.

## ELECTRICAL UNIT PREFIXES

The simple units of electrical measure—volts, amperes, ohms, and watts—are rather clumsy when very large or very small quantities are involved. Therefore, a system of electrical unit prefixes is used to indicate large and small quantities. They are kilo, milli, mega, and micro. (See Figure 10-69.)

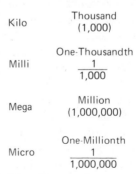

| Kilo | Thousand (1,000) |
| Milli | One-Thousandth $\dfrac{1}{1,000}$ |
| Mega | Million (1,000,000) |
| Micro | One-Millionth $\dfrac{1}{1,000,000}$ |

**Figure 10-69.** Electrical unit prefixes (*Courtesy of Frigidaire Division, General Motors Corp.*)

For example, a generator which delivers 1000 watts is called a one-kilowatt (kW) generator. Instead of saying the flow of current is one thousandth ($\frac{1}{1000}$) of a volt, we say it is one millivolt. If we are testing insulation, we would use the term megohms instead of so many million ohms. The term *farad* is used as a unit of measure for electrical capacity but, due to its size, the capacity of capacitors is given in microfarads, or so many millionths of a farad.

## TRANSFORMERS

There are many applications in which transformers play a significant part in the operation of the equipment. Because of the many uses to which they can be put, transformers are manufactured in hundreds of sizes, shapes, and current characteristics. They range in size from several tons in weight for power transmission purposes to a few ounces for radio and television sets. Regardless of the size or shape, they all have the same basic components and function on the same basic principles—self-induction and mutual induction.

### SELF-INDUCTION (COUNTER EMF):
It was learned earlier that a flow of current through a solenoid coil produces a magnetic field of flux which changes polarity as the direction of current changes. It was also learned that a moving magnetic field produces a voltage and current in a conductor. There is a law of physics which states that for every force there is an opposing force. Whenever ac voltage and current flow through a solenoid coil, each cycle produces an alternating expanding and contracting magnetic field of flux over the adjacent loops of wire in the coil. This induces a voltage in the coil which opposes the applied voltage and current. This voltage is known

as counter emf; its value is always less than that of the applied voltage.

In Figure 10-70, the large white arrows at each end of the coil indicate the direction of the applied voltage. The small, black arrows indicate the direction of the counter emf (induced voltage). Roman numeral I illustrates the relationship between the voltages during the first quarter cycle when the applied voltage is trying to build up. The counter emf opposes its build-up. Roman numeral II illustrates the second quarter cycle, when the applied voltage is dropping back to zero. The counter emf now is working in the same direction as the applied voltage and is opposing the drop in voltage. Roman numerals III and IV also illustrate that the counter emf is always opposing changes in the applied voltage. The current flow resulting from the inner action of these opposing forces is smaller than it would otherwise be. Actually, the counter emf controls and regulates the amount of current through the coil, thus preventing the coil from overheating and burning. It also is interesting to note that because of the opposition between the applied voltage and induced voltage, the resulting flow of current is slightly delayed so that its changes lag behind the changes in the applied voltage. This lag of current behind the voltage introduces a timing factor which is utilized in the operation of self-starting electric motors.

### MUTUAL INDUCTION:
The word mutual means sharing. Many years ago it was discovered that, by induction, a solenoid coil could share its current and voltage with another coil of wire connected only to a load. In other words, if a coil of wire is connected to a source of current, the current also will flow through an adjacent coil of wire which is connected only to a load.

The principle of mutual induction can be

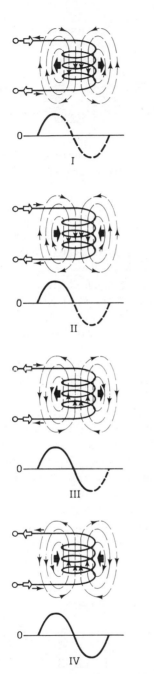

**Figure 10-70.** Self-induction (counter emf) (*Courtesy of Frigidaire Division, General Motors Corp.*)

**Figure 10-71.** Mutual induction (*Courtesy of Frigidaire Division, General Motors Corp.*)

readily demonstrated and explained (see Figure 10-71). The coil which is connected to the switch and to the battery is called the primary coil. The adjacent coil, which is connected only to the galvanometer, is called the secondary coil. The galvanometer represents the load.

The iron core around which the coils are wound is not essential to the principle of mutual induction; it merely provides a stronger field. Mutual currents and voltage also can be induced across an air gap.

When the switch is closed, the primary coil and core become an electromagnet. The build-up of the magnetic flux lines, which cut across the coils of the secondary winding, causes electrical current to flow through the secondary coil and circuit as indicated by the galvanometer. The flow of current in the secondary circuit occurs, however, only during the instant that the magnetic field is formed. When the switch is opened, the core becomes demagnetized instantly and the collapse of the magnetic flux causes current to flow in the opposite direction. Thus, it can be concluded that mutual induction depends entirely upon the alternate expanding and collapsing of the magnetic field around the adjacent coils of wire. With dc current, this expanding and collapsing of the magnetic field is accomplished by rapidly making and breaking the circuit. With ac current, the same effect is produced by rapid changes in the magnetic flux as the current alternates.

When used with ac current, the device is called a transformer. A transformer is unique in that it can remain connected to an ac supply and almost no current will flow in the primary as long as its secondary circuit is opened. The reason is that the self-induction which takes place in the primary produces a counter voltage almost equal to the applied voltage. Almost no current results from these opposing voltages. When the secondary circuit is closed, however, a current is induced in the secondary coil and produces magnetism which is opposite in polarity to that produced in the primary circuit. This reduces the counter voltage, which is produced in the primary circuit by self-induction and allows a current to flow in the primary circuit. Only because of ac current is a transformer made possible, and only because of transformers is modern transmission of ac current possible.

**CONSTRUCTION:** Having discussed self- and mutual induction, let us see how they apply to transformers. A transformer core is shaped to form a continuous circular magnetic path (see Figure 10-72). Instead of being made of solid metal, the core is made of laminations of thin sheets of soft iron, silicon steel, or other suitable metal. The sheets are insulated from each other by a

**Figure 10-72.** Transformer construction (*Courtesy of Frigidaire Division, General Motors Corp.*)

thin coating of shellac or other nonconductive material. This insulation is done to reduce the strength of *eddy* currents which would otherwise cause excessive heat and considerable loss of efficiency in the core. Eddy currents are induced currents which flow within the core. They are wasteful, have no purpose, and should be reduced to a minimum.

The primary and secondary coils are usually enameled copper wire. The relationship or ratio between the number of turns in the two coils is almost direct. For example, if the primary coil is connected to a 120-volt source of current and it is desired to supply the load with 240 volts, the secondary coil must have approximately twice the number of turns as the primary coil. This is called a step-up transformer. (See Figure 10-73.) On the other hand, if the source is 240 volts and the load requires only 120

**Figure 10-73.** Step-up transformer (*Courtesy of Frigidaire Division, General Motors Corp.*)

**Figure 10-74.** Step-down transformer (*Courtesy of Frigidaire Division, General Motors Corp.*)

**Figure 10-75.** Typical transformer (*Courtesy of Frigidaire Division, General Motors Corp.*)

## SAFETY PROCEDURES

volts, the secondary coil must have approximately half the number of turns as the primary coil. This is called a step-down transformer. (See Figure 10-74.) The number of turns in the illustration are but a few as compared to the number actually required.

While a transformer can increase or decrease the voltage and current, it cannot generate energy. For example, if the voltage is stepped up from 115 volts to 230 volts and the load requires 2 amperes and consumes 470 watts, the primary coil must be designed to carry 4 amperes so that the wattage remains the same.

A typical transformer is illustrated in Figure 10-75. This type differs from the closed-core type shown in Figures 10-73 and 10-74 in that it has a shell-type core. The shell-type core further increases efficiency because it provides two parallel paths for the magnetic field.

Although some heat losses occur in any transformer, they are considered to be one of the most efficient electrical devices made. They are usually more than 90 percent efficient.

From the preceding explanation, it should be understood how an unlimited variety of voltage combinations can be obtained by varying the number of turns in the primary or secondary coils.[1]

[1] See Frigidaire Home Study Course No. 3 for further reference.

Because approximately 85% of all service calls involve electrical trouble, the service technician is constantly in danger of receiving an electrical shock. Safety precautions must be observed to avoid serious personal injury or even death.

1. Always be certain that the electricity is disconnected from the unit before attempting to service electrical components.

2. If necessary to work on a system while the electricity is on, be certain that the service technician is not grounded.

3. Never jump a fuse or install a fuse larger than what is recommended. If a unit continues to blow fuses, a problem exists and must be corrected.

4. Always be certain that the electrical connections are tight and are not causing a high resistance to current flow.

5. Make certain that the equipment is properly grounded.

6. Always be sure that the correct voltage is provided before connecting the electricity to the unit.

7. Replace wires whose insulation is frayed or broken.

8. Never attempt to check resistance when electrical power to the unit is on.

9. In case of serious electrical shock, treat the victim for physical shock.

10. Always use the proper size wire for the current being used.

## SUMMARY

• The basis of the electron principle is the fact that all known matter or substance regardless of its form—solid, liquid, or gas —is fundamentally composed of very small particles called atoms.

• Atoms are made of smaller particles called electrons and protons.

• Electrons have a negative electrical charge.

• Protons have a positive electrical charge.

• A proton weighs almost 2000 times more than an electron.

• Each atom has a core or center which is called the nucleus.

• The nucleus contains protons and neutrons.

• A neutron is made up of one electron and one proton.

• Electrons rotate at high speeds in orbits or shells around the nucleus.

• Matter is generally defined as anything which occupies space or has mass.

• When matter has been reduced to its simplest form, each component is known as an element.

• An element may be defined as a substance containing only one kind of atom.

• As the number of electrons in the elements increase, the number of orbits and their spacing also increase.

• In a compound, two or more elements combine chemically to form a distinct substance with the same proportion of elements always present in its composition.

• A molecule may be defined as the smallest particle of a compound which is of the same chemical composition as the compound itself.

• Many compounds are composed of more than two elements. These compounds would include all the refrigerants commonly used in refrigeration and air conditioning.

• Positive charges repel each other, negative charges repel each other, and positive charges attract negative charges.

• Free electrons are those electrons which can easily move from one atom to another.

• Electricity is the movement of free electrons from one atom and molecule to another.

• Any substance with a great number of free electrons will make a good electrical conductor.

• Substances which have few free electrons are called insulators.

• Except for the splitting of atoms, energy can neither be created nor destroyed.

• The six sources of electricity are: (1) friction, (2) chemical action, (3) magnetism, (4) heat, (5) pressure, and (6) light.

• In an uncharged state, each atom has an equal number of electrons and protons.

• A statically charged insulator or dielectric has a field of force and the laws of attraction and repulsion apply.

• An ion may be defined as an electrically charged atom or a group of atoms.

• Any substance containing charged ions is capable of conducting an electrical current and is called an electrolyte.

• Copper and zinc are excellent for use as electrodes. This is due to the fact that zinc has an abundance of negatively charged ions while copper contains an abundance of positively charged ions.

• The flow of regular dc current can be compared with the flow of water produced by a centrifugal pump.

• Electrical pressure is called electromotive force (emf) and causes the free electrons to flow through the circuit.

• The direction of electrical force is always from the negative electrode to the positive electrode.

• Any particular electron does not move from one end of the conductor to the other. Thus, electron movement may be compared to a switch engine bumping a standing string of boxcars on a track. When the engine bumps one end of the string, each car progressively bumps the next car.

• Volt is the electrical unit of measure of force.

• The greater the difference between the concentration of free electrons in the electrodes, the greater the potential difference will be.

• Emf or voltage is the pressure produced throughout an electrical circuit while potential difference indicates a ratio of electron potency or concentration between the electrodes of a battery or certain components of other current-producing devices.

• Coulomb is the unit of measure of the quantity of electricity.

• Ampere is the unit of measure of current strength or rate of flow of electrons.

• All substances offer resistance to the flow of electrons.

• If the force is strong or the resistance weak, a strong current will flow.

• If the force is weak or the resistance great, only a small current will flow.

• Electrical resistance of wires is dependent on four factors: (1) diameter of wire, (2) length of wire, (3) kind of wire, and (4) temperature of the wire.

• If the resistance of a conductor is increased by any one of the four factors, the current flow is decreased.

• Thermistors are conductors whose resistance to the flow of electricity decreases as their temperature increases.

• Thermistors are being used in the control circuits of refrigeration and air conditioning equipment to provide more accurate control of the equipment.

• The ohm is the unit of measure of resistance of a conductor to the flow of electrons.

• One ohm is the resistance offered by a column of mercury to the flow of one ampere of current having an electromotive force of one volt.

• Ohm's Law states that the current in an electrical circuit is equal to the pressure divided by the resistance.

• The three general types of circuits are: (1) series, (2) parallel, and (3) series-parallel.

• The series circuit is a one-path circuit and can be recognized in two ways: (1) it will never have more than one conductor connected to one terminal, and (2) there will be only one path from the source to the load and back to the source.

• In a parallel circuit, there are two or more paths between the terminals of the source of current.

• A series-parallel circuit, as the name implies, is a combination of both the series and parallel circuits.

• The attraction of an unmagnetized piece of iron to a magnet is due to magnetic induction.

• Every magnet has two poles, a north-seeking pole and a south-seeking pole.

• In magnets, the like poles repel and unlike poles attract.

• The earth is a magnet.

• The area surrounding a magnet in which its magnetic force can be detected is known as the magnetic field of force.

• None of the lines of force cross each other.

• All of the flux lines are complete.

• All of the flux lines leave the magnet at right angles to the magnet.

• Every atom and molecule of a magnetic substance has a north pole and a south pole.

• When inducing magnetism, more strokes will produce more magnetism.

• It is the effect of magnetic fields on nonmagnetic metals that makes the generation of electric current possible.

• If a conductor is passed through a magnetic field, an electrical voltage will be induced in the conductor.

• The direction of movement of the conductor and the direction of the field of flux will determine the direction of induced current.

• The strength of the current in a conductor is determined by the position of the conductor in the magnetic field and the strength of the magnetic field.

• Three ways to induce a higher voltage are: (1) speed up the movement of the conductor, (2) increase the magnetic field strength, and (3) coil the conductor so that more turns of the wire will cut the magnetic field.

• In principle, generators and motors are identical. They are both classified as dynamos.

• The three elements involved in any generator are: (1) a magnetic field, (2) an armature, and (3) a prime mover to turn the armature.

• An electromagnet is like a natural magnet, but differs in its control. It can be turned on and off with a switch.

• Electromagnets are used to lift heavy pieces of metal from one place to another, to open or close contacts in relays, to open orifices in solenoid valves, and to turn rotors in dynamos or motors.

• If a solenoid coil is wound around a piece of iron or other easily magnetized metal, the core becomes temporarily magnetized and an electromagnet has been produced.

• An electromagnet can always produce a stronger field than a permanent one.

• The proper polarity of the pole pieces of the generator is established and maintained by winding the poles in the opposite direction.

• The two types of alternators are: (1) stationary field, and (2) rotating field.

• Power factor may be defined as the difference between the amount of power supplied by the power company for a device and that consumed by this device, for which the company receives no payment.

• It should be remembered that the rate of current flow returning to the source is the same as that entering a circuit from the source.

• When ac voltage and current flow through a solenoid coil, each cycle produces an alternating expanding and contracting magnetic field of flux over the adjacent loops of wire in the coil. This induces a voltage in the coil which opposes the applied voltage and current. This voltage is known as counter emf.

• By induction, a solenoid coil will share its current and voltage with another coil of wire connected to a load.

• A transformer can increase or decrease the voltage and current, but it cannot create energy.

## REVIEW EXERCISES

1. What is the basis of the electron principle of electricity?

2. How heavy is a charged proton in relation to a charged electron?

3. Of what is a neutron made?

4. Briefly define matter.

5. Briefly define an element.

6. Briefly define a compound.

7. What is the force that holds an atom together?

8. What makes a good electrical conductor?

9. What are the six sources of electricity?

10. What type of electricity is produced by use of dialectrics?

11. What electrical charge does an electron have?

12. When does an atom have an equal number of electrons and protons?

13. State the laws of charges.

14. Can a substance be charged either positively or negatively?

15. Where may electricity which is produced by chemical action be used in refrigeration work?

16. What is an electrolyte?

17. What metals make good electrodes?

18. What limits the life of a battery cell?

19. What type of electricity flows at an unfluctuating rate?

20. What is another name for electromotive force?

21. Does a particular electron flow the complete length of the conductor?

22. What is the definition of a volt?

23. When are two substances said to have an electrical potential difference?

24. Briefly define a coulomb.

25. Briefly define an ampere.

26. If an electrical circuit has a high resistance, will the current flow be small or large?

27. On what four factors is the resistance of a circuit dependent?

28. How does temperature affect a thermistor?

29. Briefly define an ohm.

30. State Ohm's Law.

31. Name the three general types of circuits.

32. What will be the reading on a voltmeter when connected across the total resistance in a series circuit?

33. How is the amperage in a parallel circuit determined?

34. What is the greatest and most common source of electrical power?

35. What was the earliest practical use of the magnet?

36. What is the reason an unmagnetized piece of iron is attracted to a magnet?

37. What are the poles on a magnet named?

38. What is the magnetic field of force?

39. How is the direction of the induced current in a conductor determined?

40. In what three ways can a higher voltage be induced in a conductor?

41. What is the name of the metal around which a solenoid coil is wound?

42. Which magnet can produce a stronger field, a natural or an electromagnet?

43. What are the two basic types of alternators?

44. Briefly define the power factor.

45. What induces the voltage and current from one coil to another in a transformer?

46. What does counter emf oppose?

47. The current in the secondary of a transformer is in what relation to the current in the primary?

48. Why is a transformer core laminated?

# Test Equipment

Test instruments are valuable aids to service technicians. The use of test instruments seems to frighten many service technicians, however, and they will not make any effort to use these valuable tools of their trade. The use of test instruments is of great benefit to the customer as well as the service technician. Test instruments should be just as familiar to the service technician as any other tool in his tool box.

## PURPOSE

Test instruments, when used, can help the service technician to perform service operations more accurately and faster, thereby providing a savings to the customer. There are' two very strong arguments for the use of test instruments: first, prevention of the unnecessary changing of parts which usually end up in the scrap box; second, that time saved is money. Customer satisfaction after the sale is not promoted by the parts changer service technician, who costs the customer and his employer thousands of dollars each year in wasted parts and effort.

## SELECTION

Quality tools may represent a sizable investment, but they are not expensive in the long run if care is exercised in their selection, use, and maintenance. When the task of selecting test equipment is approached properly, the errors made will be kept to a minimum. It is desirable to make a list of the features that are required, those features that would be nice but not essential, and those that are not needed. Careful shopping will probably produce a test instrument that is fairly close to what is wanted.

The cost of these instruments also should be considered. One instrument with several functions will be less expensive than an instrument for each individual purpose. It will be easier to handle, if it is not too bulky, but it will be complicated to use and will require more time in learning to use the instrument to its full capacity. The purchase of several instruments has these advantages: they are easier to use, offer lower initial cost per unit, and offer lower additional availability in that should one instrument need repairs another can take over.

## REPAIR

The field repair of test instruments should not be attempted. A few instruments have components that can be replaced in the field by any competent technician. Any major, internal repairs should be left for those qualified to make them.

Most instruments have some means of adjustment or calibration which should be checked periodically to assure dependable and accurate measurement. The manufacturer's recommendations for calibration and adjustment for any instrument should be followed.

## CARE

Some meter movements are sensitive to position. This type of instrument is more accurate when placed in a horizontal position. Any instrument should be in the position in which it is to be used when calibration is done. All instruments have a specific point at which the needle should rest when it is not in use. This point is called the *null point*.

Instrument test leads should be checked regularly for wear or deterioration of the insulation. Leads with cracked or worn insulation should be replaced. Some leads are replaceable with standard probes that can be purchased at electronic supply houses. However, some instrument leads can only be replaced with leads which are supplied by the instrument manufacturer.

A firm solid contact between the lead and the instrument is required. Loose-fitting leads can cause the meter to give an improper reading.

We can consider the mechanism that moves the needle across the meter face as being a motor. Improper lubrication can damage this mechanism as easily as rough handling or overloading. Use the instrument for its intended purpose and it will last forever.

Most test instruments are equipped with fuses or circuit breakers which prevent damage to the instrument. Be sure to have a few spare fuses of the proper type on hand.

Test instruments should be treated with care. When they are stored in a tool box, they are subjected to a beating. Extremely hot or cold temperatures are not good for test instruments. They should be protected from oil, dust, and moisture. They should not be carried or stored out of their protective cases.

The batteries used in test instruments should be of good quality. Be sure to check batteries frequently for corrosion because some batteries will leak acid even before all the power is used. This acid will ruin test instruments. It is good practice to remove the batteries from instruments which are not used frequently and store them separately.

When storing a test instrument, be sure that the selector switch is not left in a position that calls for battery output. If a selector switch is not used, the leads should be disconnected, except when the meter is being used.

## APPLICATION

The proper application of test instruments will provide more accurate trouble diagnosis in the shortest possible time, providing a savings to the customer and effecting better customer relations.

**VOLTMETERS:** The voltmeter is probably one of the most frequently used instruments. It is also the easiest instrument to read (see Figure 11-1). The service technician may encounter voltages ranging from millivolts to 5000 volts or more. Therefore, reasonable caution should be used.

In use, the selector switch should be set on the proper function for which the meter is to be used. When the correct voltage is not known, start checking the electrical circuit with the meter set on the highest scale possible and work down until one is found that provides a reading at approximately midscale. Midscale readings are the most accurate obtained.

**OHMMETER:** The ohmmeter is used for measuring resistances, checking continuity,

making quick checks on capacitors, and testing some solid state components. (See Figure 11-2.) Most ohmmeters are ruined because they are connected to a circuit or a component that has not been disconnected from the source voltage. Ohmmeters have their own power source and cannot be safely used with any other power source.

**Figure 11-2.** Ohmmeter (*Courtesy of Robinair Manufacturing Company*)

It is best to start with the lowest resistance scale when measuring an unknown resistance or continuity. Work up through the scales until a midscale reading is obtained.

When checking resistance, first select the proper resistance scale. Install the test probes in the proper jacks. Touch the ends of the probes together and turn the adjust knob until the meter reads zero resistance while the ends of the probes are touching. If the scale is changed, the meter must be recalibrated to zero. If the meter cannot be calibrated to zero, the batteries should be replaced.

When a switch or wire is to be tested for continuity, the needle should deflect to zero,

**Figure 11-1.** Multimeter (*Courtesy of A. W. Sperry Instruments, Inc.*)

which indicates a closed circuit. The part being checked has continuity. If, on the other hand, the needle does not move from the at-rest position, the switch is open. The part being tested does not have continuity.

If the needle stops between infinity and zero, the meter is measuring the resistance of the part being tested. Infinity means that the resistance is extremely high; probably an open circuit is being measured.

**AMMETER:** The clamp-on type ammeter is the most popular current-measuring instrument used for ac circuits because the current flow can be checked without interruption of the electrical circuit. (See Figure 11-3.) All that is necessary to check the current flow is to clamp the tongs around one of the wires supplying electrical power to the circuit being tested.

Clamp-on ammeters are designed primarily to measure the current flow in ac circuits only. These instruments measure the current flow in a wire by making it serve as the primary side of a transformer. The circuit to the meter movement serves as the secondary side of the transformer (see Figure 11-4). These instruments are most accurate when the current-carrying wire is in the center of the tongs. For best accuracy of the instrument, the mating surfaces of the tongs should be kept clean.

When testing a circuit that has a low current flow, the wire can be wrapped around one tong of the ammeter. The sensitivity of the meter will be multiplied by the number of turns used. To obtain the correct amperage reading, divide the meter reading by the number of turns taken on the tong.

**WATTMETER:** The wattmeter can be used to determine the voltage to the unit and the total wattage draw of the unit. (See Figure 11-5.) Be sure to set the meter on the high-

**Figure 11-3.** Clamp-on ammeter (*Courtesy of Amprobe Instrument Division of SOS Consolidated Inc.*)

est range possible because the starting wattage of most motors is very high. After the unit is running, the proper scale can be selected so that the needle indicates a mid-scale reading.

**CAPACITOR ANALYZER:** The capacitor analyzer is a very important instrument to the service technician. It can be used to check for open or shorted capacitors, for the microfarad capacity, and for the power factor. (See Figure 11-6.) Scales are provided for both the voltage at which the instrument will be used and the microfarad rating of the capacitor. It may be used to check both the running and starting capacitors.

**Figure 11-6.** Capacitor analyzer (*Courtesy of Robinair Manufacturing Company*)

**TEMPERATURE TESTER:** Most electronic temperature testers (Figure 11-7) have thermistor sensing leads that are used primarily to measure temperatures below 200°F (93°C). Do not attempt to alter these

**Figure 11-4.** Use of clamp-on ammeter (*Courtesy of Amprobe Instrument Division of SOS Consolidated Inc.*)

**Figure 11-5.** Wattmeter (*Courtesy of Robinair Manufacturing Company*)

**Figure 11-7.** Electronic temperature tester (*Courtesy of Mechanical Refrigeration Enterprises*)

leads in any way. If the leads do not function properly, they should be replaced. By using these leads, temperatures at several different locations may be taken without interruption of the normal operating cycle of the unit. Almost any number of leads can be used on instruments equipped with multiple connections.

Temperature testers also are available that will record both the time and the temperature on a piece of paper (see Figure 11-8). These instruments are valuable for helping to locate intermittent temperature-related problems. When this type of instrument is desired be sure that the chart paper and inking supplies are available.

**Figure 11-9.** Micron meter (*Courtesy of Airserco Manufacturing Company*)

and use a sensing probe to measure the thermal conductivity of the gases in the system being evacuated. These instruments are designed to measure the last part of an inch of vacuum and indicate this measurement in microns. There are 25,400 microns in one inch.

The micron-sensing element is not designed to measure pressures above atmospheric pressure. Therefore, some means must be provided to protect the sensor when charging the system with refrigerant.

**MILLIVOLT METER:** The millivolt meter is an instrument used to measure voltage in thousandths of a volt (see Figure 11-10). In air conditioning and heating service work, these instruments are used to check the thermocouple output voltage on pilot safety devices. It is often necessary to check the closed circuit voltage of a thermocouple. A special adapter can be obtained that will permit checking the thermocouple under loaded conditions. (See Figure 11-11.)

**ELECTRONIC LEAK DETECTOR:** Electronic leak detectors (Figure 11-12) are very sensitive instruments. They operate by moving refrigerant vapor across an ionizing

**Figure 11-8.** Recording temperature-time meter (*Courtesy of Airserco Manufacturing Company*)

**MICRON METER:** When it is desirable to have an accurate indication of the vacuum inside a system, a micron meter is the instrument most often used. (See Figure 11-9.) These instruments are electrically powered

**Figure 11-12.** Electronic leak detector (*Courtesy TIF Instruments, Inc.*)

**Figure 11-10.** Millivolt meter (*Courtesy of Airserco Manufacturing Company*)

element. There are several ways to warn the user that a leak is present. Some instruments produce a high-pitched screech, some use a blinking light in the probe, and others use a buzzing noise. These instruments will not function when either ammonia or sulphur dioxide is the refrigerant.

**THERMOSTAT TESTER:** Thermostat testers are used to calibrate refrigeration thermostats (see Figure 11-13). Accurate results are obtained by using this instrument. The well of the tester is partially filled with alcohol. Liquid refrigerant is bled into the

**Figure 11-11.** Adapter for testing thermocouple closed-circuit voltage

**Figure 11-13.** Thermostat tester (*Courtesy of Robinair Manufacturing Company*)

**Figure 11-14.** Cross-section of thermostat tester

**Figure 11-15.** Hermetic analyzer (*Courtesy of Mechanical Refrigeration Enterprises*)

surrounding shell to cool the alcohol. A thermometer and the thermostat feeler bulb are placed in the alcohol (see Figure 11-14). The cut-out temperature of the thermostat can be determined from the thermometer. The cut-in temperature can be found by stopping the flow of liquid refrigerant and warming the tester. As the alcohol temperature increases, the cut-in temperature can be seen on the thermometer.

**HERMETIC ANALYZER:** Hermetic analyzers (Figure 11-15) can be used for many functions, including: (1) to rock free stuck or frozen compressors by reversing motor rotation; (2) as a bank of starting capacitors; (3) as an auxiliary capacitor; and (4) to check motor windings for shorts, opens, continuity, and grounds. Some can be used as ammeter, voltmeter, capacitor analyzer, ohmmeter, and relay analyzer.

**POTENTIAL RELAY ANALYZER:** The potential relay analyzer, when used properly,

can save the service technician much time and trouble. (See Figure 11-16.) Some of the uses of these instruments are: (1) to check the relay for open, closed, or short cir-

**Figure 11-16.** Potential relay analyzer (*Courtesy of Mechanical Refrigeration Enterprises*)

cuit; (2) to indicate relay pick-up and drop-out voltage; (3) to operate the relay manually; (4) as a continuity tester; and (5) to identify defective relays.

**SLING PSYCHROMETER:** The sling psychrometer (Figure 11-17) is used to determine relative humidity. It is equipped with two mercury thermometers. Both thermometers are of the ordinary dry bulb type. One has a wick placed over the bulb end; the other is left bare. The wick is wet with water. The sling psychrometer is then twirled around until a constant reading is obtained on both thermometers. The relative humidity can then be obtained from relative humidity charts.

**Figure 11-17.** Sling psychrometer (*Courtesy of Robinair Manufacturing Company*)

**HEATING SYSTEM ANALYZER:** This analyzer was designed specifically for the heating and air conditioning service technician to thoroughly and rapidly check all types of heating systems and controls and to measure temperatures throughout the system. (See Figure 11-18.) A complete analysis for a typical heating system takes only a few minutes. The instrument functions as: (1) ac–dc voltmeter, (2) millivoltmeter, (3) ohmmeter, (4) device to measure milliampere output, and (5) thermocouple probe that reads the temperature of the pilot flame.

*NOTE:* It is always important to use the manufacturer's instruction sheet for the instrument being used. Always exercise caution to prevent damage to test equipment.

**Figure 11-18.** Heating system analyzer (*Courtesy of Thermal Engineering Co.*)

## SAFETY PROCEDURES

Test equipment is a must for efficient service procedures. The information obtained through the use of damaged test equipment is useless. Every precaution should be exercised to keep instruments functioning properly.

1. Never connect an ohmmeter to an electrical circuit when the electrical power is on.

2. Do not drop, kick, or throw test instruments.

3. Always start (except with an ohmmeter) with a higher than necessary scale and work downward until midscale is reached to prevent damage to the instrument.

4. Always use the correct meter for the job.

5. Never apply open flame to a charging cylinder.

6. Never subject gauges to pressures that exceed their rating.

7. Protect test instruments from adverse weather conditions.

8. Never leave a refrigerant charge in a charging cylinder.

9. Never jump a protective fuse in a test instrument.

10. Never leave thermometers in direct sunlight, as the mercury may exceed the scale and ruin the thermometer.

## SUMMARY

• Test instruments can help the service technician to perform service operations faster and more accurately.

• The two things accomplished when test instruments are used are: (1) prevention of the unnecessary interchanging of parts, and (2) savings of time.

• Selecting of test equipment should be approached cautiously.

• The field repair of test equipment should not be attempted.

• Some meter movements are sensitive to position. Always use them in the correct position.

• Never repair test leads. Replace them when worn or damaged.

• Test instruments stored in tool boxes are subjected to extremely rough treatment.

• Avoid storing test instruments in extremely hot or cold temperatures.

• Test instruments, when properly applied, will provide more accurate trouble diagnosis in the shortest possible time.

• The voltmeter is probably the most often used test instrument. It is also the easiest to read.

• The ohmmeter is used for measuring resistances and continuity, for making quick checks on capacitors, and for testing solid state components.

• The clamp-on ammeter is the most popular current-measuring instrument used for ac circuits because the current flow can be checked without interruption of the electrical circuit.

• The wattmeter can be used to determine the voltage to the unit and the total wattage draw of the unit.

• The capacitor analyzer can be used to check for open or shorted capacitors, for the microfarad capacity, and for the power factor.

• Temperature testers are used to sense accurate temperatures during normal operation of the equipment being tested.

• Micron meters are used to indicate a very high vacuum which is not possible with ordinary service gauges.

• Millivolt meters are used to check the output voltage of thermocouples.

• Electronic leak detectors are very sensitive instruments used to detect refrigerant leaks.

• Thermostat testers are used to calibrate refrigeration thermostats accurately.

• Hermetic analyzers are very useful in hermetic compressor servicing.

• Potential relay analyzers are used to check the condition of potential relays.

• The sling psychrometer is used in determining relative humidity.

• Heating system analyzers are designed to aid the service technician in repairing the various components in a heating system.

## REVIEW EXERCISES

1. What is the purpose of test instruments?

2. What are two strong arguments for using test instruments?

3. What is a good method of selecting test instruments?

4. What is the fault of having all the test functions in one instrument?

5. Why should the batteries be stored separately in seldom used instruments?

6. What is the purpose of fuses and circuit breakers on test instruments?

7. In what position should the selector switch be placed when storing test instruments?

8. Which test instrument is the most often used?

9. Which test instrument is the easiest to read?

10. Which test instrument is used to check continuity?

11. Why are clamp-on ammeters popular among service technicians?

12. Name the function of a capacitor analyzer.

13. What is an advantage of using a temperature tester?

14. When is a micron meter used?

15. Under what conditions can an electronic leak detector not be used?

16. What is the fluid used in a thermostat tester well?

17. When is a hermetic analyzer especially useful?

18. What instrument is used to determine relative humidity?

19. What are the functions of a heating system analyzer?

# Electric Motors
# and
# Motor Controls

The laws of magnetism and magnetic induction discussed in Chapter 10 are the basis on which all electric motors operate. It is desirable that an understanding of these principles be obtained before the basic theory of electric motors is attempted.

## ELECTRIC MOTOR THEORY

The most simple electric motor possible is made up of a rotor, a permanent magnet mounted on a movable shaft, and two magnetic poles mounted on the outside motor shell. (See Figure 12-1.) These magnetic poles are actually coils of wire; they are stationary poles and are referred to as the *stator*. These coils of wire produce strong magnetic fields when electric current is passed through them.

The amount of starting torque built into the motor circuit is the basic difference among the single-phase motors that are used in the refrigeration and air conditioning industry. The method by which torque is

**Figure 12-1.** Simple electric motor

obtained is the major difference in these motors.

The magnetic laws of attraction state that like magnetic poles repel and that unlike magnetic poles attract. The top pole marked *N* in Figure 12-1 is magnetically attracting the *S* pole of the rotor. Because the stator is stationary, the rotor moves toward the stator poles and rotation of the shaft begins.

The electric current in use today is 60-cycle alternating current. This means that the electric motors that are used must be designed to operate on this type of current. When alternating current is used, the direction of current flow is reversed alternately from a high negative potential to a high positive potential. In modern electrical systems, this change takes place 60 times each second. Therefore, the polarity of each stator coil is reversed each time the potential changes from negative to positive.

Because of this alternating polarity of the magnetic poles of the stator, there is a push–pull action on the rotor. Therefore, before the rotor can stop in line with the stator poles, the current changes direction which also changes the magnetic polarity of the poles. The momentum of the rotor carries it past the in-line position with the pole. The south pole of the stator attracts the north pole of the rotor and repels its south pole (see Figure 12-2). At the same instant, the north pole of the stator repels the north pole of the rotor and attracts its south pole. In this way, the push–pull action is kept continuous. The rotor will continue to spin as long as the poles continue to change their magnetic polarity. The rotor will theoreti-

cally adjust itself to a speed of 60 revolutions per second (rps) or 3600 revolutions per minute (rpm). This is called the *synchronous speed* of the motor. Mathematically, the synchronous speed of a motor may be determined by the following formula:

$$\text{Synchronous Speed} = \frac{120 \times \text{Frequency}}{\text{Number of Poles}}$$

For example: The synchronous speed of a two-pole motor is

$$\frac{120 \times 60}{2} = 3600 \text{ rpm}$$

The synchronous speed of a four-pole motor is

$$\frac{120 \times 60}{4} = 1800 \text{ rpm}$$

From the foregoing explanation, it can be easily seen that as the stator poles continue to alternate between north and south, the rotor will continue to rotate. However, the major problem lies in starting the rotor, not running it.

Should the rotor be stopped in line with the stator poles, the rotor could not be started, regardless of the polarity of the two stator poles. (See Figure 12-3.) In this condition, the north pole is attracting the south pole and the rotor will not move. Even when the stator pole is changed to the south pole, the repelling forces will be at such an angle that the rotor will not turn.

However, if another pole is installed in the stator, the rotor can be caused to rotate (see Figure 12-4). The phase north pole on the left will attract the south pole on the rotor at the same time that the north pole on the right does. The stator pole on the left will provide enough magnetic attraction to pull the motor to the left and start the rotation of the rotor.

**Figure 12-2.** Rotor repulsion and attraction

**Figure 12-3.** Rotor in line with stator poles

**Figure 12-4.** Phase stator pole

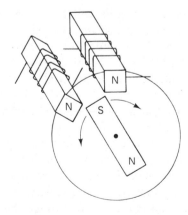

**Figure 12-5.** Equal rotor attraction between two poles

However, if the rotor should stop between these two poles the problem of starting the rotor would still exist (see Figure 12-5). As the polarity of the poles is established, the magnetic attraction of the rotor is equal in both directions.

The obvious solution would be to have a motor circuit that would build a strong magnetic strength in one stator pole before the other. In Figure 12-6 the two stator windings and the rotor have been connected to a common source of electrical current. The

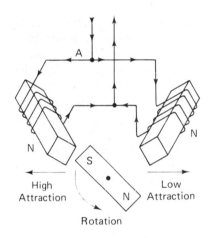

**Figure 12-6.** High and low attraction of stator poles

electricity enters the motor circuit at point A. The current is divided at this point and part flows to the left while the rest flows to the right. If the current can be made to flow to the left before it flows to the right, the left stator pole will obtain a strong north polarity earlier than the right stator pole. Therefore, the strong attraction to the left will overcome the weaker attraction to the right and the rotor will turn.

Finding a method of making the current in the left pole lead or flow ahead of the current in the right stator pole would solve the problem of effectively making the motor start. In other words, two-phase current must be available to produce the magnetic attraction and repulsion which is required to start the motor. The component used to produce this two-phase current is a capacitor.

## CAPACITORS

In order to understand how a capacitor can produce the second phase of current which is needed to start a motor, a review of some of the basic facts about ac voltage and current is necessary.

To illustrate some of these facts a pure resistance circuit will be used. In Figure 12-7, the voltage which is applied from point $A$ to point $B$ is represented in the illustration on the right by use of the sine wave. Because the current flow through a pure resistive load will be in phase with the applied voltage (solid line) the current part of the wave form (dashed line) is shown at a peak when the voltage is at a peak and at zero when the voltage is zero.

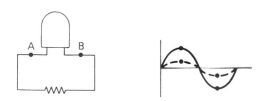

**Figure 12-7.** Pure resistive circuit

If a capacitor is connected into an alternating current circuit, the current through the load will not remain in phase with the applied voltage. In a capacitive circuit, the load current will lead the applied voltage. Figure 12-8 shows a capacitor connected in series with the resistive load. The voltage that is applied to points $C$ and $D$ has the same wave as $A$ and $B$ in Figure 12-7. The

**Figure 12-8.** Capacitive circuit

current wave form (the dashed line) is now shown leading the applied voltage.

From this it can be seen that installing a capacitor in series with one of the two stator windings will cause the current flowing through that winding to lead the current that flows through the winding that is connected directly to the power supply.

It has been stated that the capacitor is the component that provides the two-phase electrical current necessary to start a motor. Therefore, a basic knowledge of its characteristics is required.

Capacitors are rated in millionths of a farad or in microfarads (mfd). A large mfd rating is the result of using either large metal plates or a small amount of insulation between the plates. A low mfd rating is the result of using small metal plates or a large amount of insulation between the plates. (See Figure 12-9.)

A large phase shift will occur between the applied voltage and the load when a capacitor with a large mfd rating is connected in series with a load. This high phase shift is what provides the high starting torque for single-phase motors. A low phase shift and low starting torque is the result of using a capacitor with a low mfd rating.

Also, the amount of electrical current which will flow through a series load when capacitors with different mfd ratings are used is important. High mfd capacitors will provide a high current flow through the series load, while low mfd capacitors will pro-

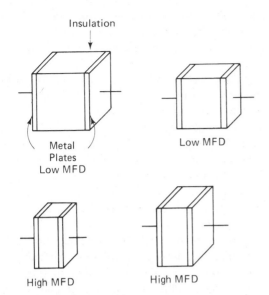

**Figure 12-9.** Effects of plates and insulation on capacity of capacitors

vide a low current flow through the series load.

Two types of capacitors are used to produce the second-phase current necessary to start single-phase motors. (See Figure 12-10.) The physical dimensions have nothing to do with the mfd rating of a capacitor. A run capacitor will usually have a low mfd

rating. It is physically larger because of the dielectric oil type insulation. This oil dissipates the internal heat from the plates to the outside atmosphere. The run capacitor provides a low starting torque and a low current flow through the series load.

The starting capacitor usually has a higher mfd rating than the running capacitor. The starting capacitor is switched out of the circuit a few seconds after the motor starts. It is necessary that the starting capacitor be switched out because it is not designed to carry the heavy current flow for an extended period of time. If the starting capacitor is left in the circuit for an extended period of time permanent damage may be done to the capacitor.

To make a graphic illustration of this principle, the motor with the auxiliary stator pole that was used in Figure 12-4 is used in Figure 12-11. Here, both the pole windings have been connected across the electrical power supply. A capacitor is connected in series with the pole on the right. The rotor has stopped in a position between the two poles. With the capacitor connected in series, the current leads the applied voltage.

Therefore, the current flowing through

Run          Start

**Figure 12-10.** Motor capacitors

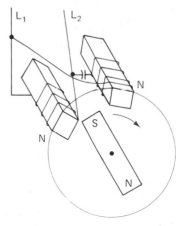

**Figure 12-11.** Use of a capacitor

the right stator pole is leading the current through the left stator pole. Thus, the north pole in the right stator pole will become strong before the pole on the left stator pole. Because of this, the south pole of the rotor will be deflected toward the pole on the right and the starting torque is created.

An instant later, the current reverses direction in the right stator pole and the polarity of the stator pole is reversed (see Figure 12-12). Now the south stator pole is repelling the south pole of the rotor. This repulsion increases the amount of work that a motor can do. A more efficient motor is the result of leaving the capacitor in the circuit.

There are five types of single-phase motors that are used on modern refrigeration and air conditioning equipment. They are: (1) split-phase, (2) capacitor-start (CSR), (3) permanent-split capacitor (PSC), (4) capacitor-start/capacitor-run (CSCR), and (5) shaded-pole.

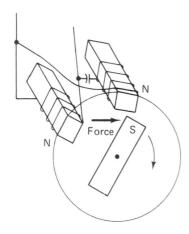

**Figure 12-12.** Rotor deflection

### SPLIT-PHASE MOTORS

These motors are popular when motor sizes ranging from $\frac{1}{20}$ to $\frac{1}{3}$ hp are required. Split-phase motors are used for such applications as oil burners, blowers, pumps, and

fans. The starting torque requirements for these applications are moderate—not a great deal more than the running torque requirements.

Split-phase motors are single-phase induction motors which use a nonwound rotor and a stator whose windings are embedded in the insulated slots of a laminated steel core (see Figure 12-13). The stator windings are comprised of two separate windings: the main or running winding and the starting or phase winding. These two windings are connected electrically in parallel (see Figure 12-14). The phase winding is physically placed in a magnetic position to the main winding to effectively produce a two-phase field which is required to start a single-phase induction type motor and help bring it up to its normal running speed. After the motor has reached approximately three-fourths of its normal running speed a centrifugal switch, which is actuated by a governor weight mechanism, opens the electrical circuit to the phase winding. (See Figure 12-15.) After the centrifugal switch removes the phase winding from the circuit,

**Figure 12-13.** Split-phase motor cutaway (*Courtesy of General Electric Co.*)

**Figure 12-14.** Split-phase motor connection diagram (starting)

**Figure 12-15.** Split-phase motor diagram in the operating position

the split-phase motor operates as a single-phase induction motor.

## CAPACITOR-START (CSR) MOTORS

These motors are similar to split-phase motors with the exception that a starting capacitor is used. The starting capacitor is connected electrically in series with the phase winding as shown in Figure 12-16. The purpose of the capacitor is to increase the starting torque and reduce the starting current. The capacitor is removed from the circuit by the centrifugal switch when the motor reaches approximately 75 percent of its normal operating speed.

Since the capacitor is in the circuit only during the starting cycle, the running characteristics of split-phase motors and capacitor-start motors of the same horsepower rating are identical.

**Figure 12-16.** Capacitor-start motor connection diagram

Capacitor-start motors are used on applications that require a high starting torque (150 percent to 350 percent of the running torque) or where low starting amperes are required. This group includes such components as compressors, large fans and blowers, and water pumps.

## TWO-SPEED MOTORS

Two-speed motors are made in both the split-phase and the capacitor-start types. Three separate windings are used in these motors. They are: the phase-winding, the low-speed main winding, and the high-speed main winding (see Figure 12-17). The low-speed and high-speed connections are determined by an external switch or relay in the control circuit. During high-speed operation, the motor functions like a single-speed motor using only the phase and high-speed main winding. During low-speed operation, the motor starts on the phase and high-speed windings. After the starting cycle is completed, the motor switches to the low-speed main winding (see Figure 12-18).

Two-speed motors are used to operate blowers in air conditioning systems that use a modulated air flow or fans, and in other

**Figure 12-17.** Two-speed electrical connection diagram

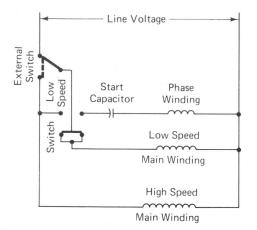

**Figure 12-18.** Two-speed motor in low speed condition

applications where more than one speed is required.

## CENTRIFUGAL SWITCH

The heart of split-phase and capacitor-start motors is the centrifugal switch (see Figure 12-20). The centrifugal switch consists of four principal components: (1) the moving contact arm, (2) the stationary contact plate, (3) the governor weight, and (4) the

weight spring. The contact arm pivots on the contacts, (8) and is held in position with two pins or screws, (5 and 6). A compression spring (7) on the top pin (5) holds the contacts (8) open when the switch is in the running position. The stationary contact plate assembly (2) includes the electrical line terminal studs and the overload protection device, when used.

**OPERATION:** When the motor is not running, the position of the switch parts is with the riding edge of the governor weight (3) against the contact arm (1) which holds the contacts (8) in a closed position and the contacts (9) open. When the motor is energized the rotor starts to rotate, and as it accelerates to the predetermined switching speed, the centrifugal forces of the governor weight (3) overcomes the force of the governor weight spring (4) and snaps outward on the pin (11), permitting the contact arm (1) to move toward the rotor, opening the contact (8) and de-energizing the phase winding. The contacts (9) close, completing an external circuit such as indicator lights, heater circuits, relays, etc.

## PERMANENT-SPLIT CAPACITOR (PSC) MOTORS

These motors are used for a large number of applications in the refrigeration and air conditioning industry. They are used on fans and blowers and on some compressors. They are rated with a medium starting torque and have excellent running efficiencies. The compressors that use this motor must be installed on refrigeration systems in which the refrigerant pressures will equalize during the off cycle.

Permanent-split capacitor motors incorporate a running capacitor connected in series with the phase winding (see Figure

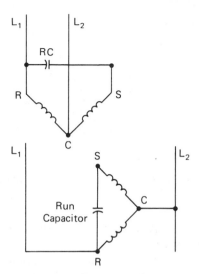

**Figure 12-19.** Permanent-split capacitor connection diagram

12-19). The running capacitor serves two important functions. First, the capacitor provides the split-phase electrical characteristics necessary to start the motor turning. Second, because the capacitor is not removed, this split-phase provides efficient motor operation. The start (phase) winding is made of much smaller wire than the run winding. It also has many more turns than the main (run) winding. Because of these differences: (1) the phase winding has a higher resistance than the main winding, and (2) the start winding cannot withstand a very high current draw for a long period of time.

Because the start winding cannot withstand a continuous high current draw, the running capacitor must have a low mfd rating to limit the flow of current through the start winding. When using running capacitors with a low mfd rating, only a small phase angle will result, which allows the motor to have only a moderate starting torque. The running capacitor is always connected between the run and start terminals of the motor.

## CAPACITOR-START/CAPACITOR-RUN (CSCR) MOTORS

These motors make use of the best characteristics of both the permanent-split capacitor motor and the capacitor-start motor. The permanent-split capacitor has excellent running efficiency but only moderate starting torque. The capacitor-start motor has a high starting torque but a lower running efficiency than the PSC motor.

Therefore, the high starting torque of the CSR motor and the higher runing efficiency of the PSC motor are combined to make the CSCR motor. The PSC electrical circuit remains the same as with the PSC motor. The capacitor-start circuit is combined with a starting relay to take the place of the centrifugal switch that was discussed earlier. The relay is added so that the circuit can be used on hermetic and semi-hermetic compressors. The centrifugal switch is eliminated on this type of equipment because the repair of the switch would be impossible. The relay is accessible for replacement.

In operation, when the electrical circuit is completed to the motor, it starts as a capacitor-start unit. As the motor reaches approximately 75 percent of the operating speed, the relay removes the starting capacitor from the circuit. The motor then operates as a permanent-split capacitor motor.

## SHADED-POLE MOTOR

These motors develop a very low torque. Therefore, they are very small in size and are usually less than ½ hp. They are used on small fans, pumps, and timer motors. They are low in cost and comparatively dependable.

The stator poles are made of laminated steel as in the other type motors. However,

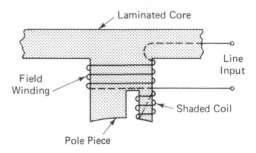

**Figure 12-22.** Windings on a shaded-pole motor

**Figure 12-20.** Capacitor-start capacitor-run connection diagram

the pole pieces are quite different because each pole piece has a slot cut in the face (see Figure 12-21). Even though it is a form of an induction motor, its windings are distributed differently than those of other induction type motors. The slot in the face of the pole piece contains the shading coil. This shading coil may be made of a single- or a multiple-strand wire (see Figure 12-22). The shading pole coil must form a closed circuit or loop. This loop is not connected to the source of electricity. The main field

**Figure 12-21.** Shaded-pole motor pole piece

windings are wound around the remainder of the pole piece.

When electrical current is applied the magnetic force around the wire is constantly changing direction, creating a magnetic pole with a force changing from $N$ to $S$. The magnetic lines of force around the pole are built up to the maximum in first one direction and then the other. The shaded-pole banding changes the rate of speed with which the building up and collapsing of the magnetic lines occur.

Thus, a condition is created where the main pole has a magnetic field that is building up and collapsing at a normal rate while the magnetic field of the smaller shaded pole is building up and collapsing at a different rate.

The main pole and the shaded pole peak in magnetic magnitude at different rates of speed. Therefore, at some time on the way to that peak, one of the poles will momentarily get ahead of the other. At that moment the magnetic strength is not the same; a torque is developed, and the rotor will turn.

### SINGLE-PHASE MOTOR PROTECTORS

Most single-phase motor circuits include a switch to protect the motor against overheating, overcurrent, or both types of conditions. These overload devices may be either internal or external types depending on the equipment design.

**INTERNAL OVERLOAD:** These overloads may be of the line break type or the thermostat type. Both types of internal overloads are located precisely in the center of the *heat sink* portion of the motor windings and protect the motor from excessive temperatures and excessive current draw. (See Figure 12-23.) The line break type of internal overload will interrupt the electrical power line to the common connection of the motor winding if either of these conditions occurs (see Figure 12-24).The internal thermostat overload protector is also mounted in the motor windings. However, the switch interrupts the control circuit to the contactor or starter (see Figure 12-25). The electrical connections are shown in Figure 12-26. If this overload becomes faulty, the compressor must be replaced.

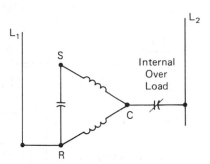

**Figure 12-24.** Line break internal overload connection

**Figure 12-25.** Internal thermostat overload (*Courtesy of Tecumseh Products Company*)

**EXTERNAL OVERLOAD:** This type of overload protector is mounted externally and may be replaced when faulty. It is mounted in contact with the motor housing or the hermetic compressor housing to sense the temperature at the point of contact.

External overloads may also be of the line break or the thermostat type. These

**Figure 12-23.** Internal overload location (*Courtesy of Tecumseh Products Company*)

## SINGLE PHASE

TO CONTROL CIRCUIT
24v ·115v OR 230 VOLTS
(300 VOLTS MAX.)

HERMETIC
TERMINAL
(MOTOR WINDING
THERMOSTAT)

SUPPLEMENTARY
OVERLOAD

CONNECTOR BLOCK

2 POLE CONTACTOR

RUN CAP

FOR PSC OPERATION ELIMINATE START
CAPACITOR & RELAY. HOOK UP RUN CAPACITOR
BETWEEN START & RUN OF CONNECTOR BLOCK.

BLEEDER RESISTOR          START
CAP

**Figure 12-26.** Internal overload thermostat connections (*Courtesy of Tecumseh Products Company*)

TO: LINE

TO: S. CAP

TO: LINE

TO: R. CAP.

**Figure 12-27.** Location of external overload device (*Courtesy of Tecumseh Products Company*)

overload devices must be placed where the manufacturer has designated. (See Figure 12-27.) They must always be replaced with an exact replacement or proper protection will not be provided.

The line break external thermostat is operated by a bimetal disc (see Figure 12-28). The switch contacts are normally closed and when an overload condition occurs the contacts open (see Figure 12-29). They are electrically connected into the electrical circuit to the common terminal of the motor winding. (See Figure 12-30.) When the resistance heater has provided sufficient heat indicating an overload condition, the contacts open and interrupt the electrical circuit to the motor. The motor then stops until the overload has cooled sufficiently to allow the contacts to reset and complete the electrical circuit to the motor.

The thermostat type of external overload is mounted in the same way as the line break type. However, the contacts do not

**Figure 12-28.** External line break overload (*Courtesy of Tecumseh Products Company*)

Open

Closed

**Figure 12-29.** External overload contacts

open the main electrical power circuit. (See Figure 12-31.) Instead, the control or pilot circuit is interrupted which causes the contactor or starter to interrupt the main power supply. (See Figure 12-32.)

## TESTING SINGLE-PHASE MOTORS

Many service technicians make the false assumption that troubleshooting motors is difficult, especially the capacitor-start/capacitor-run type motors. If the service technician is familiar with the separate electrical circuits that cause the motor to function, a systematic approach to troubleshooting each component will usually yield any troubles in the motor.

Also, certain safety precautions should be observed when attempting to troubleshoot any electrical system. Always be certain that the electrical system being serviced is disconnected from the electrical supply.

Capacitors can hold a high voltage charge on their plates long after the electrical supply has been disconnected. Be sure to discharge capacitors before working on them with a 20,000 ohm resistor touched to each terminal.

The counter emf on the start winding can be much higher than the applied voltage. Therefore, all personnel and meters should be protected against injury or damage.

**LOCATING COMPRESSOR MOTOR TERMINALS:** It is a simple process to determine the motor winding terminals on a compressor. When the compressor is not marked or the manufacturer's specifications are not readily available, a good ohmmeter will provide accurate information. The first step in determining the common ($C$), start ($S$), and run ($R$) terminals is to draw the exact location of the terminals on a piece of

# RESISTANCE START INDUCTION RUN

**Figure 12-30.** External line break overload connections

PILOT CIRCUIT
CONTACTS

BIMETAL DISC
FLEXES UPWARD

NOTE:
BIMETAL DISC DOES NOT
BREAK LINE CURRENT

**Figure 12-31.** External thermostat overload (*Courtesy of Tecumseh Products Company*)

paper (see Figure 12-33). Use the ohmmeter to determine the resistances between each terminal and mark them on the paper in the proper place. The following rule will then apply: The greatest resistance will be between the start and run terminals; the medium resistance will be between the start and common terminals; the least resistance will be between the run and common terminals. In Figure 12-33 the greatest resistance is between terminals 1 and 2; the medium resistance is between 2 and 3, and the least resistance between 1 and 3. Therefore, terminal number 3 is common ($C$), terminal number 2 is start ($S$), and terminal number 1 is run ($R$). The sum of the resistances between common to start and common to run should equal the resistance between run to start.

# THREE PHASE

**Figure 12-32.** External thermostat overload connection

**GROUNDED WINDINGS:** To check for grounded motor windings, the ohmmeter should be set on the highest resistance value, $R \times 10,000$, if possible. Zero the ohmmeter and attach one lead firmly to the motor housing, then firmly touch each terminal separately (see Figure 12-34). Any hermetic motor of 1 hp or less should have at least one million ohms resistance between the

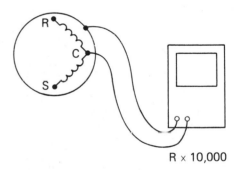

**Figure 12-34.** Checking for grounded windings

Therefore

**Figure 12-33.** Locating compressor terminals

motor windings and the housing, unless the manufacturer's specifications indicate otherwise. Motors larger than 1 hp should have a minimum resistance of 1000 ohms per volt. The motor should be warm. If possible, run the motor for a few minutes before checking the resistance. A warm winding will be more likely to show a ground than a cold winding.

**OPEN OR SHORTED WINDINGS:** An open winding indicates that the winding has separated, leaving the motor inoperative, and will be indicated by an infinite resistance reading. A shorted winding indicates that the insulation has broken down, allowing part of the winding to be left out of the circuit. This condition is probably the most difficult condition to identify. The ohmmeter will show a lower than normal resistance on a shorted winding. To test for these conditions, the $R \times 1$ scale on the ohmmeter should be used. Any reading different from the manufacturer's specifications indicates a bad motor winding. First, zero the ohmmeter. Be sure that the motor terminals are clean and a good connection can be made. Connect one lead of the ohmmeter to one terminal and alternately touch firmly the other lead to the other terminals. (See Figure 12-35.)

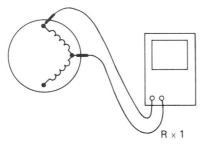

R x 1

**Figure 12-35.** Checking for open or shorted windings

**REPLACING CAPACITORS:** Because the service technician will not always have an exact replacement capacitor, he must know how to make substitutions. There are several rules that can be applied in capacitor replacement.

1. The voltage rating of any capacitor replacement must be equal to or greater than the capacitor being replaced.

2. When replacing starting capacitors, the capacitance must be equal to but not greater than 20 percent more than the capacitor being replaced.

3. When replacing a running capacitor, the capacitance may vary no more than $\pm$ 10 percent of the capacitor being replaced.

Rules for capacitors in parallel:

1. All capacitors must have a voltage rating equal to or greater than the capacitor being replaced.

2. The total mfd rating is the sum of the mfd ratings of each capacitor.

Rules for capacitors in series:

1. The sum of the voltages must at least equal the voltage of the capacitor being replaced.

2. The total capacitance of the capacitors in series is found by using the formula

$$C_t = \frac{C_1 C_2}{C_1 + C_2}$$

**STARTING RELAYS**

Starting relays are used to remove the starting circuit from operation when the motor reaches approximately 75 percent of its normal running speed. Its function is basically the same as the centrifugal switch discussed earlier. There are four types of starting relays in common usage today. They are (1) amperage (current) relay, (2) hot wire relay, (3) solid state, and (4) potential (voltage) relay. The horsepower size and design of the equipment regulate which type of starting relay is used.

**AMPERAGE (CURRENT) STARTING RELAY:** This type of starting relay is normally used on ½ hp units and smaller. This is an electromagnetic relay. The coil of the relay is connected in series with the motor run winding. Therefore, the coil wire must be large enough to carry the amperage draw of the motor. (See Figure 12-36.) The contacts are normally open. Amperage relays are positional type relays. That is, they must be installed so that the weighted armature that carries the contacts will function by gravity pull.

**Figure 12-37.** Amperage relay connections

GENERAL ELECTRIC
3ARR2
BRACKET TYPE

**Figure 12-36.** Amperage relay (*Courtesy of General Electric* Co.)

In operation, when the motor control completes the electrical circuit to the motor, the current flow through the relay coil is at its peak. The contacts are pulled together by the electromagnetic field in the relay coil (see Figure 12-37). When the contacts are closed, an electrical circuit is completed to the starting circuit of the motor providing the electrical phase shift required to start the motor. As the motor reaches approximately 75 percent of its full speed, the counter emf built up in the start winding opposes the applied voltage and the amperage draw is reduced. This reduced amperage causes a reduced magnetic field in the coil of the starting relay and the contacts are pulled apart by the weight of the armature, stopping operation of the starting circuit. The

motor is now in its normal operating condition. When the motor control interrupts the electrical power to the motor, it will stop and the relay contacts will remain open until the electrical power is again applied to the motor.

Amperage starting relays must be sized for each motor horsepower and amperage rating. A relay rated for too large a motor may not close the relay contacts, which will leave out the much-needed starting circuit. The motor will not start under these conditions. A relay which is rated for too small a compressor may keep the contacts closed at all times while electrical power is applied, leaving the starting circuit engaged continuously. Damage to the starting circuit may occur under these conditions. A motor protector must be used with this type relay.

**HOT WIRE RELAY:** This type is a form of current relay, but it is not operated by an electromagnetic coil. It is designed to sense the heat produced by the flow of electrical current. The current flowing through a resistance (hot) wire provides heat to a bimetal which operates the contacts. (See Figure 12-38.) There are two sets of contacts in this relay, a set for starting and a set

Running Contacts

Hot Wire

L

M

S

3  2  1

Starting Contacts          Bimetal Spring Blades

**Figure 12-38.  Hot wire relay**

for running the motor. The contacts are normally closed.

In operation, when the motor control completes the electrical circuit to the motor, the current is automatically supplied to both the running winding and the starting winding (see Figure 12-39). The current flowing through the resistance wire to the main winding causes the wire to get hot. This heat causes the bimetal to warp and open the starting contacts. The main or running contacts remain closed and the relay stays in the normal operating position. Should the motor continue to draw an excessive amount

Hot Wire Relay

$L_1$  L

R

M

S

C

S

$L_2$

**Figure 12-39.  Hot wire relay connections**

of current, the hot wire will provide enough heat to cause the bimetal to warp further and open the running contacts. The electrical circuit to both the running winding and starting winding is interrupted. The contacts will remain open until the bimetal has cooled sufficiently to close the contacts. Both the running and starting contacts close at the same instant to start the motor operating again.

No overload is necessary when this type of starting relay is used. However, some motor manufacturers will use an internal overload in conjunction with the hot wire relay. Hot wire relays are nonpositional relays. They also must be sized to each motor. If the relay is too large, the starting contacts may remain closed for too long a time or not open at all, thus possibly damaging the motor windings. If the relay is too small, the relay may stop the motor as if it were overloaded and sufficient running time will not be accomplished.

**POTENTIAL (VOLTAGE) STARTING RELAY:** Potential starting relays operate on the electromagnetic principle. They incorporate a coil of very fine wire wound around a core. These starting relays are used on motors of almost any size. The contacts on this relay are normally closed and are caused to open when a plunger is pulled into the relay coil. These relays have three connections to the inside in order for the relay to perform its function. These terminals are numbered 1, 2, and 5. Other terminals numbered 4 and 6 are sometimes used as auxiliary terminals. (See Figure 12-40.) The relay is installed with the number 5 terminal connected to both the electrical line to the motor and to the common terminal on the motor. Terminal number 2 is connected to the motor start winding. Terminal number 1 is connected to the starting capacitor.

**Figure 12-40.** Potential starting relay (*Courtesy of General Electric Co.*)

In operation, when the motor controller completes the electrical circuit to the motor, electricity is also supplied to the starting winding through the relay contacts between terminals 1 and 2. (See Figure 12-41.) As the motor reaches approximately 75 percent of its rated speed, the counter emf in the start winding has increased sufficiently to cause the relay to "pick up," opening the contacts and removing the starting components from the circuit. Remember, it is the voltage in the start winding that causes the relay to function. The relay is in the normal operating position. When the controller interrupts the electrical circuit to the motor,

the motor begins slowing down. The counter emf in the start winding is reduced and when a certain voltage is reached the relay "drops out," closing the contacts. The relay is now in the motor starting position.

Potential starting relays are nonpositional. The sizing of these relays is not as critical as with the amperage and hot wire relays. A good way to determine what relay is required is to manually start the motor and check the voltage between the start and common terminals while the motor is operating at full speed. Multiply the voltage obtained by .75 and this will be the pick-up voltage of the required relay.

**SOLID-STATE STARTING RELAY:** These relays use a self-regulating conductive ceramic, developed by Texas Instruments, Inc., which increases in electrical resistance as the compressor starts, thus quickly reducing the starting winding current flow to a milliamp level. The relay switches in approximately .35 second. This allows this type of relay to be applied to refrigerator compressors without being tailored to each particular system within the specialized current limitations. These relays will start virtually all split-phase 115 volt hermetic compressors up to ⅓ hp. These relays are designed for push-on installation on the compressor pin connector. (See Figure 12-42).

**Figure 12-41.** Potential starting relay connections

**Figure 12-42.** Solid-state motor start relay (*Courtesy of Klixon Controls Division, Texas Instruments, Inc.*)

In operation, this relay is connected in the motor circuit with the ceramic material between the line and the starting terminal of the compressor. (See Figure 12-43.) That is, it is connected in series with the starting winding. When electrical power is applied to the relay, the ceramic material heats up and turns the relay off in approximately .35 second. This reduces the current flow through the start winding to a milliamp level until the electric power is removed from the circuit. After the power is removed, the ceramic material requires a few seconds to cool down before the next start cycle.

The 4EA relay is not suitable for replacement on capacitor start compressors, or for those systems with unusually rapid cycle rates.

**Figure 12-43.** Solid-state relay connections (*Courtesy of Klixon Controls Division, Texas Instruments Incorporated*)

**SOLID-STATE HARD START KIT:** This is a way to provide the additional starting torque necessary to solve PSC motor starting problems. This is done through the use of ptc ceramic materials. This material increases its resistance with an increase in temperature. At a temperature known as its anomaly temperature, the resistance increase is very sharp. (See Figure 12-44.)

One of the objectives of a ptc start assist device is to provide a surge of current that lasts for a period of time sufficient to start the compressor. This additional current then decreases, permitting the motor to run in the normal PSC mode. When electrical power is applied to the PSC motor, current flows through the start winding and through the

**Figure 12-44.** Solid-state hard start kit (*Courtesy of Klixon Controls Division, Texas Instruments Incorporated*)

L₁           L₂

Important:

This solid-state device requires a 6 minute cool-down period if initial attempt to start is not successful.

Contactor

Run Capacitor

3EA1

Main Winding

Start Winding

Cont.

Motor/Compressor

**Figure 12-45.** Solid-state hard start kit connections (*Courtesy of Klixon Controls Division, Texas Instruments Incorporated*)

parallel combination of the run capacitor and the low resistance ptc. (See Figure 12-45.)

The low resistance during starting not only increases the start winding current, but also reduces its angular displacement with the main winding current. In most PSC motors, this is advantageous since the angle between these two currents is usually greater than 90 percent.

While the surge current increases motor starting torque, it is also flowing through the ptc and heating it to its high resistance region. The time it takes the ptc to heat to its high resistance state is, unlike the operation of a potential relay, independent of when the motor starts. Instead, it is a function of the ptc's mass, anomaly temperature, resistance, and the voltage applied to the ptc. When a 230-volt compressor having a 3 EA

start assist is initially energized at nominal voltage, the 3 EA switching time is 16 electrical cycles. Starting the same compressor with 25 percent less voltage increases the 3 EA switching time to 32 electrical cycles, providing added assistance under these more difficult starting conditions.

The above switching times are ideal for starting PSC motors because the ptc will only be in its low resistance state long enough to overcome the initial inertia of the motor compressor. When the ptc switching time is longer than normal motor start times, the low resistance ptc effectively shunts out the run capacitor. The excessive *on* time retards the motor speed while the motor is trying to overcome the increasing load.

Once the ptc heats up and reaches its anomaly temperature, its resistance increases dramatically to around 80,000 ohms, and effectively switches itself out of the circuit without requiring an electromechanical relay. While the motor is running, the 3 EA continues to draw only 6 milliamps of current, which is less than $\frac{1}{1000}$ of the surge current provided by the 3 EA during the motor starting period. This low current has no effect on the start winding or on motor running performance.

When the motor is shut off, the ptc is also de-energized and starts cooling to ambient temperature. Should the motor be restarted before the ptc cools below the anomaly temperature, the motor will try starting in the standard PSC mode. Cool-down times in excess of one minute will usually provide sufficient time for the ptc to cool below its anomaly temperature and provide a start assist.

It is recommended that the 3 EA 1 be used on compressors up to 48,000 Btus or motors up to 4 hp. However, its use is not limited to this size range.

## THERMOSTATS

A thermostat is a device that acts to make or break an electrical circuit in response to a change in temperature. There are many types of thermostats in use today. They range from simple bimetallic switches to multiple contact controls which operate from remote sensing bulbs. They may have fixed control points or multiple variable adjustments.

Cooling thermostats will make on a rise in temperature and break on a fall in temperature. Heating thermostats will make on a fall in temperature and break on a rise in temperature. The type used will depend on the particular installation.

## PRESSURE CONTROLS

There are two types of pressure controls: (1) low pressure and (2) high pressure. These controls are switching devices that are used to stop the motor compressor when the pressures reach a predetermined point.

**LOW PRESSURE CONTROLS:** In general, low pressure controls are connected to the low side of the compressor and are adjusted to stop the compressor if the low side pressure drops below the desired level. (See Figure 12-46.)

The main purposes of the low pressure control in air conditioning are to prevent the temperature of the evaporator coil from falling below the temperature at which frost would occur and to prevent compressor overheating due to a refrigerant shortage. It may have either manual or automatic reset.

In refrigeration work, low pressure controls are used as temperature controls, and sometimes as defrost controls, depending on the cut-in and cut-out settings. (See Table 12-1.)

CONTACTS CLOSE HIGH; OPEN LOW

B  CUT-OUT SETTING
(Changes Cut-out
Point only)

A  RANGE ADJUSTING SCREW
Set CUT-IN point first with
this adjustment.(Changes both
cut-in and cut-out points.)

**Figure 12-46.** Low pressure control (Courtesy of Penn Controls, Inc.)

**HIGH PRESSURE CONTROL:** The high pressure control is connected into the high side of the refrigeration system so that it will stop the motor compressor when the discharge pressure exceeds the control setting. (See Figure 12-47.) These high pressures could be caused by insufficient condenser cooling, excessive refrigerant charge, air in the refrigerant lines, or other abnormal conditions. The high pressure control is connected to the high side of the system, preferably where the compressor discharge service valve cannot be back-seated far enough to render the control inoperative (see Figure 12-48).

When both the high and low pressure

Table 12-1

Approximate Pressure Control Settings*

| APPLICATION | REFRIGERANT | | | | | | | |
|---|---|---|---|---|---|---|---|---|
| | 12 | | 22 | | 502 | | 717 | |
| | Out | In | Out | In | Out | In | Out | In |
| Ice Cube Maker—Dry Type Coil | 4 | 17 | 16 | 37 | 22 | 45 | — | — |
| Sweet Water Bath—Soda Fountain | 21 | 29 | 43 | 56 | 52 | 66 | 33 | 45 |
| Beer, Water, Milk Cooler, Wet Type | 19 | 29 | 40 | 56 | 48 | 66 | — | — |
| Ice Cream Trucks, Hardening Rooms | 2 | 15 | 13 | 34 | 18 | 41 | 5 | 24 |
| Eutectic Plates, Ice Cream Truck | 1 | 4 | 11 | 16 | 16 | 22 | 4 | 8 |
| Walk In, Defrost Cycle | 14 | 34 | 32 | 64 | 40 | 75 | 23 | 55 |
| Reach In, Defrost Cycle | 19 | 36 | 40 | 68 | 48 | 78 | 30 | 57 |
| Vegetable Display, Defrost Cycle | 13 | 35 | 30 | 66 | 38 | 77 | — | — |
| Vegetable Display Case—Open Type | 16 | 42 | 35 | 77 | 44 | 89 | — | — |
| Beverage Cooler, Blower Dry Type | 15 | 34 | 34 | 64 | 42 | 75 | 24 | 55 |
| Retail Florist—Blower Coil | 28 | 42 | 55 | 77 | 65 | 89 | 44 | 67 |
| Meat Display Case, Defrost Cycle | 17 | 35 | 37 | 66 | 45 | 77 | — | — |
| Meat Display Case—Open Type | 11 | 27 | 27 | 53 | 35 | 63 | — | — |
| Dairy Case—Open Type | 10 | 35 | 26 | 66 | 33 | 77 | — | — |
| Frozen Food—Open Type | 7 | 5 | 4 | 17 | 8 | 24 | — | — |
| Frozen Food—Open Type—Thermostat | 2°F. | 10°F. | — | — | — | — | — | — |
| Frozen Food—Closed Type | 1 | 8 | 11 | 22 | 16 | 29 | — | — |

* Courtesy, Sporlan Valve Co.

controls are incorporated into a single control housing, the device is called a high-low pressure control or dual-pressure control. This type of construction provides a savings in space, original cost, and electrical wiring. These controls are wired into the electrical system in two ways: (1) the control circuit may be interrupted, allowing the contactor to open and stop the motor; or (2) main power to the compressor motor may be interrupted, stopping the motor. The preferred method is to interrupt the control circuit. (See Figure 12-49.)

## OIL SAFETY CONTROLS

Oil safety controls are designed to provide compressor protection by guarding against low lubricating oil pressure. (See Figure 12-50.) A built-in time delay switch allows for oil pressure pickup on starting and avoids nuisance shutdowns on oil pressure

**Figure 12-47.** High pressure control (*Courtesy of Penn Controls, Inc.*)

**Figure 12-48.** Pressure control connections

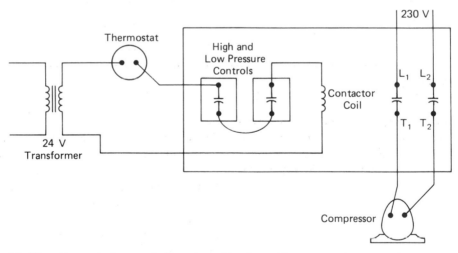

**Figure 12-49.** Typical electrical diagram of high and low pressure controls

drops of short duration during the running cycle.

In operation, the total oil pressure is the combination of crankcase pressure and the pressure generated by the oil pump. The net oil pressure available to circulate the oil is the difference between total oil pressure and refrigerant pressure in the crankcase:

total oil pressure − refrigerant pressure
= net oil pressure

This control measures that difference in pressure, referred to as *net oil pressure*.

**Figure 12-50.** Oil safety control (*Courtesy of Penn Controls, Inc.*)

When the compressor starts, a time delay switch is energized. If the net oil pressure does not increase to the control cut-in point within the required time limit, the time delay switch trips to stop the compressor. If the net oil pressure rises to the cut-in point within the required time after the compres-

sor starts, the time delay switch is automatically de-energized and the compressor continues to operate normally.

If the net oil pressure should fall below the cut-out setting during the running cycle, the time delay switch is energized and, unless the net oil pressure returns to the cut-in point within the time delay period, the compressor stops.

The time delay switch is a trip-free thermal expansion device. The time delay unit is compensated to minimize the effect of ambient temperatures from 32°F to 150°F (0°C to 66°C). Timing is also affected by voltage variations. The manufacturer's recommendation should be followed during installation and adjustment. Normally, the oil pressure line is connected to the pressure connection labeled OIL and the crankcase line to the pressure connection labeled LOW. Oil safety controls are rated for pilot duty only. Wire as suggested for specific equipment. (See Figure 12-51.)

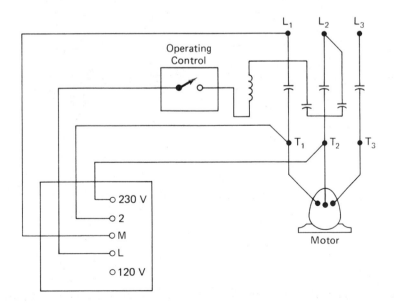

**Figure 12-51.** Penn control wiring diagram used on 230-volt system

## DEFROST TIMERS

Defrost timers are devices that cause a refrigeration system to go into a defrost cycle at predetermined periods of time. There are a variety of times available and each system has its own requirements. Therefore, the manufacturer's recommendations should be followed during replacement. Defrost timers are usually driven by a synchronous motor which runs while the unit is supplied with electrical power. In Figure 12-52, electrical power to the motor is connected to terminals 1 and 6. Terminal 2 is connected to the compressor and terminal 3 is connected to the defrost device.

In operation, electrical power is made to the compressor from terminal 1 to terminal 2 (see Figure 12-53). When the timer motor has run the required amount of time, the switch changes from terminal 2 to terminal 3, stopping the compressor and energizing the defrost device. This is known as the defrost period. After the predetermined period of time has lapsed, the switch returns to terminal 2. The defrost device, either an electric heater or a hot gas solenoid, is removed from the circuit and the compressor is again energized and producing refrigeration.

## STARTERS AND CONTACTORS

In refrigeration and air conditioning work, the compressor motor represents the largest switching load for the control system. The various fan motors, water pumps, and other machinery wired in parallel with the compressor motor also add to the load requirements. There are several methods of con-

**Figure 12-52.** Typical defrost timer connections (*Courtesy of Gem Products, Inc.*)

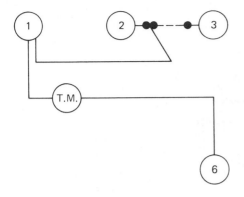

Terminal 1—Common
Terminal 2—N.C. to Compressor
Terminal 3—N.O. to Defrost Device
Terminals 4 and 5 (When Used) and Terminal 6
           are junctions only and
           have no internal connection
           to the timer

**Figure 12-53.** Internal diagram of a defrost timer (*Courtesy of Gem Products, Inc.*)

**Figure 12-54.** Three-pole contactor (*Courtesy of Arrow-Hart, Inc.*)

trolling the various loads but we will discuss only the contactor and starter at this time.

**CONTACTOR:** A contactor is a device for repeatedly establishing and interrupting an electric power circuit. Contactors have many common features. Each has electromagnetically operated contacts which make or break a circuit in which the current exceeds the operating circuit current of the device. Each may be used to make or break voltages which differ from the controlling voltage. A single device may be used to switch more than one circuit. A contactor is used for switching heavy current, high voltage, or both. (See Figure 12-54.)

In operation, the armature is pulled in when the controller completes the control circuit (see Figure 12-55). Electrical current flows through the coil and causes electromagnetism which attracts the armature. The armature carries a set of movable contacts which mate with a set of stationary contacts. The electrical circuit to the compressor is completed through these contacts and the motor begins to operate. When the controller is satisfied, the control circuit is interrupted. The electromagnetic force in the coil stops and the armature drops out, taking with it the movable contacts. The motor stops and awaits a signal from the controller for the next on cycle.

**STARTER:** A motor starter may consist of a contactor used as a means of switching electrical power to a motor. A starter, however, usually has additional components, such as overload relays and holding contacts. These components may also include step resistors, disconnects, reactors, or other hardware required to make a more sophisticated starter package for large motors. (See Figure 12-56.)

The operation of a starter is the same as that of a contactor. The armature is caused to move a set of movable contacts by electromagnetism.

Frame

Iron core

Armature spring

Armature

Coil

Coil terminals

Contact blades

Contact terminals

**Figure 12-55.** Contactor components (*Courtesy of Arrow-Hart, Inc.*)

**Figure 12-56.** Motor starter (*Courtesy of Arrow-Hart, Inc.*)

## SAFETY PROCEDURES

Electric motors provide the power to drive fans, compressors, pumps, etc., on refrigeration and air conditioning equipment. They are often needlessly abused and mistreated. An electric motor will provide many years of service if properly cared for.

1.   Motor compressors are sometimes very heavy and help should be obtained when moving one of these components.

2.   Be careful to keep hands, feet, and clothing free of all belts.

3.   Always disconnect the electrical power when checking the electrical circuit.

4.   Be sure to discharge capacitors with a resistor to prevent personal injury or damage to the capacitor.

5.   Do not tighten belts excessively.

6.   Do not overload an electric motor; it may result in burnout.

7.   Do not replace fuses with a higher rating than that recommended.

8.   Be sure that the supplied voltage corresponds with the motor rating.

9.   Be sure that an electric motor has sufficient means for cooling.

10.   Never stall an electric motor because damage to the winding will usually occur.

## SUMMARY

• The laws of magnetism and magnetic induction are the basis of operation for electric motors.

• The basic components of an electric

motor are the rotor, the permanent magnet, and two magnetic poles.

• A phase stator pole is installed to help the motor start regardless of where the rotor stops.

• The phase winding allows the electric current to flow in one winding before it flows to the other, producing an out-of-phase current to start the motor.

• Capacitors are used to produce an out-of-phase electrical current to the phase winding.

• The two types of capacitors are: (1) start and (2) run.

• Start capacitors usually have a higher mfd rating than run capacitors.

• High mfd rated capacitors will permit high current flow through a series load.

• Running capacitors are physically larger than start capacitors because of the oil that is used as a dielectric.

• Starting capacitors are to be used only for a short duration of time.

• When capacitors are connected in series with a load, the current leads the voltage.

• The five different types of single-phase motors are: (1) split-phase, (2) permanent-split capacitor (PSC), (3) capacitor-start (CSR), (4) capacitor-start/capacitor-run (CSCR), and (5) shaded-pole.

• Split-phase motors have moderate starting torque for use on blowers, pumps, fans, and oil burners.

• Capacitor-start motors have high starting torque.

• The starting capacitor is removed from the circuit after the motor reaches ap-

proximately 75 percent of its speed by the centrifugal switch.

• Capacitor-start motors are used on compressors, large fans, blowers, and water pumps.

• Two-speed motors are made in both split-phase and capacitor-start types.

• Two-speed motors are used where a modulated air flow is required and in other applications where more than one speed is required.

• The centrifugal switch is the heart of the split-phase and capacitor-start motors.

• As the motor reaches 75 percent of its normal operating speed, the centrifugal switch removes the starting components from the circuit.

• Permanent-split capacitor motors are used on compressor motors, fans, and blowers.

• A running capacitor is connected in series with the phase winding to help in starting and running.

• Permanent-split capacitor motors have only a moderate starting torque.

• Capacitor-start/capacitor-run motors have both high starting torque and efficient running characteristics.

• Shaded-pole motors have a very low torque.

• Shaded-pole motors use a winding on a part of the pole to provide an out-of-phase characteristic.

• The shaded portion of the pole in a shaded-pole motor is not connected to the electrical circuit.

• There are two general types of single-phase motor protectors: (1) internal type and (2) external type.

• The internal overload is imbedded into the heat sink of the motor winding.

• The external overload is mounted securely on the outside of the housing at the hottest point.

• Overloads are also of the line-break and thermostatic types.

• Always be certain that the electrical power is off when servicing electrical motors.

• Always discharge capacitors before handling them.

• The rule to apply when locating compressor motor terminals is: The greatest resistance will be between the start and run terminals; the medium resistance will be between the start and common terminals; the least resistance will be between the run and common terminals.

• Motor windings should have at least one million ohms resistance between the winding and the housing.

• An open winding will show infinite resistance.

• A shorted winding will show less resistance than the specifications indicate.

• Certain rules should be applied when replacing capacitors.

• Starting relays are used on hermetic and semi-hermetic compressors to remove the starting components from the circuit when 75 percent of the normal operating speed is reached.

• The types of starting relays used are: (1) amperage (current), (2) hot-wire, (3) solid-state, and (4) potential (voltage).

• Amperage relays are used on ½ hp motors and smaller.

• Hot-wire relays are a form of amperage relay because current flowing through a wire causes heat which warps a bimetal blade which operates the relay.

• Potential (voltage) relays operate on the electromagnetic principle and are used on motors of almost any size.

• Solid-state relays make use of ptc materials and triacs to provide an easy-to-use motor-starting relay.

• Solid-state hard start kits provide the additional starting torque necessary to solve PSC motor-starting problems.

• Two types of pressure controls are: (1) low pressure and (2) high pressure.

• Oil safety controls are designed to provide compressor protection by guarding against low lubricating oil pressure.

• Defrost timers are devices that cause a refrigeration system to go into a defrost cycle at predetermined times.

• A contactor is a device for repeatedly establishing and interrupting an electric circuit.

• A motor starter may consist of a contactor used as a means of switching electrical power to a motor.

• A starter usually has additional components such as overload relays and holding contacts.

## REVIEW EXERCISES

1. What are the stationary poles in an electric motor called?

2. What is the difference among single-phase motors used in refrigeration and air conditioning?

3. What causes the polarity of stator poles to change?

4. What is used in a single-phase motor to ensure that it will start each time?

5. How are capacitors rated?

6. Why is oil used as a dielectric in a running capacitor?

7. With which winding is the running capacitor connected in series?

8. Why is the starting capacitor removed from the starting circuit after the motor has started?

9. What is the purpose of a capacitor?

10. What is the purpose of a centrifugal switch?

11. At approximately what speed does the centrifugal switch open?

12. What is the purpose of the running capacitor on a PSC motor?

13. How is the capacitor electrically connected on a PSC motor?

14. Why can the run winding withstand a higher continuous current than the start winding?

15. Does the shaded-pole motor develop a high or low torque?

16. What does the term shaded-pole mean?

17. What is the purpose of motor protectors?

18. In what location should the motor protector be installed?

19. What safety precautions should be followed when handling capacitors?

20. List the steps used in locating compressor motor terminals.

21. What should the minimum resistance between the motor windings and the housing on a 1-hp motor be?

22. What resistance will be indicated on an open winding?

23. What is the rule governing the replacement of starting capacitors?

24. What is the purpose of a starting relay?

25. Are the contacts normally open or closed on an amperage starting relay?

26. What causes the hot wire to operate?

27. Between what two terminals is the coil on a potential relay?

28. Where is the voltage that operates the potential relay produced?

29. Does the solid-state starting relay need to be sized for each size motor?

30. When is a hard start kit used?

31. What is the purpose of a thermostat?

32. Where should pressure switches be connected into the refrigeration system?

33. What are oil safety controls designed to do?

34. What is the purpose of defrost timers?

35. What is the difference between starters and contactors?

# Domestic Refrigeration

Domestic refrigeration is usually considered as comprising three areas: (1) refrigerators, (2) freezers, and (3) room air conditioners. In most modern equipment, the refrigerator and freezer are combined into one cabinet. In early times these cabinets were fitted with one door. Today, however, it is more common to have two or three doors. These cabinets may be free-standing or built into a wall.

The freezer section of most refrigerator-freezers is small and is designed to hold frozen foods for only a short duration. The separate freezer is in a cabinet much like the refrigerator except that the inside temperature is much colder. This colder temperature allows foods to be stored for a much longer period of time.

Room air conditioning units are designed primarily to cool a single room within a structure. However, larger units have been designed to cool a complete home or several rooms where installation of a central air conditioning unit is impractical or not desired.

## REFRIGERATOR CABINET

A refrigerator that is used for household units consists of a specially designed insulated storage box. The modern household cabinet is constructed of sheet steel. The external surfaces of these cabinets are finished with porcelain, enamel, or some other special finish. The food storage compartment is provided with wire shelves and lined with some moisture-proof material such as porcelain or enameled steel. (See Figure 13-1.) The space between the lining and the external surface is filled with insulation.

The refrigerator cabinet has hardware fittings which enable the door to be tightly closed, minimizing leakage of heat into the box. Arrangements are made in many models for special features such as foot-controlled door openers. There are lights which automatically turn on when the door is opened. There may be a number of other features along these lines but they are all a part of the cabinet and do not directly have anything to do with the refrigeration unit.

A-7150

**Figure 13-1.** Typical domestic refrigerator (Courtesy of Frigidaire Division, General Motors Corp.)

**INSULATION:** Insulation is considered to be anything that will retard the flow of heat. There are a large number of substances which have been used for insulation in domestic refrigerators and freezers. At present, fiberglass and foamed plastic are the most widely used insulators for this purpose.

Fiberglass is made from silicon dioxide (glass sand) and is expanded with refrigerant or carbon dioxide. This material is formed into bats approximately 3 inches (7.62 cm) thick when used in domestic refrigerators and freezers. When working with fiberglass insulation, rubber gloves and protective clothing should be worn to guard against fiberglass penetration of the skin. At

one time, fiberglass was the most widely used insulating material, but in recent years foamed plastic has taken over the domestic refrigerator insulation duties.

Foamed plastic is made by expanding the plastic with refrigerant, carbon dioxide, or some other highly volatile agent. It is more expensive than the fiberglass insulation, however only half as much is required to obtain the same insulating effect. Foamed plastic is light in weight, easily worked, and is manufactured in various forms and shapes. Therefore, it makes a useful insulating material which is used in thin-walled refrigerators and freezers. The foam density ranges from approximately 1.8 pounds per cubic foot to 2.2 pounds per cubic foot (.029 to .035 $g/cm^3$). The higher the density, the greater the insulating value.

**VAPOR BARRIER:** The average kitchen has a much higher relative humidity than any other place in the home except the bathroom. Condensation of moisture from the air will occur on any surface that is at a temperature below the dew point temperature of the surrounding air. Because of this, moisture must be kept out of the insulated space between the outer shell and the inner liner. This is especially important around the freezer section. Several substances are used for this purpose. The most important substance used in domestic refrigerators with fiberglass insulation is a thin sheet of plastic placed over the outside of the insulation. The holes where the electrical or the refrigerant lines go through the cabinet are filled with a water-resistant putty. Also, when fiberglass insulation is used, a tar is spread over the metal surfaces and placed in cracks to prevent the moisture from penetrating the insulation. Foamed plastic insulation does not have the number of moisture problems that fiberglass has because of its inherent vapor barrier qualities.

**Figure 13-2.** Domestic Refrigerator Door Storage (*Courtesy of Frigidaire Division, General Motors Corp.*)

**Figure 13-3.** Breaker Strip

**Figure 13-4.** Refrigerator Door Seal

**DOOR CONSTRUCTION:** In order that the entire interior of a refrigerator cabinet be accessible to the user, the door (or doors) must cover as much as is practical of the front side of the refrigerator. The outside panel of the door is made of sheet steel. The inside of the door is usually made of high impact plastic which is molded with shelves and storage areas (see Figure 13-2). The inside and outside panels are separated by the proper thickness of insulation. The four edges of the door are finished with a non-heat-conducting material called a *breaker strip,* usually made of plastic (see Figure 13-3). A rubber seal or gasket is placed around the inside edges of the door so that when the door is closed the rubber gasket is pressed between the door and the cabinet, thus effecting an air-tight seal (see Figure 13-4). The amount of heat leakage past the

door seal into the cabinet is an extremely important factor. It has been found that a single door, even though it has a total larger opening than two smaller doors, actually allows less heat to enter the cabinet. This is because one large door requires less gasket than two or more doors would require. Heat leakage is more dependent upon the length of the gasket than upon the area of the door.

Domestic refrigerator and freezer doors use a magnetic type of gasket. This type of gasket allows for a better seal, as well as allowing the door to be opened from the inside or the outside. They are also equipped with adjustable hinges so that the door can be adjusted to obtain the best fit possible. (See Figure 13-5.)

**Figure 13-5.** Adjustable Door Hinges (*Courtesy of Frigidaire Division, General Motors Corp.*)

**Figure 13-6.** Breaker Strip Location (*Courtesy of Frigidaire Division, General Motors Corp.*)

**BREAKER STRIP:** Between the inner liner and the outer shell, a breaker strip must be used to reduce the flow of heat into the refrigerator by conduction. The lining of the refrigerator is the same temperature as the inside of the cabinet, while the outer shell is at ambient temperature. The breaker strip acts as an insulator between these two parts (see Figure 13-6). Each manufacturer installs the breaker strip in a different manner. Some use screws, some use clamps, and others just snap them in.

These breaker strips are made of plastic and are easily broken when cold. Therefore, before attempting removal of a cold breaker strip, bathe the breaker strip with a towel that has been wetted with warm water. Before installation of a breaker strip, dip it in warm, *not hot,* water. This will make it more pliable and reduce the possibility of breakage.

**MULLION HEATER:** The mullion heater is a strand of high resistance wire that is attached to a strip of aluminum foil. These heaters are installed around the cabinet doors beneath the breaker strips. Any time the refrigerator is plugged in, electricity is supplied to the mullion heater, which produces a small amount of heat. (See Figure 13-7.) This heat around the door area prevents the condensation of moisture. If moisture is allowed to accumulate around the freezer door, it could possibly freeze and prevent opening of the door. Also, condensation may enter the area between the inner

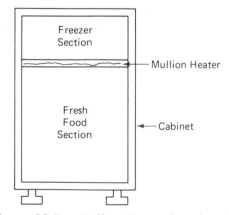

**Figure 13-7.** Mullion Heater Location

liner and outer shell, which will deteriorate the insulation and cabinet.

**SHELVING:** The shelves in a domestic refrigerator or freezer cabinet must be of the open grill type so that the air can circulate through them (see Figure 13-8). They may be of several types, such as woven wire grill, rod or flat bar grill (formed by placing rods or bars side by side), or diamond mesh expanded metal grill. The shelves are usually supported by pegs or hooks which are secured to the lining, or they may rest on lugs which are pressed into the lining. The various refrigerator manufacturers have their own special shelf features and arrangements, each of which has certain advantages. Aluminum, stainless steel, and heavily tinned steel are all used in the manufacture of shelves.

**Figure 13-8.** Refrigerator Shelves (Courtesy of Frigidaire Division, General Motors Corp.)

**INSTALLATION:** The refrigerator may be installed wherever practical in the home, but several factors should be considered. The floor should be firm, level, and capable of supporting the loaded cabinet's weight without vibration. If the floor is not strong enough, it should be reinforced with ¾-inch (19.05 mm) plywood under the refrigerator for added support.

Refrigerators are designed to operate efficiently at room temperatures between 60°F and 110°F (15.6°C and 43°C). If the refrigerator is installed where the temperature will be below 60°F the operating efficiency is affected. This may result in user complaints. Therefore, installations in ambient temperatures below 60°F are not recommended.

If possible, avoid locating the refrigerator next to a stove, radiator, or hot air register. Similarly, an installation which exposes the refrigerator to the direct rays of the sun is undesirable. The user should be advised that although the refrigerator is capable of compensating for these handicaps, it will run longer and consequently use more electricity.

Level the refrigerator by adjusting either the leveling glides or the adjustable rollers after the cabinet is located. (See Figure 13-9.)

All domestic refrigerators are equipped

BRAKE　　ROLLER　　PIN

**Figure 13-9.** Cabinet Leveling Glides (Courtesy of Frigidaire Division, General Motors Corp.)

with a three-wire service cord with a standard three-prong plug. The electrical outlet should have a grounding-type receptacle to accommodate this type of plug. Do not cut off the ground terminal of the service cord plug. The wall outlet box and the receptacle itself must be properly grounded. If a properly grounded outlet is not available, one should be installed by a qualified electrician. The refrigerator should be on a separate outlet. The voltage must be checked with the compressor running to be sure that it is adequate. When the voltage is not within 10 percent of the 115 volts required, it can cause improper operation. Extension cords should never be used. They are a source of trouble and often cause the grounding feature to be lost.

With the cabinet properly located and connected to its power source, the refrigerator interior light should come on when the door is opened. With the temperature control set at mid-range, the compressor should run. Check the operation of the fan in the freezer section of the frost-free models.

A slight gurgling or bubbling sound which may be heard when the cabinet door is open is caused by the liquid refrigerant boiling in the evaporator. This is normal and should not be misconstrued as an indication of a malfunction.

There will be some evidence of refrigeration within five minutes after starting, but the temperature may not start to drop until after 30 minutes or more of operation. Explain to the customer that the first few running cycles will be quite long, but that they will shorten after the normal operating temperatures are reached. Allow 24 hours for the cycles to settle out.

## DOMESTIC REFRIGERATION SYSTEM

Before any meaningful analysis can be made, the service technician must fully understand the function of the various refrigeration sys-

**Figure 13-10.** Domestic Refrigeration System—Static Condenser (*Courtesy of Frigidaire Division, General Motors Corp.*)

tem components. Figure 13-10 is an illustration of the static or gravity type condenser system. A refrigeration system consists of an evaporator, compressor, condenser, and filter-drier all connected with tubing through which the refrigerant circulates. Refrigeration is really the process of removing heat from within an insulated cabinet. Briefly stated, the refrigerant within the sealed system absorbs heat from inside the cabinet and through the various components of the refrigeration system and then gives up or transfers this heat to the air surrounding the refrigerator. The air surrounding the refrigerator is referred to as the ambient air. The refrigerant is circulated through the system by a pressure differential created by the compressor. The refrigeration system is divided into two parts—high pressure and low pressure sides. The dividing points are at the compressor and at the outlet of the capillary tube. The low pressure side of the refrigeration system consists of the evaporator, the suction line, and part of the compressor. The high pressure side of the system consists of the discharge line, filter-drier, capillary tube, and part of the compressor.

**EVAPORATOR:** The evaporator, interchangeably called the cooling coil, freezer coil, or just plain freezer, is where the action is. The heat in the cabinet is transferred to the refrigerant that is circulating through the evaporator by conduction, convection currents, or forced air. As the refrigerant absorbs the heat from the foods, it begins to boil, changing its state from a low pressure liquid to a low pressure heat-laden vapor. This vapor is then drawn through the accumulator and suction line to the compressor. The accumulator acts as a reservoir to prevent liquid refrigerant from flooding out into the suction line during certain portions of the refrigeration cycle, long running periods, or after the cooling coil accumulates a frost build-up. In a normal operating system with a capillary tube, approximately two-thirds of the liquid refrigerant is in the evaporator and accumulator during the running cycle.

**COMPRESSOR:** The compressor and motor are combined in a hermetically sealed steel casing. There are no exposed moving parts. All the parts are continuously lubricated internally, eliminating the need for periodic service lubrication. The motor operates on a 115-volt, 60-Hz, single-phase alternating current. It is a split-phase motor having two sets of windings, a start (phase) winding and a run winding. Some compressor motors run at 1725 rpm, while others run at 3450 rpm. The speed is determined by the type of stator used in the motor.

The function of the compressor is to compress the low pressure vapor (gas) from the evaporator, changing it into a high pressure vapor. As it is compressed, the temperature of the refrigerant vapor is increased. The temperature of the vapor must be raised so that it is warmer than the ambient air. The ambient air will then pick up the heat from the refrigerant within the condenser.

**CONDENSER:** The function of the condenser is to transfer the heat absorbed by the refrigerant to the ambient or room air. As described in the preceding paragraphs, the high pressure gas is routed to the condenser where, as the gas temperature is reduced, it condenses into a high pressure liquid state. As described before, this heat transfer takes place because the discharged gas is at a higher temperature than the air that is passing over the condenser. It is very important that an adequate air flow over the condenser be maintained.

Any accumulation of lint or other material covering the condenser should be removed to assure normal air flow. Remember, as the temperature of the condensing medium increases, the condensing temperature of the refrigerant increases. In other words, as the room temperature increases from about 70°F to 90°F (21.1°C to 32.2°C), the head pressure and temperature must also be higher to transfer the heat effectively.

**FILTER-DRIER:** A high side filter-drier is located at the outlet of the condenser. It filters foreign matter that might restrict the refrigerant flow through the system and it removes moisture that might freeze in the capillary tube, causing a restriction. When installing a filter-drier be sure that it is installed with the direction of refrigerant flow as indicated by the arrow and be certain that the outlet is always lower than the inlet.

## USE OF ALCOHOL IN REFRIGERATION SYSTEMS:

The practice of using alcohol as an antifreeze in refrigeration systems when the compressor motor is located internally in the system is not recommended. Alcohol is not compatible with aluminum or some other types of insulation used on motor windings and compressor parts in modern refrigeration systems. In addition, if alcohol is added and moisture is present in these systems, the chemical stability of the system is substantially affected. It is, however, recommended that a good filter-drier be added to the refrigeration system for the control of moisture in the system.

## CAPILLARY LINE:

The capillary line or *cap tube,* as it is commonly called, is a refrigerant metering device that determines the amount of high pressure liquid refrigerant that flows from the condenser to the evaporator. This flow is dependent upon the length of the cap tube as well as its diameter.

## DOMESTIC REFRIGERATOR OPERATION

The air flow inside a refrigerator cabinet is very important. A lack of air circulation inside the cabinet will provide poor air temperatures and allow spoilage of food.

## SIDE-BY-SIDE FROST-FREE MODELS:

The air in the freezer section is pulled through the front grill and over the entire width of the cooling coil that is located at the rear of the freezer compartment (see Figure 13-11). The heat and moisture picked up from the food and air inside the cabinet are given up to the evaporator. This results in the discharge of cold, dry air from the air

**Figure 13-11.** Air Circulation in Top Freezer Frost-Free Refrigerator (*Courtesy of Frigidaire Division, General Motors Corp.*)

duct into the freezer section. An air deflector is used to divert the air directly over the ice service area. The blower housing is designed so that some of the air is diverted through the manual damper to the refrigerator section. The freezer control or manual damper, located in the top rear of the refrigerator section, will determine the amount of air that flows into the refrigerator section (indicated by the open arrowheads in Figure 13-11). The air from the refrigerator section returns to the cooling coil area through the slots in the divider located on each side of the control housing. It mixes with the freezer return air and passes across the evaporator and back to the blower motor housing.

**SIDE-BY-SIDE FROST-FREE MODELS:**
These models use a single coil evaporator which is mounted in a vertical position in the freezer compartment to cool both the freezer and the food compartment (see Figure 13-12). Only one fan is used for air distribution. Starting in the freezer compartment, air is drawn through the grill at the bottom of the compartment and through the coil where the heat and moisture are removed. The blower housing and air ducts provide direct cooling in this area. This air flows downward, cooling the lower section on its way to the return air grill. Part of the cold, dry air entering the fan housing is diverted upward through ducts to the top of the freezer compartment. A deflector, mounted on the freezer liner top, directs air over the ice service area. From here, the air filters down through the shelves to the return air grill.

The air duct from the blower housing is so designed that some of the cold, dry air is diverted to the top of the food compartment. The amount of air entering the food compartment is controlled by a manual damper. The operation of this control is explained under Temperature Controls on p. 324. As the air enters the food compartment, the cold air flows to the bottom of the compartment. In the bottom left rear corner of the food compartment an opening is provided in the foamed divider for air to return to the cooling coil in the freezer section. From here the air returns to the cooling coil and the air cycle is repeated.

**DEFROST SYSTEMS:** Frost that forms on the cooling coil must be removed periodically to maintain cooling efficiency, air flow, and product temperatures. On the nonfrost-

**Figure 13-12.** Air Flow and Water Disposal (*Courtesy of Frigidaire Division, General Motors Corp.*)

free models, defrosting and cleaning of the evaporator is the responsibility of the user. On frost-free refrigerators this is accomplished by a timer-controlled defrost system, usually an electric heater.

As stated before, there is a wide variety of defrost systems. In this example, we will use the Frigidaire 8-hour defrost system. Regardless of the amount of time involved, the defrost cycle will be basically the same for all types.

**Defrost Heaters:** Frost-free refrigerators utilize a 500-watt infrared heater enclosed in a glass tube. On top freezer models, one of these heaters is located in a recess provided at the bottom of the evaporator. An additional drain heater prevents the drain from freezing. Some side-by-side models use two heaters wired in series. (See Figure 13-13.) They are located in recesses provided between the fins of the coil. The lower heater directs enough heat to the drain area so that a separate drain heater is not required.

**Figure 13-13.** Side-by-Side Defrost Heater Location (*Courtesy of Frigidaire Division, General Motors Corp.*)

**Defrost Timers:** An 8-hour timer de-energizes the compressor and freezer blower and energizes the defrost system for 20 minutes, three times each day. At the end of each defrost period the timer will automatically cut off the heater circuit and energize the compressor and freezer blower motor. (See Figure 13-14.)

**Frigure 13-14.** Defrost Timer (*Courtesy of Frigidaire Division, General Motors Corp.*)

**Defrost Limiter:** The defrost timer determines when the defrost period will occur, but the duration of the defrost period, not to exceed 20 minutes, is normally determined by a thermostatic control called the *defrost limiter switch.* The defrost limiter is located near the evaporator and is connected electrically in series with the heater element (see Figure 13-15). This switch opens at a temperature of 50°F ± 5°F (10°C ± 2.78°C) and disconnects the defrost heater. The timer terminates the defrost cycle after 20 minutes and energizes the refrigeration circuit.

The defrost limiter switch resets at a temperature of 30°F ± 6°F (−1.1°C ± 3.33°C) and is then ready for the next defrost period.

**Defrost Water Disposal:** On top freezer models, the defrost water is removed from the freezer section through the drain tube

**Figure 13-15.** Defrost Limiter Switch Location (*Courtesy of Frigidaire Division, General Motors Corp.*)

and deflector mounted to the top of the refrigerator section. The water flows down the rear wall of the liner to a trough which is formed in the liner bottom (see Figure 13-16). The defrost water then goes through a drain trap into a disposal pan located in the machine compartment where the water evaporates.

On the side-by-side models, a drain trough is riveted to the evaporator housing at the bottom of the freezer compartment. The defrost water is then channeled to the freezer drain tube that extends through the liner and cabinet shell at the rear of the machine compartment (see Figure 13-17). From there, a tube directs the water forward to the drain trap and disposal pan. The water in the disposal pan is then evaporated by the air that moves across the pan.

The user should be advised that periodic cleaning of the drain is necessary. The user should also be instructed on reinstalling the drain trap after cleaning. Improper placement of the drain trap can cause an air leak in the drain opening.

**Figure 13-16.** Water Disposal in Top Freezer Frost-Free Models (*Courtesy of Frigidaire Division, General Motors Corp.*)

**Figure 13-17.** Side-by-Side Water Disposal (*Courtesy of Frigidaire Division, General Motors Corp.*)

**TEMPERATURE CONTROLS:** The temperatures within a refrigerator are controlled by a thermostatic switch that regulates the running time of the compressor and freezer blower motor, where applicable.

Keeping in mind that refrigeration of the fresh food and freezer compartments results from the cooling effect of the evaporator in the freezer compartment and the circulation of cold air by a blower, we can determine how the temperature in each compartment can be regulated.

The running time of the unit is determined by the thermostatic switch bulb, located at the top of the fresh food compartment, and by the volume of cold air that is provided through the manual damper (freezer control) from the freezer compartment (see Figure 13-18).

When the fresh food compartment warms

THERMOWELL    MANUAL DAMPER

A-6302

**Figure 13-18.** Thermostat Bulb and Manual Damper Location (*Courtesy of Frigidaire Division, General Motors Corp.*)

to the cut-in temperature of the thermostat, the refrigeration unit begins to run. The evaporator in the freezer gets colder and the cold air is blown, through the manual damper, into the food compartment. When the manual damper is set at the midpoint, enough air will be diverted to the food compartment to cool it to the cut-out temperature without affecting the freezer temperature appreciably.

When a lower freezer compartment temperature is required, the manual damper is closed a little by turning the freezer control to a colder setting, restricting the volume of cold, dry air that is provided to the food compartment. Since more cold, dry air is circulating within the freezer compartment, the temperature will fall. The temperature in the fresh food compartment will not be greatly affected because the unit will run until the cut-out temperature in the fresh food compartment is reached. Simply stated, the longer running time compensates for the reduced volume of air that is provided for the fresh food compartment.

When a lower fresh food compartment temperature is required, the thermostat is set lower, which lowers the cut-in and cut-out temperatures of the thermostat. Setting the manual damper one increment higher will provide more air to the fresh food compartment with minimal effect on the freezer temperatures.

## SEALED SYSTEM REPAIR

When servicing domestic refrigerators and freezers, the malfunction is often found to be in the controls and not in the refrigeration unit. Many times the problem is solved by simply instructing the user on the proper use. At other times, it is a matter of restoring proper air flow from the cooling coil to the compartments, and the foods stored in them, and back to the cooling coil again. A lack of proper air flow is sometimes caused by fan motors not running at the proper speed. This problem also could be a result of improper setting of the controls by the user. A malfunction of the defrost system causing the evaporator to be blocked with ice could restrict the air flow. It is strongly urged that the sealed refrigeration unit be checked as a last resort. If gauges are installed, be sure that the procedures are followed properly to avoid contamination of the refrigeration system which could cause problems later on.

**SAFETY:** Always of primary importance is safety: safety for yourself as you work, safety for the surroundings in the area in which you are working, and safety for the customer after you leave the job.

Safety for yourself includes using the approved eye protection, using grounded tools and instruments, proper lifting and moving practices, and being aware of and prepared for any emergency that might arise as you work. You may use shortcuts or unsafe practices for years with no adverse effects, but the one time that you are caught can make all the time you have saved very expensive. *There is always time for safety.*

Safety for the surroundings means respect for the property around which you are working. Cover furniture and rugs that could be damaged. Do not track the dirt on your shoes into the house. Direct the discharge of refrigerant out of doors, being sure that shrubs, flowers, or grass are not damaged by the refrigerant and oil. Use extreme caution when moving the appliance. With the rugs and carpets used in kitchens today, moving an appliance can become troublesome. The good will derived from a successful service call can be nullified when property has been damaged or the work area left dirty.

Before leaving the job, be certain that the product is properly grounded and that the outlet into which the refrigerator is to be plugged is properly grounded. If there is any doubt about the receptacle being as it should be, advise the customer. Note on the work order that the customer has been advised and have it signed, if possible. Advise the customer to contact an electrician who can wire the receptacle in accordance with local codes.

Safety does not cost; it pays. Give electricity the respect that it demands.

**REFRIGERATION PERFORMANCE COMPLAINTS:** A refrigeration performance complaint is a complaint involving food storage temperatures, percentage of running time, or both. Such complaints are much more frequent during the summer months when the refrigerator is required to work at or near full capacity. Indeed, many refrigerator performance complaints require only customer education or reassurance that the refrigerator or freezer is operating normally.

It is difficult to suggest a detailed step-by-step procedure to diagnose refrigeration performance complaints. Each service call requires a particular procedure depending on the customer's complaint, severity of the symptoms, room temperature, length of service, and other factors. In other words, there is no substitute for a knowledge of basic refrigeration principles and sound, analytical thinking when diagnosing these complaints. The following procedure should be used as a guide for every refrigeration performance complaint.

1. Measure the temperature of the frozen food packages and the food stored in the fresh food section. Always use a good reliable thermometer or thermocouple. (See

**Figure 13-19.** Temperature Tester (Courtesy of Robinair Manufacturing Company)

Figure 13-19.) Place the thermometer bulb or the thermocouple probe between two packages of frozen food. If the thermocouple is wet before placing it between the packages, it will freeze to the packages and result in a more accurate temperature reading. Leave the refrigerator or freezer door closed for five to ten minutes, then read the temperature as rapidly as possible without removing the thermometer. If a temperature tester is used, the temperature can be read without disturbing the probe or opening the door. The food being measured should be located near the center of the compartment to obtain an average temperature.

Frozen food temperatures in the freezer section ideally are between 0°F to +5°F (−17.8°C to −15°C) in frost-free models. Conventional models will be from +10°F to +12°F (−12.2°C to −11.1°C) depending on the control settings, food load, usage, and ambient temperatures. Sub-zero temperatures are not necessary for food preservation and one should not try to obtain lower temperatures than indicated above. Temperatures could go higher than those indicated for a brief period of time during and immediately after the defrost period on frost-free models. The temperature

of food in an extremely lightly loaded freezer will fluctuate more than in a full freezer due to door opening and defrost. It is, therefore, desirable to keep the freezer section at least half full of food.

The food temperatures in the refrigerator (fresh food) section should be between 32°F and 40°F (0°C and 4.4°C) depending upon the model, running time, usage, load, and where the temperatures are taken. Cold air drops; therefore the bottom of the refrigerator section will be colder than the top except, of course, when foods are placed directly in front of the air discharge into the refrigerator section. Take the temperatures of a food such as a jar of pickles that has been in the refrigerator for at least 12 hours and is located on top of the hydrator cover as a guide. This gives a good average temperature.

2.   Install a temperature recording instrument. Occasionally a temperature complaint cannot be settled to the satisfaction of the customer by a single temperature reading. Perhaps the customer agrees with the thermometer check but claims that the operation is erratic. In such cases, a temperature recording instrument is a very useful tool to settle the complaint, or perhaps prove that the operation is or is not erratic.

Temperature recorders that record temperatures on a 24-hour chart are available from many manufacturers. They also record the number of cycles of operation, which is useful in detecting erratic operation and estimating the percentage of running time.

3.   See Troubleshooting Chart. Almost every manufacturer will have a service diagnosis section in the equipment literature, giving the problem and possible cause, along with the remedy. Recently, block diagram charts have been used. (See Table 13-1.) Study these charts for assistance in making a proper diagnosis.

## Table 13-1
## Troubleshooting Chart

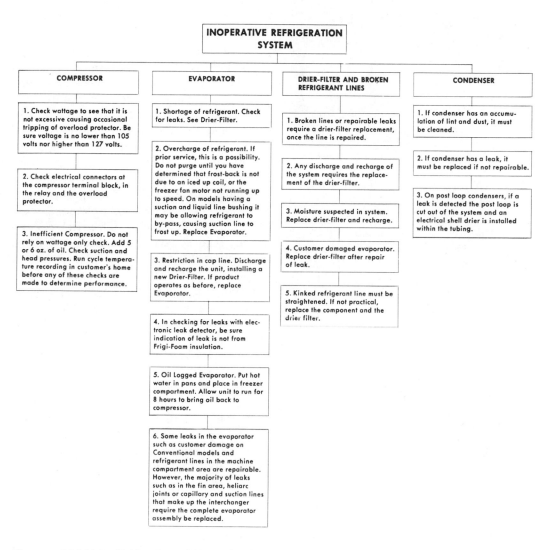

INOPERATIVE REFRIGERATION SYSTEM

**COMPRESSOR**

1. Check wattage to see that it is not excessive causing occasional tripping of overload protector. Be sure voltage is no lower than 105 volts nor higher than 127 volts.

2. Check electrical connectors at the compressor terminal block, in the relay and the overload protector.

3. Inefficient Compressor. Do not rely on wattage only check. Add 5 or 6 oz. of oil. Check suction and head pressures. Run cycle temperature recording in customer's home before any of these checks are made to determine performance.

**EVAPORATOR**

1. Shortage of refrigerant. Check for leaks. See Drier-Filter.

2. Overcharge of refrigerant. If prior service, this is a possibility. Do not purge until you have determined that frost-back is not due to an iced up coil, or the freezer fan motor not running up to speed. On models having a suction and liquid line bushing it may be allowing refrigerant to by-pass, causing suction line to frost up. Replace Evaporator.

3. Restriction in cap line. Discharge and recharge the unit, installing a new Drier-Filter. If product operates as before, replace Evaporator.

4. In checking for leaks with electronic leak detector, be sure indication of leak is not from Frigi-Foam insulation.

5. Oil Logged Evaporator. Put hot water in pans and place in freezer compartment. Allow unit to run for 8 hours to bring oil back to compressor.

6. Some leaks in the evaporator such as customer damage on Conventional models and refrigerant lines in the machine compartment area are repairable. However, the majority of leaks such as in the fin area, heliarc joints or capillary and suction lines that make up the interchanger require the complete evaporator assembly be replaced.

**DRIER-FILTER AND BROKEN REFRIGERANT LINES**

1. Broken lines or repairable leaks require a drier-filter replacement, once the line is repaired.

2. Any discharge and recharge of the system requires the replacement of the drier-filter.

3. Moisture suspected in system. Replace drier-filter and recharge.

4. Customer damaged evaporator. Replace drier-filter after repair of leak.

5. Kinked refrigerant line must be straightened. If not practical, replace the component and the drier filter.

**CONDENSER**

1. If condenser has an accumulation of lint and dust, it must be cleaned.

2. If condenser has a leak, it must be replaced if not repairable.

3. On post loop condensers, if a leak is detected the post loop is cut out of the system and an electrical shell drier is installed within the tubing.

(Courtesy of Frigidaire Division, General Motors Corp.)

**USE OF VOLT-WATTMETER:** One of the most valuable tools that a service technician can own is a volt-wattmeter. (See Figure 13-20.) This meter can be used to check line voltage and line voltage drop.

Most refrigerator compressors are designed to operate on 115 volts ± 10%. Any voltages above 127 volts or below 105 volts will have an effect on the motor rpm and the overload protection. The dropout time of

**Figure 13-20.** Volt-Wattmeter (*Courtesy of Robinair Manufacturing Company*)

the starting relay (the time required for the relay to de-energize the start winding after the initial start) would also be affected. All of this will have an adverse effect on the overall operation of the refrigerator.

Complete operating instructions accompany the volt-wattmeter. A typical procedural check is outlined below.

*Caution:* Do not exceed the current ratings specified on the wattmeter dial. Always start with the instrument set on the highest scale and then switch to the scale that gives the closest reading to mid-scale of the dial.

1. Set the toggle switch to the 130- or 260-volt position. If in doubt, use the 260-volt position.

2. Set the watts selector switch to the 5000 watts position. This will protect the instrument from surge overload.

3. Connect the male power cord of the instrument to a wall outlet. The voltmeter will now read the line voltage.

4. Connect the line cord of the refrigerator to the female receptacle on the volt-wattmeter front panel.

5. Turn on the temperature control switch of the refrigerator being tested. (The following wattage readings were taken on a Frigidaire FPCI-219VN-R refrigerator.)

You will note that the instant the compressor is energized, the overswing of the needle may go above 2000 watts depending on the damping of the meter movement within the wattmeter. The voltage is at 120 volts and the room temperature is 70°F (21.1°C). The wattage then drops off to approximately 1500 watts, which is the combined wattage draw of the start and run windings. It then drops off to approximately 475 watts. This is the running winding wattage plus whatever other electrical components, such as condensate driers, might be incorporated within the refrigerator. All of this takes place within ½ to 1½ seconds, making it difficult to establish these wattages. These wattages vary, of course, with the make and model being tested. Now switch the meter to the 500-watt scale.

It should be noted that the final running wattage settles out at approximately 400 watts, which includes the various condensate driers and fan motors. When the temperature control is turned off, the wattage draw is still approximately 60 watts. Therefore, the net compressor wattage draw is 340 watts (400 − 60 = 340).

The wattage recorded will vary slightly with the ambient temperature, make and model of the refrigerator (which determines the size of the compressor), and the other electrical components that are used, such as condensate driers, condenser fan, etc. Whether the reading is taken at the beginning or end of the running cycle (the lower wattages will be at the end of the running cycle) will make a difference. Also to be considered is the amount of air flow over the condenser and the line voltage.

Plugging the refrigerator line cord in and taking it out several times may cause the compressor overload to trip. If this should happen, wait for the compressor to cool and the overload to reset before continuing. The length of time required for it to reset will depend on the type of overload protection and whether or not it kicked off due to excessive current or because of high temperature. If a current overload is used, only a matter of minutes is required for the compressor to restart. Forced air condenser models use temperature overloads as well as current overloads which take one hour or more to reset when the cause is excessive temperature. On the other hand, these forced air units utilize starting capacitors which make it unlikely that the overload would kick out on the first few times the cord is plugged in.

**USE OF COMPRESSOR TEST CORD WITH WATTMETER:** Should a faulty compressor be suspected, all controls and wiring can be bypassed and the compressor checked directly. This may be done by using a compressor analyzer or a test cord in conjunction with a volt-wattmeter. (See Figure 13-21.) When using the test cord, the three leads can be attached to the three-lead cord from the starting relay or directly to the compressor terminals. Some service technicians make an adapter from the three-lead relay to the compressor cord that attaches to the compressor terminals. The leads are properly color-coded. The plug is always positioned with the leads upward.

Attach the test cord clips to the compressor terminals as follows: white to common terminal, red to start terminal, and black to run terminal (see Figure 13-22). If

**Figure 13-21.** Compressor Test Cord (*Courtesy of Robinair Manufacturing Company*)

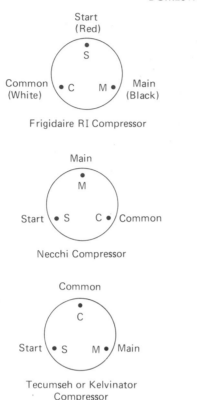

**Figure 13-22.** Compressor Terminal Identification

a terminal diagram is not available, locate them as outlined in Chapter 12. Attach the ground clip to the compressor frame. If the unit has a capacitor, connect one of the same size that is known to be good to the two capacitor lead clips. If no capacitor is used, connect these two clips together.

Be sure that the fuse in the test cord is good. A *fusetron* type fuse is preferred. Connect the male plug of the test cord into the volt-wattmeter which has already been plugged into a wall outlet. The wattmeter should be set on the 5000 watt scale. For illustration purposes, the Frigidaire model FPCI-219VN-R refrigerator is used.

The stalled or locked rotor wattage (1500 watts) of the run winding will be indicated on the wattmeter when the switch on the test cord is turned from off to run. Now switch from run to start and the combined wattage (1800 watts) of the start and run windings will be indicated. Hold the switch in the start position only momentarily and then release it to run. The wattage now is 380 watts.

*Caution:* Do not hold the switch in the start position any longer than necessary to get the compressor started or the start winding can overheat and burn out. If the compressor starts and continues to run with normal wattage, the compressor is not at fault. Normal wattages are given on the data label and wiring diagram that accompany the refrigerator. The wattage, on run settled out to 340 watts, is the compressor wattage only—no fans or condensate driers. If the compressor tries to start but does not, or if it runs with continued higher than normal wattage, the compressor must be replaced.

*NOTE:* Wattages will vary depending upon when they are taken. Wattages will be higher for a unit that has been running at normal operating temperatures and cycles than for a product that has been sitting idle. The compressor must run for 10 minutes or so to raise the oil and stator temperatures and to raise the head pressure to its normal range. Wattages will be higher if the compressor has been cycling on the overload. This could be caused by an inoperative condenser fan, for example, which will allow the compressor dome temperatures to build up.

Wattages should be used to determine if the compressor is stuck, shorted, grounded, has open windings, has a tight rotor, or has other internal malfunctions. Wattages alone should not be used to diagnose a shortage of refrigerant or an inefficient compressor.

**CHECKING CAPACITORS:** A faulty capacitor should be suspected if the customer states that the compressor will not run, has a loud hum or buzz at times, or tries to start and then kicks off.

If the capacitor shows visual evidence of liquid leakage or if the case is bulged or cracked, replace the capacitor even though it may seem to be good through testing.

Defective capacitors should always be replaced on a like-for-like rating. Substituting a capacitor with either a higher or lower mfd rating than specified will reduce the torques available from the motor. The voltage rating of the old and new capacitors should be identical.

**Testing:** There are two general methods used to accomplish this task.

A. Use of an Ohmmeter: An experienced technician can readily test a capacitor by using an ohmmeter. Testing a capacitor takes a little practice because the ohmmeter setting varies with the capacitor mfd rating. As a rule, the lower the rating, the higher the ohmmeter setting required. In refrigeration work, however, we usually check medium range capacitors. First, discharge the capacitor with a 20,000 ohm resistor. Place one probe of the ohmmeter on one terminal of the capacitor to be checked. While observing the ohmmeter, place the other probe to the other capacitor terminal. The meter needle will immediately register zero resistance and then slowly increase in resistance until it approaches infinity. This, as stated earlier, requires the proper selection of the resistance scale that is compatible with the size of the capacitor being tested. If the meter reads infinity at the beginning, the capacitor is open. If the meter continues to read zero, the capacitor is shorted. In either case, it must be replaced. To check a metal-clad capacitor for a grounded condition, set the ohmmeter on the highest scale and place one probe on the metal casing of the capacitor, and alternately touch each capacitor terminal. A high resistance (250,000 ohms) or infinity should be indicated on a good capacitor.

B. Use of a Capacitor Checker and a Volt-wattmeter to Test Capacitors:

1. Connect the capacitor to the alligator clip leads of the checker.

2. Plug the checker into a volt-wattmeter and plug the volt-wattmeter into a power supply receptacle.

3. Depress the push button of the checker momentarily and read both the voltage and the wattage. (The meter needles must stabilize before the readings can be taken.)

4. Locate the mfd rating of the capacitor in Table 13-2 and select the correct wattage found in the test under the proper test-voltage column. Values within $+20\%$ to $-0\%$ are satisfactory on electrolytic starting capacitors and within $\pm10\%$ on non-electrolytic capacitors.

For checking capacitors with mfd ratings of 325 to 550 see Figure 13-23.

1. Connect the capacitor to be checked to the alligator clipped leads of the capacitor checker.

2. Plug a volt-wattmeter into the receptacle of the capacitor checker and plug the two-wire lead of the capacitor checker into the 115-volt ac power supply receptacle.

3. Depress the pushbutton momentarily and read the voltage. The correct voltages are listed in Table 13-3.

Table 13-2

Capacitor Test Wattage at Various Test Voltages

| Capacitor Mfd. Rating | Wattage at Given Test Voltages | | | | |
|---|---|---|---|---|---|
| | 100 Volts | 105 Volts | 110 Volts | 115 Volts | 120 Volts |
| 11 | 4 | 5 | 5 | 6 | 6 |
| 15 | 8 | 9 | 10 | 11 | 12 |
| 17.5 | 11 | 12 | 13 | 14 | 15 |
| 20 | 14 | 15 | 17 | 18 | 20 |
| 28 | 26 | 29 | 31 | 34 | 37 |
| 32 | 33 | 37 | 40 | 44 | 48 |
| 35 | 40 | 44 | 48 | 52 | 57 |
| 40 | 50 | 55 | 60 | 66 | 72 |
| 42 | 54 | 60 | 65 | 72 | 78 |
| 45 | 61 | 67 | 74 | 81 | 88 |
| 50 | 73 | 80 | 88 | 96 | 105 |
| 54 | 83 | 91 | 100 | 110 | 120 |
| 60 | 97 | 107 | 118 | 128 | 140 |
| 75 | 134 | 148 | 162 | 177 | 193 |
| 85 | 157 | 172 | 189 | 207 | 225 |
| 105 | 198 | 218 | 240 | 263 | 285 |
| 115 | 214 | 236 | 262 | 187 | 311 |
| 125 | 233 | 257 | 281 | 308 | 335 |
| 130 | 240 | 265 | 290 | 318 | 346 |
| 145 | 258 | 285 | 315 | 345 | 375 |
| 155 | 270 | 298 | 327 | 361 | 392 |
| 167 | 285 | 313 | 344 | 375 | 407 |
| 190 | 305 | 336 | 370 | 404 | 438 |
| 210 | 319 | 352 | 387 | 422 | 458 |
| 240 | 334 | 369 | 405 | 442 | 480 |
| 270 | 348 | 382 | 418 | 459 | 498 |
| 315 | 360 | 398 | 437 | 479 | 520 |
| Capacitor Shorted | 400 | 440 | 485 | 530 | 575 |

**SYSTEM PRESSURE:** If a defrost cycle is not occurring, not only will the temperatures inside the cabinet be affected, but also the head and suction pressures. Such conditions should be corrected before any attempt is made to check the refrigerant system pressures. We will now concentrate on only those conditions directly associated with the unit that will adversely affect the performance of the refrigeration system.

Previously, it was established that there is a low pressure side and a high pressure side within a refrigeration system. At this time, a review of this area is in order regarding the meaning of these pressure readings.

When gauges are installed it can be determined whether or not the unit is performing normally and, if not, what the problem might be.

*NOTE:* It is again emphasized that gauges should be installed only after all other possibilities have been eliminated and it has been concluded that the fault lies in the refrigeration system. The gauges should be installed

**Figure 13-23.** Capacitor Check—325 to 550 mfd (*Courtesy of Frigidaire Division, General Motors Corp.*)

Table 13-3

Capacitor Test Voltage and mfd Rating

| CAPACITOR MFD. RATING | VOLTAGE READING |
|---|---|
| 350 | 33 |
| 400 | 30 |
| 450 | 27 |
| 500 | 24 |
| 550 | 22 |

Note: With line voltage higher or lower than 115 votes ac, the voltage drop across the capacitor will vary with the percent of change in line voltage.

on the refrigerator in the customer's home, if possible, to record the pressures under normal usage conditions. The unit should have been operating for some time before the service technician arrives. If a long or continuous running situation has existed, this is the ideal time to take pressure readings. Since the purpose of installing gauges is to detect a malfunction, do not disconnect the unit electrically to install the gauges. If the unit is already disconnected when the service technician arrives at the home, it will not be advantageous to install gauges. If the refrigerator is in the repair shop the unit should be allowed to run for several hours prior to checking. The compressor windings must reach their normal temperature range which will affect oil and refrigerant temperatures. The gauge readings should be recorded just prior to the end of the run cycle. The exception would be when checking idle head pressures, equalization pressures, or when a low side leak is suspected.

In this respect common sense and judgment should prevail. Naturally, a unit should not be operated if it has a low side leak which would allow air and moisture to be drawn into the refrigeration system. Many times this is quite obvious. It is not necessary to have the compressor running to determine if the unit is discharged. A quick gauge check will give this indication. A thorough leak check should be made to determine where the leak is located. Running the compressor with a low side leak not only pulls in air and moisture but can also result in damage to the various parts of the system and could result in making replacement of the entire system necessary.

Approximately 90 percent of sealed system entries are for repairs on the high side of the system. This includes high side leaks, restrictions, and compressor replacement.

The remaining 10 percent of entries are due either to moisture in the system or a low side leak. A leak in the low side is often subjected to a vacuum in the sealed system and air and moisture are drawn in. Therefore, it must be considered a wet system.

As a general guide, the suction pressure variation has an effect on the head pressure. (See Table 13-4.)

Head pressures and suction pressures will vary and the system performance will be adversely affected by one of the following.

**A. Use of Different Refrigerants:** When working on a refrigerator or freezer using R–12 refrigerant, different pressure readings

## Table 13-4

### Normal Conditions*

| SUCTION PRESSURE VARIATION | CAUSE | EFFECT UPON DISCHARGE PRESSURE |
|---|---|---|
| Increase | Additional heat load primarily in the freezer compartment. | Increase |
| Decrease | Decrease in heat load primarily in the freezer compartment. | Decrease |
| | Abnormal Conditions | |
| Increase | Refrigerant overcharge. | Increase |
| Decrease | Refrigerant undercharge (leak). | Decrease |
| Decrease | High Side Restriction. | Increase (See Text) |
| Increase | Inefficient compressor. | Near Normal (See Text) |
| Decrease | Low Ambients. (High ambients the Reverse.) | Decrease |
| Decrease | Low Side Restriction. | Normal to Slight Decrease |
| Near Normal | Air in System. | Increase |

* Courtesy, Frigidaire Div., General Motors Corp.

can be expected than when working on a refrigerator using R–114, for example. The use of R–22 in air conditioners will generate higher pressures than if R–12 were used.

**B. One Line of Products Versus Another:** The normal operating pressures on food freezers will be lower than those on conventional refrigerators, which in turn are lower than the pressures on air conditioners.

**C. Heat Load:** A greater heat load can result from the addition of more than a normal supply of foods, such as after doing the weekly shopping or perhaps filling all the ice cube trays at one time. The load placed in the freezer section affects the unit operation more than the load in the refrigerator section. Other items contributing to an additional heat load would be excessive door openings, poor door seals, interior lights remaining on, etc.

An increase in the amount of heat being absorbed by the refrigerant in the evaporator will affect the temperature and pressure of the refrigerant gas returning to the compressor. Compartment temperatures, power consumption, discharge and suction pressures are all affected by heat load. The operating pressures will be higher than normal under a heavy heat load condition.

**D. Ambient or Room Temperature:** Lower ambient air temperatures reduce the condensing temperature and therefore reduce the temperature of the liquid refrigerant entering the evaporator. The increase in refrigerating effect due to operation in a lower ambient temperature results in a decrease in power consumption and running time. At lower ambient temperatures, there is a reduction in cabinet heat leakage which is partially responsible for the lower power consumption and running time.

*NOTE:* An increase in refrigeration effect

cannot be expected below a certain minimum ambient temperature. This temperature varies with the type and design of the product.

Generally speaking, ambient temperatures cannot be lower than 60°F (15.6°C) without affecting the operating efficiency. Conversely, the higher the ambient temperature, the higher the head pressure must be to raise the high side refrigerant temperature above that of the condensing medium. Therefore, the head pressure will be higher as the ambient temperature rises.

Refrigerators installed in ambient temperatures lower than 60°F (15.6°C) will not perform as well because the pressures within the system are generally reduced and unbalanced. This means that the lower head pressure forces less liquid through the capillary line resulting in symptoms of a refrigerant shortage. The lower the ambient temperature, the more pronounced this condition becomes.

**E. Condenser:** If the efficiency of heat transfer from the condenser to the surrounding air is impaired, the condensing temperature becomes higher. A higher liquid temperature means that the liquid refrigerant will not remove as much heat during boiling in the evaporator as under normal conditions. Therefore, remove any paper sacks, lint accumulations, etc., that would restrict the normal air movement through the condenser. This condition would be indicated by a higher than normal head pressure.

**F. Refrigerant Charge:** Any deviation in the refrigerant charge will affect the suction pressure. As the suction pressure varies, the head pressure will vary in direct proportion. This is true because of the fixed size and length of the capillary tube that is the metering device. The refrigerant flow rate is also affected by the amount of refrigerant in the system and the head pressure forcing it through the capillary tube. An undercharge of refrigerant will cause a decrease in the flow rate (less heat is removed) and, therefore, a decrease in the desired refrigerating effect. This, in turn, causes a reduction in normal suction pressures. The wattage draw of the compressor will also decrease.

**G. Incorrect Refrigerant Charge—Overcharge:** An incorrect charge rarely occurs in manufacture. Various conditions could indicate an overcharge. For example, if the cooling coil is not defrosted at regular intervals due to a failure of the defrost system, the refrigerant will flood out and cause the suction line to frost or sweat. (See Figure 13-24.) The cause of this problem should be corrected rather than purging refrigerant from the system. The indication of an overcharge of refrigerant may result from running the freezer section colder than necessary (0°F to +5°F [−17.8°C to −15°C] are considered normal package temperatures, except in conventional models), or from continuous running of the compressor for a variety of reasons, or even from the freezer fan motor not running.

In high humidity conditions, sweating of the suction line may occur at various stages of the running cycle. This is usually temporary and should cause little or no problem.

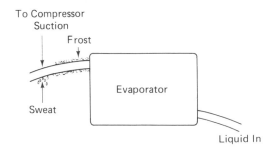

**Figure 13-24.    Indications of a Refrigerant Overcharge**

If need be, the suction line can be wrapped with insulation tape or a similar material to prevent moisture from dripping from the lines.

## SYMPTOMS OF AN OVERCHARGE

1.    Above normal freezer temperature.

2.    Frost on the suction line indicating refrigerant flooding into the suction line.

3.    High head and suction pressures.

4.    High running wattage.

If it becomes necessary to remove refrigerant from a system, it can be accomplished by accurately removing one or two ounces at a time. Thus, the service technician knows precisely the condition with relation to the refrigerant charge.

*Caution:* This procedure should not be used on systems which have had previous service for leaks, compressor burnout, or other system-contaminating service. Those systems should be completely discharged and recharged with the correct amount of refrigerant.

A.    Install the gauges to the compressor. Use a portable charging cylinder with 3 or 4 ounces of liquid R–12 charge and a short charging line connected to the bottom shutoff valve. Start the threads of the charging line onto the service valve fitting.

B.    Open the charging cylinder valve slightly to purge the air from the charging line. Tighten the charging line at the service valve fitting. The line is now filled with liquid R–12. (See Figure 13-25.)

C.    Mark the refrigerant level in the charging cylinder; then with the compressor running, open the discharge service valve and condense one ounce of R–12 from the

A-6965

**Figure 13-25.** Measuring Refrigerant Removal (*Courtesy of Frigidaire Division, General Motors Corp.*)

system into the charging cylinder by opening the bottom shut-off valve for one-second intervals. Observe the charging cylinder until the liquid level rises to the desired level.

*NOTE:* The compressor must be operating while the liquid is being condensed into the cylinder. The small amount of refrigerant in the cylinder should be discarded and not reused. If for some reason too much refrigerant has been removed from the system, it is best to discharge, evacuate, and recharge with the specified amount.

The refrigeration system is designed so that there is enough refrigerant for adequate heat removal in a 110°F (43°C) ambient temperature. However, there should not be an excess of refrigerant that would result in liquid refrigerant flooding the suction line and entering the compressor. An overcharge of refrigerant flooding through could cause damage to the compressor or could possibly keep the suction pressure and temperature high enough to prevent the thermostat switch from cycling the unit.

**H. Incorrect Refrigerant Charge—Undercharge:** This is a common occurrence insofar as component problems are concerned. The first thing noticed by the user is a rise in temperature of the foods. In the case of conventional refrigerators and freezers, a partial frosting of the evaporator is noticed. The refrigerant can leak out to the point that no frost is apparent anywhere on the evaporator. On some models, a slight shortage will result in very low freezer temperatures and a high food compartment temperature (see Figure 13-26).

This is almost always accompanied by continuous compressor running because the thermostatic control for the unit is never satisfied. On frost-free models, the cooling coil is behind a cover panel and cannot be observed, but the customer usually complains that the fresh food compartment temperatures are too warm. Although the tempera-

tures will rise in both the freezer section and the food compartment, the frozen meats and vegetables will not thaw immediately. The ice cubes will still be frozen. The customer doesn't associate the problem with the freezer section and will first notice that milk, vegetables, and even beer are not cold enough. Some older models use a fan cycling switch for the food compartment fan motor while others use an automatic air flow control. This section deals with a refrigerant shortage and it is assumed that the controls are functioning properly.

Under some circumstances, such as in the case of the conventional refrigerator or freezer with a slight shortage of refrigerant, freezing in the food compartment may be experienced due to the additional running time. With a refrigerant leak, however, it always gets worse and as the refrigerant charge decreases, the temperature will continue to rise.

With a shortage of refrigerant, the capillary tube will not have a full column of liquid refrigerant available. As a result, there is a noticeable hissing sound in the evaporator. This should not be mistaken for the regular refrigerant boiling sounds that would be considered normal.

### SYMPTOMS OF A REFRIGERANT SHORTAGE

1. A rise in the food product temperatures in both compartments.

2. Longer or continuous running time.

3. Obvious traces of oil on the floor that would occur due to a broken or cracked refrigerant line. Oil dripping from the charging screw. Do not misinterpret the rust inhibiting oil placed on the compressor charging port threads at the factory on some units as an oil leak.

**Figure 13-26.** Indication of a Refrigerant Undercharge

4.  Lower than normal wattage.

5.  The compressor feeling hot to the touch because of the heat generated by the motor windings from long continuous running. It will not be as hot as it would be with a full charge accompanied by long running times for some other reason such as a dirty condenser.

6.  The condenser will not be hot, but will be close to room temperature since it is not doing much work. This depends, of course, on the amount of refrigerant shortage. The capillary tube will be warmer than normal from a slight shortage.

7.  If the leak is on the high side of the system, both gauges will show lower than normal readings and will show progressively lower readings as the charge decreases. The suction pressure gauge will probably be in a vacuum.

8.  If the leak is on the low side of the system, the suction pressure gauge reading will be lower than normal—probably in a vacuum—and the head pressure gauge will be higher than normal. It will probably continue to become higher because of the air being drawn in through the leak and being compressed by the compressor. This air accumulates in the high side of the system. Air and R–12 do not mix so their pressures are additive.

**I. Restriction:** Always remember that refrigeration (cooling) occurs on the low pressure side of a partial restriction (obviously a total restriction will completely stop the circulation of refrigerant and no cooling will take place).

An experienced refrigeration service technician will physically feel the refrigeration lines when he suspects a restriction. The most common places for a restriction are at the filter-drier or at the capillary tube outlet

or inlet. If the restriction is not total, there will be a temperature difference at the point of restriction; the area on the evaporator side will be cooler. In many cases, frost and/or condensation will be present. Also, a longer time will be required for the system pressures to equalize (see Figure 13-27).

Any kinked line will cause a restriction, so the entire system should be visually checked. It is very difficult to diagnose a partially restricted capillary tube.

A slight restriction would give the same indication as a certain given refrigerant shortage with a lower than normal suction pressure, head pressure, and wattage, and warmer than normal product temperatures. If a total restriction is on the discharge side of the compressor, higher than normal head pressure and wattage will result. This is true only while the low side is being pumped out and if the restriction is between the compressor and the first half of the condenser.

To diagnose for a restriction versus a refrigerant shortage, discharge the system, replace the filter-drier, and recharge the unit with the specified amount of refrigerant. If

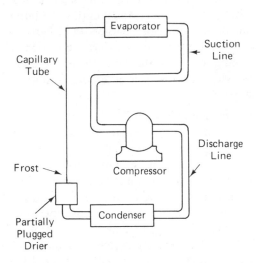

**Figure 13-27.  Partial Restriction**

the unit performs normally, three possibilities exist: (1) refrigerant loss, (2) partially restricted filter-drier, or (3) moisture in the system.

If the unit performs as it did previously, there may be a restricted capillary tube or condenser or a kinked line. Find the point of restriction and correct it.

A restriction reduces the flow rate of the refrigerant and consequently reduces the heat removal. A complete restriction may be caused by moisture, solid contaminants in the system, or a poorly soldered joint. Moisture freezes at the evaporator inlet end of the capillary tube and solid contaminants collect in the filter-drier. The compressor wattage drops because the compressor is not circulating the usual amount of refrigerant.

As far as pressure readings are concerned, if the restriction (such as a kinked line or a joint soldered shut) is anywhere on the low side, the suction pressure will probably be in a vacuum while the head pressure will be near normal. If the restriction is on the high side, the suction pressure, again, will probably be in a vacuum while the head pressure will be higher than normal during the pump-out period described earlier. In either case, it will take much longer than the normal 10 minutes or so for the head pressure to equalize with the low side after the compressor stops.

**J. Air in the System:** This condition can result from a low side leak or from improper servicing. If a leak should occur on the low side, the temperature control would not be satisfied; thus, continuous running of the compressor would result. The compressor would eventually pump the low side into a vacuum, drawing air and possibly moisture into the refrigeration system. Air and R–12 do not mix so the air pressure would be added to the normal head pressure, resulting in a higher than normal head pressure.

The surest way to determine if air is in the system is to read the head pressure gauge while the compressor is idle and then take the temperature on the condenser outlet tube. This temperature should be within 2°F or 3°F (1.11°C or 1.67°C) of what the pressure temperature relation chart shows for the given idle head pressure. If the temperature of the condenser outlet is considerably lower than the idle head pressure on the gauge, it indicates that there is air in the system.

A thorough leak checking is necessary. Correct the source of the leak. Do not attempt to purge off the air because this could result in the system being undercharged. It is best to discharge the refrigerant, evacuate, and recharge the system with the specified refrigerant charge.

**K. Inefficient Compressor:** This classification is used too frequently when the only sure things known are that the product temperatures are high and the compressor runs too much. A frostback, for example, or air in the system will affect the performance of the compressor in a way that could cause the service technician to suspect the compressor.

Before finally deciding to change the compressor, the technician should make one more check. Since it is necessary to cut the refrigerant lines to replace the compressor anyway, either cut the suction line and solder it shut or pinch it tightly shut just ahead of the low side fitting and gauge. Run the compressor for a couple of minutes to see if it will pull a 26- to 28-inch (9.95- to 4.87 cm) vacuum. If it does, it certainly is not an inefficient compressor. If it doesn't, try adding about 6 oz (170 g) of 525 viscosity oil and repeat the check. Any reading less than 26-inch vacuum indicates that the compressor should be replaced.

**Special Note on Refrigeration Oil:** The oil within the compressor plays a vital role. An insufficient amount of oil can give the indication of an inefficient compressor because the oil coating between the parts reduces friction and results in better compression without a bypass of refrigerant vapor.

**BURNT OIL CHECK:** A burnt oil check is made by taking a sample of oil from the inoperative compressor and comparing it to the darkest section of an *oil sample indicator* (see Figure 13-28). If the oil is darker than the darkest sample, it is an indication that the oil has undergone a chemical breakdown and has contaminated the system to the extent that the complete sealed system must be replaced. The sample should be collected and checked in a clear, clean glass container approximately 2 in. (5.08 cm) in diameter.

*NOTE:* If water has entered the system from a low side leak on the evaporator, the oil indicator will not show it. The oil will be gray, depending upon the amount of water that has been pulled into the system. Usually the amount of water is small and is more in the form of moisture mixed with the oil.

**Figure 13-28.** Oil Sample Indicator (*Courtesy of Frigidaire Division, General Motors Corp.*)

**LEAK TEST PROCEDURE:** Any time that a system is diagnosed as being low on refrigerant, the leak must be found and repaired. There are several types of leak detectors and each is effective and has outstanding points. The halide torch is best used where large leaks are present. Electronic leak detectors should be used when small leaks occur. Liquid leak detectors are good for locating small leaks when a particular joint or area is suspected. However, liquid leak detectors are the only type that can be used successfully around the urethane foam insulation that is used in most modern refrigerators and freezers.

The following procedure may be used when testing a refrigeration system for leaks:

1. If the system appears to be short of refrigerant, clean the charging ports and observe for traces of oil. If oil is found, leak test the suspected port with a leak detector.

2. If no leak is found on the charging ports, install the gauges. Crack the charging screws to permit pressure to be registered on the gauges.

3. If the system has a gauge reading of 25 lbs (1.76 kg/cm²) or higher, the system has sufficient pressure to proceed with the leak test. The pressure in the low side should be as high as possible. Therefore, turn the temperature control to the off position and warm up the evaporator by placing pans of hot water directly in contact with the evaporator. On frost-free models, manually initiate a defrost cycle.

4. If the head pressure is less than 25 lbs, charge refrigerant into the system until sufficient pressure is obtained so that an effective leak test can be made (see Figure

A-6968

**Figure 13-29.** Adding Refrigerant for a Leak Test (*Courtesy of Frigidaire Division, General Motors Corp.*)

13-29). The unit should be electrically disconnected for about 10 minutes before testing.

*NOTE:* Avoid breathing fumes from the leak detector flame.

5. Check all lines, freezer, charging ports, etc., with a leak detector (see Figure 13-30).

6. If a leak is found at one of the above mentioned points, discharge the system, repair the leak, evacuate the system, and recharge with refrigerant using a new filter-drier.

**Figure 13-30.** Leak Testing Using a Halide Leak Detector (*Courtesy of Frigidaire Division, General Motors Corp.*)

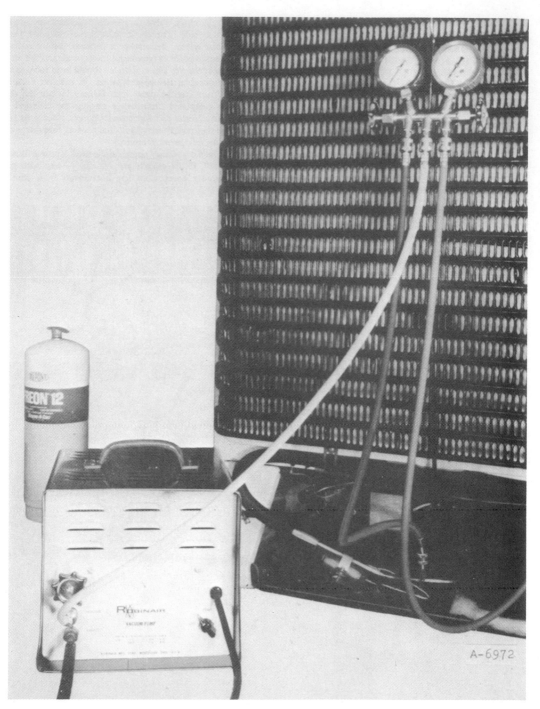

**Figure 13-31.** Evacuation Through Both Gauges (*Courtesy of Frigidaire Division, General Motors Corp.*)

*Caution:* On refrigerators and freezers with foam insulation, do not puncture the foam with a detector prod or remove plastic shelf pegs from areas to be leak-checked. This releases trapped particles of refrigerant from the foam which will falsely indicate a leak.

**EVACUATION PROCEDURE:** Evacuation is necessary before recharging the system in order to remove all traces of air and the previous refrigerant charge. The evaporator should be at room temperature, if possible, before evacuation. If not, on frost-free models, turn the defrost timer to defrost. When the defrost cycle is completed, immediately unplug the electrical cord. On conventional models, place pans of hot water in the freezer section with the electrical cord unplugged to raise the refrigerant pressure.

On units with service ports on both sides, the evacuation can be accomplished through both ports of the gauge manifold (see Figure 13-31). This requires that the vacuum pump be disconnected before recharging can take place. This is not difficult and air will not get into the system if precautions such as purging the charging line from the charging cylinder to the gauge set are followed. There, of course, must be no leaks on the gauge set or in any of the hose connections.

A preferred method is to evacuate the charging cylinder while evacuating the system (see Figure 13-32). Notice the way the charging lines are connected. In this fashion, the system can be evacuated and recharged without disconnecting any lines.

Be sure that the evaporator is warmed before evacuating so the system can be purged of refrigerant by attaching a line from the charging port to the outdoors. Adequate ventilation should be provided so that accurate leak tests can be conducted

later in the procedure. After the system is completely discharged, assemble the equipment as outlined above and as shown in Figure 13-32. Open both gauge hand valves and the charging ports. Start the vacuum pump by following the instructions that accompany the pump. Be sure that the hand valve at the base of the charging cylinder is closed during evacuation. Evacuate the system for approximately 5 minutes, shut off the valve at the vacuum pump, and unplug the pump. Observe the manifold gauges for a few minutes. If the gauge readings rise, there is a leak in the system or line hook-up. Recheck and correct as necessary. When this check is completed, continue evacuating for at least 20 minutes at 28-inch (4.87 cm) vacuum or lower.

**RECHARGING PROCEDURE:** The secret to recharging is good evacuation. The compressor should not be at a higher temperature than the ambient as would be the case when it has been running for hours prior to the service call. On some compressors, the refrigerant enters through the high side of the system and vaporizes more quickly if the compressor is hot, slowing down entry of the liquid refrigerant into the system. The pressure in the charging cylinder may be increased by applying hot towels or warm air to the charging cylinder. Some charging cylinders have their own heat source. *Never apply an open flame to a charging cylinder.* When the refrigerant charge is measured from the charging cylinder, it is in the liquid state. Liquid refrigerant should never be charged through the low side because this tends to *scrub out* the oil from the compressor. Be safe and charge through the high side. If the occasion requires that refrigerant be added through the low side, be sure the refrigerant is in the vapor state and add it slowly while the compressor is running. This

**Figure 13-32.** Preferred Method for Evacuation (*Courtesy of Frigidaire Division, General Motors Corp.*)

is also advisable for clearing the refrigerant from the charging hook-up in order to reduce the pressure to that of the low side.

Whenever the refrigerant charge is questionable, discharge and evacuate the system and recharge with the correct amount of re-

frigerant as shown on the model and serial number plate.

It should be noted that a reasonable amount of time should be allowed after recharging the unit for it to begin functioning properly. Since the refrigerant is charged into the high side of the system, it must flow through the small capillary tube to the evaporator. If the evaporator is warm, the refrigerant changes to a vapor instantly due to the absorption of heat. The evaporator must be reduced in temperature considerably before the system pressures begin to balance out.

If a service technician is expecting immediate cooling results or normal operating pressures on his gauges, he could condemn a component as being faulty when it is not. This leads to unqualified component replacements. The time required to obtain normal operating temperatures depends on many factors. The best method is to remove the charging equipment, thoroughly check for leaks, and let the unit run. Recheck the unit the next day to determine if it is performing normally.

## FRIGIDAIRE REFRESHMENT CENTER REFRIGERATOR

The refreshment center refrigerator (Figure 13-33) features a new method of automatically dispensing ice cubes, two beverages, and chilled water. But the refrigeration system is basically the same as that discussed before, and a discussion of that unit will be omitted here.

**CUBE DISPENSING:** When the user desires ice cubes, an actuator button that is labeled ICE CUBES is pushed. This raises the door of the cube chute. As the door opens to its full open position, it engages an arm that closes the contacts of a microswitch located in the divider assembly between the fresh food and the freezer compartments (see Figure 13-34). When the microswitch contacts close, a drive motor located behind the cube container in the freezer compartment is energized. The motor is rated as a 2-pole, shaded-pole type, $\frac{1}{12}$ hp, 115 volts, 60 Hz, and rotates counterclockwise as viewed from

**Figure 13-33.** Refreshment Center Refrigerator (*Courtesy of Frigidaire Division, General Motors Corp.*)

the front or lead end. It is permanently lubricated and rotates at approximately 3000 rpm. It is reduced to a 30 to 1 ratio by a gear reduction system that drives a conveyor located in the bottom of the ice cube container. (See Figure 13-35.) This conveyor pushes the cubes forward through an opening at the front of the container which has a spring-loaded trap door that leads into the cube chute. The cube chute, which is located in the divider, leads into the food

**Figure 13-34.** Ice, Juice, and Water-Dispensing System—Exploded View (*Courtesy of Frigidaire Division, General Motors Corp.*)

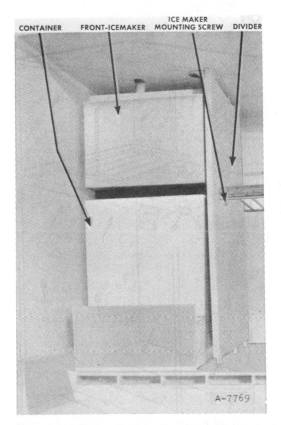

CONTAINER   FRONT-ICEMAKER   ICE MAKER MOUNTING SCREW   DIVIDER

A-7769

**Figure 13-35.** Ice Service Compartment (*Courtesy of Frigidaire Division, General Motors Corp.*)

compartment door and dispenses into the user's glass. A bellows and valve arrangement is used in conjunction with the cube actuator to slow the closing action of the cube door. This prevents cubes from being lodged in the chute when the user releases the cube actuator.

**CHILLED WATER DISPENSING:** When the user depresses the WATER dispenser push buttom, a spring is raised off of the valve seat taking the pressure off the diaphragm. This is done so that full water pressure is not placed on the dispenser valve. Full wa-

ter pressure could result in splashing in the user's glass. When the push rod is depressed to its fullest extent a microswitch closes. When this switch closes, the electrical circuit to the black solenoid of the water valve is completed, causing the water to flow by pressurizing the system. Water from the supply line flows through the water valve at a controlled rate of 334–366 cc per 10 seconds to the bottom (inlet) of the storage tank, which is mounted to the rear wall of the food compartment liner. The tank is designed as a series flow but has crossovers at the ends to prevent air from being trapped within the tank. The air is forced to the top of the tank and is purged when the customer pushes the dispenser button. As a matter of fact, it is recommended that this be done by the person installing the water line to the refrigerator. The tank holds ¾ gallon (2.84 l) of water. As the WATER push button is depressed and all the air is purged, water flows from the top of the storage tank through a plastic tube that connects the tank to the dispenser in the door and into the user's glass. The temperature of the chilled water will be between +40°F and +48°F (4.4°C and 8.9°C), depending on the quantity of water dispensed at one time and the temperature in the food compartment.

**BEVERAGE DISPENSING:** The flow of water through the system to the dispenser valve is exactly the same as it is when the user pushes the WATER button. However, when one of the BEVERAGE buttons is depressed, water leaving the dispenser valve passes through the *venturi tube* (see Figure 13-36). The stream of water passing through the venturi tube creates a negative pressure in the storage container and the beverage concentrate is drawn from the storage container. The beverage concentrate and water mix in the dispenser outlet tube. The amount

**Figure 13-36.** Beverage-Dispensing Valve (*Courtesy of Frigidaire Division, General Motors Corp.*)

of water added to the concentrate is controlled by the user through a manually operated richness control valve. When this valve is set on RICH, no additional water is added to the mixture. As the valve is turned to MILD, additional water is added to the mixture. After a few days of trial period, the user will determine the consistency of the beverage he desires.

**ICEMAKER OPERATION:** When the water line is connected to the refrigerator and the water shut-off valve is turned on, the icemaker will begin after the unit has reached zero zone operating temperatures.

At the proper temperature, the icemaker will call for a fill. The green solenoid of the two-part water valve is energized, which causes water to flow to the icemaker. The fill portion of the commutator gear allows the water to flow for 12 seconds into the cubelet tray. The flow washer in this portion of the water valve meters the water flow through an orifice so that 105 to 130 cc of water is dispensed in this 12-second fill time.

The polypropylene party ice tray consists of 3 rows of 7 cavities each for a total of 21 ice cubes per harvest. The cubes are square on top and rounded on the bottom to make automatic ejection easier. It should be noted that air must be in circulation and at the desired temperatures over this tray for proper operation. Too often the icemaker is blamed for a problem when improper air flow and temperature are the cause. The best way to determine this is by taking freezer product temperatures which should be between 0°F and +5°F (−17.8°C and −15°C) before any other analysis is attempted.

The bulb of the icemaker thermostat senses the front, middle cube cavities and when it is satisfied—13°F (−10.6°C)—the icemaker mechanism starts the harvest cycle providing, of course, that the cube container is in place. The icemaker mechanism rotates the front of the cube tray to the right (clockwise) a few degrees, creating a gradual twist to break the cubes loose from the molds. It returns to the horizontal position, then rotates to the left (counterclockwise). The cubes begin to fall out of the tray cavities into the container. As the tray reaches a vertical position, it engages a restrainer spring at the rear of the icemaker frame which causes the tray to get an additional twist as the motor continues to drive the gear and slide rack that turns the tray. The twist of the tray eventually overcomes the tension of the restrainer spring and allows the tray to snap off the spring and hit the stop at the rear of the icemaker frame. This sudden jar causes any cubes still remaining in the tray to be released from the tray cavities. After this action, the tray starts to rotate back again until it reaches the horizontal position. At this time the tray is filled with water again. The overall travel time for the complete harvest cycle is approximately 3 minutes.

This icemaker model has no weight switch or adjustable cube level control because of the automatic dispensing feature. It does have a heavy gauge metal wire sensing arm that moves up and down with each harvest cycle. (See Figure 13-37.) As the rack moves, the end of it engages a gauge arm and switch shaft to which the sensing arm is attached. When the sensing arm comes back down, it senses the level of the cubes in the container. If the level of the cubes is high enough to raise the sensing arm, an actuator opens a microswitch called the gauge switch and cube production stops.

In addition, the automatic shut-off lever works in conjunction with the container and ice-sensing arm. If the container is removed, the automatic shut-off lever raises the ice-sensing arm which in turn opens the gauge switch, shutting off the icemaker. The container itself holds approximately 500 cubes, but automatic operation allows somewhat less.

**Freeze Cycle:** Figure 13-38 shows the complete cycle diagram as well as the electrical diagram. The numbers to the left of the sequence indicate the cycle time in seconds. This cycle of operation can be followed from this diagram, which is located on the inside of the icemaker cover.

In Figure 13-39, the water is freezing into ice. This cycle requires most of the time in the cube-making process. The tray is in the horizontal position and is full of water. The thermostat has warmed up above 16°F (−8.9°C). The gauge arm is down but not touching ice because the container is not full. The container is in place, however.

The electrical power comes in through the black and white leads at the icemaker terminals marked D and E respectively. The blue lead from the water valve solenoid is attached to terminal F. The icemaker motor is not running simply because the icemaker thermostat internal contacts are in the warm position (1 and 2).

**Start of Harvest Cycle:** In Figure 13-40, the harvest cycle is started because the icemaker

**Figure 13-37.** Icemaker Mechanism Assembly (Courtesy of *Frigidaire Division, General Motors Corp.*)

ELECTRICAL MOUNTING
PLATE ASSEMBLY

GAUGE ARM &
SWITCH SHAFT

ICE LEVEL
SENSING ARM

CUBELET TRAY

FRAME

AUTO SHUT
OFF WIRE

RESTRAINER
SPRING

STOP

A-7802

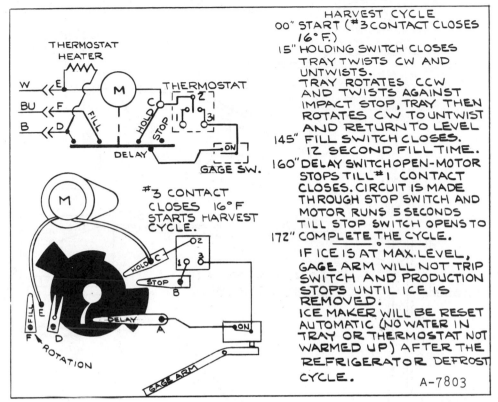

HARVEST CYCLE
00" START (#3 CONTACT CLOSES 16° F.)
15" HOLDING SWITCH CLOSES TRAY TWISTS CW AND UNTWISTS.
TRAY ROTATES CCW AND TWISTS AGAINST IMPACT STOP, TRAY THEN ROTATES CW TO UNTWIST AND RETURN TO LEVEL
145" FILL SWITCH CLOSES. 12 SECOND FILL TIME.
160" DELAY SWITCH OPEN—MOTOR STOPS TILL #1 CONTACT CLOSES. CIRCUIT IS MADE THROUGH STOP SWITCH AND MOTOR RUNS 5 SECONDS TILL STOP SWITCH OPENS TO
172" COMPLETE THE CYCLE.

IF ICE IS AT MAX. LEVEL, GAGE ARM WILL NOT TRIP SWITCH AND PRODUCTION STOPS UNTIL ICE IS REMOVED.
ICE MAKER WILL BE RESET AUTOMATIC (NO WATER IN TRAY OR THERMOSTAT NOT WARMED UP) AFTER THE REFRIGERATOR DEFROST CYCLE.

A-7803

THERMOSTAT HEATER

THERMOSTAT

#3 CONTACT CLOSES 16° F STARTS HARVEST CYCLE.

**Figure 13-38.** Icemaker Cycle Diagram *(Courtesy of Frigidaire Division, General Motors Corp.)*

thermostat has reached its cut-out setting of 13°F ($-10.6$°C), causing the internal contacts to flip to positions 2 and 3. This completes the circuit from the black power lead at terminal $D$ to the commutator, to the delay contact through the gauge switch, to the icemaker thermostat, through contacts 3 and 2 to the hold contact and the drive motor. The other side of the drive motor coil is connected to the common side of the electrical power line, white. This causes the drive motor to run. As the commutator and gear advance, the hold contact comes into play by contacting the commutator. This gives direct contact to the drive motor from the commutator and bypasses the icemaker

thermostat and gauge switch. It is at this point that the tray begins its first twist (clockwise) and now cannot be interrupted by the gauge switch or icemaker thermostat.

The harvest cycle continues as the drive motor turns the commutator and gear assembly. The thermostat cold contacts, across 2 and 3, are still closed although the hold contact is making on the commutator which bypasses the thermostat and gauge switch. This is why the drive motor continues to run even though the ice sensing arm (gauge arm in Figure 13-41) has started to rise out of the container breaking the microswitch (gauge switch) contacts. The tray is in its final twist.

FREEZE CYCLE

1. Tray horizontal and filled with water.
2. Thermostat warm (closed 2-1).
3. Gauge arm down, not touching ice.

**Figure 13-39.** Freeze Cycle (*Courtesy of Frigidaire Division, General Motors Corp.*)

START HARVEST CYCLE

1. Thermostat contacts 2 to 3 close.
2. Tray entering pretwist position.
3. Circuit hold contact has engaged commutator.
4. Gauge arm has not yet started to rise.

**Figure 13-40.** Icemaker Start of Harvest Cycle Diagram (*Courtesy of Frigidaire Division, General Motors Corp.*)

A-7806

## HARVEST CYCLE

1. Thermostat contacts 2 to 3 still closed.
2. Gauge arm has moved out of container.
3. Tray in final twist & dispensing ice.
4. Gauge switch to off position.

**Figure 13-41.** Harvest Cycle Continued (*Courtesy of Frigidaire Division, General Motors Corp.*)

MOTOR

THERMOSTAT

HOLD

STOP

2

1

3

DELAY

ON

0 OFF

GAGE SWITCH.

FILL

FILL
VALVE
SOLENOID

GAGE ARM

ICE

CONTAINER

115 V.

A-7807

FILL CYCLE

1. Tray horizontal and filling with water.
2. Thermostat contact 2-3 open contacts 2-1 close.
3. Gauge arm enters container. Does not reach ice.
4. Gauge switch returns to on for next cycle.

**Figure 13-42.** Fill Cycle (*Courtesy of Frigidaire Division, General Motors Corp.*)

**Fill Cycle:** The cube tray has returned to the horizontal position and the tray is filling with water because the commutator and gear have advanced to the portion of the commutator that energizes the fill contact and then to the water valve solenoid through the blue electrical lead (see Figure 13-42). The incoming water will raise the temperature—above 16°F (−8.9°C)—of the icemaker thermostat, causing the internal warm contacts 1 and 2 to be made. The ice-sensing arm enters the container but does not touch ice so the gauge switch returns to on —ready for the next harvest which will occur when the water is frozen and the icemaker thermostat is satisfied at 13°F (−10.6°C).

**Full Container:** When the level of the ice cubes in the container reaches a point that will cause the ice-sensing arm to rise, it will open the gauge switch contacts. (See Figure 13-43.) Even though the ice in the tray is ready to be harvested, there is no electrical circuit to the drive motor—thus no harvest. The ice remains in the tray until such time as the level of the ice is reduced to the point that the ice-sensing arm will complete the circuit.

**Container Removed:** When the container is removed for any reason the automatic shut-off lever causes the ice level sensing arm to move up out of the container. This causes the gauge switch contacts to open, preventing a harvest (see Figure 13-44). When the container is replaced in its normal position, the system returns to a normal operating condition. As stated previously, removal of the container should be discouraged except for general cleaning purposes.

**Delay Feature:** The *delay feature* puts the icemaker mechanism into a hold position when there is a possibility that water might be harvested before it is frozen into cubes. A raised portion designed into the plastic drive gear protrudes through the metal com-

mutator. This protrusion causes a momentary electrical break, sometimes referred to as a *dead spot,* that interrupts electrical contact to the delay contact or to the on contact of the gauge switch. If for any reason the icemaker thermostat fails to warm up after the tray is filled and if the internal contacts do not flip back to 1 and 2, it is possible to harvest the water that just filled the tray. (See Figure 13-45.) Should the commutator and gear assembly stop in this delay position, a reset of the cycle will automatically occur after the next defrost period. Package temperatures do not rise appreciably during a defrost cycle but the air temperature over the cooling coil does warm for a short time period. This air should be warm enough to raise the temperature of the icemaker thermostat the few degrees necessary to reset the thermostat, especially if no water is in the tray. If this fails to reset the thermostat, check the electrical hook-up or maybe replace the thermostat to correct the condition. If you get no fill then, it isn't the thermostat. Check the fill system. See Figure 13-46 for an exploded view of the icemaker.

## DOMESTIC ELECTRICAL CIRCUITS

When troubleshooting a domestic refrigeration unit, it is necessary for the service technician to be able to properly read the wiring diagrams which accompany the unit. These wiring diagrams range from very simple to very complex. However, if one has the ability to read these diagrams, he or she will realize that the more complex diagrams only have more components. Also, a basic knowledge of unit operation is very helpful.

There are basically two types of wiring diagrams in use today. They are the *pictorial* and the *schematic* diagrams. Both are very useful. The pictorial is helpful in locating components on the unit. The schematic is helpful in determining the unit's operating sequence.

FULL CONTAINER

1. Tray of ice is ready for harvest.
2. Ice container is full.
3. Gage arm contacting ice & will not let switch complete circuit.
4. Normal harvest will occur when level of ice is reduced to point gage arm will complete circuit.

**Figure 13-43.** Full Container (*Courtesy of Frigidaire Division, General Motors Corp.*)

A-7809

CONTAINER REMOVED

1. Remove container.
2. Gage arm moves out of container.
3. Gage switch moves to off preventing ice harvest.
4. Replace container & system returns to a normal operating condition.

**Figure 13-44.** Icemaker Container Removed (*Courtesy of Frigidaire Division, General Motors Corp.*)

DELAY FEATURE

1. Delay feature prevents continuous operation of ice maker if thermostat does not reset.
2. Dead spot in commutator prevents completion of cycle and a water harvest.
3. Similar situation could occur if no water enters the tray.
4. Unless thermostat or bellows heater is faulty, reset of the cycle will automatically occur after defrost.

**Figure 13-45.** Delay Feature (*Courtesy of Frigidaire Division, General Motors Corp.*)

DETAIL "A"
THERMOSTAT BULB MUST BE
INSTALLED IN THIS OPENING -
NOT IN THE SPACE BELOW.

AUTOMATIC ICE

A-7761

**Figure 13-46.** Icemaker Detail Parts—Exploded View (*Courtesy of Frigidaire Division, General Motors Corp.*)

**SIMPLE FREEZER DIAGRAM:** When reading a wiring diagram, regardless of how simple or complex, it is less confusing to trace each circuit all the way through before starting a new one. For the following explanation, see the schematic diagram in Figure 13-47.

Beginning with the side of the electric power line labeled HOT, the first circuit that is approached is the light circuit. The light switch is shown to be a normally closed switch. The electric power progresses through the wire extending to the freezer light and to the other side of the electric power line. Therefore, any time the light switch is closed, the light will be on.

When we continue on the hot line, the next circuit approached contains only a door heater. There are no switches or other devices in this circuit. Therefore, any time the unit's electric cord is plugged in, the door heater will provide heat.

The next circuit approached is the compressor circuit. The first component encountered is the cold control (thermostat) followed by the overload protector which is in the common line to the compressor. The electric power then goes through the compressor start winding, through the start-contacts on the starting relay which are normally open, and to the other side of the line. To complete this circuit, the electric power, after leaving the common terminal of the compressor, proceeds through the compressor run winding to the starting relay coil and to the other side of the line. Any time the cold control and overload protector are closed, the compressor will attempt to run. Electric power is supplied to both compressor windings through the common terminal. The high current through the run winding will cause the contacts to close on the amperage starting relay and energize the start winding. As the compressor gains speed, the amper-age decreases, allowing the starting relay contacts to open and the run winding to remain in the circuit. The unit is now in its normal operating cycle. When the cold control contacts open, the electric power to the compressor is interrupted and the unit stops.

It should be evident from the foregoing explanation that any circuit can be easily traced and the operating sequence determined by properly reading and diagnosing the wiring diagram.

**COMPLEX REFRIGERATOR WIRING DIAGRAM:** Even though this type of diagram has more components, it can be read and the operating sequence can be determined in the same manner. (See Figure 13-48.) Each of the components is shown in the normal operating position. Simply follow through each individual circuit completely before attempting another.

## ROOM AIR CONDITIONERS

To this point our discussion of domestic refrigeration has involved systems that operate with an evaporator temperature below freezing. Room air conditioning units operate with the evaporator temperature above freezing. They are designed to handle loads between 5000 and 32,000 Btus (1260 and 8064 kcal) per hour. Refrigerators and freezers are designed to handle loads between 450 and 1000 Btus (113.4 and 252 kcal) per hour with a zero-degree evaporator. Window units operate with an evaporator temperature of about 40°F (4.4°C) and lower the circulating air temperature about 20°F (11.12°C). At the same time the temperature is being lowered, moisture is being removed, drying the air and making the room more comfortable.

There are two basic systems in any air conditioner: the refrigeration system and the air circulating system.

Wiring Pictorial

Door Heater
(Optional)

Cold
Control

Light

Light Switch

Relay

Overload

Wire Colors

Red
White
Blue
Black
Green

Cross over

Schematic

Overload
Protector

Compressor

Starting
Relay

Start (S)

(S)

White

C Common

Cold
Control

Black

Blue

Run (M)

(M)

Door Heater
~~~~~~~~~~~
(19 ± 1 Watts)

Light Switch

Black

Red

Freezer Light

White

Hot

Ground

Neutral

**Figure 13-47.** Simple Freezer Wiring Diagram (*Courtesy of Frigidaire Division, General Motors Corp.*)

**Figure 13-48.** Refreshment Center Refrigerator Wiring Diagram (*Courtesy of Frigidaire Division, General Motors Corp.*)

The sealed refrigeration system is much the same as the sealed system used in refrigerators and freezers. The refrigerant generally used in room air conditioners is R–22. The capillary tube is usually much larger in diameter than those used on refrigerators and freezers in order to operate at the higher evaporator pressure and temperature. (See Figure 13-49.) The same service procedures used for other refrigeration systems should be applied to these units.

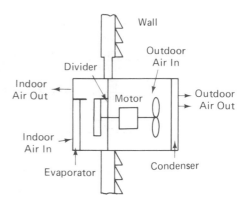

**Figure 13-50.** Types of Circulating Air

**Figure 13-49.** Room Air Conditioner Refrigeration System (*Courtesy of Frigidaire Division, General Motors Corp.*)

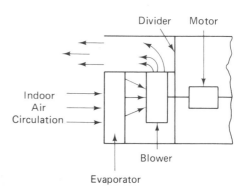

**Figure 13-51.** Indoor Air System

The air circulation system on room air conditioning units can be divided into two categories: indoor air and outdoor air (see Figure 13-50).

The indoor air is the air circulated in the space to be cooled. It is circulated over the evaporator by a centrifugal blower (see Figure 13-51). The centrifugal blower has high air-moving capabilities with low noise.

The outdoor fan is usually a blade-type fan. These fans are usually referred to as condenser fans and are mounted outside the conditioned space (see Figure 13-52). This fan moves outdoor air over the compressor and condenser. It also has a slinger ring around the outside periphery of the blades which picks up condensate from a condensate well and distributes it over the condenser. This moisture helps cool the condenser and increase the efficiency of the unit.

The fan motor on these units may be either single speed or multiple speed. When replacing a motor, one with the same speed must be used. If the rpm of the motor is changed, the air volume will also be changed. If the air volume is changed, the Btu rating also will be affected and the comfort in the conditioning space will be changed.

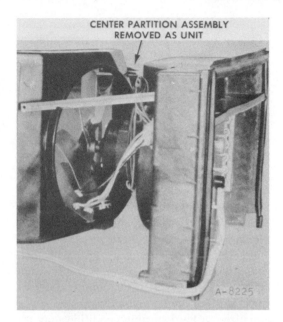

CENTER PARTITION ASSEMBLY
REMOVED AS UNIT

A-8225

**Figure 13-52.** Outdoor Fan (*Courtesy of Frigidaire Division, General Motors Corp.*)

**CHECKING CAPACITY:** Room air conditioners are rated according to the results obtained from a controlled test which is spelled out in AHAM (Association of Home Appliance Manufacturers) Standard CNI. During this test the room air is controlled at 80°F (26.7°C) with a 50% relative humidity and the condenser air is controlled at 95°F (35°C) db and 38% relative humidity. The capacity of the unit will vary as these conditions vary.

Variations in ambient temperatures and relative humidity directly affect the cooling capacity and the temperature drop across the evaporator coil. Installation in restrictive or unfavorable operating locations, improper sizing of the unit for the area to be cooled, occupancy above normal, dirty filters, and improperly directed louvers are some conditions that can contribute to complaints of unsatisfactory cooling.

It is possible, however, to check the unit against established average capacity ratings, disregarding everything but the temperature of the air entering the condenser, the dry bulb temperature drop across the evaporator, and the normal wattage draw of the unit.

**CAPACITY CHECKING PROCEDURES AND PRECAUTIONS:** When making the capacity tests, be certain that the filter, the evaporator, and the condenser are clean and that the coil fins are not bent and obstructing the air flow. Also, be sure that the adjustable air directional louvers are directing the air up and to the right or left, not down. Air that is directed downward can circle back into the air conditioner and give a false reading.

It will be necessary to take the wet bulb (WB) temperature, dry bulb (DB) temperature, and the total wattage draw. The unit must be operated on maximum cooling for about 15 minutes before the readings are taken.

The wet and dry bulb temperatures, necessary for determining the relative humidity, may be determined by use of a *sling psychrometer*.

The wattmeter should be connected in the circuit before the unit is started.

Determine the dry bulb temperature of the evaporator inlet and outlet air streams.

The inlet return air temperature should be taken at the center of the return air opening. The discharge air temperatures should be taken in front of the air discharge grill. Since the temperature will vary slightly between the grills, make at least three readings across the grill area to obtain an average reading of the discharge air temperature.

The dry bulb condenser inlet temperature can be taken with a pencil-type thermometer or with any accurate temperature tester. This temperature should be taken about ¼ to ½ in. (6.4 mm to 12.7 mm) away from the condenser air inlet louvers. The thermometer should not touch the louvers nor be exposed to the direct rays of the sun.

These readings can be compared to capacity check tables which can be obtained from the unit manufacturer or distributor. It is necessary to obtain a table for the specific model being checked.

## THERMOSTAT ANALYSIS

**1.   Operating Range of Thermostat:** The operating range of a thermostat is the difference between the cut-in temperature at the coldest thermostat setting and the cut-out temperature at the warmest thermostat setting. This difference usually represents approximately 20–21°F (11°C) temperature change in 180° rotational movement of the thermostat shaft, or a change equal to approximately 5°F (2.78°C) for every 45° of rotation.

**2.   Operating Differential of Thermostat:** Determine the cut-in and cut-out temperature of the thermostat by sensing the return inlet air temperature with an accurate temperature tester. The temperature should be sensed in a manner similar to that described in checking the capacity of the room air conditioner.

*NOTE:* The air conditioner must be tested in an environment where it has the capacity of cooling to the cut-out temperature of the thermostat setting.

Before and during the operating temperature check, be sure that the discharge air is not being directed or deflected directly back to the return air inlet by louvers, drapes, curtains, adjacent objects such as furniture, or by standing in front of the air conditioner.

For proper performance, it is important that the sensing bulb of the thermostat be positioned ¼ to ⅜ in. (6.4 to 9.5mm) in front of the evaporator. If the thermostat capillary line or bulb is too close or touching the evaporator, it will sense the coil temperature directly, instead of the air temperature, and cause short cycling of the compressor.

**SHORT-CYCLING CONDITION:** Most short-cycling is caused by the discharge air from the grill that hits a nearby object such as furniture, walls, or curtains and is deflected back into the air inlets. This cool air satisfies the thermostat and it shuts off the compressor before the room is really cooled. Warm air entering the inlet grill will soon cause the thermostat contacts to close and start the compressor. If the compressor motor is still hot or the unit pressures have not equalized, a hard starting condition may result, causing the motor overload protector to trip, a circuit breaker to trip, or the circuit fuse to blow.

If short-cycling or premature compressor shut-off persists, it may be desirable to place a small piece of electrical tape on the back side of the clip to reduce the conductivity between it and the coil. It also may be desirable to place a small piece of sleeving over a portion of the bulb to reduce its sensitivity.

Outside air coming in the bottom of a poorly installed unit may strike the bulb and cause poor temperature control or short-cycling. In a poorly sealed building or where the unit is mounted close to the floor, it may be desirable to move the bulb clips higher up on the coil so that it is better isolated from drafts.

**ICED COIL CONDITION:** Icing is usually caused by an insufficient air flow over the evaporator. This could be due to a dirt-blocked coil, dirty filter, curtains or drapes, nearby furniture, etc. Also, when the outside condensing air becomes quite cool, the air conditioner cooling capacity may increase to the point at which the evaporator coil will ice up. To guard against icing on some units, the clips holding the thermostat bulbs are configured so that they touch the evaporator. If the ice touches the clips, it will cool the bulb to the point where the thermostat will shut off the compressor, allowing the ice to melt.

**COOL AMBIENT TEMPERATURE OPERATION:** Occasionally an air conditioner will not run when the outside ambient temperature is less than 60°F (15.6°C) even though the room is warm. The metal in the air conditioner can become so cold that the liquid or vapor actuating medium in the thermostat power element condenses to the extent that its existing pressure cannot close the thermostat contacts. The volume of vapor or liquid in the power element bellows is so great as compared to the volume of vapor in the bulb that the thermostat operation will be controlled by the temperature of the thermostat body and not by the roomside thermobulb capillary line. Even heating the bulb will not expand the power element charge enough to operate the switch.

*Note:* If the compressor is operated at very low outside ambient temperatures to cool a room, it will probably overheat and possibly damage the compressor motor due to the lack of sufficient refrigerant passing through the capillary tube and evaporator to cool the motor windings.

**LOW VOLTAGE:** Low voltage is a common cause of trouble in the operation of a room air conditioning system. It therefore becomes doubly important, because of the motor size, that the service technician check the voltage when servicing room air conditioners.

Improper voltage may result in one or more of the following complaints:

1.  The unit will not start.

2.  Compressor motor cycling on the motor protector.

3.  Premature failure of the motor protector caused by burned or pitted contact points.

4.  Blown fuses.

5.  Noticeable brightening of the lights when the compressor cycles off.

6.  Evaporator icing—a low voltage may reduce the fan speed enough to give an inadequate air flow over the evaporator, thereby allowing it to ice up.

Low voltage often can be attributed to the use of extension cords or an inadequately wired circuit, but low voltage into the building and loose fuses or connections should not be overlooked.

**HIGH VOLTAGE:** High voltage can be equally as troublesome by causing motors to overheat and cycle on their overload protectors or break down electrically. Extremely high voltages can create enough back voltage across the capacitor to cause it to fail. It will cause resistance heaters on heating models to deliver more than their rated capacity and open the safety protectors or fuse links.

**COMPRESSOR:** An inoperative compressor should be isolated electrically and checked before a decision is made to replace it. Winding resistance can be checked against the values on the schematic wiring diagrams. The resistances on some types of compressors will vary according to the stator manufacturer.

When testing the compressor, use a start cord and a capacitor of the proper rating. Connect the start cord as shown in Figure 13-53.

All PSC motors are designed to run with

**Figure 13-53.** Compressor Test Cord Connections (*Courtesy of Frigidaire Division, General Motors Corp.*)

the capacitor in the circuit. Consequently, the start button must be held in or the compressor may stop when it is released.

**FAN AND BLOWER ALIGNMENT:** If the condenser fan and evaporator blower are not located in proper relation to the shroud orifice, reduction of air flow may result. This could give high condenser head pressures in hot weather or icing on the evaporator in cold weather.

The slinger ring on the condenser fan

picks up water from the sump and "flings" it onto the condenser. This serves two purposes: (1) it disposes of the water that has condensed on the evaporator coil and (2) it increases the capacity of the unit by removing heat from the condenser as the water evaporates.

*Note:* Water should stand in the base pan; do not drain it.

**CONDENSER AIR BAFFLES:** It is extremely important that the condenser air baffle be installed on the end of the condenser and that it be installed so that it will keep hot condenser air from recirculating. (See Figure 13-54.)

It is also important that the aluminum rear grill be installed on the sleeve so that exhaust condenser air is not drawn back into the condenser air inlet grill.[1]

The following charts indicate the problem, or complaint, and the probable solution. They are designed for quick and easy reference.

[1] See Frigidaire "Tech-Talk," Vol. 74, No. 8 for further reference.

**Figure 13-54.** Sleeve—Air Flow (*Courtesy of Frigidaire Division, General Motors Corp.*)

# TYPICAL SERVICE DIAGNOSIS CHART

CONDITION: No Refrigeration
Unit Does Not Run

CAUTION: Throughout these procedures, product is plugged in for voltage checks.

**START HERE**
Turn cold control to cold position. Open food door—observe if interior light is on or not.

Check voltage at wall receptacle. Should be 115 volts ± 10%

Gain access to relay and using a test light or voltmeter, check for voltage at terminals 1 and 2 of the relay. On models using a 4 CR relay check at the black and white leads of the terminal board.

Voltage present at terminals 2 and L of the relay or on models with 4 CR relay it would be at relay terminal 4 and white of the terminal board.

If there is no voltage at 1 and 2 or at ther terminal board but voltage at the wall receptacle, replace service or line cord.

FORCED AIR CONDENSER MODELS USING 4 CR RELAY
Check between the pink and the white lead terminals of the terminal board for proper voltage.

If voltage is at the pink and white lead terminals and not at relay terminal 4 and white then the motor protector is open. If compressor is cool replace protector. If compressor is hot proceed to . . .

Check fuse box. Replace fuse or reset breaker. Restore power.

Check for voltage on the load side of the temperature control (Pink lead) and any white connector. If voltage is indicated there is a broken lead between the temperature control and the terminal board.

Repair or replace broken lead.

Determine nature of overload. Advise customer.

Check for voltage on the live side of the temperature control. If voltage is indicated, short across the control. Unit should run.

Replace temperature control

Insert probes of voltmeter into the ends of terminals 1 and 2 of the timer plug. Line voltage should be indicated. Be sure timer is not in the defrost position.

If line voltage is indicated but there is no voltage on the line side of the temperature control, there is an open lead between the timer and temperature control. Replace lead.

Advise customer to call electrician. Provide temporary power supply to product.

Insert probes of the voltmeter into the ends of terminals 1 and 3 of the timer plug. Line voltage should be indicated. Be sure timer is not in the defrost position.

If voltage is at 1 and 3 terminals and not at 1 and 2, replace the defrost timer.

No voltage at 1 and 3 of timer, but voltage is at the black and white terminals of the terminal board. There is an open lead between the timer and the terminal board.

Repair or replace broken lead.

371

HOW TO USE THIS DIAGRAM:

Follow arrow to the right or left if statement in the block is true.
Follow arrow downward if statement in the block is not true.

Disconnect plug at compressor. Install starter cord with capacitor, if required, through Volt-Wattmeter. Product runs with normal wattage.

Check the 3-prong plug attached to compressor for good connections. Check the wiring harness from Relay to compressor for open circuit. Follow color coding Replace harness if necessary.

Check the capacitor, by one of several methods. If it is open or shorted, replace it.

If all checks to this point are good, the relay is at fault and should be replaced.

If compressor fails to run and is hot to the touch, allow to cool and repeat. If wattage is too high, will not start or blows fuses, replace the Compressor.

Clean condenser if it has an accumulation of lint and dust.
If compressor is hot because it has been cycling on the overload protector, do not change the protector. Check condenser fan motor to see that there is no interference with the blade.
Check motor winding. If it is open, rotor is stuck or it has bad bearing, replace the condenser fan motor.

STATIC CONDENSER MODELS

Check for voltage on load side of temperature control (Pink lead) and terminal 2 of relay.

If there is a poor or loose connection within the plug, disconnect and check visually. Reconnect to insure good electrical contact.

Unit runs. Recheck by flexing leads and connectors to insure continuing good electrical contact. No other checks necessary. Reassemble product.

There is an open circuit between cold control and relay. Probably at the connector block in the freezer on ABS liner models. Gain access to connector block in right rear corner of freezer compartment.

Check for voltage on line side of temperature control (Brown lead) and terminal 2 of relay. If voltage is present, short across control. Unit should run.

Separate plugs and disconnect line or service cord. Use an ohmmeter and check the circuits to determine the continuity of each lead. Follow color coding and appropriate wiring diagram.

Repair or replace leads and/or contacts within the connector block and necessary.

Replace Temperature Control.

Insert probes of voltmeter into the ends of the defrost timer plug terminals 1 and 2. Line voltage should be indicated. Be sure timer is not in defrost position.

If voltage is present there is an open circuit between timer and temperature control.

Insert probes of voltmeter into the ends of the timer plug terminals 1 and 3. Be sure timer is not in defrost position.

If voltage is at terminals 1 and 3 and none at terminals 1 and 2, replace defrost timer.

No voltage at 1 and 3 of timer but voltage at 1 and 2 of relay. There is an open circuit, probably at the connector block in the freezer (ABS liner models) or black lead from relay to timer. The white lead comes from the connector block.

Repair or replace black lead from relay to timer.

COMPRESSOR RUNS—TEMPERATURE COMPLAINT

Temperature of the foods store should be recorded. Desired range of temperatures in the freezer is $0°$ to $+5°F$. and $+32°$ to $40°F$. in the food compartment. Temperatures vary due to ambients, usage (door openings), food loads, control settings, time of day temperatures taken, etc.

FREEZER COMPARTMENT
TOO WARM

Reset temperature control dials colder. Freezer control on models so equipped. Allow 24 hours for temperatures to settle out.

Control dial(s) is set on extreme position but temperatures are OK now.

Adjust cutin and cutout or range screw of temperature control as outlined in Tech-Talk for particular model given. Return control dials to mid-position.

Temperature still high with insufficient run time of the unit. Replace Cold Control if broken or discharged.

Freezer fan motor not running. Air duct misaligned. Foreign object restricting air flow. Correct as necessary.

Check for voltage at fan motor leads. If voltage is present, check for interference. Realign blower wheel or motor mounting.

Disconnect line cord and check motor windings with ohmmeter. If open windings or tight bearing, replace motor.

If no voltage, refer to Procedure "Unit does not run."

Condenser fan motor not running.

Check for voltage at motor leads. If voltage is present, check for interference. Realign fan or motor mounting.

Disconnect line cord and check motor windings with ohmmeter. If open windings or tight bearing, replace motor.

If no voltage, refer to Procedure "Unit does not run."

Coil iced up.

Food packages holding door open or very poor door seal. Conventional and Cycla-matic models need defrosting.

Re-position packages or align door for proper sealing. Defrost coil if needed; on Conventional and Cyclamatic models.

Observe defrost timer motor. If not running, replace timer.

If timer motor is running—see Procedure on "Defrost Circuit."

Light stays on.

Check light bulb and push switch plunger several times, Replace switch if necessary. Check to see that switch is wired properly.

Poor door seal.

Food packages holding door open or door not properly aligned. Adjust door. If door seal is torn or "rolled" badly, replace.

Moist-minder. Check to see that trap is in place and working properly.

If a malfunction of the refrigerating system is suspected, see "Inoperative Refrigeration System."

COMPRESSOR RUNS—TEMPERATURE COMPLAINT (Cont'd)

Temperature of the foods stored should be recorded. Desired range of temperatures in the freezer is 0° to +5°F, and +32° to 40°F. in the food compartment. Temperatures vary due to ambients, usage (door openings), food loads, control settings, time of day temperatures taken, etc.

374

## COMPRESSOR RUNS—TEMPERATURE COMPLAINT Cont'd

Temperature of the foods stored should be recorded. Desired range of temperatures in the freezer is 0° to +5°F. and +32° to 40°F. in the food compartment. Temperatures vary due to ambients, usage (door openings), food loads, control settings, time of day temperatures taken, etc.

```
┌─────────────────────────┐
│   FOOD COMPARTMENT      │
│     TOO COLD           │
└─────────────────────────┘
            │
            ▼
┌─────────────────────────┐
│ Reset temperature control dials for│
│ better air distribution within food│
│ compartment. Allow 24 hours for    │
│ temperatures to settle out.        │
└─────────────────────────┘
            │
            ▼
┌─────────────────────────┐        ┌─────────────────────────┐
│ Control dial(s) is set on extreme │──────▶│ Adjust cutin and cutout or range │
│ warm end. Temperatures are now    │       │ screw of cold control as outlined in │
│ satisfactory.                     │       │ Tech-Talk for particular model. Re- │
│                                   │       │ turn control dial to mid-position.  │
└─────────────────────────┘        └─────────────────────────┘
            │
            ▼
┌─────────────────────────┐        ┌─────────────────────────┐
│ Temperatures still too cold. If tem-│──────▶│ On models equipped with an Auto- │
│ perature control is providing too   │       │ matic Air Flow Control, adjust con- │
│ long run time of the unit or control│       │ trol for warmer food compartment   │
│ is damaged, replace control. Check  │       │ temperature. If control is damaged, │
│ control switch bulb mounting. If    │       │ replace. If an ice interference in  │
│ loose, it could cause long run cycles.│     │ holding control open, correct.     │
│ Tighten clamp.                      │       │                                   │
└─────────────────────────┘        └─────────────────────────┘
```

375

## COMPRESSOR RUNS—TEMPERATURE COMPLAINT Cont'd

Temperature of the foods stored should be recorded. Desired range of temperatures in the freezer is 0° to +5°F. and +32° to 40°F. in the food compartment. Temperatures vary due to ambients, usage (door openings), food loads, control settings, time of day temperatures taken, etc.

IN OPERATIVE DEFROST CIRCUIT

Power to Product and Refrigeration Cycle O.K.

Observe if defrost timer motor is running by using flashlight and mirror. Motor does not run.

Check to seet that voltage is supplied to motor at timer terminals 1 and 3. If voltage is present, motor cover may be binding rotor. Adjust cover.

If voltage is at motor and cover is not binding, disconnect power and check the motor winding with an ohmmeter. If open winding, replace the defrost timer.

If no power to timer motor, check timer connector plug.

Install volt-wattmeter and advance timer to defrost position. Should read between 500-600 watts on most models. Others could read between 600-700 watts.

Only approximately 50 watts indicated on wattmeter, depending on model tested. These are usually condensate driers. Be sure doors are closed or wattage of light bulbs will also be indicated. With product still plugged in, use a voltmeter and insert probes into end of defrost timer plug at terminals 1 and 3. Voltage shoudl be indicated. Check at terminals 1 and 4, voltage should be indicated.

No voltage at terminals 1 and 4. Replace defrost timer.

Disconnect defrost timer plug and disconnect product from wall outlet. Use an ohmmeter and check defrost circuit from terminals 1 and 4 of plug. Should read approximately 25 ohms of defrost heaters.

If an open circuit is indicated, it will be necessary to disassemble the freezer compartment. Disconnect product from source of power. Use an ohmmeter for the following.

Replace defrost limiter switch.

DEFROST HEATER CHECK.

Should read approximately 25 ohms. Some models have two heaters connected in series. Each will read 12½ ohms. These use all red leads.

If either or both heaters have an open circuit, they must be replaced.

DEFROST LIMITER SWITCH CHECK.

If coil is frosted the limiter should be closed–zero resistance. If a 47,000 ohm resistance is read, the switch is open.

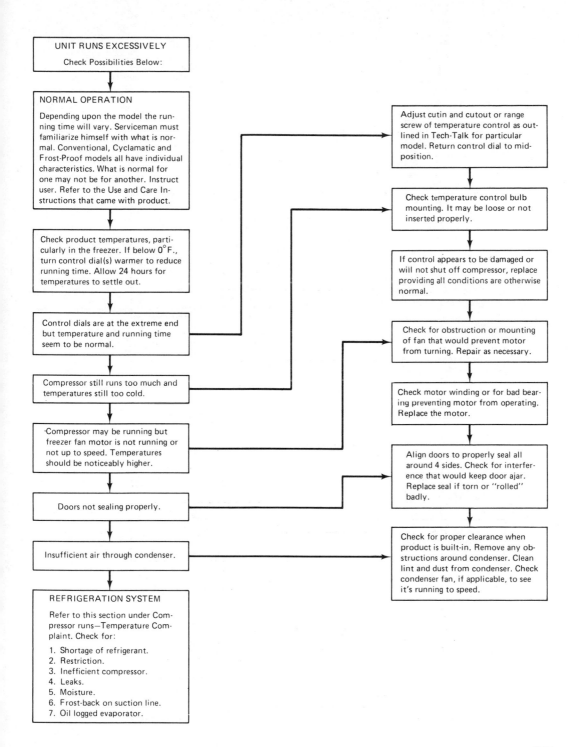

**UNIT RUNS EXCESSIVELY**

Check Possibilities Below:

**NORMAL OPERATION**

Depending upon the model the running time will vary. Serviceman must familiarize himself with what is normal. Conventional, Cyclamatic and Frost-Proof models all have individual characteristics. What is normal for one may not be for another. Instruct user. Refer to the Use and Care Instructions that came with product.

Check product temperatures, particularly in the freezer. If below 0°F., turn control dial(s) warmer to reduce running time. Allow 24 hours for temperatures to settle out.

Control dials are at the extreme end but temperature and running time seem to be normal.

Compressor still runs too much and temperatures still too cold.

Compressor may be running but freezer fan motor is not running or not up to speed. Temperatures should be noticeably higher.

Doors not sealing properly.

Insufficient air through condenser.

**REFRIGERATION SYSTEM**

Refer to this section under Compressor runs—Temperature Complaint. Check for:

1. Shortage of refrigerant.
2. Restriction.
3. Inefficient compressor.
4. Leaks.
5. Moisture.
6. Frost-back on suction line.
7. Oil logged evaporator.

Adjust cutin and cutout or range screw of temperature control as outlined in Tech-Talk for particular model. Return control dial to mid-position.

Check temperature control bulb mounting. It may be loose or not inserted properly.

If control appears to be damaged or will not shut off compressor, replace providing all conditions are otherwise normal.

Check for obstruction or mounting of fan that would prevent motor from turning. Repair as necessary.

Check motor winding or for bad bearing preventing motor from operating. Replace the motor.

Align doors to properly seal all around 4 sides. Check for interference that would keep door ajar. Replace seal if torn or "rolled" badly.

Check for proper clearance when product is built-in. Remove any obstructions around condenser. Clean lint and dust from condenser. Check condenser fan, if applicable, to see it's running to speed.

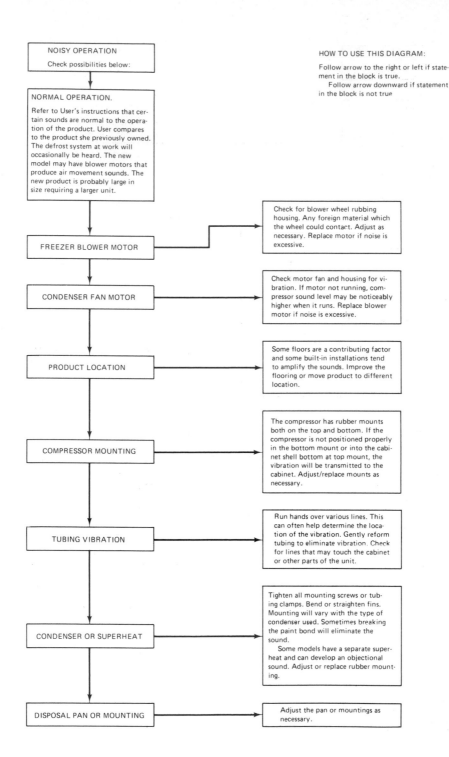

**NOISY OPERATION**

Check possibilities below:

HOW TO USE THIS DIAGRAM:

Follow arrow to the right or left if statement in the block is true.

Follow arrow downward if statement in the block is not true

**NORMAL OPERATION.**

Refer to User's instructions that certain sounds are normal to the operation of the product. User compares to the product she previously owned. The defrost system at work will occasionally be heard. The new model may have blower motors that produce air movement sounds. The new product is probably large in size requiring a larger unit.

**FREEZER BLOWER MOTOR**

Check for blower wheel rubbing housing. Any foreign material which the wheel could contact. Adjust as necessary. Replace motor if noise is excessive.

**CONDENSER FAN MOTOR**

Check motor fan and housing for vibration. If motor not running, compressor sound level may be noticeably higher when it runs. Replace blower motor if noise is excessive.

**PRODUCT LOCATION**

Some floors are a contributing factor and some built-in installations tend to amplify the sounds. Improve the flooring or move product to different location.

**COMPRESSOR MOUNTING**

The compressor has rubber mounts both on the top and bottom. If the compressor is not positioned properly in the bottom mount or into the cabinet shell bottom at top mount, the vibration will be transmitted to the cabinet. Adjust/replace mounts as necessary.

**TUBING VIBRATION**

Run hands over various lines. This can often help determine the location of the vibration. Gently reform tubing to eliminate vibration. Check for lines that may touch the cabinet or other parts of the unit.

**CONDENSER OR SUPERHEAT**

Tighten all mounting screws or tubing clamps. Bend or straighten fins. Mounting will vary with the type of condenser used. Sometimes breaking the paint bond will eliminate the sound.

Some models have a separate superheat and can develop an objectional sound. Adjust or replace rubber mounting.

**DISPOSAL PAN OR MOUNTING**

Adjust the pan or mountings as necessary.

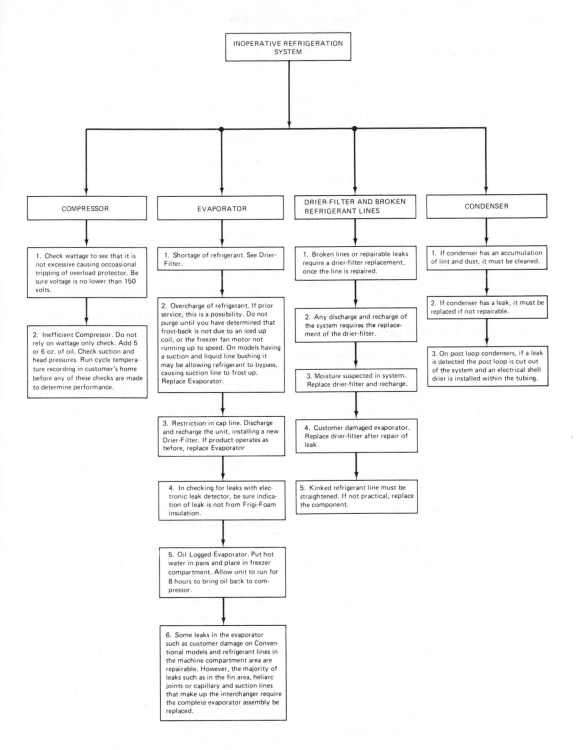

**INOPERATIVE REFRIGERATION SYSTEM**

**COMPRESSOR**

1. Check wattage to see that it is not excessive causing occoasional tripping of overload protector. Be sure voltage is no lower than 150 volts.

2. Inefficient Compressor. Do not rely on wattage only check. Add 5 or 6 oz. of oil. Check suction and head pressures. Run cycle temperature recording in customer's home before any of these checks are made to determine performance.

**EVAPORATOR**

1. Shortage of refrigerant. See Drier-Filter.

2. Overcharge of refrigerant. If prior service, this is a possibility. Do not purge until you have determined that frost-back is not due to an iced up coil, or the freezer fan motor not running up to speed. On models having a suction and liquid line bushing it may be allowing refrigerant to bypass, causing suction line to frost up. Replace Evaporator.

3. Restriction in cap line. Discharge and recharge the unit, installing a new Drier-Filter. If product operates as before, replace Evaporator

4. In checking for leaks with electronic leak detector, be sure indication of leak is not from Frigi-Foam insulation.

5. Oil Logged Evaporator. Put hot water in pans and place in freezer compartment. Allow unit to run for 8 hours to bring oil back to compressor.

6. Some leaks in the evaporator such as customer damage on Conventional models and refrigerant lines in the machine compartment area are repairable. However, the majority of leaks such as in the fin area, heliarc joints or capillary and suction lines that make up the interchanger require the complete evaporator assembly be replaced.

**DRIER-FILTER AND BROKEN REFRIGERANT LINES**

1. Broken lines or repairable leaks require a drier-filter replacement, once the line is repaired.

2. Any discharge and recharge of the system requires the replacement of the drier-filter.

3. Moisture suspected in system. Replace drier-filter and recharge.

4. Customer damaged evaporator. Replace drier-filter after repair of leak.

5. Kinked refrigerant line must be straightened. If not practical, replace the component.

**CONDENSER**

1. If condenser has an accumulation of lint and dust, it must be cleaned.

2. If condenser has a leak, it must be replaced if not repairable.

3. On post loop condensers, if a leak is detected the post loop is cut out of the system and an electrical shell drier is installed within the tubing.

379

## SAFETY PROCEDURES

Domestic refrigeration is a very large part of the refrigeration industry. It is a must that proper safety procedures be followed in order to protect the service technician, the equipment, and the products.

1.   Always get help when moving a domestic refrigerator or freezer.

2.   Always lift with the legs, never with the back.

3.   Do not remove frost and ice from an evaporator with sharp instruments such as knives, ice picks, etc.

4.   Always be certain that the unit is properly grounded electrically.

5.   Always practice good housekeeping. Keep oil and water off the floor.

6.   Take proper precautions to prevent electrical shock because of jewelry such as watches, rings, etc.

7.   When charging or purging refrigerant always wear goggles.

8.   Be sure that proper ventilation is provided when repairing leaks, purging refrigerant, etc.

9.   Never use carbon tetrachloride for cleaning because continued use can be fatal to the user.

10.   Always remove the doors from discarded refrigerators and freezers to prevent children from possibly suffocating during play.

## SUMMARY

• The areas comprising domestic refrigeration are: (1) refrigerators, (2) freezers, and (3) room air conditioners.

• Fiberglass and foamed plastic insulation are the types of insulation used in modern refrigerators and freezers.

• The density of foam insulation ranges from 1.8 to 2.2 pounds (.81 to .99 kg/m³) per cubic foot.

• A vapor barrier is used to keep moisture out of the insulated space between the shell and the liner.

• Foamed plastic has an inherent vapor barrier.

• Magnetic door seals are used today because they provide a better seal and the door is easily opened from either side.

• Breaker strips are used to reduce the flow of heat into the cabinet by conduction.

• Breaker strips should be warmed with a warm, wet cloth to prevent breaking before attempting to remove them.

• A mullion heater is a strand of high resistance wire which is attached to a strip of aluminum foil.

• Mullion heaters are installed around cabinet doors beneath the breaker strip.

• Mullion heaters prevent condensation of moisture around doors.

• The shelves in a domestic unit must be of the open grill type to allow for the circulation of air inside the cabinet.

• Avoid locating a domestic unit next to a stove, radiator, or hot air register.

• Domestic units should be installed on a separate electric circuit.

• The temperature inside a domestic unit will not start to drop until after about 30 minutes of operation.

• The refrigeration system is divided into two parts—the high pressure and the low pressure sides.

- The evaporator is sometimes called the cooling coil or freezer.

- The evaporator is where the action is in a refrigeration system.

- The compressor and motor are a hermetically sealed unit.

- The function of the compressor is to compress the low pressure vapor into a high pressure vapor.

- The temperature of the vapor must be raised above the ambient temperature so that it can give up the heat absorbed in the evaporator.

- The function of the condenser is to transfer the heat absorbed by the refrigerant to the ambient air.

- It is very important that an adequate air flow be maintained over the condenser.

- The filter-drier removes foreign matter that might restrict the flow of refrigerant through the system, and it removes moisture that might freeze in the capillary tube causing a restriction.

- The practice of using alcohol in hermetic systems is not recommended.

- The capillary tube meters the flow of liquid refrigerant from the condenser to the evaporator.

- Poor circulation of air inside a domestic unit will cause higher air temperature and spoilage of food.

- An air deflector is used to divert the cold air directly over the ice service area.

- Side-by-side frost-free refrigerators use a single coil evaporator mounted in a vertical position in the freezer compartment to cool both the freezer and fresh food compartments.

- The air that cools the fresh food section in refrigerator-freezer combination units is controlled by a damper.

- Most defrost systems use electric defrost heaters.

- The defrost timer puts the unit into defrost at the prescribed time.

- The duration of the defrost limiter normally determines the duration of the defrost cycle.

- Defrost water is disposed of by draining into a disposal pan located in the machine compartment where the water evaporates.

- The running time of the unit is determined by the thermostat switch bulb located at the top of the fresh food compartment and by the volume of cold air that is provided from the freezer compartment.

- When a lower freezer compartment temperature is required, the manual damper is closed a little by turning the freezer control to a colder setting, restricting the volume of cold, dry air that is provided to the food compartment.

- In servicing domestic units the malfunction, if any, is often found to be in the controls, not in the refrigeration unit.

- A lack of proper air flow is sometimes caused by the fan motor not running at the proper speed.

- Many domestic unit performance complaints require only customer education or reassurance that the unit is operating normally.

- The temperature in the freezer section is 0°F to +5°F (−17.8°C to −15°C) in frost-free models and from +10°F to +12°F (−12.2°C to −11.1°C) in other models, depending on the control setting.

• The temperature in the fresh food section should be between 32°F and 40°F (0°C to 4.4°C).

• Plugging the unit line cord in and taking it out several times may cause the compressor overload to trip.

• A test cord can be used to determine whether or not a compressor is bad.

• Wattages should be used to determine if the compressor is stuck, shorted, grounded, has open windings, has a tight rotor, or other internal malfunctions.

• Wattages alone should not be used to diagnose a shortage of refrigerant or an inefficient compressor.

• A faulty capacitor should be suspected if the compressor will not run, if there is a hum or buzz at times, or if it tries to start and then kicks off.

• Defective capacitors should be replaced on a like-for-like basis.

• Refrigeration gauges should be installed only after all other possibilities have been eliminated and it has been concluded that the fault is in the refrigeration system.

• Approximately 90 percent of the entries to a sealed system are for repairs on the high side.

• As a general guide, the suction pressure variation has an effect on the head pressure.

• Generally speaking, domestic units cannot operate efficiently in ambient temperatures lower than 60°F (15.6°C).

• A domestic unit is designed so that there is enough refrigerant for adequate heat removal in a 110°F (43°C) ambient temperature.

• The first thing noticed when a shortage of refrigerant occurs is a rise in the temperature of the foods.

• Under some circumstances, such as in the case of the conventional refrigerator or freezer with a slight shortage of refrigerant, freezing in the food compartment may be experienced due to the additional running time.

• Refrigeration occurs on the low pressure side of a partial restriction. An experienced service technician will physically feel the refrigeration lines when he or she suspects a restriction.

• A restriction reduces the flow rate of the refrigerant and consequently reduces the heat removal.

• Since air and R–12 do not mix, the air pressure would be added to the normal head pressure, resulting in a higher than normal head pressure.

• Before changing a compressor, the service technician should seal off the suction line and pump a vacuum with the compressor to see if it will pull a 26- to 28-inch (9.95- to 4.87-cm) vacuum.

• An insufficient amount of oil in a compressor can give an indication of an inefficient compressor because the oil coating between the parts results in better compression.

• Any time a system is low on refrigerant, the leak must be found and repaired.

• Liquid type leak detectors are the only type which can be used successfully around urethane insulation.

• Evacuation is necessary before recharging the system in order to remove all traces of air and the previous refrigerant charge.

- The secret to recharging a system with refrigerant is good evacuation.

- Liquid refrigerant should be charged into the system through the high side to prevent washing oil out of the compressor.

- Before an icemaker can operate, the freezer must reach the zero zone operating temperature.

- Two types of electrical wiring diagrams in use today are pictorial and schematic.

- When reading a wiring diagram, it is less confusing to trace each circuit all the way through before starting a new one.

- Room units operate with an evaporator temperature above the freezing point.

- The two basic systems in a room air conditioner are: (1) the refrigeration system and (2) the air circulating system.

- The outdoor fan has a slinger ring around the periphery of the blade which picks up condensate from the evaporator and splashes it on the condenser to increase the unit's capacity.

- Most room unit short-cycling is caused by air from the grill being deflected back into the air inlets.

- Icing is usually caused by an insufficient air flow over the evaporator.

- If a compressor is operated in low ambient temperatures, it will probably overheat and possibly burn up the compressor motor due to a lack of refrigerant to cool the windings.

- Low voltage is a common cause of trouble in the operation of a room air conditioning unit.

- An inoperative compressor should be isolated electrically and checked before a decision is made to replace it.

- If the condenser fan and evaporator blower are not located in proper relation to the shroud orifice, reduction of air flow may result.

## REVIEW EXERCISES

1. What is insulation?

2. What is the density of foam insulation?

3. Is a vapor barrier necessary when polyurethane insulation is used?

4. What type of door seals are used on modern refrigerators and freezers?

5. Where is the mullion heater installed on domestic units?

6. What is the purpose of breaker strips?

7. What is the purpose of a mullion heater?

8. Why must the shelving in a domestic unit be the open grill type?

9. In what ambient temperatures are domestic units designed to operate?

10. What does a slight gurgling or bubbling sound when the door is opened indicate?

11. How much time will usually lapse after plugging in a domestic unit before a drop in temperature will be noticed?

12. What is refrigeration?

13. Where is the action in a refrigeration system?

14. What is the function of a compressor in a refrigeration system?

15. Why must the temperature of the refrigerant vapor be increased?

16. What is the function of the condenser?

17. What is the purpose of the filter-drier?

18. Is it recommended to use alcohol in a refrigeration system? Why?

19. What is the purpose of the capillary tube?

20. What device is used to divert the air over the ice service area?

21. What is the purpose of the defrost system?

22. What does a defrost limiter do?

23. How is the defrost water disposed of?

24. What component limits the running time of the refrigeration unit?

25. In what components will a malfunction be most likely to occur in a domestic unit?

26. What should be the temperature range of a fresh food compartment?

27. On what voltage are most refrigerator compressors designed to operate?

28. Will the ambient temperature affect the wattage of a compressor?

29. What can be used to test a compressor to see if it or some external components are bad?

30. Why should electrical power not be applied to the start winding for long periods of time?

31. What conditions would cause a capacitor to be suspected of being faulty?

32. When should the gauges be installed on a refrigeration system?

33. What is the most common cause of seal system entry?

34. What will cause a high condensing temperature?

35. What are the symptoms of a system overcharge?

36. What is indicated by a hissing sound in the evaporator using a capillary tube?

37. Will low running wattage always be an indication of a low charge?

38. How can a restriction be located?

39. What would indicate a mixture of air and R–12?

40. What should be the vacuum reading of a good compressor?

41. Will insufficient oil indicate an inefficient compressor?

42. Why should foam insulation not be punctured when leak testing a system?

43. Why is evacuation of a system necessary?

44. What is the secret to recharging a system?

45. When will an icemaker begin operation?

46. What should be done when attempting to read an electrical diagram?

47. At what evaporator temperatures does a room air conditioner operate?

48. What are the two basic systems in room air conditioners?

49. What is the purpose of the slinger ring on the condenser fan blade?

50. What is the cause of most short-cycling conditions?

# Commercial Refrigeration

There are several classifications of commercial refrigeration cabinets depending on their size and design. These classifications are dependent on the intended use of the cabinet. Grocers, butchers, and others who sell perishable food require refrigerated cabinets suitable for the preservation of these foods. In general, the size and construction of the cabinet that best meets the needs of the merchant depends on the particular line of business in which he or she is engaged and the quantity of perishable food that must be kept in stock. Because of this, commercial cabinets may be found in almost every shape, size, and description.

## DEFINITION

Commercial refrigeration units are those used in businesses for the preservation of perishable merchandise. Commercial refrigeration units normally use high starting torque compressor motors and thermostatic expansion valves. Some of the smaller units, however, may use capillary tubes as the metering device.

## REQUIREMENTS OF COMMERCIAL REFRIGERATION EQUIPMENT

The store owner is extremely interested in the operating and maintenance cost of his or her refrigeration equipment. A quality commercial refrigeration installation must provide satisfactory service economically. Usually this type of equipment is installed to be used as a profitable investment because every place of business is operated for a profit. Properly designed and installed refrigeration equipment will keep operating costs low and remarkably improve the preservation of stored merchandise. It can, therefore, be seen that it is necessary to estimate the requirements for the cabinet as closely as possible. It is also necessary that the refrigeration unit be sized properly—not too small, not too large to do the work. If the capacity of the refrigeration unit is too large, the operating costs will increase. If, however, the unit is too small, it will be overloaded with longer periods of operation, more frequent starting, and more wear. The maintenance and repairs will increase and the performance probably will not be satisfactory. Therefore, it can be concluded that: (1) careful estimating of the refrigeration load is necessary and (2) the refrigeration unit chosen should be sized according to the manufacturer's specifications. Allowances should be made for emergencies and a proper safety factor should be applied. Properly engineered and installed refrigeration equipment is less expensive than that which is not. At the same time profits are increased due to the elimination of loss and spoilage while permitting a more effective display of the refrigerated goods.

Although there are many and varied applications of commercial refrigeration, the principles applied are the same as those that apply to domestic refrigeration. In most types of commercial refrigeration, the success or failure in performance of the system depends on the following conditions:

1.   Proper sizing for the heat load.

2.   Correct installation of the equipment.

3.   Insulation of the proper thickness at all points where the entrance of heat is to be retarded.

4.   Whether the system is installed in a refrigerator cabinet, display case, or freezer case of any type, the construction and location of the evaporator, and baffles when required, is of vital importance for successful operation.

5.   The size of the openings and locations of the warm and cold air passages determine the efficiency of the cooling unit in certain types of large size cabinets.

6.   The general design of the interior of the cabinet should allow for free and rapid circulation of the air through the food storage and over the cooling unit to facilitate the transfer of heat from the food to the cooling unit.

## TYPES OF COMMERCIAL CABINETS

Commercial refrigeration cabinets are designed and constructed to fit the needs of the user. The shell and the liner are usually made of metal, although the liner may be plastic. The finishes are designed for easy cleaning and upkeep. The insulation is either polystyrene or urethane which may be installed in pieces or, in more modern cabinets, foamed in place.

The refrigeration unit is designed to provide adequate refrigeration under the most severe usage. To prevent condensation electric heaters are installed around the doors or other openings, like the mullion heaters used in domestic units.

Commercial refrigeration units are usually designed to be used with remote condensing units. Most commercial refrigeration units use air-cooled condensing units because they are used in freezing weather. These condensing units may be connected to several commercial cabinets.

**WALK-IN COOLERS:** Walk-in coolers generally are larger than 100 ft³ (2.83 m³) in size. These cabinets are constructed so that access to the interior is gained by entering through a door. They are usually installed in large meat markets, larger restaurants, dairies, etc.

These coolers are generally constructed of metal exterior sections and can be expanded by using additional stock sections. (See Figure 14-1.) The entire exterior surface is also metal, thus giving protection against rats and vermin. The interior sur-

**Figure 14-1.** Walk-In Cooler (*Courtesy of Tyler Refrigeration Corp.*)

faces are metal as well except for the hardwood floor.

These units are also available with self-service display doors. These units allow for the storage of merchandise in the same cabinet from which it is selected, thus saving the labor of having to move the merchandise from one cabinet to another. The units may be either remote-installed or self-contained.

**Insulation:** Refrigeration efficiency is obtained with foam-sandwich construction. With this construction technique, polyurethane foam is pumped under pressure into a form which holds both the inner and outer case walls of all sections. As the foam expands and hardens in place, it bonds the steel outer and inner walls. (See Figure 14-2.) The result is a wall that is exceptionally strong and solid, yet thin and lightweight. The solid wall of foam is fire-resistant and waterproof. It retains the cold much longer and requires less compressor capacity to maintain proper temperature levels than old-fashioned fiberglass insula-

**Figure 14-2.** Insulation of Walls (*Courtesy of Tyler Refrigeration Corp.*)

tion. There is no way for moisture to cause rotting, molding, or insulation sag.

**Door Systems:** Completely foamed doors are rugged but lightweight. They close automatically on heavy-duty cam lift hinges. Magnetic vinyl door gasketing gives positive seal against heat leakage. When magnetic gaskets are used, the door does not require an inside safety release.

A lighting system is usually provided for the inside of the cooler. There is a pilot light on the outside to indicate when the light is on.

**REACH-IN CABINET:** These cabinets usually fall between 20 and 100 ft³ (.566 m³ and 2.83 m³) of storage space. They may be of the single- or multiple-door design. (See Figure 14-3.) The glass doors permit effective display of the merchandise when desirable.

**Figure 14-3.** Reach-In Cabinet (*Courtesy of Tyler Refrigeration Corp.*)

A blower coil is usually installed in the top or on the rear inside of the cabinet. The merchandise should be placed so that the air flow to and from the coil will not be restricted. Reach-in cabinets are usually self-contained units which require only that drain lines be installed.

Reach-in cabinets have a variety of uses such as: bottled beverage storage, dairy product storage, food storage, and dough storage in bakeries. The temperatures inside these cabinets are about the same as those for domestic units—ranging from about 32°F to 40°F (0°C to 4.4°C).

**FLORIST CABINET:** The size, type of construction, and the interior arrangement of florist cabinets vary widely. They are used largely for display and are, therefore, constructed with glass doors and often have glass ends. In no case should there be less than two thicknesses of glass, and three thicknesses are commonly used, particularly for lower temperature storage. (See Figure 14-4.) A blower coil may be placed on the back wall or regular cooling coils may be used. Many florist cabinets are tall and have the coil installed across the top allowing the entire lower portion to be devoted to display purposes.

The temperatures inside a florist cabinet vary from 47°F to 54°F (8.3°C to 12.2°C). The humidity is kept as high as possible to prevent moisture evaporating from the flowers. This is done with a coil designed to have a large cooling surface.

**Figure 14-4.** Florist Cabinet (*Courtesy of Uniflo Manufacturing Co.*)

**DISPLAY CASES:** In addition to the reach-in cabinet and the walk-in cooler, most businesses have one or more refrigerated display cases which are also made in various sizes and designs. Display cases are used almost exclusively for the preservation and display of meats and dairy products. Although they are generally used in combination with larger cases, there are designs and sizes of display cases where the case is the sole means of refrigerated storage.

Such cases as top-display, full-vision, combination top-display, and bottom-storage or double-duty, delicatessen, candy, dairy, vegetable, multiple-deck, etc. have more than proven their worth in enticing customers to purchase more of the products which are so invitingly displayed. This type of display accentuates impulse purchasing and the customer can either point to the merchandise desired or, when in self-service establishments, pick out the item personally.

**Double-Duty Case:** As the name suggests, the case has two compartments, an upper compartment for display and a compartment underneath for storage purposes which is also refrigerated. (See Figure 14-5.) The storage compartment is located on the rear side of this cabinet. The refrigeration coil is located in the bottom of the cabinet where a fan is used to circulate the refrigerated air throughout the case to provide proper refrigeration for the stored foods.

**Open Type Self-Service Display Cases:** The open type refrigerated display case is the latest milestone in merchandising products which must be kept constantly under refrigeration. Foods displayed in this type of case are refrigerated by a blanket of cold air and are stored below the top level of that cold blanket. Foods stored in these cases are usually wrapped in clear, transparent material such as cellophane which protects them from dust and germs in the air and from contamination through handling by the cus-

**Figure 14-5.** Double-Duty Case *(Courtesy of Tyler Refrigeration Corp.)*

tomer. While this type of display case has introduced some refrigeration problems, as compared to the completely enclosed case that has doors or covers, the main problem has been essentially that of providing the additional refrigeration capacity which is required to take care of the higher heat gain.

As an example, a 6-ft (1.8-m) enclosed top-display case can be adequately refrigerated by a ¼-hp condensing unit with a matching evaporator coil, while a 6-ft open-type display case will require a ½-hp unit with a correspondingly larger evaporator.

Even though the cold air is heavier than warm air and the design of the case should keep the air inside the case, the losses are so great that twice the amount of refrigeration capacity is required. While the cold air does tend to remain within the fixture, since it is in constant circulation past the evaporator and over the stored products, the top cold air comes in contact with the warm air above the case and some of it is warmed. This top layer consequently rises and leaves the case, being replaced by new warm air from outside the fixture. Since warm air contains moisture, it is necessary not only to cool the warm air down to case temperature but also to condense out the moisture it is carrying. For every pint of moisture condensed, approximately 1000 Btus (252 kcal) must be removed, and this process is occurring continuously. The moisture condensed passes out of the case through the drain.

Even where the movement of air in the cabinet is slight, warm, moist air from outside the case will mix with the cold, dry air in the cabinet and add substantially to the refrigeration load required to cool the air and condense the moisture. Customers reaching into the fixture to remove stored products also cause air movement and mixing of cold and warm air. It is therefore most important to locate these units away

from all externally induced air circulation possibilities such as fans, entrance doors which might permit blasts of air to blow past the fixture, heating duct openings, air blasts from unit heaters, and similar causes of air movement.

Modern open-type display cases usually have the evaporator located in the bottom of the cabinet (see Figure 14-6). A fan is used to circulate the air over the coil and through the case. The food stored in these cases must be placed within a certain zone to prevent spoilage. This case has an optional canopy with mirrors that can be used to improve the appearance of the stored merchandise.

**Frozen Food Display Cases:** The successful merchandising of frozen foods and packaged ice cream has proven to require the open display of such items in fixtures which attract the customer and are readily accessible for the choice of the product or flavor desired. While the open display case is most suitable for this purpose, the required low temperature of 0°F (−17.8°C) or lower adds problems to the use of the open-type case. (See Figure 14-7.)

Because of the required low temperature, the difference in temperature between the room air and the cold air in the case is substantially increased with the result that the heat absorbed by the interchange of warm air is also greatly increased. The lower temperatures induce greater infiltration of moisture-laden air with the resultant higher refrigeration load. Both the moisture-laden air and the lower temperatures required combine to increase the condensing unit size. Therefore, for a 6-ft (1.8-m) open-type frozen food storage cabinet, approximately a ¾-ton refrigeration unit would be required. To obtain efficient loading of the compressor motor for such cabinets, it is important that the unit be equipped with the

**Figure 14-6.** Top-Display Open Meat Case (*Courtesy of Tyler Refrigeration Corp.*)

proper size pully if it is the conventional open-type compressor. If it is a hermetic compressor, the proper size cylinder is essential for low temperature work.

**Figure 14-7.** Tyler Frozen Food and Ice Cream Cabinet (*Courtesy of Tyler Refrigeration Corp.*)

Some open-type low temperature cabinets use vertical plate type evaporators which are placed about 12 in. (.304m) apart along the full length of the storage compartment with the packaged goods stored between the plates. The warm air penetration occurs through the top 2- or 3-inch (5- or 7.6-cm) layer of cold air just below the top edge of the refrigerated plates so that frozen foods cannot satisfactorily be stored in this area. All cabinets of this type have indicated levels beyond which the products should not be stored. Some equipment manufacturers provide a supplementary means of flowing a cold air blanket over the top of the evaporator plates to permit loading of the cabinet to the top edge of the plates. This may be done by a forced air fan, an auxiliary coil built into the structure above the case, or an

Lightweight,
durable,
interlocking deck pans

Solid steel
wire shelves

10-in-1
ticket molding

Rear sill
for easy stocking

White plastic
air discharge grid

2" water
seal outlet

All-steel
construction
with urethane
insulation

Full-length
evaporator coil
for efficient operation

Hinged
plenum cover
and fan assembly

Dual-contact
stainless steel
bumper

Speed-align locks

Reinforced fiberglass
end frame
for added rigidity
and maximum
thermal break

Electrical raceway
is accessible
without tools

Pass-through
for running
drain lines
case to case

Kick plate

Two optional locations
for refrigeration lines—
inside or outside
refrigerated area

**Figure 14-8.** Method of Circulating Air (*Courtesy of Hill Refrigeration Division, Emhart Industries, Inc.*)

extra plate placed lengthwise above the top edge of the compartment plates.

Regardless of the method used, a blanket of cold air at least 2 or 3 in. (5 or 7.6 cm) thick must separate the frozen foods and the warm moist air above the storage compartment (see Figure 14-8).

## OPERATING CHARACTERISTICS OF SELF-SERVICE OPEN-TYPE CASES

The operating characteristics of self-service open-type refrigerated cases differ from those of the closed type. Particular attention

must be given to the proper control adjustments along with other considerations which must be taken into account if these cases are to provide the user with the desired service. Three main principles which must be kept in mind during the installation and maintenance of these units are:

1.  The evaporator must be completely filled with refrigerant.

2.  The cycling of the unit must be different. A longer running time is necessary to obtain the desired refrigeration.

3. Complete defrosting of the evaporator during the defrost period is necessary for the best operation of the equipment.

The one sure way to determine whether or not the evaporator is fully refrigerated is to observe the refrigerant line from the thermostatic expansion valve to the evaporator and the suction line leading from the evaporator. Both lines should be equally fed with refrigerant and fully frosted. A condition that may cause the service technician to believe the evaporator is fully refrigerated is that the suction line frosts periodically for several feet out of the case, even though the superheat setting of the thermostatic expansion valve may be too high. The "hunting" action of the thermostatic expansion valve serves to refrigerate the evaporator past the thermostatic bulb and out of the case, causing the suction line to frost several feet. The hunting action results in the evaporator being flooded and then starved of refrigerant, so that generally the evaporator is only partially refrigerated. However, during much of the running cycle the suction line will have little or no frost when compared to the evaporator coil.

The only positive check to see whether the evaporator is being fully refrigerated is to determine if the suction line and the thermostatic expansion valve outlet line to the evaporator are approximately equally frosted during the running cycle.

An evaporator that is fully refrigerated will cause the suction line to frost and sweat for several feet out of the case on humid days. To prevent this condition, a heat exchanger can be installed inside the case. The frost line should then extend through the heat exchanger but not out of the case. To obtain the best results, it is recommended that a heat exchanger be installed on cases that are maintained below 40°F (4.4°C).

Concerning the second principle—the long running time—it must be understood that the condensing unit is normally required to run 40 minutes or longer to maintain the desired temperatures.

The problem of completely defrosting the evaporator demands special attention when open-type cases are involved. Because cabinet temperatures of 28°F to 36°F (−2.2°C to 2.2°C) are required by the store operator to properly display pre-packaged meats in an open-type case, it is often necessary to operate a system using R–12 with a 4- to 6-pound (.28- to .42-kg/cm$^2$) cut-out point, and a 26- to 28-pound (1.83- to 1.96-kg/cm$^2$) cut-in point on the low pressure control. These settings result in a lower and more consistent cabinet temperature.

The question may arise as to whether or not the evaporator will build up an excess of frost and lose its efficiency at such operating pressures. This will not occur within a 24-hour period and that is the reason that the time-clock method of evaporator defrosting is in wide general use for this purpose. This type of defrost system uses an electric time clock wired into the control circuit.

## METHODS OF HEAT TRANSFER

Commercial refrigeration systems make use of four primary methods of heat transfer, as follows:

1. *Direct contact cooling* with the cooling coils hung from the ceiling or walls of the cooler, thus cooling the air directly. The air in turn circulates naturally, carrying the heat from the products to the evaporator.

2. *Indirect contact cooling* by means of a brine tank which is located at the ceiling or on the walls of the cooler.

3. *Combination direct and indirect cooling* where half of the evaporator is exposed

while the other half is immersed in the tank. Thus, the heat storage capacity of the brine tank minimizes temperature variation.

4. *Forced draft cooling* by means of a blast of air which is blown over the evaporator into the cooler.

## COMMERCIAL REFRIGERATION CYCLE

The commercial refrigeration cycle is basically the same as the refrigeration system used in domestic refrigerators. The major differences are in the types of controls used. These controls will vary with the design and requirements of the particular system. The refrigerant pressures also will vary with the type of system. On low temperature systems, extreme caution must be exercised to prevent the entrance of moisture into the refrigeration system. Moisture in these systems will probably be noticed first with freezing at the flow control device. When freezing at the flow control device is experienced, refrigeration will be at least partially reduced, depending upon the severity of freeze-up. Refrigeration can be completely stopped because of moisture freezing at the flow control device.

## COMMERCIAL REFRIGERATION COMPRESSORS

It was shown earlier in the discussion of the refrigeration cycle that the compressor has one purpose. That is to raise the refrigerant gas pressure from evaporator pressure to condenser pressure. The compressor must deliver the compressed refrigerant gas to the condenser at a pressure and temperature at which the condensing process can be accomplished.

**SINGLE-STAGE LOW TEMPERATURE SYSTEMS:** Low temperature single-stage refrigeration systems have become increasingly critical from a design and application

standpoint as the desired evaporating temperature is decreased. The combination of high compression ratios, low operating temperatures, and thin return gas can cause lubrication and overheating problems, thus making the compressor more vulnerable to permanent damage due to moisture and contaminants in the refrigeration system.

The selection of the compressor, the suction gas temperature, and the application must be made so that the discharge line temperature taken within 1 to 6 in. (2.54 to 15.2 cm) of the compressor discharge service valve does not exceed 230°F (110°C) for refrigerants R–12, R–22, and R–502. Under these conditions, the estimated average temperature at the discharge port—measured at the discharge valve retainer on the valve plate—will be approximately 310°F (154°C) for refrigerants R–12 and R–502, and 320°F (160°C) for refrigerant R–22.

The compressor displacement, pressure limiting devices, and quantity of cooling air or water must be selected to prevent the compressor motor winding temperature from exceeding the following limits:

1. 210°F (99°C) when protected by inherent protectors affected by line current and motor temperature.

2. 190°F (88°C) when protected by motor starters.

The compressor motor temperature should be determined by the resistance method and should be measured when the motor is tested in the highest ambient temperature in which it is expected to operate, at 90 percent of the rated voltage, with 90°F (32.2°C) suction gas temperature. For longer motor life, the maximum operating temperatures of 170°F to 190°F (77°C to 88°C) are recommended.

In order to prevent the discharge gas temperature and motor temperature from exceeding the recommended limits, it is desirable, and in some cases absolutely necessary, to insulate the suction lines which will allow the suction gas to return to the compressor at a lower than normal temperature. This is especially important when R–22 is used with suction-cooled compressors.

Most suction-cooled compressors require some auxiliary cooling by an air blast on the compressor when the compressor is used for operating temperatures below 0°F (−17.8°C).

To prevent compressor motor overload during pulldown periods or after a defrost period, either the evaporator must be properly designed or a pressure limiting device such as an expansion valve or a crankcase valve must be used.

Some compressor manufacturers recommended that R–502 refrigerant be used in all single-stage low temperature applications where the evaporator temperature is −20°F (−28.9°C) or below. Now that R–502 is readily available, the use of R–22 should be avoided in single-stage low temperature compressors of 5 hp and larger. The lower refrigerant gas discharge temperatures encountered when using R–502 have resulted in much more trouble-free operation.

An adequate supply of oil must be maintained in the compressor crankcase at all times to ensure continuous lubrication of the parts. If the refrigerant velocity in the system is so low that rapid return of the oil is not assured, an adequate oil separator must be used. The normal oil level should be maintained at or slightly above the center of the compressor oil sight glass. An excess of refrigerant or oil in the system must be avoided because it may result in liquid slugging of the compressor and damage to the compressor valves, pistons, or cylinders. (See Figure 14-9.)

The refrigerant lines must be designed and installed to prevent oil trapping. The

**Figure 14-9.** Oil Level in Sight Glass

highest refrigerant velocity possible without excessive pressure drop is recommended.

Care must be exercised to prevent the evaporating temperature from dipping so far below the normal system operating temperature that the refrigerant velocity becomes too low to return the oil to the compressor. The low pressure control cut-out setting should not be below the lowest published rating for the compressor.

The smallest practical size tubing should be used in evaporators and condensers in order to hold the system refrigerant charge to a minimum. When large refrigerant charges are unavoidable, recycling pump-down control methods should be used.

When air-cooled condensing units are required for operation in low ambient temperatures, some means of head pressure control to prevent the condensing pressure from falling too low is highly recommended to maintain the required refrigerant velocities.

A filter-drier of generous size must be installed in the liquid line, preferably in the cold zone. The desiccant used must be able to remove moisture to a low end point and be capable of removing a reasonable quantity of acid. It is most important that the filter-drier be equipped with an excellent filter to prevent the circulation of carbon and foreign particles inside the system. A permanent suction line filter is highly recommended to protect the compressor from any contaminants which may have been left in the system during installation.

After installation has been completed, all

systems should be thoroughly evacuated with a high grade vacuum pump and dehydrated to ensure that no air or moisture is left in the system. The triple-evacuation method should be used, breaking the vacuum each time with dry refrigerant charges into the system through a drier. The compressor motor must not be operated while the high vacuum pump is in operation. If the motor is run, permanent damage to the windings could occur.

The system should be charged with clean, dry refrigerant through a drier (see Figure 14-10). Other substances such as liquid driers or alcohol must not be used.

**Figure 14-10.** Gauge Manifold Hook-Up Using a Drier

**Two-Stage Low Temperature Systems:** Because of their basic design and operation, two-stage systems are inherently more efficient and encounter fewer operating difficulties at the lower operating temperatures than do single-stage units. The two-stage compressor, however, has its limitations. When evaporator temperatures below $-80°F$ ($-62.2°C$) are encountered, they lose their efficiency. Compressor motor overheating also becomes an increasing problem. At evaporating temperatures below $-80°F$, a cascade system is recommended. However, for applications with evaporating temperatures in the $-20°F$ to $-80°F$ ($-28.9°C$ to $-62.2°C$) range, the two-stage compressor efficiency is high, the refrigerant discharge temperatures are low, and field experience with properly applied two-stage compressors has been excellent.

Two-stage refrigeration systems are somewhat more complex and sophisticated than simple single-stage systems. Many of the operating problems encountered on two-stage systems stem from the fact that too often their application has been made without sufficient appreciation of the safeguards which must be taken in proper system design.

**Volumetric Efficiency:** Three definitions which were given previously are of importance in analyzing two-stage systems:

The compression ratio is the ratio of the absolute discharge pressure (psia) to the absolute suction pressure (psia).

The absolute pressure is gauge pressure plus atmospheric pressure, which at sea level is standardized at 14.7 pounds per square inch ($1.03 kg/cm^2$).

Volumetric efficiency is defined as the ratio of the actual volume of the refrigerant gas pumped by the compressor to the volume displaced by the compressor pistons.

A typical single-stage volumetric efficiency curve is shown in Figure 14-11. Note that as the compression ratio increases, the volumetric efficiency decreases.

Two factors that cause a loss of efficiency with an increase in compression ratio are:

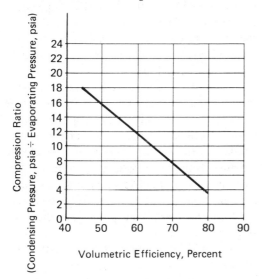

**Figure 14-11.** Typical Single-Stage Low Temperature Compressor Efficiency Curve (*Courtesy of Copeland Corp.*)

1. The density of the residual gas remaining in the cylinder clearance space after the compression stroke is determined by the discharge pressure—the greater the discharge pressure, the greater the density. Since this gas does not leave the cylinder on the discharge stroke, it re-expands on the suction stroke, thus preventing the intake of a full cylinder of vapor from the suction line. As the compression ratio increases, the added space in the cylinder on the intake stroke is filled by the residual gas.

2. The high temperature of the cylinder walls resulting from the heat of compression is a factor in the loss of efficiency. As the compression ratio increases, the heat of compression increases, and the cylinders and compressor head become very hot. The suction gas which enters the cylinder on the suction stroke is heated by the cylinder walls and expands, resulting in a reduced weight of refrigerant gas entering the compressor.

Therefore, a single-stage compressor has its limitations as the compression ratio increases. The effective low limit of even the most efficient single-stage system is approximately −40°F (−40°C) evaporating temperature. At lower evaporating temperatures, the compression ratio becomes so high that the compressor capacity falls rapidly. The compressor may no longer be handling a sufficient weight of return refrigerant gas for proper motor cooling. Also, because of the decreased gas density, oil may no longer be properly circulated throughout the system.

**TWO-STAGE COMPRESSION AND COMPRESSION EFFICIENCY:** In order to increase the compressor's operating efficiency at low evaporating temperatures, the compression ratio can be done in two steps or stages. For two-stage operation, the total compression ratio is the product of the compression ratio of each stage. In other words, for a total compression ratio of 10 to 1, the ratio of each stage might be 4 to 1; or compression ratios of 4 to 1 and 5 to 1 in separate stages will result in a total compression ratio of 20 to 1.

Two-stage compression may be accomplished with the use of two compressors with the discharge of one pumping into the suction of the second. However, because of the difficulty of maintaining the proper oil levels in the two crankcases, it is more satisfactory to use one compressor with multiple cylinders. A greater volume of the low-stage cylinders is necessary because of the difference in specific volume of the refrigerant vapor at low and interstage pressures. While the compression ratios of the two stages are seldom exactly equal, they will be approximately the same. Two-stage compressors generally use an external manifold and a de-superheating expansion valve. (See Figure 14-12.) A typical three-cylinder two-stage compressor with its external manifold is shown in Figure 14-13.

**Figure 14-12.** Typical Six-Cylinder Two-Stage Compressor (*Courtesy of Copeland Corp.*)

**Figure 14-13.** Typical Three-Cylinder Two-Stage Compressor (*Courtesy of Copeland Corp.*)

The volumetric efficiency of compressors can be entered on graphs or curves to indicate their performances. The three straight lines in Figure 14-14 are typical single-stage curves—one for an air conditioning compressor, one for a typical multipurpose compressor, and one for a low temperature compressor. There are some variations in the compressor design involved, but the primary difference in characteristics is due to the clearance volume.

The two vertical curved lines represent the comparative efficiency of a two-stage compressor. Actually, each separate stage would have a straight line characteristic similar to the single-stage curves to enable a comparison with single-stage compressors. The overall volumetric efficiency has been computed on the basis of the total displacement of the compressor, not just the low-stage displacement.

The solid black curve represents the efficiency of a two-stage compressor without a liquid subcooler. Note that the efficiency is relatively constant over a wide range of total compression ratios, and that the crossover in efficiency with the best low temperature single-stage compressor is at a compression ratio of approximately 13 to 1. In other words, at compression ratios lower than 13 to 1, a single-stage compressor will have more capacity than a two-stage compressor of equal displacement without liquid sub-cooling.

The dotted curve represents the efficiency of the same two-stage compressor with a liquid subcooler. In the subcooler, the liquid refrigerant being fed to the evaporator is first subcooled by liquid refrigerant fed through the interstage de-superheating expansion valve, and a much greater share of the refrigeration load has been transferred to the high-stage cylinders. Since the high-stage cylinders operate at a much higher

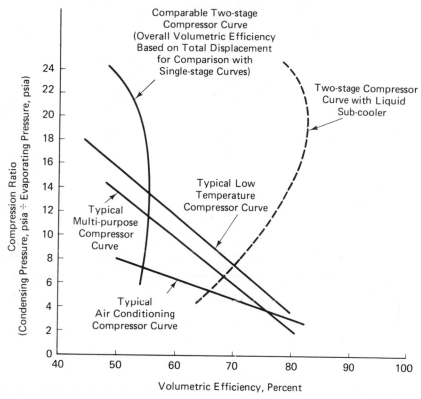

**Figure 14-14.** Typical Compressor Volumetric Efficiency Curves (*Courtesy of Copeland Corp.*)

suction pressure, the refrigeration capacity there is far greater per cubic foot (.0283 m³) of displacement than in the low-stage cylinders. In effect, the capacity of the compressor has been greatly increased without having to handle any additional suction gas returning from the evaporator. Note that with the liquid subcooler, the crossover point in efficiency as compared with a single-stage compressor is at a compression ratio of approximately 7 to 1. In other words, at compression ratios lower than 7 to 1, the single-stage compressor will have more capacity for an equal displacement, but at compression ratios higher than 7 to 1, the two-stage compressor will have more capacity.

Table 14-1 lists comparative operating data at varying evaporating temperatures for a Copeland compressor which is available either as a single-stage or a two-stage compressor. Although the displacement, refrigerant, and motor are the same, the rapidly increasing advantage of two-stage operation as the evaporating temperature decreases is plainly shown.

**COMPRESSOR OVERHEATING AT EXCESSIVE COMPRESSION RATIOS:** In addition to efficiency, the extremely high temperatures created by operation at abnormally high compression ratios make the use of single-stage compressors impractical for ultralow temperature applications. A valve plate can be ruined with carbon for-

Table 14-1

Efficiency Comparison of Single-Stage versus Two-Stage Compression*

|  | Evaporating Temperature | | |
|---|---|---|---|
|  | -30° F. | -40° F. | -50° F. |
| Condensing Temperature | 120° F. | 120° F. | 120° F. |
| Condensing Pressure, Psig | 280.3 | 280.3 | 280.3 |
| Condensing Pressure, Psia | 295 | 295 | 295 |
| Evaporating Pressure, Psig | 9.40 | 4.28 | 0.04 |
| Evaporating Pressure, Psia | 24.1 | 18.97 | 14.74 |
| **Single Stage** | | | |
|    Compression Ratio | 12.5/1 | 15.6/1 | 20/1 |
|    Capacity, BTU/hr | 46,000 | 32,000 | 23,000 |
|    BTU/watt | 3.42 | 2.86 | 2.3 |
| **Two Stage with Subcooler** | | | |
|    Compression Ratio—Low Stage | 3.74 | 3.68 | 4.05 |
|    Compression Ratio—High Stage | 3.26 | 4.23 | 4.95 |
|    Capacity, BTU/hr | 61,000 | 50,000 | 38,500 |
|    BTU/watt | 4.15 | 3.84 | 3.35 |
| Increase in Capacity Two Stage vs. Single Stage | 32% | 56% | 67% |

* Capacity Data Based on Equal Displacement (Courtesy of Copeland Refrigeration Corp.)

mation due to lubricating oil breakdown from excessive heat. Excessive cylinder temperatures also can cause rapid piston and cylinder wear, cylinder scoring, and early failure of the compressor. With two-stage compressors, the interstage expansion valve maintains safe operating temperatures and this type of damage is prevented.

**BASIC TWO-STAGE SYSTEM**

The basic flow of refrigerant in a six-cylinder two-stage compressor is different from the single-stage system. (See Figure 14-15.) The suction gas returning from the evaporator enters the four low-stage cylinders directly from the suction line. Since the discharge gas from the first-stage cylinders is heated from compression, it must be cooled by the de-superheating expansion valve before entering the compressor motor cham-

ber. The de-superheated refrigerant vapor, now at the interstage pressure, enters the high-stage cylinders, is compressed, and then discharged to the condenser.

Figure 14-16 is a schematic view of a typical two-stage system showing the various components that are necessary for operation.

**DE-SUPERHEATING EXPANSION VALVE:** Expansion valves which are currently supplied as original equipment with two-stage compressors are of the nonadjustable superheat type. In the event of failure, standard replacement valves with adjustable superheat which have been approved by the manufacturer may be used. Improper valve selection can result in the compressor overheating and possible damage due to improper liquid refrigerant control.

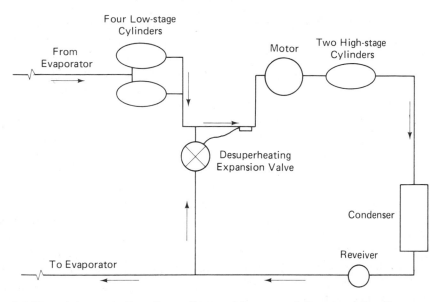

**Figure 14-15.** Schematic Two-Stage System (*Courtesy of Copeland Corp.*)

**Figure 14-16.** Two-Stage System with Six-Cylinder Compressor (with Liquid Subcooler) (*Courtesy of Copeland Corp.*)

**LIQUID REFRIGERANT SUBCOOLER:** Two-stage systems may be operated either with or without liquid subcoolers. The function of the subcooler is to cool the liquid refrigerant being fed to the evaporator; this is accomplished by the evaporation of refrigerant which is fed through the de-superheating expansion valve. This transfers a greater portion of the refrigeration load to the high-stage cylinders, and because of the greater compressor capacity at higher suction pressures, the system capacity is greatly increased.

The temperature of the liquid refrigerant being fed to the evaporator is reduced in the subcooler to within approximately 10°F (5.56°C) of the interstage saturated evaporating temperature. The increase in system capacity can only be realized if the subcooled liquid is maintained at this low temperature and heat transfer into the liquid line is prevented. Normally this requires insulation of the liquid line.

When selecting expansion valves for two-stage systems with liquid subcoolers, the designer must keep in mind that the expansion valve will have greatly increased capacity due to the low temperature of the refrigerant entering the valve. Unless this is taken into consideration, the increased refrigerating effect per pound of refrigerant may result in an oversized expansion valve with resulting erratic operation.

## CHARGING COMMERCIAL REFRIGERATION SYSTEMS

The proper performance of any refrigeration or air conditioning system is greatly dependent on the proper charge of refrigerant. An undercharged system will result in a starved evaporator, resulting in excessively low compressor suction pressures, loss of capacity, and possible compressor motor overheating.

On the other hand, overcharging can flood the condenser, resulting in high discharge pressures and in potential compressor damage caused by liquid refrigerant flooding the compressor. Most refrigeration systems have a reasonable tolerance for some variation in charge. However, some small systems will actually have a critical refrigerant charge that is essential for proper operation.

Each refrigeration system must be considered separately. Systems with the same capacity or horsepower rating may not necessarily require the same refrigerant charge. Therefore, it is important first to determine the type of refrigerant required for the system. The unit nameplate normally identifies both the type and weight of refrigerant required.

**LIQUID CHARGING:** Charging a system with liquid refrigerant is much faster than charging with vapor. Because of this factor, liquid charging is almost always used on large field-installed systems. Liquid charging requires either a charging valve in the liquid line, a process fitting in the high pressure side of the system, or a receiver outlet valve with a charging port. It is recommended that liquid charging be done through a filter-drier to prevent any contaminants from entering the system. Never charge liquid into the compressor suction or discharge service valve ports because this can damage the compressor valves.

On original installations, the entire refrigeration system should be thoroughly evacuated. Weigh the refrigerant drum and attach the charging line from the refrigerant drum to the charging valve. If the approximate weight of refrigerant required is known, or if the charge must be limited, the refrigerant drum should be placed on a scale so that the weight of the refrigerant remaining in the drum can be checked frequently.

Purge the charging line of any air and open the cylinder liquid valve and the charging valve. The vacuum in the system will cause the liquid to flow through the charging connection until the system pressure is equalized with the pressure in the refrigerant cylinder. (See Figure 14-17.)

**Figure 14-17.** Liquid Charging of a Commercial Refrigeration System. (*Courtesy of E. I. DuPont de Nemours & Co.*)

Close the receiver outlet valve and start the compressor. Liquid refrigerant will now flow from the refrigerant cylinder to the liquid line through the evaporator and will be collected in the condenser and receiver.

To determine if the charge is approaching the amount required by the system, open the receiver outlet valve, close the charging valve, and observe the operation of the system. Continue charging until the proper charge has been introduced into the system.

Again weigh the refrigerant drum, and make a record of the weight of refrigerant charged into the system.

Watch the discharge pressure gauge closely. A rapid rise in discharge pressure indicates that the condenser is filling with liquid and the system pumpdown capacity has been exceeded. If this occurs, stop charging from the cylinder immediately and open the receiver outlet valve.

On factory assembled package units using welded compressors, charging is normally accomplished by drawing a deep vacuum on the refrigeration system, and introducing the proper refrigerant charge by weight into the high side of the system by means of a process connection which is later sealed and brazed closed. To field-charge such systems, it may be necessary to install a special process fitting or charging valve, and weigh in the exact charge required for that unit.

**VAPOR CHARGING:** Vapor charging is normally used when only small amounts of refrigerant are to be added to a system, possibly up to 25 pounds (11.34 kg), although it can be more precisely controlled than liquid charging. Vapor charging is usually accomplished by use of a charging manifold connected to the compressor suction service valve port. If no valve port is available—for example, on welded compressors—it may be necessary to install a piercing valve or fitting in the suction line.

Service gauges are connected to read both the suction and discharge pressure. (See Figure 14-18.) When adding refrigerant to a system the discharge pressure should be observed to be sure the system is not being overcharged and that the refrigerant is not being added too rapidly. A higher than normal discharge pressure indicates that either the condenser is filling with liquid or the compressor is being overloaded by too rapid

**Figure 14-18.** Vapor Charging Refrigerant Through Compressor Suction Service Valve (*Courtesy of E. I. DuPont de Nemours & Co*).

charging. The charging manifold permits throttling of the vapor from the cylinder. The cylinder is mounted on a scale to measure the amount of refrigerant being charged into the system. An approved valve wrench is used to operate the cylinder valve.

The refrigerant cylinder must remain in the upright position with the refrigerant being withdrawn only through the vapor valve to ensure that only vapor is reaching the compressor. The vaporizing of the liquid refrigerant in the cylinder will chill the remaining liquid and reduce the cylinder pressure. To maintain cylinder pressure and expedite charging, warm the cylinder by placing it in warm water or by using a heat lamp. *Do not apply heat with a torch.*

To determine if sufficient charge has been introduced, close the refrigerant cylinder valve and observe system operation. Continue charging until the proper charge has been added.

Watch the compressor discharge pressure closely during the charging operation to be certain that the system is not overcharged.

**HOW TO DETERMINE THE PROPER CHARGE:** There are several methods that can be used to determine whether or not a system contains enough refrigerant. The method used will depend to a great extent on the system design and the personal preference of the service technician. The following is a list of these methods.

**1. Weighing the Charge:** The most accurate charging procedure is to actually weigh the refrigerant which is charged into the system. This method can only be used when the system requires a complete charge and the amount of that charge is known. Normally, this data is available on packaged unitary equipment. If the charge is small, it is common practice to purge the charge to the atmosphere when repairs are required. The system is completely recharged after the repairs have been made.

**2. Using a Sight Glass:** The most common method of determining whether or not the system is properly charged is by use of a liquid indicator (sight glass) in the liquid line. Because a solid head of liquid refrigerant is required for proper operation of the expansion valve, the system can be considered properly charged when a clear stream of liquid refrigerant is visible in the sight glass. Bubbles or flashing usually indicate a shortage of refrigerant. Keep in mind that if there is only vapor and no liquid in the sight glass, it will also appear clear. However, the service technician should be aware of the fact that at times the sight glass may show bubbles as flash gas even when the system is fully charged. A restriction in the liquid line ahead of the sight glass may cause sufficient pressure drop to cause flashing of the refrigerant. If the expansion valve feed is erratic or surging, the increased flow when the expansion valve is wide open can create enough pressure drop to create flashing at the receiver outlet. Rapid fluctuations in the condensing pressure can be a source of flashing. For example, in a temperature-controlled room, the sudden opening of shutters or the cycling of a fan can easily cause a change in the condensing temperature of 10°F to 15°F (5.56°C to 8.34°C). Any liquid refrigerant in the receiver may then be at a temperature higher than the saturated temperature equivalent to the changed condensing pressure and flashing will occur until the liquid temperature is again below the saturation temperature.

Some systems may have different refrigerant charge requirements under different operating conditions. Low ambient head pressure control systems for air-cooled applications normally depend on partial flooding of the condenser to reduce the effective surface area. Under these conditions, a system operating with a clear sight glass may possibly require twice the amount of refrigerant for proper operation under low ambient conditions.

While the sight glass can be a valuable aid in determining the proper charge, the system performance must be carefully analyzed before placing full reliance on it as a positive indicator of the system charge.

**3. Using a Liquid Level Indicator:** On some systems, a liquid level test port may be provided on the receiver. The proper charge can then be determined by charging until liquid refrigerant is found when the test port is cracked. (See Figure 14-19.) With less than a full charge, only vapor will be available at the test port.

Larger receiver tanks may be equipped with a float indicator to show the level of liquid in the receiver in much the same manner as a gasoline gauge on an automobile.

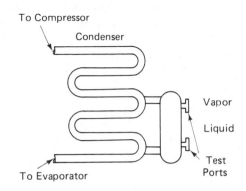

**Figure 14-19.   Liquid Level Indicator**

**4. Checking Liquid Subcooling:** On smaller systems, if no other means of checking the refrigerant charge is available, a test of the liquid subcooling at the condenser outlet can be used. With the unit running under stabilized conditions, compare the temperature of the liquid line leaving the condenser with the saturation temperature equivalent to the condensing pressure. (See Figure 14-20.) This provides an approximate comparison of the condensing temperature and the liquid temperature leaving the condenser. Continue charging the unit until the liquid line temperature is approximately 5°F (2.78°C) below the condensing temperature

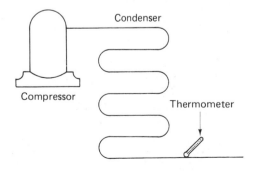

**Figure 14-20.   Checking Liquid Subcooling**

under maximum load conditions. This charging procedure is normally used only on factory-packaged systems. However, it does provide a means of emergency field checking which should indicate proper system operation.

**5. Charging by Superheat:** On small self-contained systems equipped with capillary tubes, the operating superheat may be used to determine whether or not the system contains the proper charge.

If a service port is available so that the suction pressure can be determined, the superheat may be calculated by determining the difference between the temperature of the suction line approximately 6 in. (15.24 cm) from the compressor and the saturation temperature equivalent of the suction pressure. If no means of determining the pressure is available, the superheat can be taken as the difference between the suction line temperature 6 in. from the compressor and a temperature reading on the evaporator tube (not a fin) at the midpoint of the evaporator. (See Figure 14-21.)

With the unit running at its normal operating condition, continue charging the unit with refrigerant until the superheat is found to be approximately 20°F to 30°F (11.2°C to 16.68°C). A superheat approaching 10°F

(5.56°C) indicates an overcharged condition. A superheat approaching 40°F (22.24°C) indicates an undercharge.

**6. Charging by Manufacturer's Charging Charts:** Some manufacturers of self-contained equipment provide charging charts so that the proper charge can be determined by observing the system operating pressures. Follow the manufacturer's directions for determining the proper charge if the unit is to be charged in this fashion.

## CRANKCASE PRESSURE REGULATING VALVES (CPR)

These valves are installed in the suction line ahead of the compressor. The valve controls the maximum pressure at the compressor suction (see Figure 14-22). This provides overload protection for the compressor motor. The valve is adjusted for the maximum pressure which is specified by the condensing unit manufacturer.

At conditions of overload, the valve modulates to prevent suction gas pressures at the compressor suction greater than the pressure for which the valve is adjusted. When the overload has passed and the pressure drops below the valve setting, the valve assumes a wide open position. The valve setting is determined by a pressure spring, as the valve modulates from fully open to fully closed in response to the outlet pressure, closing on a rise in outlet pressure. (See Figure 14-23.)

These valves assure positive protection against motor overload on systems using hot gas defrosting. For example, high coil temperatures following defrosting plus an accumulation of liquid in the evaporator results in high pressures at the start of the normal refrigerating cycle. Also, high system heat loads may result in low side pressures above that for which the compressor motor is designed.

**Figure 14-21.** Checking Superheat

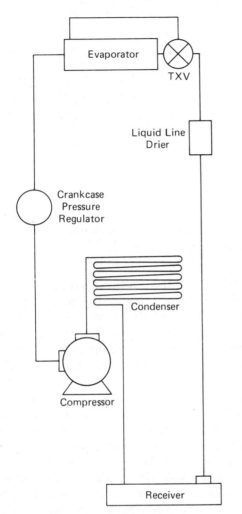

**Figure 14-22.** Crankcase Pressure Regulator Installation

**Figure 14-23.** Cross-Section of Crankcase Pressure-Regulating Valve (*Courtesy of Singer Controls Division*)

## EVAPORATOR PRESSURE REGULATOR (EPR)

These valves are installed in the suction line between the evaporator and compressor. They maintain the desired minimum evaporator pressures within close limits. (See Figure 14-24.)

On systems with multiple evaporators operating at different temperatures or on systems where the evaporating temperature cannot be allowed to fall below a given temperature, an evaporator pressure-regulator valve is frequently used to control the evaporating temperature. This valve, often called an EPR valve, acts like the crankcase pressure regulator, except that it is responsive to the inlet pressure. (See Figure 14-25.)

Since the valve responds to the inlet pressure, it opens when the inlet pressure equals or exceeds the opening point for which the valve is set. As the evaporator continues to rise, the valve stroke increases, permitting an increased flow of refrigerant.

Close-off is caused when the inlet pressure drops below the valve setting, preventing the evaporator temperature from dropping below this point.

## AIR-COOLED CONDENSER PRESSURE REGULATOR

The function of this valve is to maintain adequate high side (receiver) pressure during low ambient temperatures. Sufficient high side pressure must be available at all times to provide an adequate pressure drop across the system expansion valve.

**Figure 14-25.** Cross-Section of Evaporator Pressure-Regulating Valve (*Courtesy of Singer Controls Division*)

**Figure 14-24.** Two-Temperature System with EPR Valve

The recommended location for the condenser pressure regulator valve is at the condenser outlet between the condenser and liquid receiver. In this location the valve handles liquid and will perform as a high capacity device. (See Figure 14-26.)

It is also very important to note that these valves are not recommended for installation at the condenser inlet or other locations where compressor discharge pulsations may reach the valve. This regulator is responsive to the valve inlet pressure, opening on an inlet pressure increase and closing on an inlet pressure decrease. If the inlet pressure changes in rapid sequence (compressor discharge pulsations, etc.) the valve will be adversely affected. When a condition develops to cause the valve to open and close with each revolution of the compressor, the life of the valve bellows will be extremely short. The addition of a dampener weight assembly in this case will not totally protect the valve, but it will assist in prolonging valve life.

**HIGH TO LOW SIDE BYPASS VALVE**

These valves are installed between the system high and low sides and respond to the valve outlet (system low side) pressure. The valve opens when the low side pressure is reduced to the valve setting. The machine capacity is thus reduced and the evaporator temperature ceases to drop as the suction pressures are maintained. (See Figure 14-27.)

**Figure 14-26.** Typical Low Ambient Control System for Air-Cooled Condensers

**Figure 14-27.** High to Low Side Bypass Valve Location (*Courtesy of Singer Controls Division*)

The bypass valve is an effective way to control system capacity as well as to limit a minimum evaporator temperature. Continuous machine operation is made possible without a reduction of evaporator temperatures below predetermined limits.

The bypass gas may be piped through the valve to either the suction line ahead of the compressor or to the inlet of the evaporator downstream from the expansion valve. When the bypass gas is circulated through the evaporator, compressor protection against high superheat is provided by an expansion valve de-superheating effect.

## BYPASS CONTROL VALVE FOR AIR-COOLED CONDENSERS

These valves function as condenser bypass devices on air-cooled condensers with a winter start control. The valves are installed in the condenser bypass line between the compressor discharge and the liquid receiver (see Figure 14-28).

**Figure 14-28.** Bypass Control Valve for Air-Cooled Condensers

Operating in response to receiver (valve outlet) pressure, the valve is open when the receiver pressure is below the valve set point. At the start of each machine on cycle, high pressure discharge gas is admitted directly to the receiver and quickly builds up adequate receiver pressure. When the valve opening pressure is reached, the valve closes and the compressor discharges directly into the condenser.

Were it not for the bypass control, some delay in reaching adequate receiver pressure would occur due to the time required for the larger condenser mass to reach operating temperatures.

## CAPACITY CONTROL DEVICES AND METHODS

On many refrigeration and air conditioning systems, the refrigeration load will vary over a wide range. This may be due to differences in product load, ambient temperatures, usage, occupancy, or other factors. In such cases, some type of compressor capacity control is necessary for satisfactory system performance.

**ON-OFF OPERATION:** The simplest form of capacity control is on–off operation of the compressor. This method is acceptable for small compressors, but it is seldom satisfactory for larger compressors because of the fluctuations in the controlled temperature. When light load conditions are experienced, the compressor may short-cycle. On refrigeration applications where the formation of ice on the coil is not a problem, users frequently lower the low pressure cut-out setting to a point below the design limits of the system in order to prevent compressor short-cycling. As a result, the compressor may operate for long periods at extremely low evaporating temperatures. Since the compressor capacity decreases rapidly with a reduction in suction pressure, the reduced refrigerant density and velocity is frequently inadequate to return the oil to the compressor. Operation of the system at temperatures below those for which it was designed may also lead to overheating of the motor-compressor. Both of these conditions can cause compressor damage and ultimate failure.

**COMPRESSOR CYLINDER UNLOADERS:** In order to provide a means of changing compressor capacity under fluctuating load conditions, larger compressors are frequently equipped with cylinder unloaders.

Unloaders which are used on reciprocating compressors are of two general types. On some compressors, the suction valves on one or more of the cylinders are held open by some mechanical means in response to a pressure control device. With the suction valves open, refrigerant vapor is forced back into the suction chamber rather than the discharge chamber during the compression stroke, and the cylinder performs no pumping action.

Copelametic ® compressors with unloaders have a bypass valve arranged so that the unloaded cylinder, or cylinders, are isolated from the discharge pressure which is created by the loaded cylinders. A 3-way valve connects the discharge ports of the cylinder either to the normal discharge line when loaded, or to the compressor suction chamber when unloaded. (See Figure 14-29.) Since the piston and cylinder do no work when in the unloaded cycle other than pumping vapor through the bypass circuit (and only suction vapor), the problem of cylinder overheating while unloaded is practically eliminated. At the same time the electrical power consumption of the compressor motor is greatly reduced because of the re-duction in the work performed. The reduced electrical power consumption and better temperature characteristics of this type of compressor unloading, when properly applied, are major advantages over external hot gas bypass unloading where all the cylinders of the compressor are working against the condensing pressure.

Because of the decreased volume of available suction vapor returning to the compressor from the system for compressor motor cooling, the operating range of unloaded compressors must be restricted. In general, compressors with capacity control are recommended only for high temperature applications, but in some instances they can be satisfactorily applied in the medium temperature range. Because of the danger of overheating the compressor motor on low temperature systems, either cycling the compressor or the hot gas bypass method of capacity control is recommended.

**HOT GAS BYPASS:** Compressor capacity control by means of using hot gas bypass is recommended where normal compressor cycling or the use of unloaders may not be satisfactory. Basically, this is a system of bypassing the condenser with the compressor discharge gas to prevent the compressor suction pressure from falling below a desired setting.

All hot gas bypass valves operate in a similar manner. They open in response to a decrease in downstream (low side) pressure, and modulate from fully open to fully closed over a given range of pressures. The introduction of the hot, high pressure gas into the low pressure side of the system at a metered rate prevents the compressor from lowering the suction pressure further.

The control setting of the valve can be varied over a wide range by means of an adjusting screw. Because of the reduced power

**DE-ENERGIZED**

**ENERGIZED**

**Figure 14-29.** Compressor Unloading Valve (*Courtesy of Copeland Corp.*)

consumption at lower suction pressures, the hot gas valve should be adjusted to bypass gas at the minimum suction pressure within the operating limits of the compressor which will result in acceptable system performance.

If a refrigeration system is properly designed and installed, field experience indicates that maintenance may be greatly reduced if the compressor operates continuously within the design limitations of the system as opposed to frequent cycling of the compressor. Electrical problems are minimized, compressor lubrication is improved, and liquid refrigerant migration is avoided.

Therefore, on multiple evaporator systems where the refrigeration load is continuous but may vary over a wide range, hot gas bypass may not only provide a convenient method of capacity control, but may also result in more satisfactory and more economical system operation.

**Bypass into Evaporator Inlet:** On close connected systems using a single evaporator, it is sometimes possible to introduce the hot gas into the evaporator immediately after the expansion valve (see Figure 14-30). Refrigerant distributors are available with side

openings for this purpose. An artificial cooling load is created by the bypass gas entering at this point. Because the regular system thermostatic expansion valve will control the flow of refrigerant required to maintain its superheat setting, the refrigerant returns to the compressor at normal operating temperatures. Thus, no compressor motor heating problem is encountered. Because high refrigerant velocities are maintained in the evaporator, the oil return is also aided. Because of these advantages, this type of control is the simplest, least costly, and most satisfactory bypass system.

**Bypass into the Suction Line:** Where multiple evaporators are connected to a single compressor, or where the condensing unit is located some distance from the evaporator, it is sometimes necessary to bypass the hot gas into the refrigerant suction line. (See Figure 14-31.) The suction pressures can be satisfactorily controlled by using this method. However, a de-superheating expansion valve is required to meter the liquid refrigerant into the suction line in order to keep the temperature of the refrigerant gas which is returning to the compressor within

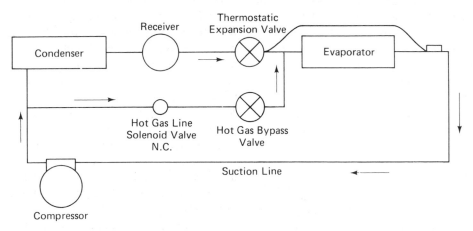

**Figure 14-30.** Bypass into Evaporator Inlet (*Courtesy of Copeland Corp.*)

**Figure 14-31.** Typical Hot Gas Bypass into Suction Line (*Courtesy of Copeland Corp.*)

the allowable temperature limits. It is necessary to thoroughly mix the bypassed hot gas, the liquid refrigerant, and the suction gas from the evaporator so that the refrigerant mixture entering the compressor is at the correct temperature. A mixing chamber is recommended for this process. A suction line accumulator can serve as an excellent mixing chamber and at the same time protect the compressor from liquid floodback.

Another commonly used method of mixing the refrigerant is to arrange the piping so that a mixture of discharge gas and liquid refrigerant is introduced into the suction line at some distance from the compressor, in a suction header if possible. (See Figure 14-32.)

## PUMPDOWN CONTROL

Refrigerant vapor will always migrate to the coldest part of the system. If the compressor crankcase can become colder than the other parts of the system, the refrigerant in the condenser, receiver, and evaporator will vaporize and travel through the system. This vaporized refrigerant will condense in the compressor crankcase.

Because of the difference in vapor pressures of the lubricating oil, the refrigerant vapor is attracted to the lubricating oil. Even though no pressure or temperature difference exists to cause the flow, the refrigerant vapor will migrate through the system and condense in the oil until the oil is saturated. During any equipment off-cycles which extend over several hours, it is possible for liquid refrigerant to almost fill the compressor crankcase completely. Because of the attraction to equalize at an ambient temperature of 70°F (21.1°C), the oil-refrigerant mixture in the crankcase will be about 70 percent refrigerant before pressure equilibrium is reached.

The most positive and dependable method of preventing refrigerant migration is to use

**Figure 14-32.** Alternate Method of Bypass into Suction Line (*Courtesy of Copeland Corp.*)

a system pumpdown cycle. By closing a liquid line solenoid valve, the refrigerant can be pumped into the condenser and receiver. The compressor operation can be controlled by use of a low pressure control (see Figure 14-33). The refrigerant can then be isolated during periods when the compressor is not in operation. Therefore, refrigerant migration to the compressor crankcase is prevented.

System pumpdown control can be used on all systems equipped with a thermostatic expansion valve with the addition of a liquid line solenoid valve, provided, of course, that adequate receiver capacity is available. A slight refrigerant leakage may occur through the solenoid valve during the off cycle which will cause the suction pressure to rise gradually. Because of this rise in suction pressure, a recycling type low pressure control is recommended to repeat the pumpdown cycle as needed. The occasional short-cycle usually is not objectionable.

A pumpdown cycle is highly recommended whenever it can be used. If a non-recycling pumpdown cycle is required, then consideration should be given to the use of a crankcase heater in addition to the pumpdown for more dependable compressor protection.

## COMMERCIAL REFRIGERATION DEFROST SYSTEMS

Refrigerated cases are equipped with evaporator coils in which liquid refrigerant is evaporated. This evaporating liquid absorbs heat from the case. Relatively warm, moist air which passes through and over the coil gives up its heat and as a result loses its moisture in the form of frost being deposited on the fins and tubes of the evaporator. This frost gradually builds up until the air movement is reduced and eventually completely stopped if the evaporator is not defrosted. Since proper refrigeration of the case is dependent upon air circulation, it is necessary to defrost the evaporator periodically so that refrigeration may continue.

Evaporator defrosting is accomplished by the following basic methods.

**Figure 14-33.** Typical Pumpdown System

**Figure 14-34.** Low Pressure Control (Courtesy of Penn Controls, Inc.)

**OFF-CYCLE DEFROST:** This method uses a low pressure control to control the compressor operation. By setting the control cut-in point at a pressure and hence a temperature above freezing, the evaporator will be completely clear of frost before the compressor starts again. This method of defrost is used for closed meat cases, produce cases, and coolers. (See Figure 14-34.)

**TIMED DEFROSTS:** This type uses a defrost control, usually a time clock, to provide the desired time and duration of the defrost cycle.

**A. No Supplemental Heat:** This method is used on open and multishelf meat and dairy cases, multishelf produce cases, and coolers. Since the evaporator does not completely defrost between compressor cycles or when the compressor does not cycle at all, an enforced off period alows a natural warming to clear the evaporator of any accumulated frost.

**B. With Electric Heaters:** This method is sometimes used with meat and dairy cases. It is usually a standard method used with frozen food and ice cream cases.

**C. With Hot Gas:** This method is used on some unit cooler evaporators but almost never on display cases. This method uses

**Figure 14-35.** Typical Hot Gas Defrost System

valves operated by a defrost control to circuit the compressor discharge gas directly to the evaporator (see Figure 14-35). Usually some method must be used to avoid flushing liquid refrigerant from the evaporator to the compressor. Extra heat is often used to supplement the available sensible heat of the hot gas since only the heat of friction and some motor heat is added after the first few minutes of defrosting.

**D. Latent Heat Defrosting:** The latent heat defrost method is used sometimes where there are multiple evaporators. In its simplest form, there would be three evaporators with one defrosting while the other two refrigerate. (See Figure 14-36.) The coil being defrosted actually acts as the condenser for the refrigerant that is being evaporated in the two refrigerating evaporators. This procedure is sequenced usually by a programmed time clock so that each of the three evaporators is defrosted as often as needed in rotation.

**DIRECT ELECTRIC DEFROST:** When supplemental heat is required for defrost, some manufacturers equip their cases with supplemental electric heat strips and a defrost limit switch. The advantages of electric defrost are:

1. Low initial cost
2. Simple installation
3. Efficient operation
4. Easy maintenance and service.

All of these factors result in the lowest total operating cost on a year-to-year basis.

Additionally, the defrost limit switch affords a built-in safety device which is also used to reliably terminate the defrost cycle by using electric current sensing relays.

**METHODS OF TERMINATING DEFROST:** All time clock defrost periods are initiated on a time basis. Termination is accomplished mainly by three methods.

**1. Straight Time Termination:** The system is controlled entirely by the time clock. The length of the defrost cycle, a constant preset value, must be determined by past experience with similar systems or by adjustment

## LATENT HEAT DEFROST AS USED ON A COMBINED SYSTEM

(Do not confuse this with Reverse Cycle Defrost with all cases being defrosted at once.)
Only one-third to one-fourth of the total Btuh load can be defrosted at any one time.

REFRIGERATION CYCLE - Liquid refrigerant is expanded through the expansion valve, absorbs heat in the coil and is drawn as a vapor back to the condensing unit.

The MULTI-CIRCUIT TIME CLOCK CONTROL operates solenoid pilot operated Three-Way valves when a defrost is scheduled.

DEFROST CYCLE - Three way valve (at condensing unit or near case line-up sends Hot Gas from compressor thru suction line. Drain pan is heated first and frost is melted from the coil as the hot gas cools and condenses. Condensed liquid is returned to liquid line through the by-pass and check valve around the expansion valve. EPR's (when used) are located either down stream from the three way valves or upstream. An ACCUMULATOR prevents condensed liquid from harming the compressor.

CASE LINE UP ON DEFROST CYCLE

By-Pass & Check Valve

CASE LINE UP ON REFRIGERATION CYCLE

CASE LINE UP ON REFRIGERATION CYCLE

EPR

Liquid Line

THREE-WAY VALVES

Valve

Normal Position

Suction Line

Locations

Defrost Position

(When Used)

Normal Position

Receiver

Compressor

Liquid Line Solenoid, N.O. Closes during a defrost of any part of the system. Pressure control on Liquid Line over-rides Solenoid, opening it when pressure drops too low.

Condenser

Accumulator

INSTALLATION NOTICE - Since suction lines (used as hot gas lines during defrost) may lengthen up to 2" in 100 feet when the hot gas heats the line, vibration absorbers are recommended to prevent line damage. The lines must also be allowed to "float free" in the trenches to allow full freedom of movement.

NOTE: These lines must be INSULATED.

Vibration Eliminators

**Figure 14-36.** Latent Heat Defrosting Cycle (*Courtesy of Tyler Refrigeration Corp.*)

**Figure 14-37.** Pressure-Terminating Defrost Control Clock (*Courtesy of Paragon Electric Corp.*)

after installation. While this is a reliable method, it has the disadvantage of making no allowance for changing conditions of the service load, humidity, or other variables. To ensure proper operation at all times, the length of the defrost period must be set for the most severe conditions, which could result in excessively long defrost periods much of the time.

**2. Pressure Termination:** A pressure switch which is incorporated in the control is connected to the low side of the refrigeration system. (See Figure 14-37.) At a predetermined and preset pressure—most often around 45 psig (3.16 kg/cm²) for low pressure systems and 40 psig (2.8 kg/cm²) for nonelectric medium temperature systems using R–12 as the refrigerant—the defrost is terminated. The controls are usually provided with a "failsafe" time termination in

case of failure of the pressure switch. The pressure-terminated control has the advantage of adjusting automatically to the changing conditions of service load and humidity. This results in defrost periods of only such duration as are needed to clear the evaporator of frost.

*NOTE:* Cold suction lines or cold machines—outdoor condensing units—will prevent this type of termination from being fully reliable. It works best when the pressure rise is consistent throughout the year.

**3. Solenoid Reset Termination:** This type of defrost clock usually has a "failsafe" dial and can be used in a number of ways. (See Figure 14-38.)

A. Close on rise of thermostats on evaporators in the cases. A series pilot circuit must be installed to all cases. The

**Figure 14-38.** Solenoid Defrost Control Clock (*Courtesy of Paragon Electric Corp.*)

solenoid resets when all the thermostats are satisfied completing this pilot circuit. It can be used on either electric or hot gas defrost systems.

B.   Open on rise of thermostats on the evaporators in series with electric heater elements. The electrical supply to the heaters from the defrost contactor or time clock is wired through a current relay. As long as the current flows through the defrost heaters the relay holds open an electric circuit to the solenoid. (See Figure 14-39.) When all of the case thermostats in the cases have opened, zero current draw to the electric heaters allows the relay contacts to close and reset the solenoid. The advantages here are that each case is protected individually from excessive defrost heat, and no extra wiring is required to the cases.

C.   Close on rise—pressurestat activates the solenoid. This is just a variation of the pressure termination method of defrost.

Current Relay
(RBM Controls)

Limit Switch
(Texas Instrument)

Portion of control panel wiring diagram showing use of a current relay. Defrost limit switches are sometimes mounted to return bends of case coils.

**Figure 14-39.** Tyler TG Control Defrost (*Courtesy of Tyler Refrigeration Corp.*)

**DEMAND DEFROST:** Solid state components have made accurate monitoring of simple devices possible. One method monitors the case air velocity, initiating a defrost cycle. A lowered air velocity indicates that the evaporator has accumulated all the frost it can hold and still maintain the desired case temperature.

Another method employs the *fluidic principle* by monitoring the air pressure in the front and the back of the evaporator. The fluidic principle amplifies small pressure changes so that a defrost can be initiated when necessary. Termination for either method is usually controlled by a thermostat.

### DEFROST CONTROLS

There are basically two types of controls used for defrost systems. These are the low pressure switch and the time clock. Most variations will have a combination of these two types of controls.

**PRESSURE CONTROL:** Since a pressure-temperature relationship is known for each type of refrigerant, regulating the suction pressure of the compressor will also regulate the operating temperature of the evaporator in the case. (See Table 14-2.)

The dual pressure control used by most equipment manufacturers has three adjustable settings. One setting is the high pressure cut-out, a safety device which is set according to the type of refrigerant used and the manufacturer's recommendation. The other two settings are the cut-in pressure—the pressure at which the compressor starts—and the differential—the number of pounds of pressure that the compressor suction will pull the system down to before stopping. The cut-out is the value determined by subtracting the differential setting

Table 14-2
Pressure-Temperature Chart*

| °F | R-12 | R-22 | R-502 | °F | °C |
|----|------|------|-------|----|----|
| -60 | 19.0 | 12.0 | 6.9 | -60 | -51.1 |
| -55 | 17.3 | 9.1 | 3.6 | -55 | -48.3 |
| -50 | 15.4 | 6.0 | 0 | -50 | -45.0 |
| -45 | 13.2 | 2.6 | 2.0 | -45 | -42.7 |
| -40 | 11.0 | 0.6 | 4.3 | -40 | -40.0 |
| -35 | 8.3 | 2.6 | 6.7 | -35 | -37.2 |
| -30 | 5.5 | 5.0 | 9.4 | -30 | -34.4 |
| -28 | 4.2 | 6.0 | 10.5 | -28 | -33.3 |
| -26 | 3.0 | 7.0 | 11.7 | -26 | -32.2 |
| -24 | 1.6 | 8.0 | 12.9 | -24 | -31.1 |
| -22 | 0.2 | 9.1 | 14.2 | -22 | -30.0 |
| -20 | 0.6 | 10.2 | 15.5 | -20 | -28.9 |
| -18 | 1.3 | 11.3 | 16.9 | -18 | -27.8 |
| -16 | 2.1 | 12.6 | 18.3 | -16 | -26.7 |
| -14 | 2.8 | 13.9 | 19.7 | -14 | -25.6 |
| -12 | 3.7 | 15.2 | 20.2 | -12 | -24.4 |
| -10 | 4.5 | 16.6 | 22.8 | -10 | -23.3 |
| -8 | 5.4 | 18.0 | 24.4 | -8 | -22.2 |
| -6 | 6.3 | 19.3 | 26.0 | -6 | -21.1 |
| -4 | 7.2 | 21.0 | 27.7 | -4 | -20.0 |
| -2 | 8.2 | 22.5 | 29.4 | -2 | -18.9 |
| 0 | 9.2 | 24.1 | 31.2 | 0 | -17.8 |
| 2 | 10.2 | 25.7 | 33.1 | 2 | -16.7 |
| 4 | 11.2 | 27.5 | 35.0 | 4 | -15.6 |
| 6 | 12.3 | 29.2 | 37.0 | 6 | -14.4 |
| 8 | 13.5 | 31.1 | 39.0 | 8 | -13.3 |
| 10 | 14.7 | 33.0 | 41.1 | 10 | -12.2 |
| 12 | 15.9 | 34.9 | 43.2 | 12 | -11.1 |
| 14 | 17.1 | 36.9 | 45.4 | 14 | -10.0 |
| 16 | 18.4 | 39.0 | 47.7 | 16 | -8.9 |
| 18 | 19.7 | 41.0 | 50.1 | 18 | -7.8 |
| 20 | 21.1 | 43.3 | 52.4 | 20 | -6.7 |

| °F | R-12 | R-22 | R-502 | °F | °C |
|----|------|------|-------|----|----|
| 22 | 22.5 | 45.5 | 54.9 | 22 | -5.6 |
| 24 | 24.0 | 47.9 | 57.4 | 24 | -4.4 |
| 26 | 25.4 | 50.1 | 60.0 | 26 | -3.3 |
| 28 | 27.0 | 52.7 | 62.7 | 28 | -2.2 |
| 30 | 28.5 | 55.2 | 65.4 | 30 | -1.1 |
| 32 | 30.1 | 57.8 | 68.2 | 32 | 0 |
| 34 | 31.8 | 60.5 | 71.1 | 34 | 1.1 |
| 36 | 33.5 | 63.3 | 74.1 | 36 | 2.2 |
| 38 | 35.2 | 66.1 | 77.1 | 38 | 3.3 |
| 40 | 37.0 | 69.0 | 80.2 | 40 | 4.4 |
| 45 | 41.7 | 76.6 | 87.7 | 45 | 7.2 |
| 50 | 46.8 | 84.7 | 96.9 | 50 | 10.0 |
| 55 | 52.1 | 93.3 | 107.9 | 55 | 12.8 |
| 60 | 57.8 | 102.4 | 115.6 | 60 | 15.6 |
| 65 | 63.9 | 112.1 | 125.8 | 65 | 18.3 |
| 70 | 70.3 | 122.5 | 136.6 | 70 | 21.1 |
| 75 | 77.1 | 133.4 | 147.9 | 75 | 23.9 |
| 80 | 84.3 | 145.0 | 159.9 | 80 | 26.7 |
| 85 | 92.0 | 157.2 | 172.5 | 85 | 29.4 |
| 90 | 100.0 | 170.0 | 185.8 | 90 | 32.2 |
| 95 | 108.4 | 183.6 | 199.7 | 95 | 35.0 |
| 100 | 117.4 | 197.9 | 214.4 | 100 | 37.8 |
| 105 | 126.8 | 212.9 | 229.7 | 105 | 40.6 |
| 110 | 136.7 | 228.6 | 245.8 | 110 | 43.3 |
| 115 | 147.1 | 245.2 | 266.1 | 115 | 46.1 |
| 120 | 158.0 | 262.5 | 280.3 | 120 | 48.9 |
| 125 | 169.5 | 280.7 | 298.7 | 125 | 51.7 |
| 130 | 181.5 | 299.7 | 318.0 | 130 | 54.4 |
| 135 | 194.1 | 319.6 | 338.2 | 135 | 57.2 |
| 140 | 207.2 | 340.3 | 359.2 | 140 | 60.0 |
| 145 | 221.0 | 362.0 | 381.1 | 145 | 62.8 |
| 150 | 235.4 | 384.6 | 404.0 | 150 | 65.6 |
| 155 | 250.4 | 406.3 | 427.8 | 155 | 68.3 |
| 160 | 266.1 | 443.3 | 452.6 | 160 | 71.1 |

Vacuum (shaded region at top)

* Vacuum in inches of mercury, Pressure in psig. (Courtesy of Tyler Refrigeration Corp.)

from the cut-in setting. (See Figure 14-40.) Do not confuse this cut-out setting with the high pressure cut-out setting.

**DEFROST TIME CLOCKS:** The 24-hour time clock operated devices have both normally open (NO) and normally closed (NC) contacts. Up to six defrost periods can be scheduled per 24 hours with the pins provided. The defrost controls used by most manufacturers initiate a defrost period when a pin on the main dial trips the switches (see Figure 14-41). The normally closed contacts allow electric current to pass through the defrost control to the dual pressure control, which in turn starts and stops the condensing unit. During a defrost period

**Figure 14-40.** Dual Pressure Control (*Courtesy of Penn Controls, Inc.*)

**Figure 14-41.** Defrost Clock Dial (*Courtesy of Tyler Refrigeration Corp.*)

the condensing unit is shut off and the normally open contacts close, energizing the defrost heater circuit, if one is used. (See Figure 14-42.)

Defrost methods are named by their mode of defrost initiation and termination. The following are the basic types.

1. TC (Straight time clock defrost—*Time Clock* Termination) The length of the defrost period is set on the inner dial and can be set from 2 to 110 minutes.

2. TP (*Time Pressure* Termination) The length of the defrost period varies with the time required for the suction pressure to raise the cut-in setting. (See Table 14-3.) TP controls can be used successfully only when pressure-limiting expansion valves are not used. These valves prevent a normal pressure rise during the defrost period.

3. TG (*Temperature Guard–Temperature* Termination) This is a Tyler Refrigeration Company innovation that utilizes a current relay which holds the time clock in defrost until all of the case limit switches are open, or the failsafe dial setting takes effect.

*NOTE:* All cases which are equipped for electric defrost systems have defrost limit switches attached to the evaporator tubing to cut off the electric current to the defrost heater(s) when the evaporator temperature reaches 50°F (10°C). Thus, each case is individually protected from overheating dur-

ing the defrost period. When all the limit switches have opened, the electric current ceases to flow, allowing the current relay to close. This, in turn, activates a solenoid, tripping the time clock back to normal operation. As in the first two controls, the inner dial is set for maximum or failsafe time. (See Figure 14-43.)

**SETTING DEFROST CONTROLS:** When the defrost period is dependent on the off-cycle defrost method, the pressure switch cut-in setting must be at 36 or 37 psig (2.5 or 2.6 kg/cm²) when R–12 refrigerant is used. This is most accurately determined with a pressure gauge because the indicator on the pressure control is not that accurate. The use of a pressure gauge when setting this control is also recommended when a timed defrost control is used.

Setting the timed defrost control involves setting the length, in minutes, of each defrost period and setting the number of defrost periods each day. Most manufacturers publish their recommendations as follows:

1. On a case setting label inside each case.

2. In a case setting card packed with the condensing unit.

3. In the instructions applicable to the case.

These recommendations may not always serve the need of a particular installation. However, they do serve as a starting point. They are based on laboratory and field experience.

**VARIABLE FACTORS:** Factors that affect the defrost cycles are:

1. Store humidity and temperatures which vary throughout the year. The amount

On a TC Control — No Extra Devices
On a TP Control — Tubing to Low Side
On a TG Control — Current Relays Would
be in the Control Box.

Tubing to Low Side of Compressor

Tubing to High Side

Dual Pressure Control

Defrost Wiring to Cases

Circuit Breaker
(or Pull-Fuses)

Motor Contactor

To Compressor

Defrost Control

**Figure 14-42.** Typical Control Box with Controls (*Courtesy of Tyler Refrigeration Corp.*)

Table 14-3
Reset Pressure Settings for TP Controls*

|  | LOW TEMP CASES | NORMAL TEMP CASES | |
|---|---|---|---|
|  |  | W/O extra heat | W/Elect. Heaters |
| R 12 | 45# | 40# | 52# |
| R 502 | 95# | 84# | 100# |

* Courtesy of Tyler Refrigeration Corp.

**Figure 14-43.** TP Defrost Control (*Courtesy of Paragon Electric Corp.*)

of frost to be removed, therefore, varies throughout the year.

2.   Voltage to the defrost heaters may vary causing the amount of heat available also to vary.

3.   The use of doors (coolers and reach-ins) varies.

4.   The operating temperature of the case (particularly with multishelf cases, low and normal temperatures) may be lower than necessary. This will cause an extra heavy frost load on the evaporator.

5.   Drafts across the case permit too much warm, moist air to enter the case, thereby loading the evaporator excessively.

**PRINCIPLES OF DEFROST:** The duration of defrost must be long enough to clear the evaporator completely each defrost period, while the defrost frequency must be often enough to keep the refrigerated air circulating freely throughout the case.

A recording temperature gauge is the most efficent method of determining when a case evaporator is becoming choked with frost. When the temperature "sags" before a defrost period, another defrost period per day is required. Make sure, however, that the factors causing the temperature sag (case or ambient temperature, humidity, usage, or drafts) have been controlled as much as possible.

## MULTIPLE (COMBINED) SYSTEMS

Not too long ago almost everyone subscribed to the theory that a single refrigeration unit was required for each case. Some still think that satisfactory operation will be obtained only if not more than two display cases per condensing unit are installed. In general, however, this type of thinking has advanced to the point that more people are combining more systems. Most store planners and operators are finding it entirely satisfactory and desirable to combine not only similar cases but cases of differing temperature requirements as well.

Combining systems is certainly not new. The combination of refrigeration systems on very large, continuously operating condensing units has been in use since the 1920s. However, the advantages of these large systems are still somewhat nullified by the complexities caused by involved combining.

The choices the design engineer must make when designing a system are many and varied. He (or she) can use from 1 to 20 compressors, depending on compressor size. If 1 or 2 compressors are chosen, the designer must combine systems extensively. If 20 compressors are chosen, they are small and less efficient compressors, along with 20 small refrigeration systems. Most experienced designers agree that the best route lies between these extremes.

Basically, combining is the grouping of

more than one refrigeration function on one compressor. This procedure usually requires the use of larger condensing units. Some of the reasons for combining are:

1. Good, relatively inexpensive, larger condensing units are now available.

2. Improved and varied controls are now on the market.

3. Refrigeration technology has improved.

4. The prepackaging of machinery can now take more of the installation burden from the local level.

However, the most important pros are economic, saving the user money in:

1. Original costs and installation costs.

2. Operating costs.

3. Maintenance costs.

Motor efficiency is one of the attributes of a larger compressor. Above 5 hp, however, there isn't too much efficiency to be gained. (See Table 14-4.) It also must be considered that the larger the compressor, the more complicated the total system will tend to be.

**SIMPLE SYSTEMS VERSUS COMBINED SYSTEMS:** A simple system is one case connected to one condensing unit as shown in Figure 14-44. The electrical wiring for this system is also simplified.

A simple multiple system is two or more cases of the same suction pressure—evaporator or suction temperature—and defrost requirements connected to one condensing unit. (See Figure 14-45.) The electrical wiring and the refrigerant system are more complicated than in the simple system.

Table 14-4
Motor Efficiency Comparison

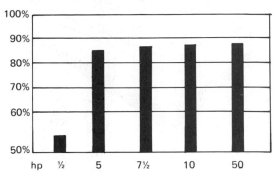

hp    ½      5     7½     10     50

* Courtesy of Tyler Refrigeration Corp.

**Figure 14-44.** Simple System (*Courtesy of Tyler Refrigeration Corp.*)

A combined system involves cases of different suction pressures and/or defrost requirements on one condensing unit. (See Figure 14-46.)

One thing that must be considered when combining cases and/or coolers of different suction pressures is the operating premium which must be paid. A rule of combining is that the entire system must be sized to

**Figure 14-45.** Simple Multiple System (*Courtesy of Tyler Refrigeration Corp*).

Ice Cream Cases

Frozen Food Cases with Ice Cream Defrost Heaters and an Evaporator Pressure Regulator on the Suction Line.

**Figure 14-46.** Simple Combined System (*Courtesy of Tyler Refrigeration Corp.*)

match the Btus which are figured at the lowest suction pressure required in the system. The operating costs rise in this situation. (See Table 14-5).

Complex combined systems are designed to handle all low temperature (or all normal temperature) cases and coolers on one condensing unit.

Ultracomplex combined systems include all of the refrigeration equipment in the store on one condensing unit (usually with an air conditioning or other unit available as an emergency spare).

All of these systems are mechanically feasible. The equipment and controls to make them work are all practical and available. The equipment which will serve the customer best through the years that follow the installation should be given first priority in making the selection.

The maintenance of commercial refrigeration equipment is also of prime importance and is usually dependent on competent local service technicians. The equipment to be used should be within their ability to service properly.

In general, the following statements serve well for most supermarket installations:

1. The use of 5-, 7½-, 10- and sometimes 20-hp units will reduce the number of condensing units in an average supermarket to 6 or 8. A saving of space will be realized. The condensing units can even be put in a prepackaged room outside the store. In addition, these units are fairly common, with parts or entire replacements available

Table 14-5
Total Heat (Btu in 1000s) Rejected from Condensing Units*

| HORSEPOWER | 3 HP | 5 HP | 5 HP | 7½ HP | 7½ HP | 10 HP | 15 HP | 20 HP |
|---|---|---|---|---|---|---|---|---|
| **HIGH BACK PRESSURE COMPRESSORS   R 12** | | | | | | | | |
| Designation | 300H | | 500H | 750H | | 1000H | | 2000H |
| Displ'ment CFH | 698 | | 1080 | 1620 | | 2380 | | 3580 |
| +35° | 39.2 | | 58.4 | 90.0 | | 134.0 | | 202.0 |
| +20° | 28.2 | | 43.4 | 68.0 | | 103.0 | | 156.0 |
| +10° | 25.4 | | 34.0 | 52.5 | | 85.0 | | 128.0 |
| **MEDIUM BACK PRESSURE COMPRESSORS   R 12** | | | | | | | | |
| Designation | 300S | | 500S | | 760S | 1000S | 1500S | 2000S |
| Displ'ment CFH | 815 | | 1375 | | 2120 | 2380 | 3020 | 3580 |
| +20° | 34.0 | | 55.2 | | 88.0 | 103.5 | 188.0 | 155.0 |
| +10° | 27.8 | | 45.0 | | 72.3 | 85.0 | 145.0 | 128.0 |
| **LOW BACK PRESSURE COMPRESSORS   R 502 ONLY** | | | | | | | | |
| Designation | 300L | 450L | 500L | 750L | 760L | 1000L | 1500L | 2000L |
| Displ'ment CFH | Marginal | 1080 | 1375 | 1620 | 2120 | 2380 | 3020 | 3580 |
| −25° | | 31.2 | 39.0 | 45.0 | 58.4 | 70.8 | 88.4 | 94.5 |
| −35° | | 24.7 | 30.8 | 36.8 | 48.0 | 56.0 | 69.5 | 79.4 |

* Courtesy of Tyler Refrigeration Corp.

on an over-the-counter basis. (See Figure 14-47.)

2. The resultant refrigeration systems will range between the simple multiple type and the simple combined type.

3. Service on these sizes of compressors and on these types of systems is generally available locally.

It can be seen that every mechanical selection is made by weighing positive and negative factors. When the positive factors outweigh the negative, then it is a good selection and a good compromise. Negative factors do, nevertheless, exist and have to be recognized for what they are.

**MAINTAINING CASE TEMPERATURE DIFFERENCES:** To separate cases and coolers of different temperatures, there are two commonly used methods. One is the use of liquid line solenoid valves controlled by thermostats (see Figure 14-48). This method is acceptable in situations where ups and downs of temperature (due to the thermostat differential) inside the case will not be critical. However, this method introduces a secondary responsibility for the performance of the whole system. Extra electrical circuits must be installed (an extra installation cost) and these circuits must be maintained.

The second method is to use evaporator pressure regulating (EPR) valves. These are nonelectric, pressure-actuated devices that

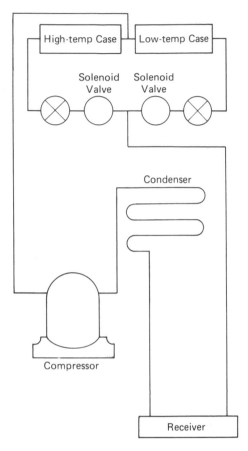

**Figure 14-47.** Solenoid Valve Location

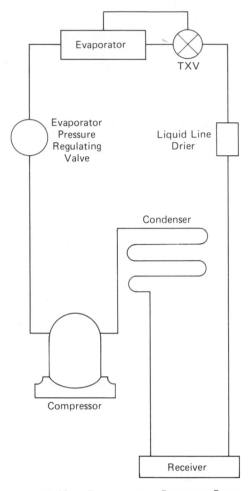

**Figure 14-49.** Evaporator Pressure-Regulating Valve Installation

**Figure 14-48.** Composite Chart—Typical Thermostat and Solenoid Valve Action (*Courtesy of Tyler Refrigeration Corp.*)

control the refrigerant pressure in the evaporator (see Figure 14-49). This method has proven to be highly reliable. EPR valves are installed in the suction line of a refrigeration system. Initially, they must be adjusted to the desired pressure (hence the desired temperature). Once they are adjusted, they will maintain a constant evaporator temperature. With a constant refrigeration load, as is the case with open supermarket equipment,

**Figure 14-50.** Composite Chart—EPR Operation (*Courtesy of Tyler Refrigeration Corp.*)

constant evaporator temperatures will maintain almost perfectly constant case temperatures. (See Figure 14-50.)

**MAINTAINING DEFROST REQUIREMENT DIFFERENCES:** Sometimes these defrost differences can be resolved by making the defrosts alike. For instance, a top display produce case can tolerate the four defrosts per day required for a multishelf produce case. However, this is not generally done. Another equalizing method is that of equipping frozen food cases with an extra defrost heater so that it can be defrosted in the same way as ice cream cases.

An ordinary defrost time clock is not always suitable for defrosting the combined system. A multicircuit time clock has been developed which is capable of handling six different functions with its six pairs of normally open and normally closed switches. Each pair of switches has its own 24-hour clock dial with initiating tab slots every hour and a terminating dial timing up to 110 minutes. This clock makes it possible to delay the cooler fans after defrost. It can also be used to permanently synchronize defrosts for case lineups on more than one condensing unit.

Compressor unloaders could be required when most of the load on a system is iso-

lated during defrost while the compressor is still running. Actually, all compressors have a built-in unloading feature. As case requirements are satisfied and the suction pressure drops, the amount of refrigerant gas pumped is reduced. The cut-out on the pressure control should be set low enough to permit this. Avoid the unnecessary use of unloaders because they make a compressor inefficient. A half-unloaded compressor still draws 70 percent of full load amps.

## HEAT RECOVERY SYSTEMS

People in most businesses today must get the maximum out of every dollar of operating cost. The food store operator is certainly no exception. This is the reason that so many people have been so interested in the technique of heat recovery when considering new stores or planning an extensive remodeling program.

Heat recovery is a most complex subject and, if it is discussed in its entirety, would leave the uninitiated completely baffled. To fully understand it, one must have more than a nodding acquaintance with basic electricity, refrigeration, heat transfer, and cost accounting.

Let's suppose that a food store operator or a chain store group has decided to use heat recovery systems. Possible yearly projected savings indicate that their investment can be paid for in five years, which is acceptable.

The savings that justify the expense of added equipment come from the simple fact that a refrigeration system is made to move heat. It removes heat from display cases and discharges the heat elsewhere—either directly to air or water. The refrigeration compressor is the pump, and the refrigerant is the fluid that acts as a vehicle for the heat.

The heat extracted from the display cases

comes from the store in the beginning, so if the heat is discharged to the outside air during the heating season, it will have to be replaced by conventional heating sources. By discharging the heat into the store, some or most of the normal heating costs can be saved. There is an added bonus in that the heat generated by normally operating electric motors which are powering the refrigeration compressors is also recovered at the condenser. With well-engineered and installed heat recovery systems, there have been amply documented reports that a large portion of the winter heating cost is eliminated. In some areas, particularly the southern states, no make-up heat is necessary in an average year.

The amount of heat available for use in a heat recovery system is tremendous. For instance, a 15-hp unit operating dairy cases at 10°F (−12.2°C) suction temperature has 145,000 Btus (36,540 kcal) of heat to reject, all of which is available to heat the store when properly channeled. (See Table 14-5.) This single unit rejects more heat than is required to heat a large home! The heat from this and other units found in a typical supermarket can quickly add up enough heat to maintain the store at comfortable levels on all but the coldest days of a normal heating season in the midwestern United States.

Care must be taken in the design of heat recovery systems to eliminate every unnecessary cost. These costs, though often hidden, must come out of the budget. A primary example is unnecessary electrical consumption for moving air or water in transferring the heat.

**BASIC APPROACHES TO HEAT RECOVERY:** No one can state with any certainty the origin of recovering heat in food stores, but it is easy for the "old timers" to recall homemade arrangements that did the job. Some of these date back close to 20 years.

**Direct Air Method:** In the summertime, the air from an air-cooled condenser was discharged to the atmosphere, but in the winter that air was utilized to supplement store heating. These systems worked. Even so, they were cumbersome, moved great quantities of unfiltered air, and often required blower fans with high initial and operating costs. In these arrangements, even in a store that used no natural fuel for heating, the electrical costs for moving the air around the store could not be ignored. They frequently involved 10-hp motors or larger, whose monthly operating costs would be approximately $100 at .02 cents per kWh.

Recognizing the shortcomings of the first approach, various other schemes have been used or tried. One variation of the direct-air approach used a single multiple circuit condenser with face and bypass dampers on both sides, Some even used two-speed fans so that in the wintertime the large volume of air was not required. The basic flaw in that particular scheme lay in the fact that the building had to be specifically prepared to utilize this type of approach (see Figure 14-51).

**Closed-Circuit Water Method:** The northwest area of the country has for years utilized closed-circuit water-cooling for the condensing side of refrigeration units. This scheme led some, quite normally, to simply circuit the hot water to a coil in an air handler, specially put there to discharge the heat from the water into the store air during the wintertime. In the summertime, a 3-way valve would isolate that coil and allow the water to be cooled in the evaporative condenser where the heat was discharged to the atmosphere.

On first examination, this scheme has merit, but further study discloses the fact that once again pumping horsepower must be considered. With these arrangements frequently using 5-, 7½-, or 10-hp water pumps, the cost of operating the water pump must be taken into account on the debit side. Furthermore, the closed-circuit arrangement must utilize closed-circuit water running at 100°F (37.8°C) or cooler. A 100°F heat source is not particularly good for the store because it is far too cool. The temptation has been to raise the head pressure in order to raise the water temperature so that more comfortable and reasonable conditions may be maintained in the store. However, this leads to other concealed costs that come from higher maintenance because of the higher pressures and the higher cost of getting the refrigeration job done using higher discharge pressures.

**Dual Condenser Method:** Any mechanical alternative has faults; it is, in its own way, a compromise. However, the dual condenser or alternate condenser technique seems to have the fewest shortcomings.

This is an arrangement whereby air or remote air condensers may be used for summertime cooling, Thus, maximum effectiveness of the refrigeration equipment is achieved during the most trying season. A refrigerant gas valve bypasses the primary condenser and causes the gas to be circulated in an appropriate coil which is located in the store air-handling system.

Figure 14-52 illustrates a typical arrangement showing the relationship of the air conditioning cooling coil, fans, filter, the heat recovery coil, and the make-up furnace. The heat recovery coil can be single- or multiple-circuited depending upon the number of condensing units from which heat is recovered. Obviously, the piping of six condensing units to such a coil is comparatively simple. The piping of eight or ten might be quite acceptable, but the use of more than a dozen condensing units with a heat recovering technique might be deemed impractical for an installation.

Nevertheless, regardless of the number of units employed, the heat is recovered for the least possible cost. Head pressures are still maintained at an optimum level—very low, but not so low as to cause refrigeration problems. The temperature of the gas being circulated through the recovery coil is high —most of it around 150°F (66°C) or higher. Furthermore, there is no cost for pumping the heat in any fashion except that which would normally be required because the fan in the air conditioning section would be required to distribute heat even without heat recovery equipment. Even better, while the heat recovery system is operating, condenser fans and/or water pumps are turned off on the alternate outdoor condensers.

Illustrated in Figure 14-51 is the whole scheme as it relates to a food store, showing the relationship of make-up air to the return ducting as well as the entire air handler package to the store. In this connection, it is important to mention two significant points. In most situations, it can be flatly stated that the best heating and air conditioning and distribution of these throughout the store will be achieved with constant running of the fan in the air handler.

The second point is one being adopted by one of the companies promoting doorless openings for food stores. Their contention that "You can't heat a vacant lot" seems to make sense. This leads to the second significant statement. Allowing the air from a building to escape through various uncontrolled openings can lead to loss or control of air distribution throughout the store. It can allow cold winter air to find its own entrances into the building and let shoppers be subjected to it before the air is warmed to a comfortable level. Tightening the building by eliminating unnecessary openings seems a reasonable requirement with the idea that controlled make-up air will come into the building as it is required. The above thoughts would seem to apply equally to a store with conventional doors as to one with a doorless arrangement.

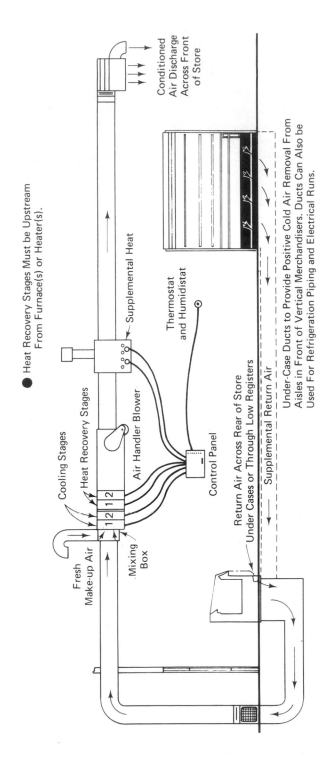

● Heat Recovery Stages Must be Upstream From Furnace(s) or Heater(s).

Conditioned Air Discharge Across Front of Store

Supplemental Heat

Thermostat and Humidistat

Cooling Stages

Heat Recovery Stages

Air Handler Blower

Control Panel

Fresh Make-up Air

Mixing Box

Return Air Across Rear of Store Under Cases or Through Low Registers

Supplemental Return Air

Under-Case Ducts to Provide Positive Cold Air Removal From Aisles in Front of Vertical Merchandisers. Ducts Can Also be Used For Refrigeration Piping and Electrical Runs.

**Figure 14-51.  Direct Air Heat Recovery System (Courtesy of Tyler Refrigeration Corp.)**

Furnace Bypass (as Required)

Air Duct to Store

Optional Gas (or Oil) Make-up Heat Section

Optional Electric Make-up Heat Section

Blower Isolating Section

Blower Section (Blower Should Operate Continually to Maintain Proper Pressure Balance in the Building.)

This Condenser Used When Heat is Not Required in the Store.

When Heat is Required, the Hot Gas Goes Directly to the H.R. Coil to Heat Store Air.

Other H.R. Circuits

Multi-Circuit Heat Recovery Coil

Air Conditioning Coil Section

Filter Section

Pilot Operated Hot Gas Diverting Valve

Check Valves

Make-up Air

Return Air

To Cases

**Figure 14-52. Dual Condenser Heat Recovery System (Courtesy of Tyler Refrigeration Corp.)**

433

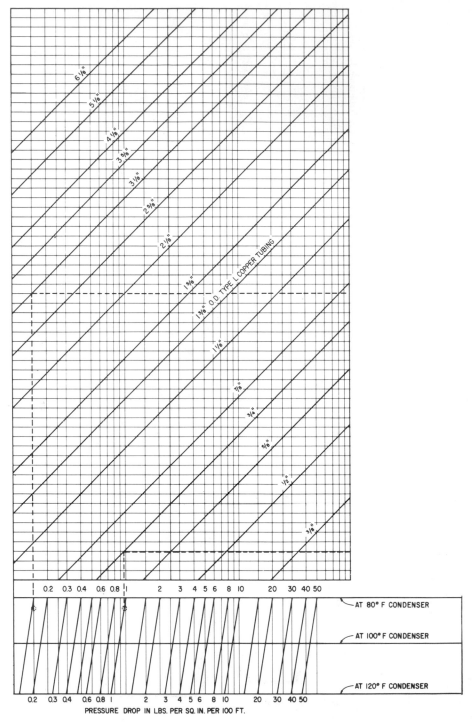

**Figure 14-53.** Freon—12 Refrigerant Pressure Drop in Lines (65° Evaporator Outlet)

TONS OF REFRIGERATION

NOTE: Pressure drops do not allow for pulsating flow. If flow
is pulsating, use next larger pipe size.
Liquid line determined at 0°F evap. & 80°F. cond.
Discharge lines at 0°F evap. Other conditions do
not appreciably change result. Vapor at evap. outlet
assumed to be at 65°F.

EXAMPLE:

5.5 Tons at −40°F Evap. and 85°F Cond.

3 1/8" Suction Line Pressure Drop = 0.155 p.s.i./100 Ft.

3/4" Liquid Line Pressure Drop = 1.0 p.s.i/100 Ft.

(Courtesy of Copeland Corp.)

## COMMERCIAL REFRIGERATION PIPING

Refrigeration piping design procedures are a series of compromises, at best. A good piping system will have a maximum capacity, be economical, provide proper oil return, provide minimum power consumption, require a minimum amount of refrigerant, have a low noise level, provide proper refrigerant control, and allow perfect flexibility in system performance from 0 to 100 percent of unit capacity without lubrication problems. Obviously, it is impossible to obtain all of these goals because some of them are in direct conflict with others. To make an intelligent decision as to what type of compromise is most desirable, it is essential that the piping designer have a thorough understanding of the basic effects on the system performance of the piping design in the different points of the refrigerant system.

**PRESSURE DROP:** Pressure drop, in genaral, tends to decrease system capacity and increase the amount of electric power required by the compressor. Therefore, excessive pressure drops should be avoided. The amount of the pressure drop allowable will vary depending on the particular segment of the system that is involved. Each part of the system must be considered separately. There are probably more charts and tables available which cover refrigeration line pressure drop and capacities at a given pressure, temperature, and pressure drop than any other single subject in the field of refrigeration and air conditioning.

The piping designer must realize that there are several factors which govern the sizing of refrigerant lines and that pressure drop is not the only criterion to be used in designing a system. It is often required that refrigerant velocity, rather than pressure drop, be the determining factor in system design. Also, the critical nature of oil return can produce many system difficulties. A rea-

sonable pressure drop is far more preferable than oversized lines which may hold refrigerant far in excess of that required by the system. An overcharge of refrigerant can result in serious problems of liquid refrigerant control, and the flywheel effect of large quantities of liquid refrigerant in the low pressure side of the system can result in erratic operation of the refrigerant flow control device.

The size of the refrigerant line connection on a service valve which is supplied with a compressor, or the size of the connection on an evaporator, condenser, or some other system accessory does not determine the correct size of the refrigerant line to be used. Equipment manufacturers select a valve size or connection fitting on the basis of its application to an average system, and other factors such as the application, length of connecting lines, type of system control, variation in load, and a multitude of other factors. It is entirely possible for the required refrigerant line size to be either smaller or larger than the fittings on various system components. In such cases, reducing fittings must be used.

**OIL RETURN:** Since oil must pass through the compressor cylinders to provide the proper lubrication, a small amount of oil is always circulating through the system with the refrigerant. The oils used in refrigeration systems are soluble in liquid refrigerant, and at normal room temperatures they will mix completely. Oil and refrigerant vapor, however, do not mix readily, and the oil can be properly circulated through the system only if the velocity of the refrigerant vapor is great enough to carry the oil along with it. To assure proper oil circulation, adequate refrigerant velocities must be maintained not only in the suction and discharge lines, but in the evaporator circuits as well.

Several factors combine to make oil return most critical at low evaporating temperatures. As the suction pressure decreases and the refrigerant vapor becomes less dense, it becomes more difficult for the refrigerant to carry the oil along. At the same time, as the suction pressure decreases, the compression ratio increases and the compressor capacity is reduced. Therefore, the weight of the refrigerant which is circulated decreases. Refrigeration oil alone becomes as thick as molasses at temperatures below 0°F (−17.8°C). As long as it is mixed with a sufficient amount of liquid refrigerant, however, it flows freely. As the percentage of oil in the mixture increases, the viscosity increases.

At low temperature conditions several factors start to converge and can create a critical condition. The density of the refrigerant vapor decreases, the velocity decreases, and as a result more oil starts to accumulate in the evaporator. As the oil and refrigerant mixture becomes thicker, the oil may start logging in the evaporator rather than returning to the compressor. This results in wide variations in the compressor crankcase oil level in poorly designed systems.

Oil logging in the evaporator can be minimized with adequate refrigerant velocities and properly designed evaporators even at extremely low evaporating temperatures. Normally, oil separators are necessary for operation at evaporating temperatures below −50°F (−45.6°C), in order to minimize the amount of oil in circulation.

**EQUIVALENT LENGTH OF PIPE:** Each valve, fitting, and bend in a refrigerant line contributes to the friction pressure drop because of its interruption or restriction of smooth flow. Because of the detail and complexity of computing the pressure drop of each individual fitting, normal practice is to establish an equivalent length of straight tubing for each fitting. This allows the consideration of the entire length of line, including the fittings, as an equivalent length of straight pipe. Pressure drop and line sizing tables and charts are normally set up on a basis of a pressure drop per 100 ft (30.4 m) of straight pipe, so the use of equivalent lengths allows the data to be used directly. (See Table 14-6.)

Table 14-6
Equivalent Lengths in Feet of Straight
Pipe for Valves and Fittings*

| O.D., IN. LINE SIZE | GLOBE VALVE | ANGLE VALVE | 90° ELBOW | 45° ELBOW | TEE LINE | TEE BRANCH |
|---|---|---|---|---|---|---|
| ½ | 9 | 5 | .9 | .4 | .6 | 2.0 |
| ⅝ | 12 | 6 | 1.0 | .5 | .8 | 2.5 |
| ⅞ | 15 | 8 | 1.5 | .7 | 1.0 | 3.5 |
| 1⅛ | 22 | 12 | 1.8 | .9 | 1.5 | 4.5 |
| 1⅜ | 28 | 15 | 2.4 | 1.2 | 1.8 | 6.0 |
| 1⅝ | 35 | 17 | 2.8 | 1.4 | 2.0 | 7.0 |
| 2⅛ | 45 | 22 | 3.9 | 1.8 | 3.0 | 10.0 |
| 2⅝ | 51 | 26 | 4.6 | 2.2 | 3.5 | 12.0 |
| 3⅛ | 65 | 34 | 5.5 | 2.7 | 4.5 | 15.0 |
| 3⅝ | 80 | 40 | 6.5 | 3.0 | 5.0 | 17.0 |

For accurate calculations of pressure drop, the equivalent length for each fitting should be calculated. As a practical matter, an experienced piping designer may be capable of making an accurate overall percentage allowance unless the piping system is extremely complicated. For long runs of piping of 100 ft (30.4 m) or greater, an allowance of 20 to 30 percent of the actual length may be adequate. For short runs of piping, an allowance as high as 50 to 75 percent or more of the lineal length may be necessary. Judgment and experience are necessary in making a good estimate, and estimates should be checked frequently with actual calculations to ensure reasonable accuracy.

For items such as solenoid valves and pressure-regulating valves, where the pressure drop through the valve is relatively large, data is normally available from the manufacturer's catalog so that items of this nature can be considered independently of lineal length calculations.

**PRESSURE DROP TABLES:** There are pressure drop tables available which show the combined pressure drop for all refrigerants. Figures 14-53, 14-54, and 14-55 are for refrigerants R–12, R–22, and R–502. Pressure drops in the discharge line, suction line, and the liquid line can be determined from these charts for condensing temperatures ranging from 80°F to 120°F (26.7°C to 49°C).

To use the chart, start in the upper right hand corner with the design capacity. Drop vertically downward on the line representing the desired capacity to the intersection with the diagonal line representing the operating condition desired. Then move horizontally to the left. A vertical line dropped from the intersection point with each size of copper tubing to the design condensing temperature line allows the pressure drop in psi ($kg/cm^2$) per 100 ft (30.4 m) of tubing to be read directly from the chart. The diagonal pressure drop lines at the bottom of the chart represent the change in pressure drop due to a change in condensing temperature.

For example, in Figure 14-57 for R–502, the dotted line represents a pressure drop determination for a suction line in a system having a design capacity of 5.5 tons or 66,000 Btus/hr (16,632 kcal) operating with an evaporating temperature of –40°F (−40°C). The 2⅝ in. (66.7 mm) suction line illustrated has a pressure drop of 0.22 psi (.015 $kg/cm^2$) per 100 ft (30.4 m) at 85°F (29.4°C) condensing temperature. The same line with the same capacity would have a pressure drop of 0.26 psi (.018 $kg/cm^2$) per 100 ft at 100°F (37.8°C) condensing temperature, and 0.32 psi (.022 $kg/cm^2$) per 100 ft at 120°F (49°C) condensing temperature.

In the same manner, the corresponding pressure drop for any line size and any set of operating conditions within the range of the chart can be determined.

**SIZING HOT GAS DISCHARGE LINES:** Pressure drop in discharge lines is probably less critical than in any other part of the system. Frequently, the effect on capacity of the discharge line pressure drop is overestimated since it is assumed that the compressor discharge pressure and the condensing pressure are the same. In fact, these are two different pressures. The compressor discharge pressure is greater than the condensing pressure by the amount of the discharge line pressure drop. An increase in pressure drop in the discharge line might increase the compressor discharge pressure materially, but have little effect on the condensing pressure. Although there is a slight increase in the heat of compression for an increase in discharge pressure, the volume of gas pumped is decreased slightly due to a decrease in volumetric efficiency of the compressor. Therefore, the total heat to be dissipated through the condenser may be relatively unchanged, and the condensing temperature and pressure may be quite stable, even though the discharge line pressure drop and, because of the pressure drop, the compressor discharge pressure might vary considerably.

The performance of a typical compressor, operating at air conditioning conditions with R–22 and an air-cooled condenser indicates that for each 5 psi (.35 kg/cm²) pressure drop in the discharge line, the compressor capacity is reduced less than ½ of 1 percent, while the electrical power required is increased about 1 percent. On a typical low temperature compressor operating with R–502 and an air-cooled condenser, approximately 1 percent of the compressor capacity will be lost for each 5 psi pressure drop, but there will be little or no change in electrical power consumption.

As a general guide, for discharge line pressure drops up to 5 psi, the effect on the system performance would be so small it would be difficult to measure. Pressure drops up to 10 psi (.703 kg/cm²) would not be greatly detrimental to system performance provided the condenser is sized to maintain reasonable condensing pressures.

Actually, a reasonable pressure drop in the discharge line is often desirable to dampen compressor discharge pulsation, and thereby reduce noise and vibration. Some discharge line mufflers actually derive much of their efficiency from a pressure drop through the muffler.

Discharge lines on a factory built condensing unit usually are not a field problem, but on systems installed in the field with remote condensers, line sizes must be selected to provide proper system performance.

Because of the high temperatures which exist in the discharge line, the oil circulation through both horizontal and vertical lines can be maintained satisfactorily with reasonably low refrigerant velocities. Since oil traveling up a riser usually creeps up the inner surface of the pipe, oil travel in vertical risers is dependent on the velocity of the gas at the tubing wall. The larger the pipe diameter, the greater will be the required velocity at the center of the pipe to maintain a given velocity at the wall surface. Figures 14-56 and 14-57 list the maximum recommended discharge line riser sizes for proper oil return for varying capacities. The variation at different condensing temperatures is not great, so the line sizes shown are acceptable on both water-cooled and air-cooled applications.

If horizontal lines are run with a pitch in the direction of refrigerant flow of at least ½ in. (12.7 mm) in 10 ft (3.04 m), there is normally little problem with oil circulation at lower velocities in horizontal lines. However, because of the relatively low velocities required in vertical discharge lines, it is recommended wherever possible that both horizontal and vertical discharge lines be sized on the same basis.

To illustrate the use of the chart we will assume a system operating with R–12 at 40°F (4.4°C) evaporating temperature has a capacity of 100,000 Btus/hr (25,200 kcal). The intersection of the capacity and evaporating temperature lines at point X on Figure 14-58 indicates the design condition. Since this is below the 2⅛ in. OD (53.97 mm) line, the maximum size that can be used to ensure oil return up a vertical riser is a 1⅝ in. OD (41.27 mm) line.

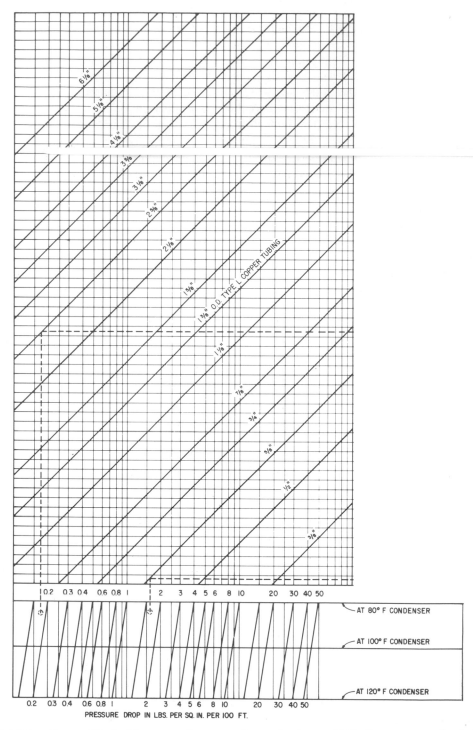

**Figure 14-54.** Freon—22 Refrigerant Pressure Drop in Lines (65° Evaporator Outlet)

**TONS OF REFRIGERATION**

0.2  0.3 0.4  0.6 0.8  1    2   3  4   6  8  10    20   30 40   60  80 100

BTU'S PER HR.

2000 3000 4000 6000 8000 10,000

-60°F EVAPORATOR TEMPERATURE
-40°F
-20°F
0°F
20°F
40°F

DISCHARGE LINES

80°F COND.
100°F COND.
120°F COND.

LIQUID LINE

NOTE: Pressure drops do not allow for pulsating flow. If flow
is pulsating, use next larger pipe size.
Liquid line determined at 0°F evap. & 80°F cond.
Discharge lines at 0°F evap. Other conditions do not
appreciably change result. Vapor at evap. outlet
assumed to be at 65°F.

EXAMPLE:

5.5 Tons at −40°F Evap. and 85°F Cond.

2 ⅝" Suction Line Pressure Drop = 0.18 p.s.i./100 Ft.

⅝" Liquid Line Pressure Drop = 1.7 p.s.i./100 Ft.

*(Courtesy of Copeland Corp.)*

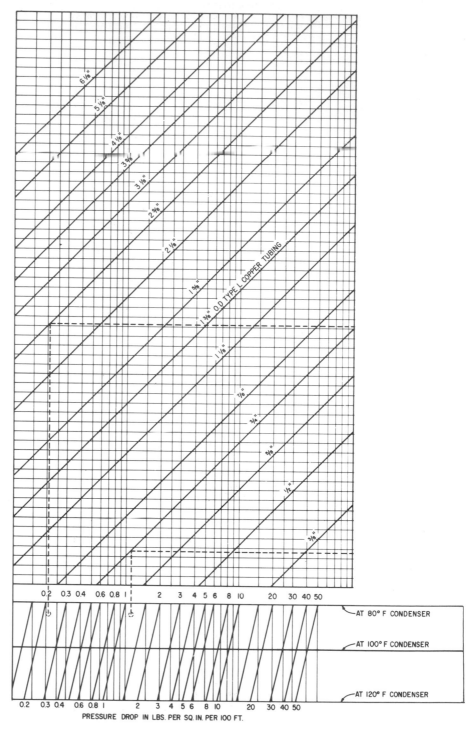

**Figure 14-55.** Freon–502 Refrigerant Pressure Drop in Lines (65° Evaporator Outlet)

PRESSURE DROP IN LBS. PER SQ. IN. PER 100 FT.

442

TONS OF REFRIGERATION

NOTE: Pressure drops do not allow for pulsating flow. If flow
is pulsating, use next larger pipe size.
Liquid line determined at 0°F evap. & 80°F. cond.
Discharge lines at 0°F evap. Other conditions do
not appreciably change result. Vapor at evap. outlet
assumed to be at 65°F.

EXAMPLE:

5.5 Tons at −40°F Evap. and 85°F Cond.

2 5/8" Suction Line Pressure Drop = 0.22 p.s.i./100 Ft.

3/4" Liquid Line Pressure Drop = 1.2 p.s.i./100 Ft.

(Courtesy of Copeland Corp.)

**COMPRESSOR CAPACITY (BTU/Hr.)**

Example: At 40° F. evaporating temperature and a system capacity of 100,000 BTU/hr, the maximum size tubing that can be used to insure oil return is 1 ⅝" O.D.

Line Size O.D.

3 ⅛
2 ⅝
2 ⅛
1 ⅝
1 ⅜
1 ⅛
⅞
⅝
½

**EVAPORATING TEMPERATURE (°F.)**

**Figure 14-56.** Maximum Recommended Vertical Discharge Line Sizes for Proper Oil Return R–12 (*Courtesy of Copeland Corp.*)

444

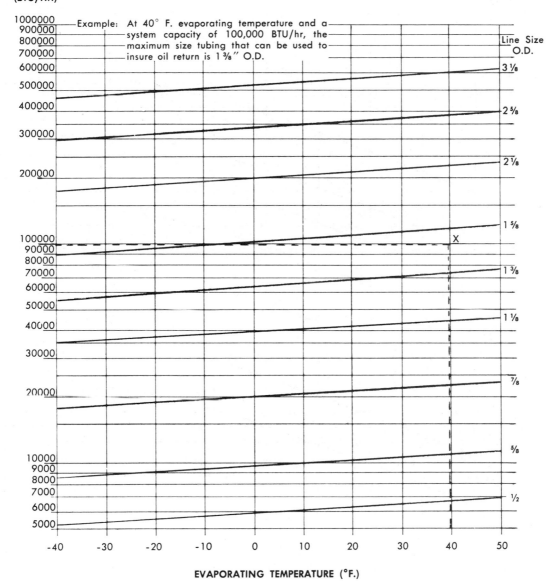

COMPRESSOR
CAPACITY
(BTU/Hr.)

Example: At 40° F. evaporating temperature and a system capacity of 100,000 BTU/hr, the maximum size tubing that can be used to insure oil return is 1⅜" O.D.

Line Size O.D.

**Figure 14-57.** Maximum Recommended Vertical Discharge Line Sizes for Proper Oil Return, R–22 and R–502 (*Courtesy of Copeland Corp.*)

445

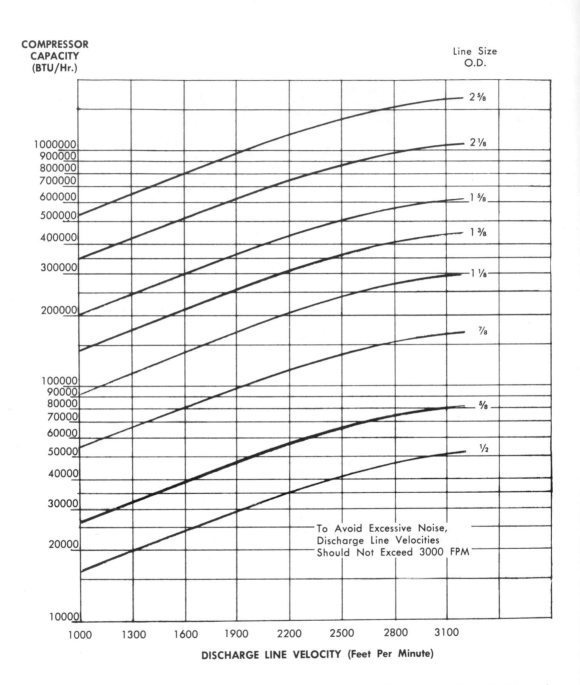

**Figure 14-58.** Discharge Line Velocities for Various Btu/Hour Capacities, R–22 and R–502 (*Courtesy of Copeland Corp.*)

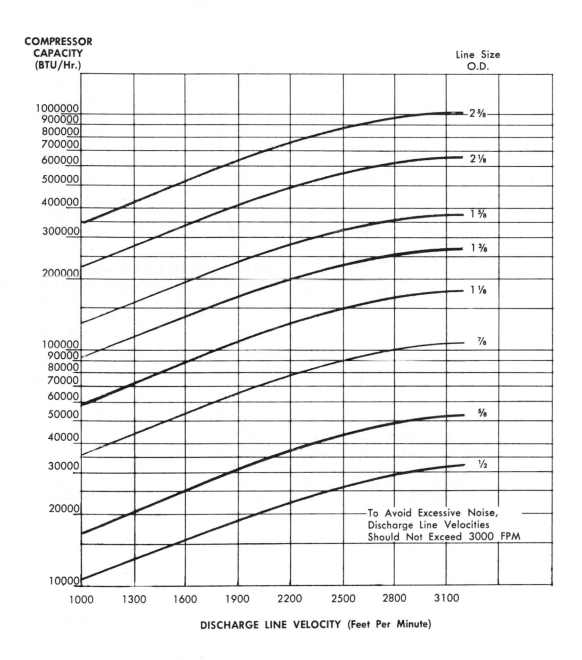

**Figure 14-59.** Discharge Line Velocities for Various Btu/Hour Capacities, R–12 (*Courtesy of Copeland Corp.*)

Oil circulation in discharge lines is normally a problem only on systems where large variations in system capacity are encountered. For example, an air conditioning system may have steps of capacity control allowing it to operate during periods of light load at capacities possibly as low as 25 percent or 33 percent of the design capacity. The same situation may exist on commercial refrigeration systems where compressors connected in parallel are cycled for capacity control. In such cases, vertical discharge lines must be sized to maintain velocities above the minimum velocity necessary to properly circulate oil at the minimum load condition.

For example, consider an air conditioning system using R–12 having a maximum design capacity of 300,000 Btus/hr (75,600 kcal) with steps of capacity reduction up to 66 percent. Although the 300,000 Btus/hr (75,600 kcal) condition could return oil up to a 2⅝ in. OD (66.67 mm) riser at light load conditions, the system would have only 100,000 Btus/hr (25,200 kcal) capacity, so a 1⅝ in. OD (41.27 mm) riser must be used. In checking the pressure drop chart (Figure 14-57) at maximum load conditions a 1⅝ in. OD (41.27 mm) pipe will have a pressure drop of approximately 4 psi (.28 kg/cm²) per 100 ft (30.4 m) at a condensing temperature of 120°F (49°C). If the total equivalent length of pipe exceeds 150 ft (45.6 m), in order to keep the total pressure drop within reasonable limits, the horizontal lines should be the next larger size or 2⅛ in. OD (53.97 mm) which would result in a pressure drop of only slightly over 1 psi (.0703 kg/cm²) per 100 ft.

Because of the flexibility in line sizing that the allowable pressure drop makes possible, discharge lines can almost always be sized satisfactorily without the necessity of double risers. If modifications are made to an existing system which result in the existing discharge line being oversized at light load conditions, the addition of an oil separator to minimize oil circulation will normally solve the problem.

One other limiting factor in discharge line sizing is that excessive velocity can cause noise problems. Velocities of 3000 feet per minute (fpm) (912 mpm) or more may result in high noise levels, and it is recommended that maximum refrigerant velocities be kept well below this level. Figures 14-58 and 14-59 give equivalent discharge line gas velocities for varying capacities and line sizes over the normal refrigeration and air conditioning range.

**SIZING LIQUID LINES:** Since liquid refrigerant and oil mix completely, velocity is not essential for oil circulation in the liquid line. The primary concern in liquid line sizing is to ensure a solid liquid column of refrigerant at the expansion valve. If the pressure of the liquid refrigerant falls below its saturation temperature, a portion of the liquid will flash into vapor to cool the remaining liquid refrigerant to the new saturation temperature. This can occur in a liquid line if the pressure drops sufficiently due to line friction or vertical lift.

Flash gas in the liquid line has a detrimental effect on system performance in several ways.

1.   It increases the pressure drop due to friction.

2.   It reduces the capacity of the flow control device.

3.   It can erode the expansion valve pin and seat.

4.   It can cause excessive noise.

5.   It can cause erratic feeding of the liquid refrigerant to the evaporator.

For proper system performance, it is essential that liquid refrigerant reaching the flow control device be subcooled slightly below its saturation temperature. On most systems, the liquid refrigerant is sufficiently subcooled as it leaves the condenser to provide for normal system pressure drops. The amount of subcooling necessary, however, is dependent on the individual system design.

On air-cooled and most water-cooled applications, the temperature of the liquid refrigerant is normally higher than the surrounding ambient temperature, so no heat is transferred into the liquid, and the only concern is the pressure drop in the liquid line. Besides the friction loss caused by the refrigerant flow through the piping, a pressure drop equivalent to the liquid head is involved in forcing liquid refrigerant to flow up a vertical riser. A vertical head of 2 ft (.6 m) of liquid refrigerant is approximately equivalent to 1 psi (.0703 kg/cm$^2$). For example, if a condenser or receiver in the basement of a building is to supply liquid refrigerant to an evaporator three floors above, or approximately 30 ft (9.12 m), a pressure drop of approximately 15 psi (1.05 kg/cm$^2$) for the liquid head alone must be provided for in system design.

On evaporative or water-cooled condensers where the condensing temperature is below the ambient air temperature, or on any application where the liquid lines must pass through hot areas such as boiler or furnace rooms, an additional complication may arise because of heat transfer into the liquid. Any subcooling in the condenser may be lost in the receiver or liquid line due to temperature rise alone unless the system is properly designed. On evaporative condensers when a receiver and subcooling coils are used, it is recommended that the refrigerant flow be piped from the condenser to the receiver and then to the subcooling coil. In critical applications, it may be necessary to insulate both the receiver and the liquid line.

On the typical air-cooled condensing unit with a conventional receiver, it is probable that very little subcooling of the liquid is possible unless the receiver is almost completely filled with liquid refrigerant. Vapor in the receiver in contact with the subcooled liquid will condense, and this effect will tend toward a saturated condition.

At normal condensing temperatures, a relation between each 1°F (.56°C) of subcooling and the corresponding change in saturation pressure applies. (See Table 14-7.)

To illustrate, 5°F (2.78°C) subcooling will allow a pressure drop of 8.75 psi (.615

### Table 14-7
### Relationship of Subcooling and Saturation Pressure

| REFRIGERANT | SUBCOOLING | EQUIVALENT CHANGE IN SATURATION PRESSURE |
|---|---|---|
| R–12 | 1°F (.56°C) | 1.75 psi (.12 kg/cm$^2$) |
| R–22 | 1°F (.56°C) | 2.75 psi (.19 kg/cm$^2$) |
| R–502 | 1°F (.56°C) | 2.85 psi (.2 kg/cm$^2$) |

* Courtesy of Copeland Refrigeration Corp.

kg/cm²) with R–12, 13.75 psi (.966 kg/cm²) with R–22, and 14.25 psi (1 kg/cm²) with (4.72°C) with R–12, 5.5°F (3.06°C) with R–502 without flashing in the liquid line. For the previous example of a condensing unit in a basement requiring a vertical lift of 30 ft (9.12 m), or approximately 15 psi (1.05 kg/cm²), the necessary subcooling for the liquid head alone would be 8.5°F (4.72°C) with R–12, 5.5°F (3.06°C) with R–22, and 5.25°F (2.92°C) with R–502.

The necessary subcooling may be provided by the condenser used, but for systems with abnormally high vertical risers a suction to the liquid heat exchanger may be required. Where long refrigerant lines are involved, and the temperature of the suction gas at the condensing unit is approaching room temperatures, a heat exchanger located near the condenser may not have sufficient temperature differential to adequately cool the liquid. Individual heat exchangers at each evaporator may be necessary.

In extreme cases, where a great deal of subcooling is required, there are several alternatives. A special heat exchanger with a separate subcooling expansion valve can provide maximum cooling with no penalty on system performance. It also is possible to reduce the capacity of the condenser so that a higher operating condensing temperature will make greater subcooling possible. Liquid refrigerant pumps also may be used to overcome large pressure drops.

Liquid line pressure drop causes no direct penalty in electrical power consumption, and the decrease in system capacity due to friction losses in the liquid line is negligible. Because of this, the only real restriction on the amount of liquid line pressure drop is the amount of subcooling available. Most references on pipe sizing recommend a conservative approach with friction pressure drops in the 3 to 5 psi (.21 kg/cm² to .35

kg/cm²) range. Where adequate subcooling is available, however, many applications have successfully used much higher design pressure drops. The total friction includes line losses through such accessories as solenoid valves, filter-driers, and hand valves.

In order to minimize the refrigerant charge, liquid lines should be kept as small as practical, and excessively low pressure drops should be avoided. On most systems, a reasonable design criterion is to size liquid lines on the basis of a pressure drop equivalent to 2°F (1.11°C) subcooling.

A limitation on liquid line velocity is possible damage to the piping from pressure surges or liquid hammer caused by the rapid closing of liquid line solenoid valves, and velocities above 300 fpm (91.2 mpm) should be avoided when they are used. If liquid line solenoid valves are not used, higher refrigerant velocities can be employed. Figure 14-60 gives liquid line velocities corresponding to various pressure drops and line sizes.

**SIZING SUCTION LINES:** Suction line sizing is more important than that of the other lines from a design and system standpoint. Any pressure drop occurring due to frictional resistance to flow results in a decrease in the refrigerant pressure at the compressor suction valve, compared with the pressure at the evaporator outlet. As the suction pressure is decreased, each pound of refrigerant that returns to the compressor occupies a greater volume, and the weight of the refrigerant being pumped by the compressor decreases. For example, a typical low temperature R–502 compressor operating at a −40°F (−40°C) evaporating temperature will lose almost 6 percent of its rated capacity for each 1 psi (.0703 kg/cm²) suction line pressure drop.

**Figure 14-60.** Liquid Line Velocities for Various Pressure Drops, R–12, R–22, R–502 *(Courtesy of Copeland Corp.)*

Table 14-8

Pressure Drop Equivalent for 2°F Change in Saturation
Temperature at Various Evaporating Temperatures*

| EVAPORATING TEMPERATURE | PRESSURE DROP, PSI (kg/cm²) | | |
| | R–12 | R–22 | R–502 |
| --- | --- | --- | --- |
| °F      (°C) | | | |
| 45°F (    7.2) | 2.0  (.14) | 3.0  (.21) | 3.3  (.23) |
| 20°F (— 6.7) | 1.35 (.09) | 2.2  (.15) | 2.4  (.17) |
| 0°F (—17.8) | 1.0  (.07) | 1.65 (.12) | 1.85 (.13) |
| —20°F (—28.9) | .75 (.05) | 1.15 (.08) | 1.35 (.09) |
| —40°F (—40  ) | .5  (.04) | .8  (.06) | 1.0  (.07) |

* Courtesy of Copeland Refrigeration Corp.

The normally accepted design practice is to use a design criterion of a suction line pressure drop equal to a 2°F (1.11°C) change in saturation temperature. Equivalent pressure drops for various operating conditions are developed and placed in tables. (See Table 14-8.)

The maintenance of adequate velocities to properly return the lubricating oil to the compressor is also of great importance when sizing suction lines. Studies have shown that oil is most viscous in a system after the suction gas has warmed up a few degrees higher than the evaporating temperature so that the oil is no longer saturated with refrigerant. This condition occurs in the suction line after the refrigerant vapor has left the evaporator. The movement of oil through the suction line is dependent on both the mass and the velocity of the suction vapor. As the mass or density decreases, higher refrigerant velocities are required to force the oil along.

Nominal minimum velocities of 700 fpm (212.8 mpm) in horizontal suction lines and 1500 fpm (456 mpm) in vertical suction lines have been recommended and used sucessfully for many years as suction line sizing design standards. The use of one nominal refrigerant velocity provided a simple and convenient means of checking velocities. However, tests have shown that in vertical risers the oil tends to crawl up the inner surface of the tubing, and the larger the tubing, the greater the velocity required in the center of the tubing to maintain tube surface velocities which will carry the oil. The exact velocity required in vertical suction lines is dependent on both the evaporating temperature and the size of the line. Under varying conditions, the specific velocity required might be either greater or less than 1500 fpm (456 mpm).

For better accuracy in line sizing, revised maximum recommended vertical suction line sizes based on the minimum gas velocities have been calculated and are plotted in chart form for easy use. (See Figures 14-61 and 14-62.) These revised recommendations supersede previous vertical suction riser recommendations. No change has been made in the 700 fpm (212.8 mpm) minimum velocity recommendation for horizontal lines. (See Figures 14-63 and 14-64.)

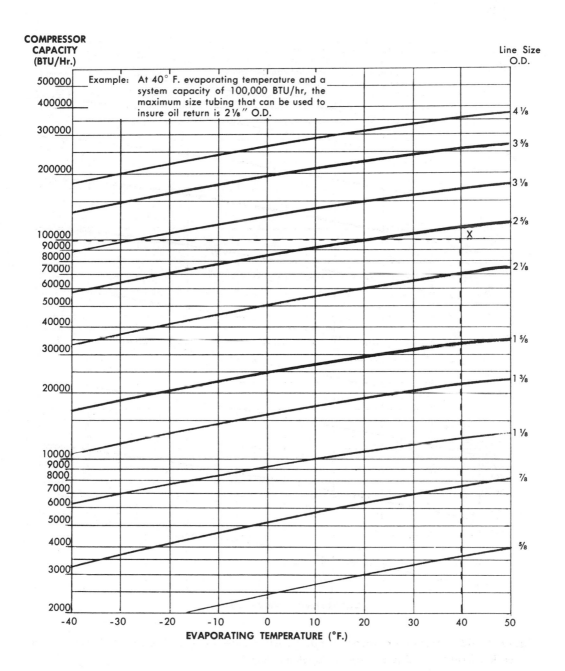

**COMPRESSOR CAPACITY (BTU/Hr.)**

Example: At 40° F. evaporating temperature and a system capacity of 100,000 BTU/hr, the maximum size tubing that can be used to insure oil return is 2⅛" O.D.

**Line Size O.D.**

4⅛
3⅝
3⅛
2⅝
2⅛
1⅝
1⅜
1⅛
⅞
⅝

**EVAPORATING TEMPERATURE (°F.)**

**Figure 14-61.** Maximum Recommended Suction Line Sizes for Proper Oil Return, Vertical Risers, R–12 (*Courtesy of Copeland Corp.*)

453

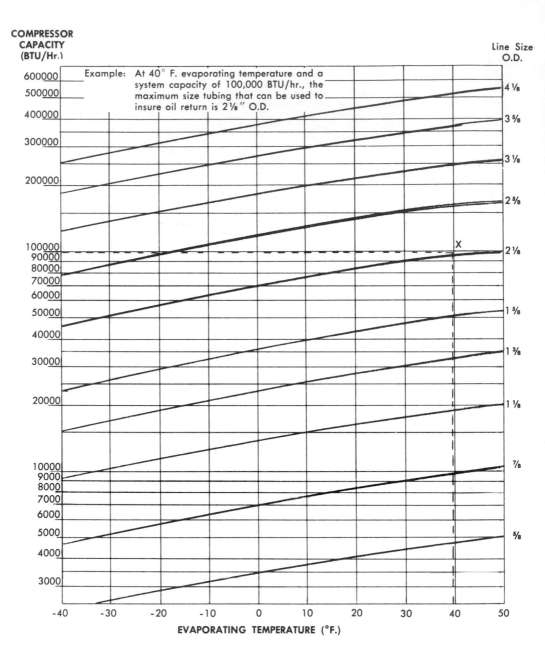

**Figure 14-62.** Maximum Recommended Suction Line Sizes for Proper Oil Return, Vertical Risers, R–22 and R–502 (*Courtesy of Copeland Corp.*)

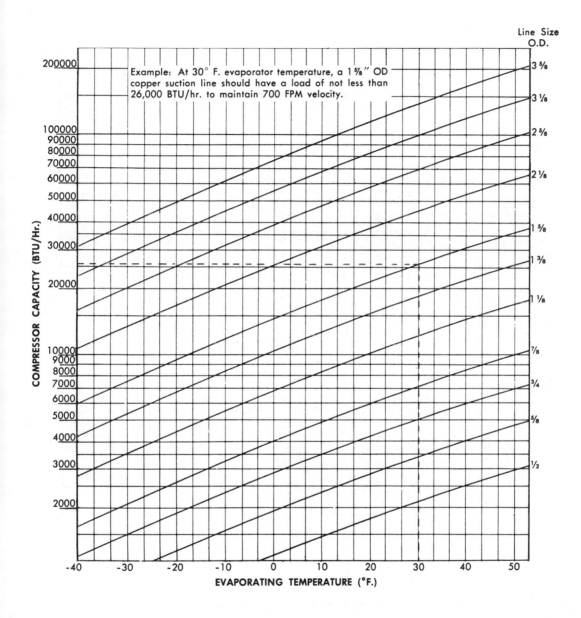

**Figure 14-63.** Maximum Recommended Horizontal Suction Line Sizes for Proper Oil Return, R–12 (*Courtesy of Copeland Corp.*)

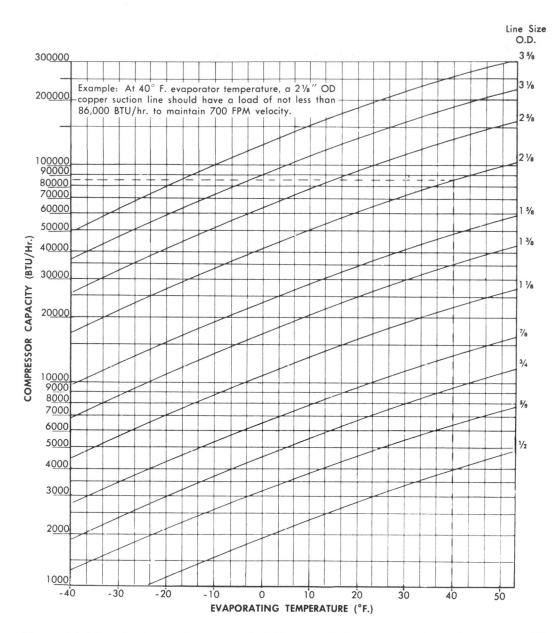

**Figure 14-64.** Maximum Recommended Horizontal Suction Line Sizes for Proper Oil Risers, R–22 and R–502 (*Courtesy of Copeland Corp.*)

To illustrate, again assume that a system operating with R–12 at a 40°F (4.4°C) evaporating temperature has a capacity of 100,000 Btu/hr (25,200 kcal). On Figure 14-61 the intersection of the evaporating temperature and capacity lines indicate that a 2⅛ in. (53.97 mm) OD line will be required for proper oil return in the vertical risers.

Even though the system might have a much greater design capacity, the suction line sizing must be based on the minimum capacity anticipated in operation under light load conditions after allowing for the maximum reduction in capacity from the capacity control, if used.

Since the dual goals of low pressure drop and high velocities are in direct conflict with each other, obviously some compromises must be made in both areas. As a general approach, in suction line design, velocities should be kept as high as possible by sizing lines on the basis of the maximum pressure drop that can be tolerated. In no case, however, should the gas velocity be allowed to fall below the minimum levels necessary to return the oil to the compressor. It is recommended that a tentative selection of suction line sizes be made on the basis of a total pressure drop equivalent to a 2°F (1.11°C) change in the saturated evaporating temperature. The final consideration always must be to maintain velocities adequate to return the lubricating oil to the compressor, even if this results in a higher pressure drop than is normally desirable.

**DOUBLE RISERS:** On systems that are equipped with capacity control compressors, or where tandem or multiple compressors are used with one or more compressors cycled off for capacity control, single-suction line risers may result in either unacceptably high or low gas velocities. A line that is sized properly for light load conditions may have too high a pressure drop at maximum load. If the line is sized on the basis of full load conditions, the velocities may not be adequate at light load conditions to move the oil through the tubing. On air conditioning applications where somewhat higher pressure drops at maximum load conditions can be tolerated without any major penalty in overall system performance, it is usually preferable to accept the additional pressure drop imposed by a single vertical riser. But on medium or low temperature applications where pressure drop is more critical, and where separate risers from individual evaporators are not desirable or possible, a double riser may be necessary to avoid an excessive loss of capacity.

The two lines of a double riser should be sized so that the total cross-sectional area is equivalent to the cross-sectional area of a single riser that would have both satisfactory gas velocity and an acceptable pressure drop at maximum load conditions (see Figure 14-65). The two lines are normally of different sizes, with the larger line being trapped as shown. The smaller line must be

**Figure 14-65. Suction Line Double Riser** (*Courtesy of Copeland Corp.*)

sized to provide adequate velocities and acceptable pressure drop when the entire minimum load is carried in the smaller riser.

In operation, at maximum load conditions, gas and entrained oil will be flowing through both risers. At minimum load conditions, the gas velocity will not be high enough to carry the oil up both risers. The entrained oil will drop out of the refrigerant gas flow, and accumulate in the $P$ trap, forming a liquid seal. This will force all of the flow up the smaller riser, thereby raising the velocity and assuring oil circulation through the system.

As an example, assume a low temperature system as follows:

| | | |
|---|---|---|
| Maximum capacity | 150,000 | (37,800 kcal) |
| Minimum capacity | 50,000 | (12,600 kcal) |
| Refrigerant | R–502 | |
| Evaporating Temperature | −40°F | (−40°C) |
| Equivalent length of horizontal piping | 125 ft | (38 m) |
| Vertical riser | 25 ft | ( 7.6 m) |
| Desired design pressure drop [equivalent to 2°F (1.11°C)] | 1 psi | (.0703 kg/cm²) |

A preliminary check of the R–502 pressure drop chart indicates for a 150-ft (45.6-m) run with 150,000 Btus/hr (37,800 kcal) capacity and a total pressure drop of approximately 1 psi (.0703 kg/cm²), a 3⅛ in. (79.37 mm) OD line is indicated. (See Figure 14-57.) At the minimum capacity of 50,000 Btus/hr (12,600 kcal) a 3⅝ in. (92 mm) OD horizontal suction line is acceptable (see Figure 14-66). However, Figure 14-64 indicates that a maximum vertical riser size is 2⅛ in. (53.97 mm) OD. Referring again to the pressure drop chart of Figure 14-57, the pressure drop for 150,000 Btus/hr (37,800 kcal) through a 2⅛ in. (53.97 mm) OD tubing is 4 psi (.28 kg/cm²) per 100 ft (30.4 m), or 1 psi (.0703 kg/cm²) for the 25-ft (7.6-m) suction riser. Obviously, either a compromise must be made in accepting a greater pressure drop at maximum load conditions, or a double riser must be used.

If the pressure drop must be kept to a minimum, then the size of the double riser must be determined. At maximum load conditions, a 3⅛ in. (79.37 mm) OD riser would maintain adequate velocities, so a

**Figure 14-66.** Suction Line Riser (*Courtesy of Copeland Corp.*)

combination of line sizes approximating the 3⅛ in. (79.37 mm) OD line can be selected for the double riser. The cross-sectional area of the line sizes to be considered are:

| | |
|---|---|
| 3⅛ in. (79.37 mm) | 6.64 sq. in. (42.82 cm²) |
| 2⅝ in. (66.67 mm) | 4.77 sq. in. (30.76 cm²) |
| 2⅛ in. (53.97 mm) | 3.10 sq. in. (19.99 cm²) |
| 1⅝ in. (41.27 mm) | 1.78 sq. in. (11.48 cm²) |

At the minimum load conditions of 50,-000 Btus/hr (12,600 kcal), the 1⅝ in. (41.27 mm) OD line will have a pressure drop of approximately .5 psi (.035 kg/cm²), and will have acceptable velocities, so a combination of 2⅝ in. (66.67 mm) OD and 1⅝ in. (41.27 mm) OD tubing should be used for the double riser.

In a similar manner, double risers can be calculated for any set of maximum and minimum capacities where single risers may not be satisfactory.

**SUCTION PIPING FOR MULTIPLEX SYSTEMS:** It is common practice in supermarket applications to operate several cases, each with a liquid line solenoid valve and expansion valve control, from a single compressor. Temperature control of individual cases is normally achieved by means of a thermostat which opens and closes the liquid line solenoid valve as necessary. This type of system, commonly called multiplexing, requires careful attention in design to avoid oil return problems and compressor overheating.

Since the cases fed by each liquid line solenoid valve may be controlled individually, and because the load on each case is relatively constant during operation, individual suction lines and risers are normally run from each case, or group of cases, which are controlled by a liquid line solenoid valve

for a minimum pressure drop and maximum efficiency in oil return. This provides excellent control so long as the compressor is operating at its design suction pressure. However, there may be periods of light load when most or all of the liquid line solenoid valves are closed. Unless some means of controlling compressor capacity is provided, this can result in the compressor short-cycling or operating at excessively low suction pressures. Either condition can result not only in overheating the compressor, but in reducing the suction pressure to a level where the gas becomes so rarified it can no longer return oil properly in lines sized for much greater gas density.

Because of the fluctuations in the refrigeration load caused by the closing of the individual liquid line solenoid valves, some means of compressor capacity control must be provided. In addition, the means of capacity control must be such that it will not allow extreme variations in the compressor suction pressure.

Where multiple compressors are used, cycling of the individual compressors provides satisfactory control. Where multiplexing is done with a single compressor, a hot gas bypass system has proven to be the most satisfactory means of capacity reduction. This system allows the compressor to operate continuously at a reasonably constant suction pressure while compressor cooling can be safely controlled by means of a de-superheating expansion valve.

In all cases, the operation of the system under all possible combinations of heavy load, light load, defrost, and compressor capacity must be studied carefully to be certain that operating conditions will be satisfactory.

Close attention must be given to piping design on multiplex systems to avoid oil return problems. The lines must be properly sized so that the minimum velocities necessary to return the oil are maintained in both horizontal and vertical suction lines under minimum load conditions. Bear in mind that although a hot gas bypass maintains the suction pressure at a proper level, the refrigerant vapor being bypassed is not available in the system to aid in returning the oil.

**PIPING DESIGN FOR HORIZONTAL AND VERTICAL LINES:** Horizontal suction and discharge lines should be pitched downward in the direction of flow to aid in oil drainage, with a downward pitch of at least ½ in. (12.7 mm) in 10 ft (3.04 m). Refrigerant lines should always be as short and should run as directly as possible.

Piping should be located so that access to system components is not hindered, and so that any components that could possibly require future maintenance are easily accessible. If piping must be run through boiler rooms or other areas where they will be exposed to abnormally high temperatures, it may be necessary to insulate both the suction and liquid lines to prevent excessive heat transfer into the lines.

Every vertical suction riser greater than 3 to 4 ft (.9 to 1.2 m) in height, should have a *P* trap at the base to facilitate oil return up the riser (see Figure 14-66). To avoid the accumulation of large quantities of oil, the trap should be of minimum depth and the horizontal section should be as short as possible.

Prefabricated wrought copper traps are available, or a trap can be made by using two street ells and one regular ell. Traps at the foot of hot gas risers are normally not required because of the easier movement of oil at higher temperatures. However, it is recommended that the discharge line from the compressor be looped to the floor prior to being run vertically upward to prevent the drainage of oil back to the compressor

**Figure 14-67.** Discharge Line Riser (*Courtesy of Copeland Corp.*)

head during shutdown periods. (See Figure 14-67.)

For long vertical risers in both suction and discharge lines, additional traps are recommended for each full length of pipe (approximately 20 ft—6.08m) to ensure proper oil movement.

In general, trapped sections of the suction line should be avoided except where necessary for oil return. Oil or liquid refrigerant accumulating in the suction line during the off cycle can return to the compressor at high velocity as a liquid slug on start-up, and can break the compressor valves or cause other damage.

**SUCTION LINE PIPING DESIGN AT THE EVAPORATOR:** If a pumpdown control system is not used, each evaporator must be trapped to prevent liquid refrigerant from draining back to the compressor by gravity during the off cycle. Where multiple evaporators are connected to a common suction line, the connections to the common suction line must be made with inverted traps to prevent drainage from one evapo-

Evaporator Trap.
Not Required With
Pumpdown System

Thermal Expansion
Valve Bulb Location

No Inverted Trap
Required Because
of Horizontal
Line Pitch

Inverted Traps

Pitch in Direction of Refrigerant
Flow—Minimum $\frac{1}{2}''$ in 10 ft

Trap to Protect Thermal
Expansion Valve Bulb
From Liquid Refrigerant

**Figure 14-68.** Typical Suction Line Piping with Multiple Evaporators (*Courtesy of Copeland Corp.*)

rator from affecting the expansion valve bulb control of another evaporator.

Where a suction riser is taken directly upward from an evaporator, a short horizontal section of tubing and a trap should be provided ahead of the riser so that a suitable mounting for the thermal expansion valve bulb is available. The trap serves as a drain area and helps to prevent the accumulation of liquid under the bulb which could cause erratic expansion valve operation. If the suction line leaving the evaporator is free-draining, or if a reasonable length of horizontal piping precedes the vertical riser, no trap is required unless necessary for oil return (see Figure 14-68).

## COMMERCIAL ICE MACHINES

There are many brands of commercial ice machines on the market. They are generally referred to as *flakers* and *cubers*. This reference is to the style of ice the machine produces. It is impossible to cover every brand and every type of ice machine in this text. Every manufacturer of ice machines will have a different operating sequence and sometimes more than one operating sequence is found on different models of the same brand. Therefore, this example of ice machine operation will deal only with the Frigidaire SCK and MCK models.

These ice machines are designed primarily for back bar installation, but can be used almost anywhere. (See Figure 14-69.) The storage bin opening is designed for easy ice cube removal. The storage bin door also is placed on an inclined plane which makes possible the full opening and closing of the door without the use of a latch. The ice bin will store approximately 70 lb (31.75 kg) of ice cubes. (See Figure 14-70.)

The unit is completely serviceable from the front. Both the front and rear machine compartment openings are the same size, permitting the use of an additional panel for free-standing installations.

**WATER RESERVOIR:** The stainless steel water reservoir is mounted on the cabinet walls of the rear right hand corner of the

**Figure 14-70.** Ice Bin *(Courtesy of Manitowac Equipment Company)*

ice bin storage compartment (see Figure 14-71).

**THERMOSTATS:** The ice thickness thermostat and the bin thermostat are located behind the access plate just above the grid on the right hand side of the bin storage compartment. The ice level in the bin can be varied by removing the two mounting screws in the ice bin thermostat well-mounting flange and rotating the well ½ turn.

**EXPANSION VALVE:** The expansion valve is located on the upper left hand wall of the cabinet liner at the side of the evaporator. The expansion valve bulb is attached to the suction line approximately in the center of the cabinet, under the evaporator.

**Figure 14-69.** Frigidaire SCK-11 Ice Machine *(Courtesy of Frigidaire Division, General Motors Corp.)*

**Figure 14-71.** Rear View of Ice Cuber with Panel Removed (*Courtesy of Manitowac Equipment Company*)

**MACHINE COMPARTMENT:** The condensing unit consists of the following components: compressor, air-cooled condenser, upright receiver, fan and motor, connecting refrigerant lines, hand shutoff valves, and a hot gas solenoid valve (see Figure 14-72). Each of these components is individually replaceable.

The entire condensing unit is easily removed through the front of the machine

**Figure 14-72.** Front View of Compressor Compartment (*Courtesy of Manitowac Equipment Company*)

compartment. The unit is fastened to the cabinet base by a hold-down clip in the rear and two hold-down bolts in the front.

The ¾-in. (19.05-mm) female drain and ¼-in. (6.4-mm) female pipe water supply connections are located in the upper left hand side of the cabinet shell to accommodate the lines to these connections.

**ELECTRICAL CONNECTIONS AND CONTROL BOX:** A choice of two openings is provided in the control box for the entrance of electrical supply lines—one in the left hand side of the box and one in the back. The two screw terminals on the terminal board are the line connections. The control box contains a terminal board, transformer, grid circuit fustat, starting relay, starting capacitor, and service switch (see Figure 14-73). The wiring diagram is located on the inside of the control box cover (see Figures 14-74 and 14-75).

**CYCLE OF OPERATION:** Water from the circulating pump in the reservoir cascades over the freezer plate or plates and freezes out until it builds up into a slab of the desired thickness. The ice thickness control then opens the electrical circuit to the water pump and energizes the hot gas solenoid valve. The hot gas from the receiver passes into the freezer plate and releases the ice slab which slides down onto the cutting grids. As soon as the water pump is shut off by the ice thickness control, all the water in the system drains back into the reservoir, resulting in a siphoning action which drains almost all the water from the reservoir. The reservoir then fills with fresh water for the next freezing cycle. After the ice slabs are released, a hinged mercury heater switch is tripped to add heat to the thickness control. The thickness control then opens the circuit to the hot gas solenoid and closes the circuit to the water pump to complete the cycle.

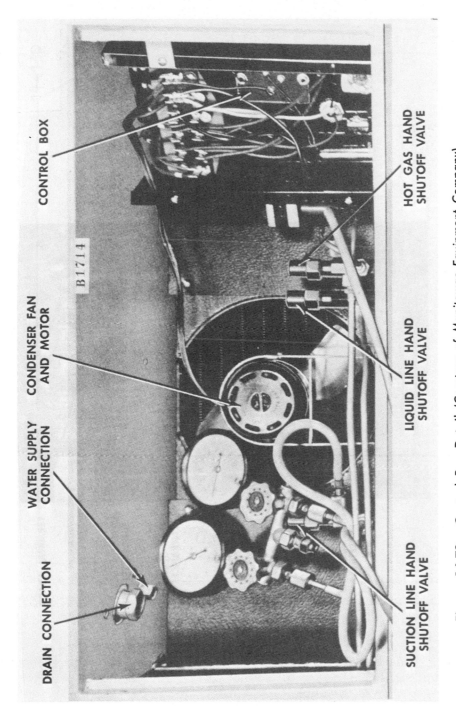

**Figure 14-73.** Control Box Detail (Courtesy of Manitowac Equipment Company)

CONTROL BOX

HOT GAS HAND
SHUTOFF VALVE

B1714

CONDENSER FAN
AND MOTOR

WATER SUPPLY
CONNECTION

LIQUID LINE HAND
SHUTOFF VALVE

DRAIN CONNECTION

SUCTION LINE HAND
SHUTOFF VALVE

465

**Figure 14-74.** Frigidaire SCK-11A Wiring Diagram (High Voltage) (*Courtesy of Frigidaire Division, General Motors Corp.*)

**Figure 14-75.** Frigidaire SCK-11A Wiring Diagram (Low Voltage) (*Courtesy of Frigidaire Division, General Motors Corp.*)

# Common Electrical Symbols

| | |
|---|---|
| Bell | ▭ ◯ |
| Buzzer | ▭ / |
| Convenience outlet duplex | ⊖ |
| Convenience outlet waterproof | ⊖ WP |
| Lighting panel | ▬ LP |
| Power panel | ▭ PP |
| Outlet, floor | ⊙ |
| Fixture, light (letter designates fixture type) | Ⓐ  or  Ⓐ  or  ▭ A |
| Motor | Ⓜ |
| Outlet, range | ⊖ R |
| Push button | ⊡ |
| Outlet, special purpose (see specs.) | ▲ |
| Switch, single pole | $ |
| Switch, double pole | $ |
| Switch, three-way | $₃ |
| Switch and convenience outlet | ⊖ S |
| Telephone | ◁ |
| Transformer | Ⓣ |

The compressor operates continuously until the storage bin thermostat is satisfied, at which time all electrical circuits are opened except for the low voltage circuit to the cutting grids which are energized as long as the ice machine is connected to the external electrical power supply.

## SAFETY PROCEDURES

A very good safety precaution is to check the unit before working on it to be certain that it is installed in accordance with local and national fire, electrical, plumbing, and building codes. The board of health has regulations that apply to all equipment used around food and beverages.

1. Check floor load limits before installing heavy commercial equipment.

2. Be sure not to trap liquid refrigerant inside a system without gas space; hydraulic pressure may build up and cause a rupture.

3. Obtain help when moving or lifting heavy cases and compressors.

4. Never purge burnt refrigerant into a building where food is stored since the food might absorb the burnt refrigerant and spoil.

5. Never use oxygen to test for leaks because oxygen and oil form an explosive mixture.

6. Be sure to disconnect the electrical service before attempting to work on the electrical components.

7. Always provide plenty of ventilation when welding, soldering, or brazing on refrigerant lines.

8. Wear goggles when charging or purging refrigerant from a system.

9. Wear rubber gloves when working with acidic materials.

10. Keep hands, feet, and clothing away from belts and moving parts.

## SUMMARY

• Commercial refrigeration units are those used in business for the preservation of perishable merchandise.

• A quality commercial refrigeration installation must provide satisfactory service economically.

• It is necessary that the refrigeration unit be sized properly.

• Commercial refrigeration cabinets are designed and constructed to fit the needs of the user.

• Walk-in coolers generally are larger than 100 cubic feet ($2.83 \text{ m}^3$) in size.

• Foam-in-place insulation is used to improve refrigeration by a foam-sandwich type construction.

• Reach-in cabinets generally have between 20 and 100 cubic feet ($.566 \text{ m}^3$ and $2.83 \text{ m}^3$) storage space.

• Reach-in cabinets are generally designed with glass doors to permit effective display of merchandise when desirable.

• Reach-in cabinets have a variety of uses such as: bottled beverage storage, dairy product storage, food storage, and dough storage in bakeries.

• Florist's cabinets are designed for display. They have glass doors and often have glass ends.

• Florist's cabinet temperatures range from 47°F to 54°F (8.3°C to 12.2°C).

• Display cases are used almost exclusively for the preservation and display of meats and dairy products.

- Double-duty cases have two compartments: an upper compartment for display and a lower compartment for storage.

- Foods displayed in the open type self-service display case are stored below a blanket of cold air.

- The main problem with open self-service cases is that of providing the additional refrigeration capacity required to take care of the greater loss of cold air.

- One sure way to determine whether or not the evaporator is fully refrigerated is to observe the refrigerant line from the expansion valve to the evaporator and the suction line leading from the evaporator. Both must be frosted.

- It is recommended that a heat exchanger be installed on cases maintained below 40°F (4.4°C) to obtain the best results.

- The commercial refrigeration cycle is basically the same as the refrigeration system used in domestic refrigerators.

- Extreme caution must be exercised to prevent the entrance of moisture into the refrigeration system.

- Low temperature single-stage refrigeration systems become increasingly critical from a design and application standpoint as the desired evaporating temperature is decreased.

- In an effort to prevent the discharge gas and motor temperatures from exceeding the recommended limits, it is desirable and in some cases absolutely necessary to insulate the suction lines which will allow the suction gas to return to the compressor at a lower than normal temperature.

- Some compressor manufacturers recommend that R–502 refrigerant be used in all single-stage low temperature applications where the evaporator temperature is −20°F (−28.9°C) or below.

- The refrigerant lines must be designed and installed to prevent oil trapping.

- When large refrigerant charges are unavoidable, a recycling pumpdown control should be used.

- To prevent possible permanent damage to the motor windings, the compressor motor must not be operated while the high vacuum pump is in operation.

- Two-stage low temperature units are more efficient and encounter fewer operating difficulties at lower operating temperatures than single-stage units.

- As the compression ratio increases, the volumetric efficiency decreases.

- Two-stage compressors generally use an external manifold and a de-superheating expansion valve.

- Below compression ratios of 7 to 1, the single-stage compressor will have greater capacity; however, two-stage compressors are more efficient above these ratios.

- De-superheating expansion valves are used to keep low temperature compressors cool.

- The function of the subcooler is to cool the liquid refrigerant being fed to the evaporator by the evaporation of refrigerant which is fed through the de-superheating expansion valve.

- The proper performance of any refrigeration or air conditioning system is greatly dependent on the proper charge of refrigerant.

- Never charge liquid refrigerant into the compressor suction or discharge valve ports because the compressor valves may be damaged.

• Vapor charging is normally used when only small amounts of refrigerant are to be added to a system.

• The refrigerant charge can be more precisely controlled by using vapor charging rather than liquid charging.

• Crankcase pressure-regulating valves (CPR) are installed in the suction line ahead of the compressor suction valve and control the maximum pressure at the compressor suction.

• The function of the condenser pressure-regulating (CPR) valve is to maintain adequate high side pressure during low ambient temperature.

• The CPR valve is located between the condenser and receiver.

• High to low side bypass valves are installed between the system high and low sides and respond to the valve outlet pressure. The valve opens when the low side pressure is reduced to the valve setting.

• Bypass control valves function as condenser bypass devices on air-cooled condensers with a winter start control.

• The simplest form of compressor capacity control is on–off operation of the compressor.

• Cylinder unloaders are used on large equipment operating under a fluctuating load condition.

• Hot gas bypass capacity control is recommended where normal compressor cycling or the use of unloaders may not be satisfactory.

• The most positive and dependable method of preventing refrigerant migration is to use a system pumpdown cycle.

• Since proper refrigeration of a case is dependent upon air circulation, it is necessary to defrost the evaporator periodically so that refrigeration may continue.

• Off-cycle defrost uses a low pressure control to control the compressor operation.

• Timed defrost systems use a defrost control, usually a time clock, to provide the desired time and duration of the defrost cycle.

• When supplemental heat is required for defrost, some cases are equipped with supplemental electric heat strips and a defrost limit switch.

• The methods used in defrost termination are: (1) straight-time termination, (2) pressure termination, and (3) solenoid reset termination.

• Demand defrost is accomplished by use of solid state components in two ways: (1) air velocity and (2) the fluidic principle.

• Pressure controls are used to control the compressor in response to the suction or discharge.

• Defrost methods are named by their mode of defrost initiation and termination.

• The duration of defrost must be long enough to clear the evaporator completely each defrost period, while defrost must occur often enough to keep the refrigerated air circulating freely throughout the case.

• Basically, combining is the grouping of more than one refrigeration function on one compressor.

• A simple system is two or more cases of the same suction pressure and defrost requirements connected to one condensing unit.

• A combined system involves cases of different suction pressures and/or defrost requirements on one condensing unit.

• Complex combined systems are designed to handle all low temperature (or all normal temperature) cases and coolers on one condensing unit.

• Ultracomplex combined systems include all of the refrigeration equipment in the store on one condensing unit.

• To separate cases and coolers of different temperatures there are two commonly used methods: (1) liquid line solenoid valves and (2) evaporator pressure regulating (EPR) valves.

• Heat recovery systems take the heat extracted from the refrigerated cases and return it to the store.

• The three approaches to heat recovery are: (1) direct-air method, (2) closed-circuit water method, and (3) dual-condenser method.

• Refrigerant line pressure drop tends to decrease system capacity and increase the amount of electrical power required by the compressor.

• Compressor lubricating oil can be circulated through the system only if the velocity of the refrigerant vapor is great enough to carry the oil along with it.

• Normally oil separators are necessary for operation at evaporating temperatures below $-50°F$ ($-45.6°C$) in order to minimize the amount of oil in circulation.

• Each valve, fitting, and bend in a refrigerant line contributes to the friction pressure drop because of its interruption or restriction of smooth flow and must be figured as part of the equivalent length of pipe.

• Pressure drop in discharge lines is probably less critical than in any other part of the system.

• The primary concern in liquid line sizing is to ensure a solid liquid column of refrigerant at the expansion valve.

• Liquid line pressure drop causes no direct penalty in electric power consumption, and the decrease in system capacity due to friction losses in the liquid line is negligible.

• The normally accepted design practice in sizing suction lines is to use a design criterion of a suction line pressure drop equal to $2°F$ ($1.11°C$) change in saturation temperature.

• Nominal minimum velocities of 700 fpm (212.8 mpm) in horizontal suction lines and 1500 fpm (456 mpm) in vertical suction lines have been recommended and used for years as suction line sizing design standards.

• The two lines of a double riser should be sized so that the total cross-sectional area is equivalent to the cross-sectional area of a single riser that would have both satisfactory gas velocity and an acceptable pressured drop at maximum load conditions.

• Close attention must be given to piping design on multiplex systems to avoid oil return problems.

• Horizontal suction and discharge lines should be pitched downward in the direction of flow to aid in oil drainage.

• If a pumpdown control system is not used, each evaporator must be trapped to prevent liquid refrigerant from draining back to the compressor by gravity during the off cycle.

• Commercial ice machines are generally referred to as flakers and cubers.

• Periodic maintenance on ice machines is necessary for satisfactory performance.

• Water from the water reservoir is circulated over the evaporator where it freezes to the desired thickness.

• The ice machine compressor operates continuously until the storage bin thermostat is satisfied.

## REVIEW EXERCISES

1. Define commercial refrigeration units.

2. Why is it necessary that the refrigeration unit be sized properly?

3. What size cabinet is considered to be a walk-in cooler?

4. For what are the glass doors used on reach-in boxes?

5. For what are reach-in cabinets used?

6. What should the temperature range be in a florist's cabinet?

7. What type of case has a storage area and a display area?

8. How are foods stored in open type self-service cases?

9. What has proven to be the main problem in open type self-service cases?

10. What should the temperature be in a frozen food display case?

11. What should be the thickness of the cold air blanket which separates the frozen food from the ambient air?

12. What is the one sure way to determine whether or not the evaporator is fully refrigerated?

13. How will moisture probably be noticed in low temperature refrigeration systems?

14. How is a decreased evaporating temperature hazardous to single-stage low temperature compressors?

15. When should compressor motor windings be tested?

16. Below what temperature do most compressor manufacturers recommend that R–502 be used in single-stage low temperature applications?

17. Where should the filter-drier be installed in the liquid line on commercial cases?

18. Why should the compressor motor not be operated when it is under a high vacuum?

19. Why are two-stage low temperature systems more efficient than single-stage low temperature systems?

20. What are the two factors that cause a loss of efficiency with an increase in compression ratio?

21. Why are two-stage compressors used in low temperature applications?

22. What is the function of a liquid sub-cooler?

23. What does a starved evaporator indicate?

24. Why should liquid refrigerant not be fed into the compressor suction?

25. Is liquid charging or vapor charging more accurate?

26. What will a higher than normal discharge pressure indicate?

27. What is the most common method of determining whether or not a system is properly charged?

28. What is the purpose of the crankcase pressure-regulating valve?

29. What is the purpose of the evaporator pressure-regulating valve?

30. Where is the condenser pressure-regulating valve located?

31. Can the high to low side bypass valve be used for capacity reduction?

32. When are bypass control valves for air-cooled condensers used?

33. When are compressor cylinder unloaders most likely to be used?

34. Name the methods used in hot gas bypass capacity control.

35. Under what conditions is a pumpdown control recommended?

36. Why do refrigerated cases need a defrost period?

37. Name the basic methods used to defrost evaporators in commercial cases.

38. What are the three settings on a dual pressure control?

39. How are defrost methods named?

40. Define a combined system.

41. What are the two methods used in maintaining case temperature differences?

42. Where is the heat obtained for a heat recovery system?

43. How can oil logging of an evaporator be prevented on a low temperature system?

44. What are the two main considerations given to refrigerant piping design?

45. What are the maximum desirable refrigerant velocities in discharge lines?

46. What is the main consideration when sizing liquid lines?

47. Why is the sizing of the suction line more important that the sizing of the other lines?

48. How should horizontal suction and discharge lines be installed?

49. How are commercial ice machines referenced?

50. When do ice machine compressors stop running?

# Air Conditioning (Heating)

The term *air conditioning* has for many years been misunderstood. Air conditioning involves treating all of the properties of air. Total air conditioning involves: circulating, cooling, heating, humidifying, dehumidifying, and cleaning the air. Therefore, heating is only a portion of total air conditioning.

The heating process involves air circulation, heating, humidification, and cleaning. The other processes are accomplished during the cooling process.

## HEAT SOURCES

A heat source is any substance that produces heat while going through a given process. The main sources of heat are in the form of solids, liquids, gases, electricity, and solar energy. The value of a source of heat is derived from the amount of heat which is released when it is processed. A good heat source must be readily obtainable in large quantities at relatively low prices.

**SOLIDS:** The main solid used for heating purposes for the general public is coal. Coal is obtained by one of two methods: open-pit and deep mining. Coal is divided into four classifications: anthracite, bituminous, subbituminous, and lignite. The anthracite and bituminous types are classified according to their carbon content when they are dry. Subbituminous, and lignite are classified according to their calorific value.

Anthracite and bituminous coals are the most popular types used for heating purposes. Lignite coal is used in electric power generating plants in some areas of the country. Anthracite coal will release from 13,000 to 14,000 Btus/lb (7150 to 7700 kcal/kg) when burned with 9.6 lb (4354.56 g) of air. Bituminous coal will release from 12,000 to 15,000 Btus/lb (6600 to 8250 kcal/kg) when burned with 10.3 lbs (4672 g) of air. Dry air at about 70°F (21.1°C) has a volume of 13.3 ft³/lb (.376 m³/g).

Coal is burned by use of several methods.

Hand firing is the oldest method. The coal is thrown onto the grates through the opening through which the primary air is drawn into the firebox. The fuel bed consists of a layer of ashes lying directly on the grates (a hot zone where the combustion process occurs), a cooler (distillation) zone where the gases are driven from the coal, and then a layer of green coal. (See Figure 15-1.) There are two reasons why it is necessary to supply secondary air over the fire to obtain complete combustion: (1) hot carbon and carbon dioxide react to form combustible carbon monoxide and (2) there is a distillation of gases from the distillation zone.

During the hand-firing process, one method was to put the green coal in the front of the firebox and push the hot coals to the back. The secondary air entering the firebox would carry the gases over the hot coals, causing them to burn more completely than if the green coal were placed on the top.

The next method used in firing coal as a fuel was the mechanical firing process. During this process, coal and air were fed simultaneously into the firebox. The coal was pushed into the firebox by means of a rotating worm or screw. The hotter part of the fire was either pushed to the rear or to the top of the firebox, depending on whether front or bottom feed was used.

**Figure 15-1. Coal Bed**

**LIQUIDS:** Fuel oil is the most common liquid being used as a heat source. Fuel oil is a petroleum product and is made up of a mixture of liquid hydrocarbons; it is produced as a by-product of the petroleum refining process. Petroleum has been known for thousands of years. Seepages of oil and gas around the Caspian and Black Seas were known and used for heating and cooking before the birth of Christ. The Chinese people drilled for oil long before the Christian era.

Fuel oil is a strong competitor of natural gas in the heating industry. It is graded and classified according to the range of distillation. The grades range from 1 to 6, omitting number 3. Grades 1, 2, and 4 are used for heating, with grades 1 and 2 being the most popular for domestic use. Number 4 is generally used in light industrial furnaces. The lower graded numbers are more expensive than the higher numbered fuel oils because of fewer impurities, such as asphalt. Grades 5 and 6 are too thick for use in domestic oil-burning equipment and require preheating to ensure a steady flow to the burner. The specifications which govern fuel oils are set forth by the U.S. Department of Commerce and conform to American Society for Testing Materials (ASTM) specifications.

The proper combustion of fuel oils can be accomplished only when the oil is prop-

Table 15-1

Heat Ratings of Fuel Oil

| GRADE | BTU | K CALORIES |
|-------|-----|------------|
| No. 1 | 132,900–137,000 | 33,490.8–34,524 |
| No. 2 | 135,800–141,800 | 34,221.6–35,733.6 |
| No. 4 | 143,100–148,100 | 36,061.2–37,321.2 |
| No. 5 | 146,800–150,000 | 36,993.6–37,800 |
| No. 6 | 151,300–155,900 | 38,127.6–39,286.8 |

erly atomized and mixed with air. The heat given off by burning fuel oil ranges from 137,000 to 151,000 Btus/gal (34,524 to 38,052 kcal/3.7 l) depending on the grade. The heat content of number 1 fuel oil is 137,000 Btus/gal (34,524 kcal/3.7 l), and number 2, the most popular domestic fuel oil, is 140,000 Btus/gal (35,280 kcal/3.7 l). (See Table 15-1.) The flash points of fuel oils will vary considerably because of the refining methods.

The storage of fuel oil is a contributing factor in the efficient operation of the heating system. In cold climates, the oil should be stored indoors or in some type of heating device. Steam or hot water pipes around the tank or electric emersion heaters can be used to preheat the oil before entering the combustion area. The fuel oil may be stored outdoors in milder climates without the problem of having it become excessively thick.

The use of fuel oil for domestic purposes is a result of the convenience and cleanliness of oil as compared to coal. Despite the competition of natural gas, fuel oil is still a leader as a source of heat in heating equipment, especially in the northeastern sections of the United States.

**GASES:** There are two major classifications of gaseous fuels used in heating equipment today. They are natural gas and liquified petroleum (LP) gas. Natural gas is distributed by means of a pipeline system, while LP gas is usually transported by truck and stored where it is to be used.

**Natural Gas:** This is the lightest of all petroleum products. It is generally found where oil is found; however, in some cases oil is not present. Theorists have long argued about the exact origin of natural gas. However, most agree that natural gas was formed during the decomposition of plant and animal remains that were buried in prehistoric times. Because these plants and animals lived during the same period as those that are presently found as fossils, natural gas is sometimes called a fossil fuel.

The gas industry may be broken down into the following areas: exploration, production, transmission, and distribution.

The exploration section of the industry actually finds the areas where the gas is located. The people who work in exploration examine new areas and make the necessary reports, purchases, and perform other essential duties prior to the actual drilling.

When all the exploration functions are completed, the production department does the actual drilling of the well. The gas, or crude oil, is brought to the earth's surface and is blocked at this point. The well remains in this state until the gas or oil is needed.

When the need arises, the transmission department receives the gas from the well at pressures from 500 to 3000 psig (35.15 to 210.9 kg/cm²). Even with these high pressures, the resistance given by the pipe and the distance to be covered necessitate the use of booster pumps to transfer the gas from the well to the refinery. At the refinery, the gas passes through a drying process that removes moisture, propane, and butane. During this process, most of the odor is also removed from the raw gas. An odorant is therefore added to aid in leak detection.

Natural gas has a specific gravity of .65, an ignition temperature of 1100°F (593°C), and a burning temperature of 3500°F (1909°C). One cubic foot (.0283 m³) of natural gas will emit from 900 to 1400 Btu/ft³ (226.8 kcal/m³ to 352.9 kcal/m³) with most of the gas emitting approximately 1100 Btus/ft³ (277.2 kcal/m³). The Btu (calorie) content of natural gas will vary from area to area. The local gas company should be consulted when the exact Btu content is desired.

Natural gas is made up of 55 to 90 percent methane ($CH_4$), 1 to 14 percent ethane ($C_2H_6$), and .5 percent carbon dioxide ($CO_2$). It requires 15 ft³ (.4245 m³) of air per ft³ (.0283 m³) of gas for proper combustion. It is lighter than air. Because methane and ethane have such low boiling points —methane −258.7°F (− 160°C) and ethane −127°F (−89°C)—natural gas remains a gas under the pressures and temperatures encountered during distribution. Because of the varying amounts of methane and ethane, the boiling point of natural gas will vary according to the mixture.

**Liquified Petroleum (LP):** Liquified petroleum is either butane or propane, and in some cases a mixture of the two. These two fuels are refined natural gases and were developed for use in rural areas.

When these fuels are extracted from raw gas at the refinery, they are in the liquid state, under pressure, and remain in this state during storage and transportation. Only after vapor is drawn from above the liquid does the liquified petroleum become a gaseous fuel.

LP gas has at least one definite advantage in that it can be stored in the liquid state and, thus, the heat content is concentrated. This concentration of heat makes it economically feasible to provide service anywhere that portable cylinders may be used. The fact that 1 gal (3.785 l) of liquid propane turns to 36.31 ft³ (1.02 m³) of gas when it is evaporated illustrates its feasibility.

When LP gas is stored in a tank, it is in both the liquid and gaseous state (see Figure 15-2). To make the action of LP gas more easily understood, let's review briefly the boiling point and pressures of water (see Figure 15-3).

When the pressure cooker is first filled with water with no heat applied and the top remaining off, there is no pressure on the surface of the water. (See Figure 15-3(a).) However, when the top is in place and heat is applied, as in Figure 15-3(b), the pressure will begin to rise after the boiling point of the water is reached. As more heat is ap-

**Figure 15-2.** LP Gas Storage Tank

70° F     0# psi     Water 70° F

70° F     Steam     15# psi     Water 250° F

(a)          (b)

**Figure 15-3.** Pressure Cooker

plied, more pressure will be created above the water by the evaporating liquid. When a constant temperature is maintained, a corresponding pressure will also be maintained. Likewise, when the pressure is reduced, the boiling point is reduced.

If we apply this principle to LP gas stored in a tank, it can readily be seen that as vapor is withdrawn from the tank, more liquid will evaporate to replace that which was withdrawn. It must be remembered that each liquid has its own boiling point and pressure.

Butane ($C_4H_{10}$), like propane ($C_3H_8$) and natural gas, is placed in the hydrocarbon series of gaseous fuels, because it is comprised of hydrogen and carbon. Butane has a boiling point of 31.1°F ($-0.6$°C), a specific gravity of 2, a heating value of 3267 Btus/ft³ (823.3 kcal/m³) of vapor. It requires 30.97 ft³ (.87 m³) of air per ft³ (m³) of vapor for proper combustion. At sea level, it has a gauge pressure of 36.9 lb (2.59 kg/cm²) at 100°F (37.8°C). Butane expands to 31.75 ft³ (.898 m³) of vapor per gal (3.785 l) of liquid. It is heavier than air. The ignition temperature is approximately 1100°F

(593°C) and the burning temperature is 3300°F (1798°C).

Propane has a boiling point of 43.8°F (6.6°C), a specific gravity of 1.52, a heating value of 2521 Btus/ft³ (635.29 kcal/m³) of vapor. It requires 23.82 ft³ (.67 m³) of air per ft³ (m³) of vapor for proper combustion. At sea level, it has a gauge pressure of 175.3 lb (12.32 kg/cm²) at 100°F (37.8°C). Propane expands to 36.35 ft³ (1.02 m³) of vapor per gal (3.785 l) of liquid. It is also heavier than air. It has an ignition temperature of 1100°F (593°C) and a burning temperature of 2975°F (1616.9°C).

When we study the physical properties of LP gases, we can see that each has both good and bad qualities. These qualities should be given a great deal of consideration when determining which fuel is to be used for any given application. The two characteristics deserving the most consideration are the Btu (cal) content and vapor pressure. See Table 15-2 for a comparison of vapor pressures.

When considering these fuels, the pressure is the major limiting factor, especially

## Table 15-2

### LP Gas Vapor Pressures psi (kg/cm²)

| TEMP °F (C°) | PROPANE | BUTANE | TEMP °F (C°) | PROPANE | BUTANE |
|---|---|---|---|---|---|
| 0 (−17.8) | 38.2 (2.68) | | 70 (21.1) | 124 ( 8.7 ) | 16.9 (1.2 ) |
| 10 (−12.2) | 46 (3.2 ) | | 80 (26.7) | 142.8 (10 ) | 22.9 (1.6 ) |
| 20 (− 6.7) | 55.5 (3.9 ) | | 90 (32.2) | 164 (11.53) | 29.8 (2.1 ) |
| 30 (− 1.1) | 66.3 (4.66) | | 100 (37.8) | 187 (13.15) | 37.5 (2.64) |
| 40 ( 4.4) | 78 (5.48) | 3 (.21) | 110 (43 ) | 212 (14.9 ) | 46.1 (3.24) |
| 50 ( 10 ) | 91.8 (6.45) | 6.9 (.49) | 120 (49 ) | 240 (16.87) | 56.1 (3.94) |
| 60 ( 15.6) | 107.1 (7.53) | 11.6 (.82) | 130 (54 ) | 272 (19.12) | 66.1 (4.65) |

in colder climates. As we study the table, we can see that when the temperature of liquid butane reaches 30°F (−1.1°C) or lower, there is no pressure in the tank. Therefore, butane would not be a suitable fuel at lower temperatures without some source of heat for the storage tank. This source of heat may be steam or hot water pipes around the tank, electrical heaters around the tank, or even burial of the tank in the ground. However, these all add to the initial cost of the equipment.

If we look at the pressures of propane, we see that they are suitable throughout a wide range of temperatures. Therefore, from the pressure standpoint, propane would be the ideal fuel. On the other hand, its low Btu (cal) rating makes it less desirable than butane. To overcome this dilemma, the two fuels may be mixed to obtain some of the desirable characteristics of each gas. An example of this may be a mixture of 60 percent butane and 40 percent propane. At a temperature of 30°F (−1.1°C), it will have a vapor pressure of approximately 24 psig (1.68 kg/cm²) and a heat content of 2950 Btus/ft³ (734.4 kcal/m³). Since these fuels are usually mixed before delivery to the local distributor, it is difficult to deter-

mine exactly the tank pressure and Btu/ft³ (calories/m³). As long as there is enough pressure in the storage tank to allow 11 in. (279.4 cm) water column of pressure to enter the house piping, there is little or nothing the service technician can do.

Before getting too involved in working with LP gases, the state and local authorities should be consulted. Some states maintain strict control over the personnel working with these fuels.

**ELECTRICITY:** Electricity is becoming more popular as a source of heat for heating applications. The early applications of the heat-producing ability of electricity were limited. It was used only in a few specialized areas, mainly in industrial processes and in portable heaters to help supplement inefficient heating systems. Today, electric heating is adaptable to almost any type of construction and any climate.

The major methods of using electricity for heating applications are: resistance heating, heat pump, and a combination of the two. There are three types of resistance elements used for electric heating. They are (1) the open wire, (2) the open ribbon, and (3) the tubular cased wire (see Figure

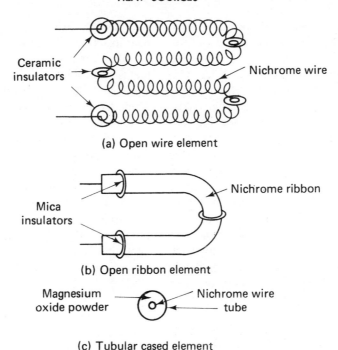

(a) Open wire element

(b) Open ribbon element

(c) Tubular cased element

**Figure 15-4.   Resistance Elements**

15-4). An electric resistance heating element may be defined as: an assembly consisting of a resistance wire, insulated supports, and terminals for connecting the electrical supply wire to the resistance wire. Resistance heating will convert electrical energy to heat energy at the rate of 3412 Btus/kW (859.8 kcal/kW). Theoretically, electrical heating is 100 percent efficient, that is, for each Btu (cal) input to the heating equipment, 1 Btu (cal) in usable heat is recovered.

**Open Wire Heating Elements:** Open wire heating elements are usually made of *nichrome* wire, which is wire made from nickel and chromium, without iron, wound in a springlike shape and mounted in ceramic insulators to prevent electrical shorting to the metal frame. The open wire elements have a longer life span than the others because they operate cooler since they release all the heat directly into the air stream. They also have a lower air pressure drop as compared to the other types of elements.

**Open Ribbon Heating Elements:** The open ribbon elements are also made of nichrome wire and are insulated in the same manner as the open wire elements. The flat design allows more intimate contact between the wire and the air because it has more surface area. Its efficiency is thereby increased over the open wire element. However, due to the increase in manufacturing costs, the open wire element is more popular.

**Tubular-Cased Heating Element:** The tubular-cased heating element is the same type as that used in cooking stoves. The nichrome wire is placed inside a tube and insulated from it by magnesium oxide powder. Thus, tubular-cased heating elements do not require external insulation as do the other two resistance elements. They are also less efficient because of the energy loss caused by the extra material that the heat must pass through before reaching the moving air. They are, however, safer to use because of the interior insulation which is used in the manufacturing process. The tubular-cased elements have a shorter life than either of the other two elements because of higher operating temperatures. The control of these elements is more difficult than with the others because of the extra material involved. Also, the air must be circulated sufficiently long to ensure the proper cooling of these elements.

**Heat Pump:** The heat pump is the most efficient method of heating in use today. A heat pump is a refrigeration unit that reverses the flow of refrigerant according to the different seasons of the year. By use of a series of valves, it cools the space in the summer and provides heat for it in the winter. Electricity is used to drive the compressor and if the system design conditions —usually 72°F (22.2°C) and 50 percent rh indoor conditions—are maintained, a heat pump will release up to eight times as much heat as could be obtained from resistance elements. Therefore, with an input of 1 kW (3412 Btu or 859.2 kcal) approximately 8 times the input or 27.280 Btu (6.8745 kcal) are released by the heat pump.

A heat pump is less efficient at lower outdoor temperatures because the evaporator temperature must be lower than the ambient temperature in order to absorb enough heat to evaporate the refrigerant. This lower temperature is accompanied by a lower suction pressure that reduces the compressor capacity. Also, at the lower temperatures, there is

not enough heat absorbed to replace the heat loss through the walls of the building.

When these units are installed in cold climates, additional heat is usually required. Because of the reduced efficiency, resistance elements are used in conjunction with these systems (see Figure 15-5).

If the cost of equipment per hour of operation is considered, a heat pump is more economical to own than a cooling system with a separate heating unit. The reason for this is that when one part (either heating or cooling) is used, the other part is not operating. Therefore, the cost per hour of operation is more.

**Figure 15-5.** Supplementary Heating Element and Heat Pump

## COMBUSTION

The available energy contained in a fuel is converted to heat energy by the combustion process. Combustion may be defined as the chemical action of a substance with oxygen resulting in the evolution of heat and some light.

There are three basic requirements for combustion. They are: (1) sufficiently high temperature, (2) oxygen, and (3) fuel (see Figure 15-6).

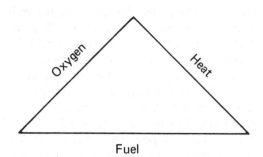

**Figure 15-6.** Combustion Triangle

**TEMPERATURE:** When the air-fuel mixture is admitted to the combustion chamber some means must be provided to bring the mixture to its flash point. This usually is done by a pilot light or an electric arc. If for any reason the temperature of the mixture is reduced below its flash point, the flame will automatically be extinguished. For example, if the temperature of a mixture of natural gas and air is reduced below its flash point of 1100°F (593°C), there will be no flame.

**OXYGEN:** Also, an ample supply of properly distributed oxygen must be supplied. The oxygen requirements governing the combustion process vary with each fuel. They also depend on whether or not the fuel and air are properly mixed in the correct proportions.

**FUEL:** The third requirement for combustion is the fuel. Properties of fuel were dealt with earlier in this chapter. The physical properties of each fuel must be considered when determining its requirements for combustion. All of the basic requirements for combustion must be met or there will be no combustion.

An important factor to keep in mind when making adjustments involving gaseous fuels is the limits of flammability, which are stated in percentages of gas in air of the mixture that would allow combustion to take place. To simplify this, if there is too much gas in the air, the mixture will be too rich to burn. If there is too little gas in the air, the mixture will be too lean to burn. The upper and lower limits are shown for the more common gaseous fuels in Table 15-3.

Table 15-3
Fuel Gas Limits of Flammability

| GAS | UPPER LIMIT | LOWER LIMIT |
| --- | --- | --- |
| Methane | 14 | 5.3 |
| Ethane | 12.5 | 3.2 |
| Natural | 14 | 3 |
| Propane | 9.5 | 2.4 |
| Butane | 8.5 | 1.9 |
| Manufactured | 29 | 4 |

Complete combustion can be obtained only when all of the combustible elements are oxidized by all the oxygen with which they will combine. The products of combustion are harmless when all of the fuel is completely burned. These products are carbon dioxide ($CO_2$) and water vapor ($H_2O$).

The rate of combustion, or burning, depends on three factors:

1. The rate of reaction of the substance with oxygen.

2. The rate at which the oxygen is supplied.

3. The temperature due to the surrounding conditions.

All the oxygen supplied to the flame is not generally used. The excess is commonly called excess oxygen or excess air. This excess oxygen is expressed as a percentage, usually 50 percent, of the air required for the complete combustion of a fuel. An example of this is that natural gas requires 10 ft$^3$ (.28 m$^3$) of air for each ft$^3$ (m$^3$) of gas. When 50 percent excess air is added to this figure, the quantity of air supplied is calculated to be 15 ft$^3$ (.42 m$^3$) of air to each ft$^3$ (m$^3$) of natural gas. There are several factors governing the excess air requirements. They are: (1) the uniformity of air distribution and mixing, (2) the direction of gas flow from the burner, and (3) the height and temperature of the combustion area. Excess air constitutes a loss and should be kept to a minimum. However, it cannot usually be less than 25 to 35 percent of the air required for complete combustion.

Excess air has both good and bad effects in the combustion area operation. It is added as a safety factor in case the 10 ft$^3$ (.28 m$^3$) of air which is required is reduced for some reason. Dirty burners, improper primary air adjustments, or a lack in the supply of primary air can cause such a reduction. The adverse effect is that the nitrogen in the air does not change chemically and tends to reduce the burning temperature and the flue gas temperature, thereby reducing the efficiency of the heating equipment. The air supplied for combustion contains about 79 percent nitrogen and 21 percent oxygen.

The products of combustion produced when 1 ft$^3$ (m$^3$) of natural gas is completely burned are: 8 ft$^3$ (.23 m$^3$) of nitrogen, 1 ft$^3$ (.0283 m$^3$) of carbon dioxide, and 2 ft$^3$ (.057 m$^3$) of water vapor. (See Figure 15-7.) These products are harmless to human beings. In fact, carbon dioxide is the ingredient added to water that makes soft drinks fizz.

The by-products of incomplete combustion are: carbon monoxide—a deadly product; aldehydes—colorless, inflammable, volatile liquids with a strong pungent odor that are irritating to the eyes, nose, and throat; keytones—used as paint removers; oxygen acids; glycols; and phenols. These by-products are harmful and must be guarded against by the proper cleaning and adjustment of heating equipment.

When considering the combustion requirement for any given installation, the air supply to the equipment must be calculated. The air supply is governed by: size of the

Figure 15-7.  Elements of Combustion

equipment in Btu (cal) rating, type of fuel, size of equipment room, building construction tightness, exhaust operation, and the local code requirements.

Heating equipment which is regulated by experienced personnel will produce clean, economical, efficient combustion. However, when the design of the equipment and the fundamentals of combustion are ignored, a potential hazard exists.

## ORIFICES

The orifice may be defined as the opening through which gas is admitted to the main burner. (See Figure 15-8.) It is mounted on the gas manifold by means of pipe threads and projects into the burner (see Figure 15-9). It is normally referred to as the *orifice spud.*

**Figure 15-8.** Orifice Location

For a burner to operate satisfactorily, it must be supplied with the proper Btu (cal) input. This is the function of the orifice. The size of the orifice, the gas manifold pressure, and the gas density are the factors which determine the rate of gas flow to the burner. The orifice and burner must be matched in Btu ratings. If the type of gas is

**Figure 15-9.** Orifice Spuds

changed, say from natural to LP gas, the orifice size must also be changed. When the orifice is oversized, an insufficient amount of primary air is drawn into the burner. Thus, incomplete combustion results and a yellow, inefficient flame is produced.

An orifice that is too small also will have an adverse effect on the burner. There will be delayed ignition accompanied by a loud boom when the burner is ignited. Also, the burner will not operate to its full capacity, resulting in poor heating.

The orifice must direct the gas stream exactly down the center of the burner (see Figure 15-10). The velocity of gas down the burner tube causes the primary air to be drawn in through the burner face and mixed with the gas. If the gas velocity is reduced, insufficient primary air will be the result. Therefore, the orifice must be drilled straight and in the exact center of the spud. This may be accomplished easily by drilling the

**Figure 15-10.** Orifice in Relation to Burner

**Figure 15-11.** Cutaway Diagram of Orifice

hole from the rear (see Figure 15-11). The V-shape will insure the desired hole position. The foregoing information should point up the fact that the orifice is a precision piece of equipment, and it should be treated as such when service is required. To clean an orifice a soft instrument, such as a broom straw, wire brush bristle, etc., must be used. Care must be taken not to change the shape or size of the orifice.

Table 15-4 can be used to ensure the proper orifice size for the job.

Table 15-4

Orifice Capacities for Natural Gas 1000 Btu/ft³ (252 kcal/m³)
Manifold Pressure 3½ in. Water Column (8.89 cm)

| WIRE GAUGE DRILL SIZE | RATE/HR FT³ (m³) | RATE/HR BTU (kcal) |
|---|---|---|
| 70 | 1.34 ( .037) | 1,340 ( 337.7) |
| 68 | 1.65 ( .046) | 1,650 ( 415.8) |
| 66 | 1.80 ( .050) | 1,870 ( 471.2) |
| 64 | 2.22 ( .062) | 2,250 ( 567 ) |
| 62 | 2.45 ( .069) | 2,540 ( 640 ) |
| 60 | 2.75 ( .077) | 2,750 ( 693 ) |
| 58 | 3.50 ( .099) | 3,050 ( 768.6) |
| 56 | 3.69 ( .104) | 3,695 ( 931.1) |
| 54 | 5.13 ( .145) | 5,125 ( 1,291.5) |
| 52 | 6.92 ( .195) | 6,925 ( 1,745.1) |
| 50 | 8.35 ( .236) | 8,350 ( 2,104.2) |
| 48 | 9.87 ( .279) | 9,875 ( 2,488.5) |
| 46 | 11.25 ( .318) | 11,250 ( 2,835 ) |
| 44 | 12.62 ( .357) | 12,625 ( 3,181.5) |
| 42 | 15.00 ( .424) | 15,000 ( 3,780 ) |
| 40 | 16.55 ( .468) | 16,550 ( 4,170.6) |
| 38 | 17.70 ( .501) | 17,700 ( 4,460.4) |
| 36 | 19.50 ( .552) | 19,500 ( 4,914 ) |
| 34 | 21.05 ( .596) | 21,050 ( 5,304.6) |
| 32 | 23.70 ( .671) | 23,075 ( 5,814.9) |
| 30 | 28.50 ( .807) | 28,500 ( 7,182 ) |
| 28 | 34.12 ( .966) | 34,125 ( 8,599.5) |
| 26 | 37.25 (1.054) | 37,250 ( 9,387 ) |
| 24 | 38.75 (1.097) | 39,750 (10,017 ) |
| 22 | 42.50 (1.203) | 42,500 (10,710 ) |
| 20 | 44.75 (1.267) | 44,750 (11,277 ) |

# MAIN GAS BURNERS

A gas burner is defined as a device that provides for the mixing of gas and air in the proper ratio to ensure satisfactory combustion.

The first burners were used for lighting purposes. The carbon in the flame was the result of incomplete combustion that produced the light. This flame was not good for heating because of the lower temperature produced.

The blow pipe was the first burner to produce a blue flame and, therefore, any great heat intensity. The blue flame was produced by blowing primary air into the base of the flame. (See Figure 15-12.) The modern blow pipe accomplishes the same thing by blowing compressed air in the same end that the gas enters. The gas and air are thus mixed in the pipe as in the burners in modern furnaces.

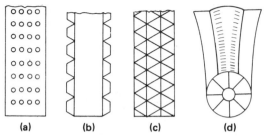

(a)    (b)    (c)    (d)

**Figure 15-13.** Types of Burners

ing on the type and purpose. Usually, the drilled port burner is made of cast iron. The others are made by forming steel into the desired shape. The type of burner used depends on the particular equipment design. Each manufacturer will use the burner that lends itself most effectively to the requirements of his equipment.

**PRIMARY AIR OPENINGS:** The control of primary air is accomplished on the face of the main burner. There are several methods used for this purpose (see Figure 15-14). The adjustable shutter, however, is the most common method used. The adjustment is made by enlarging or reducing the size of the opening through which the primary air is admitted.

The threaded shutter is less common and is used only on cast iron burners. The deflector type, or adjustable baffle is gaining acceptance in the industry. This burner has a larger air opening than necessary. This is an advantage because it will not clog up as fast as the other types. The primary air adjustment is made by pushing the deflector into the gas stream. This added restriction reduces the gas velocity and reduces the intake of primary air. Some burners have fixed openings and no adjustment can be made. These openings are sized according to the burner Btu (cal) rating and the proper flame is automatically obtained when the correct orifice size is used.

**Figure 15-12.** Basic Blow Pipe

Modern furnaces make use of four types of main gas burners. They are: (a) drilled port, (b) slotted port, (c) ribbon, and (d) inshot (see Figure 15-13). They are made of either cast iron or stamped steel, depend-

(a) Adjustable shutter          (b) Threaded shutter

(c) Nonadjustable shutter          (d) Deflector shutter

**Figure 15-14.** Primary Air Adjustment Methods

**FORCED DRAFT BURNERS:** The burners discussed previously were atmospheric types. Another type of burner that is gaining popularity, especially with the advent of the roof-mount units, is the forced draft burner. The equipment using this type of burner makes use of a small blower to provide the combustion air (see Figure 15-15). The combustion area is sealed from atmospheric conditions, and the air is forced into the combustion zone in which there is a positive pressure. The amount of air supplied and the inside pressure are controlled by an adjustment on the flue outlet. By forcing a fixed amount of air into the combustion chamber, preset combustion conditions are maintained.

Blower

Heat exchanger

**Figure 15-15.** Forced Draft Burner

Forced draft burners are ideal for outdoor heating equipment where unusual conditions prevail and where vent pipe heights are restricted. The main gas valve cannot open on these units until a definite pressure is reached inside the combustion zone. When making adjustments involving this pressure, the manufacturer's specifications must be consulted. This pressure will vary from manufacturer to manufacturer and sometimes on different models of the same manufacturer.

**POWER GAS BURNERS:** Another type of burner that is now being used is the power gas burner. (See Figure 15-16.) These burners are small, highly efficient, and suited for use with natural gas. Power burners have a very wide range of capacities. The maximum capacity is 500,000 Btus (126,000 kcal) and the minimum is 120,000 Btus (30,240 kcal). They are appropriate for any heating application where oil or an equivalent capacity is presently being used or as an integral component of a new boiler-burner system. They also can be directly installed and fired through existing doors of cast iron or

**Figure 15-16.** Power Gas Burner (Courtesy of Ray Burner Co.)

steel firebox boilers without requiring further brickwork or modification of the boiler base.

Power burners furnish 100 percent of the combustion air through the burner for firing rates up to their rated capacities. These burners have locked air shutters, with a reduced input start by means of a slow opening gas valve. Power burners are easy to install and to adjust due to the use of an orifice which meters the gas and a dial which adjusts the air shutter to match the gas orifice used. An automatic electric pilot ignition is used that incorporates a 6000-volt transformer to provide the necessary spark to ignite the pilot gas.

In operation, these burners are noisier than the atmospheric type burner. Power burners are applicable with small boilers, residential, and industrial furnaces. They are also used as antipollution afterburners. These burners are automatic and are equipped with the latest types of safety and operating controls.

## PILOT BURNERS

A pilot burner is defined as a small burner used to ignite the gas at the main burner ports and as a source of heat for the pilot safety device. Approximately 50 percent of all heating equipment failures can be attributed to this burner.

The automatic pilot is probably the most important safety device used on modern gas heating equipment. During normal operation, the pilot flame provides the heat necessary to actuate the pilot safety device. Also, the pilot is located in the combustion chamber in proper relation to the main burners so that the flame will ignite the gas admitted when the thermostat demands heating. Should something happen to cause the pilot to become unsafe, the pilot safety device will prevent any unburned gas from escaping into the combustion chamber.

There are basically three types of pilot burners. They are: (1) the millivolt, (2) the bimetal, and (3) the liquid-filled.

**MILLIVOLT PILOT BURNER:** The millivolt pilot burner can be divided into two types: the *thermocouple* and the *thermopile*. The thermocouple is an electricity-producing device made of dissimilar metals. One end of the metal is heated and is called the hot junction, while the other end remains relatively cool and is called the cold junction (see Figure 15-17). The output of a thermocouple is approximately 30 millivolts when heated to a temperature of about 3200°F (1742°C). The thermopile is a series of thermocouples wired so that the output voltage will be approximately 750 millivolts (see Figure 15-18). The greater the number of thermocouples in series, the higher will be the output voltage.

**Figure 15-17.** Thermocouple Principle

**BIMETAL PILOT BURNER:** The bimetal pilot is so named because of the sensing element. The sensing element is made of two dissimilar metals welded together so that they become one piece (see Figure 15-19). The two metals have different expansion rates and this causes them to bend when heated. As the bimetal element is heated by the pilot flame, a movable contact is pushed toward a stationary contact. When these two contacts touch, an electric circuit is completed through them (see Figure 15-20).

**Figure  15-18.**   Thermopile Principle

**Figure  15-19.**   Bimetal Element

**Figure  15-20.**   Operation  of  a  Bimetal Element

The bimetal pilot is usually employed to operate in the control or low voltage circuit.

**LIQUID-FILLED PILOT:** The liquid-filled pilot functions from the force of pressure exerted when a liquid is heated and a vapor is formed. This liquid/vapor is contained in a closed system so that more accurate con-

trol can be obtained. The bulb is located in the pilot flame (see Figure 15-21) and when the designated temperature and pressure is reached, a circuit is completed and the heating unit can function.

### FLAME TYPES

There are basically two types of flames: yellow and blue flame. There are, however, variations of these two flames that are the result of changes in the primary air adjustment. Primary air is the air that is mixed with the fuel before ignition. Secondary air is the air that is admitted to the flame after ignition (see Figure 15-22). It is also called excess air.

**YELLOW-FLAME:** This flame has a small blue-colored area at the bottom of the flame (see Figure 15-23). The outer portion (outer envelope) is completely yellow and is usually smoking. The smoke is unburned carbon from the fuel. The yellow or lumi-

**Figure  15-21.**   Liquid-Filled Element

**Figure 15-22.** Primary and Secondary Air Supporting Combustion

**Figure 15-23.** Yellow Flame

nous portion of the flame is caused by the slow burning of the carbon that is being burned. Soot is the result of the yellow flame which also provides a lower temperature than the blue flame. The yellow flame is an indication of insufficient primary air. *Incomplete combustion* is also indicated by a yellow flame—a hazardous condition that cannot be allowed.

**YELLOW-TIPPED FLAME:** This condition occurs when not quite enough primary air is admitted to the burner. The yellow tips will be on the upper portion of the outer mantle (see Figure 15-24). This flame is also undesirable because the unburned carbon will be deposited in the heat exchanger, thus eventually restricting the passage of the products of combustion.

**Figure 15-24.** Yellow-Tipped Flame

**ORANGE-COLORED FLAME:** Some orange or red color in a flame, usually in the form of streaks, should not cause any concern. These streaks are caused by dust particles in the air and cannot be completely eliminated (see Figure 15-25). When there is a great amount of red or orange in the flame, however, the combustion area location should be changed or some means provided for filtering the combustion air before it enters the equipment.

**Figure 15-25.** Orange or Red Streaks in Flame

**SOFT, LAZY FLAME:** This type of flame appears when just enough primary air is admitted to the burner to cause the yellow tips to disappear. The inner cone and the outer envelope will not be as clearly defined as they are in the correct, blue flame (see Figure 15-26). This type of flame is best for high–low fire operation. It is not suitable, however, whenever there may be a shortage of secondary air. This flame will burn blue in

**Figure 15-26.** Soft, Lazy Flame

the open air. When it touches a cooler surface, however, soot will be deposited on that surface.

**SHARP, BLUE FLAME:** When the proper ratio of gas to air is obtained, there will be a sharp, blue flame (see Figure 15-27). Both the outer envelope and the inner cone will be pointed and the sides will be straight. The flame will be resting on the burner ports and there will not be any noticeable blowing noise. The flame will ignite smoothly on demand from the thermostat. Also, it will burn with a nonluminous flame. This is the most desirable flame for heating purposes on standard gas heating units.

**Figure 15-27.** Sharp, Blue Flame

**LIFTING FLAME:** When too much primary air is admitted to the burner, the flame will actually lift off the burner ports. This flame is undesirable for many reasons. When the flame is raised from the ports, there is a possibility that the intermediate products of combustion will escape into the atmosphere. (See Figure 15-28.) The flame will be small and will be accompanied by a blowing noise, much like that made by a blow torch. The ignition will be rough, and if enough secondary air is not available, ignition may be impossible. Intermediate products of combustion are hazardous and must be avoided.

**Figure 15-28.** Lifting Flame

**FLOATING FLAME:** A floating flame is an indication that there is a lack of secondary air to the flame. In severe cases, the flame will leave the burner and have a cloudlike appearance (see Figure 15-29). Again the intermediate products of combustion are apt to escape from between the burner and the flame. The flame will actually be floating from one place to another wherever sufficient secondary air for combustion may be found.

This flame is sometimes hard to detect because as the equipment room door is opened sufficient air will be admitted to allow the flame to rest on the burner properly. However, the obnoxious odor of aldehydes will still be present.

Another possible cause of this flame is an improperly operating vent system. If the products of combustion cannot escape, fresh air cannot reach the flame. This is a hazardous situation that must be eliminated.

## HEAT EXCHANGERS

A heat exchanger may be defined as a device used to transfer heat from one medium to another. There are two types of heat exchangers: (1) primary heat exchanger and

**Figure 15-29.** Floating Flame

(2) secondary heat exchanger. For a heat exchanger to be efficient, the transfer of heat must be made with as little loss as possible.

**PRIMARY HEAT EXCHANGERS:** The heat exchanger is the heart of the heating plant. (See Figure 15-30.) It is designed to transfer heat from the combustion gases to the heating medium flowing through the passages. On direct-fired heating systems, it also serves as the combustion area. Openings in the lower section permit installation of the main burners and allow secondary air to reach the flame. As the fuel is burned, the flue gases rise through the vent passages. Restrictions are incorporated in the heat exchanger to control the flow of combustion air and reduce the amount of excess air to within the 35 to 50 percent limits. These restrictions are made in several different ways: (1) the top section is reduced in cross-sectional width, (2) baffles are placed in the exhaust openings at the top of the heat exchanger, and (3) a combination of the two (see Figure 15-31). These restrictions also permit maximum heat transfer by reducing the vent gas velocity which allows the maximum heat to be extracted from it. The vent gases then leave the heat exchanger and enter the venting system.

There are two general classifications of primary heat exchangers: (1) the barrel and (2) the sectionalized (see Figure 15-32). They may be constructed of either cast iron or steel. Due to the weight, expense, and slow response to temperature change of cast

(a)

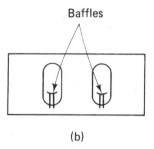

Baffles

(b)

**Figure 15-31.** Gas Vent Restrictions

iron, steel has dominated the forced air furnace heating industry in recent years.

The major problem with steel is the noise which sometimes accompanies the expansion and contraction associated with the heating and cooling of the heat exchanger. This noise can, however, be reduced by placing dimples and ribs in the metal during the stamping operation. Ceramic coating is also used to help eliminate noise as well as providing corrosion protection. This ceramic coating does not hinder the transfer of heat from the heat exchanger.

The application of the heat exchanger determines its physical shape. The heat exchanger used in a horizontal forced air furnace (see Figure 15-33) would need to be shaped differently from one used in an upflow furnace (see Figure 15-34). In most instances, there is very little difference between the upflow heat exchanger to the downflow heat exchanger.

**Figure 15-30.** Forced Air Heat Exchanger

(a) Barrel

(b) Sectionalized

**Figure 15-32.** Types of Primary Heat Exchangers: (a) Barrel; (b) Sectionalized.

**Figure 15-33.** Horizontal Heat Exchanger

**Figure 15-34.** Upflow Heat Exchanger

There are two flow paths through a heat exchanger. Both the heated air and the combustion gases pass through the unit without any mixing of the two. The heated medium flows around the outside of the heat exchanger, while the flue gases pass through the inside (see Figure 15-35).

The heat exchanger is normally a trouble-free apparatus, especially when the burners are installed and adjusted properly. From time to time, however, due to the continuous expansion and contraction, a hole will develop in the metal. This is a dangerous situation. When the flame is burning before the blower starts, carbon monoxide could be admitted to the air stream. After the blower starts, the flame will be agitated due to the greater pressure in the duct system. The amount of agitation will, naturally, depend

**Figure 15-35.** Flow Path Through a Heat Exchanger

on the size of the hole. A large hole could possibly cause the flame to be blown from the front of the furnace and set fire to anything in the immediate area, along with depositing the products of combustion in the building. The only difference between a large hole and a small hole is that the small hole is more difficult to detect. In any case, a leaking heat exchanger must be replaced.

**SECONDARY HEAT EXCHANGERS:** Hydronic heating (the use of circulated hot water or steam) enjoys a fair share of the heating industry. The boilers used for hydronic heating use both primary and secondary heat exchangers. The primary heat exchangers used with hydronic heating heat water rather than air. (See Figure 15-36.) When hydronic heating is used a secondary

heat exchanger must be used. When hydronic heating (hot water or steam) is used, the heating medium is passed from the boiler to the secondary heat exchanger, located in the area to be heated, through a series of pipes and returned to the boiler where it is reheated and recirculated. In some installations, the secondary heat exchanger is attached to a duct system and used as a central heating unit. The secondary heat exchanger is usually in the form of a finned coil (see Figure 15-37).

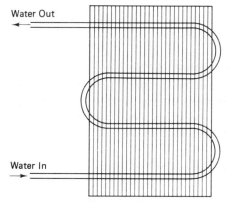

Water Out

Water In

**Figure 15-37.** Secondary Heat Exchanger

Another type of secondary heat exchanger is the embedded coil. This type of exchanger consists of a pipe embedded in the floor, wall, or ceiling (see Figure 15-38). This system of piping is connected to the boiler in the conventional manner. The coil has no fins and uses the surrounding structure to radiate the heat. It is ideal in playrooms or dens where a warm floor is desired. These exchangers, naturally, should be installed during construction of the room.

The efficient operation of any hydronic heating unit requires that all air be kept from collecting in the piping. Most installations have air vents on top of the coil as well as in the highest point of the entire system. Should these exchangers become air-locked,

Water flow

Vent gas

**Figure 15-36.** Water and Vent Gas Flow Through a Boiler (*Courtesy of Weil-McLain Co., Inc., Hydronic Division*)

Concrete slab

Water inlet

Water outlet

**Figure 15-38.** Embedded Hydronic Heat Exchanger

the heating will be reduced and sometimes completely stopped. In order to maintain maximum output from these units, the air must be continuously removed from the piping, coils, and boilers.

## ELECTRIC HEATING

Electric heating is beneficial from the standpoint of ecology because the fuel being burned in the production of electricity is consumed in one remote location. At the consumer level, there is no waste to be removed. There are no vent pipes to be installed as with gas, oil, or coal.

However, when considering electric heating, there are other factors to be weighed. The user must be willing to pay higher electric bills. There is also an increased initial cost of construction. This cost is accrued by the added electrical service and wiring, and the extra insulation which is required when electric heating is used.

When discussing electric heating, there are several facets that must be considered. They are resistance heating, heat pump, and a combination of the two.

**RESISTANCE HEATING:** Resistance heating involves a large number of operations, all of which are merely variations but are considered different in use and application. Resistance heating includes resistance elements, baseboard elements, heating cable, portable and unit type heaters, and duct heaters. However, regardless of the use or application, resistance heating wire is selected for high resistance per unit of length, and for the stability of this resistance at high temperatures. The type of wire used for this purpose is a high nickel chrome alloy resistance wire. Resistance heating does not use a centrally heated medium. The electric energy is expended directly into the air stream. Durability must also be considered when determining the type of wire to be used for resistance heating. The high nickel content in this type of wire retards oxidation and separation of the wire. If, however, the resistance wire does become broken, it may be repaired by cleaning the two ends, overlapping them, and applying borax liberally to the two pieces. Apply heat to the cleaned area until the two ends fuse together. After the wire has cooled to room temperature, remove any excess borax with a wet cloth. If this borax is allowed to remain on the wire, an immediate burnout will occur because the borax will cause excessive spot heating.

**Resistance Elements:** These elements are more commonly used in forced air control heating systems. They are so termed in order to differentiate them from duct heater strips, and are mounted directly in the air stream inside the furnace. (See Figure 15-39.) The elements are staggered for more uniform heat transfer, to eliminate hotspots, and to ensure maximum heat transfer. Each element is protected by a temperature limit

Heating elements are staggered for more uniform heat transfer, eliminates hot spots, and insures black heat. Each element is protected by an over temperature disc in one end and a fuse link in the other.

**Figure 15-39.** Resistance Element Installed in a Furnace Frame (*Courtesy of Electric Products Manufacturing Co.*)

switch and a fuse link. These elements are installed inside the furnace proper in an element compartment. (See Figure 15-40.) These elements will emit approximately 3412 Btu/kW (859.8 kcal/kW) input.

**Baseboard Elements:** Baseboard elements may be used for heating the whole building or for supplementary heating. These ele-

**Figure 15-40.** Element Compartment (*Courtesy of Electric Products Manufacturing Co.*)

ments are usually enclosed in a housing that provides safety as well as efficient use of the available heat. (See Figure 15-41.)

These units are available in a wide variety of sizes and may be joined together, usually end to end, to provide the Btu (cal) required for any given installation. The ideal location is at the point of greatest heat loss. The element in these units is normally of the enclosed type. That is, the electric resistance element has fins added to increase the surface area of the heating element, thereby increasing its efficiency. Baseboard units have a direct Btu (cal) output per watt input.

**Duct Heaters:** Electric duct heaters are factory assembled units consisting of a steel frame, open coil heating elements, and an integral control compartment. These matched combinations are fabricated to order in a wide range of standard and custom sizes, heating capacities, and control

Baseboard heater

**Figure 15-41.** Electric Baseboard Heating Units (*Courtesy of Chromalox Comfort Conditioning Division, Emerson Electric Co.*)

modes. They are prewired at the factory, inspected, and ready for installation at the job site.

The duct heater's frame is made in two basic configurations—slip-in (see Figure 15-42) or flanged (see Figure 15-43). The slip-in model is normally used in ducts up to 72 in. wide by 36 in. high (182.8 cm wide by 91.4 cm high), while the flanged model is used in larger ducts or where duct layout would make it impossible or impractical to

Air Flow

**Figure 15-42.** Slip-In Duct Heater (*Courtesy of Gould Inc., Heating Element Division*)

Air Flow

**Figure 15-43.** Flanged Duct Heater (*Courtesy of Gould Inc., Heating Element Division*)

use the slip-in type.

The slip-in heater, which is standard, is designed so that the heater frame is slightly smaller than the duct dimensions. The frame is inserted through a rectangular opening cut in the side of the duct, with the face of the heater at a right angle to the air stream. It is secured in place by sheet metal screws running through the control compartment.

The flanged heater is designed so that the frame matches the duct dimensions. The frame is then attached directly to the external flanges of the duct.

**AIR FLOW:** Electric heaters differ from steam or hot water coils in that the Btu/hr (cal/hr) output is constant as long as the heater is energized. It is, therefore, necessary that sufficient and uniform air flow be provided to carry away this heat and to prevent overheating and nuisance trapping of the thermal cut-outs. The minimum air velocity required is determined on the basis of entering air temperature and watts/ft² of the cross-sectional duct area.

## HEAT PUMP

A heat pump can be defined as an air conditioning system that moves heat both to and from the conditioned area. Through the proper use of controls, these units may be used during all seasons of the year. In the cooling season, the unit functions as a regular cooling unit. During the heating season, however, the refrigerant flow is reversed and it becomes a heating unit.

Heat pumps are classified according to their heat source and the medium to which the heat is transferred during the heating season. The classifications are: (1) air-to-air, (2) water-to-air, (3) water-to-water, and (4) ground-to-air. (See Figure 15-44.) Because of the expense of installation and equipment, the air-to-air heat pump is the most popular.

**COOLING CYCLE:** During the cooling season the heat pump operates as a normal cooling system. The refrigerant is compressed by the compressor and discharged to the outdoor coil (condenser), where it is condensed to a liquid (see Figure 15-45). It then travels to the flow control device and into the indoor coil (evaporator) where it is evaporated and absorbs heat; then back to the compressor and the cycle is repeated. Notice that the names of the coils have been changed. This is to eliminate confusion when discussing heat pumps. The evaporator is termed the *indoor coil* and the condenser is termed the *outdoor coil.* As the season changes, the function of the coils change.

**HEATING CYCLE:** During the heating season the heat pump operates because of the reversed refrigerant flow. The refrigerant is discharged by the compressor into the indoor coil where it is condensed into a liquid refrigerant. The heat removed from the re-

**Figure 15-44. Heat Pump Classifications**

frigerant is discharged into the conditioned area. The liquid refrigerant is then directed to the outdoor coil where it is evaporated before returning to the compressor (see Figure 15-46).

It should be noted that the reversed flow is accomplished by use of a 4-way valve and check valves. Some manufacturers will use different variations of this setup, but all accomplish the same thing.

The indoor coils used on heat pump units are usually designed especially for heat pumps and many times a coil is designed for a particular model unit. In most instances, the indoor coil used on a regular cooling unit will not function properly on heat pump installations. This is because of the

**Figure 15-45.** Heat Pump Cooling Cycle *(Courtesy of Carrier Corp.)*

**Figure 15-46.** Heat Pump Heating Cycle *(Courtesy of Carrier Corp.)*

reduced air flow over the indoor coil. The proper refrigerant-to-air ratio must be maintained for economical performance.

For these units to be accepted by the public, they must operate as economically as possible. Every possible Btu (cal) must be squeezed from every kW input. For each kW put into a heat pump, more than 3412 Btus (859.8 kcal) are released. A properly designed heat pump can produce three to four times the amount of heat released by resistance heating equipment. Some of the newer, more efficient units can produce as much as eight times the input Btu (cal). This is known as the coefficient of performance (COP). The COP is determined by dividing the Btu output by the Btu paid for.

Example: If we have a 10-ton heat pump which has been found to be delivering 86,000 Btus (21,672 kcal), the COP would be:

$$COP = \frac{86,000 \text{ Btu } (21,672 \text{ kcal})}{10 \text{ BHP (brake horsepower)} \times 2545 \text{ Btu/hp } (641.3 \text{ kcal/hp})} = 3.3 \text{ total heat available}$$

That is, the COP of this particular unit is 3.3 or the Btu (cal) output is 3.3 times greater than the Btu (cal) input. It should be noted here that the higher the COP, the greater the amount of heat available. In comparison, direct resistance heating has a COP of only 1.

When properly designed, installed, and maintained, a heat pump can be superior to a gas heating unit. This can be attributed to a lower discharge air temperature and a higher relative humidity that provides a more even temperature during the heating cycle.

However, these units do not operate as satisfactorily in extremely cold climates. Some manufacturers introduce an automatic means of stopping the compressor completely when the outdoor temperature drops to around 20°F (−6.7°C). In ambient temperatures below this, the COP will fall below the economical point. Also, the outdoor coil must operate at a temperature below the ambient temperature in order to absorb heat for the evaporation process. At these low temperatures, the suction pressure is low enough to cause damage to the compressor. Some units are equipped with an electric heater on the air inlet side of the outdoor coil to provide for the evaporation process at low ambient temperatures. (See Figure 15-47.) Because of the added expense in manufacturing and operation, this is done only in special cases.

**SOLAR ASSISTED HEAT PUMP:** Solar assisted heat pumps use the same principle as supplemental electric heat. The major difference is that the electric heater is replaced by a water coil (see Figure 15-48). The solution inside the coil is heated by use of a solar panel. This added heat keeps the operating efficiency as high as possible without

**Figure 15-47.** Electric Outdoor Heat for a Heat Pump

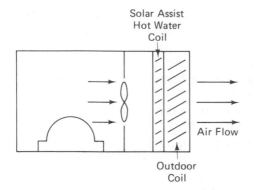

**Figure 15-48. Solar Assisted Heat Pump**

the added operating expense of using electricity.

**DEFROST CYCLE:** During the heating season, it is quite possible that the outdoor coil will operate at a temperature below freezing. At this temperature, any moisture which may be removed from the air will immediately freeze on the surface of the coil. This frost on the coil will continue to build up and cause the outdoor coil to lose its efficiency. This frost will also act as an insulator which will reduce the heat transfer and reduce the coil efficiency even more. When the coil efficiency has been reduced enough to affect the performance of the system, the frost must be removed.

There are three basic steps involved in the defrost cycle.

1. Detection of the frost build-up (defrost initiation).

2. Frost removal.

3. Detection of sufficient frost removal to allow the unit to return to the heating cycle (defrost termination).

The methods used for frost detection are many and varied. The following are some of the more common methods.

**Air Pressure:** This method uses the air pressure differential across the outdoor coil (see Figure 15-49). This type of control has a time delay to prevent false initiation. When the air pressure reaches a given drop across the coil, the defrost cycle is initiated.

**Time-Temperature:** This control uses a temperature bulb which is attached to the outdoor coil to initiate the defrost cycle at the prescribed temperature (see Figure 15-50). Incorporated in this control is a time clock that will only allow the defrost period to be initiated at given intervals. This method requires that both a low temperature be reached and a sufficient amount of time elapse before defrost initiation can occur.

**Solid State:** Some units are equipped with a solid state circuit board which initiates the defrost cycle. Thermistors within the solid state system sense the difference between

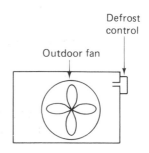

**Figure 15-49. Air Pressure Defrost Initiation**

Air flow

Defrost
control

Coil temperature
sensing bulb

**Figure 15-50.** Time-Temperature Defrost Control Method

the ambient air temperature and the refrigerant temperature. When the temperature difference exceeds the differential band, the defrost cycle is activated and continues until the defrost pressure switch terminates it.

In operation, when the ice on the coil has reached a predetermined point, the defrost cycle is initiated. The unit is returned to the cooling cycle so that the hot discharge gas now passes through the condenser. The outdoor fan is stopped so that the cold outdoor air will not flow through the coil and adversely affect the defrost efficiency. The indoor fan is kept running to evaporate the refrigerant before it returns to the compres-

sor. Also, the auxiliary heat is turned on to prevent cold air from blowing into the conditioned area. When the ice has all melted, the defrost termination control will return the unit back to the normal heating cycle.

Many of the controls that were used for defrost initiation are also used to terminate the defrost cycle.

**Air Pressure:** After the frost has been cleared from the outdoor coil, the air pressure drop across the coil is reduced. The air pressure on both sides of the coil is almost equal. The unit is then automatically returned to the heating cycle.

**Time Termination:** This type of control is usually used in conjunction with any of the previously mentioned initiation methods. In operation, after a predetermined amount of time has elapsed, the unit will return to the heating cycle if the defrost initiation control has reached the cut-in setting.

**Pressure Termination:** This type of defrost termination control functions because the refrigerant pressure increases when the frost has been melted. As the pressure inside the system increases, the control puts the unit back into the heating cycle at a predetermined setting.

Some manufacturers incorporate electric strip heaters which are attached to the outdoor coil to speed up the defrosting process. These heaters are energized by the defrost initiation control (see Figure 15-51).

Defrost heating
element

Electric connections

Outdoor coil

**Figure 15-51.** Electric Defrost Heater

**REHEATING:** A heat pump should not be used for reheating purposes, especially in the summer. The compressors used in these units have a higher compression ratio than normal air conditioning compressors. It is not difficult to obtain a 300°F (149°C) discharge gas temperature. This is the maximum allowable temperature and many manufacturers will include controls to stop the compressor before it is reached. This higher temperature is caused by too low an air flow over the indoor coil. A high indoor air temperature (90°F—32.2°C—or higher) will require more air to be moved across the indoor coil. Caution must be exercised to ensure that any supplemental heating equipment is installed in the air stream after the indoor coil.

**SUPPLEMENTARY HEAT:** These units are sized to satisfy the cooling load and supplementary heat is added to bring the unit to the desired heating capacity. Each manufacturer will provide charts indicating the cooling and heating capacity in Btu (cal) for any given unit. From these charts and the calculated heat load, the amount of supplementary heat may be determined. This supplementary heat is usually in the form of duct heaters installed in the supply duct of the equipment (see Figure 15-52).

Supplementary heat is used only when the heat pump cannot supply the desired amount of heat. It is controlled from two points: (1) the second stage on the indoor thermostat and (2) the outdoor thermostats which are mounted under the eaves of the building. When both of these thermostats demand heat, the supplementary heat will be energized. The design engineer will determine the temperature setting of the outdoor thermostats. If these thermostats are set too high, the electricity bill may be excessive. On the other hand, if they are set too low, an insufficient amount of heat may be supplied to the conditioned area. There must be a point where comfort and economy are maintained.

**REVERSING VALVE:** The reversing valve is the device that changes the direction of the refrigerant flow in a heat pump system. (See Figure 15-53.) It is installed in the discharge line of the compressor and switches the flow of refrigerant on demand

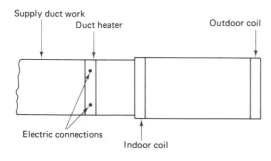

**Figure 15-52.** Location of Supplementary Duct Heaters

**Figure 15-53.** Reversing Valve (*Courtesy of Ranco, Inc.*)

COOLING END              HEATING END

HIGH PRESSURE

SLIDE BRACKET

PILOT TUBE    PISTON ASS'Y.

BODY

PILOT TUBE

END COVER    BLEED HOLE

SLIDE ASS'Y. (WITH TEFLON SLIDE)

PISTON ASS'Y.

END COVER

EVAPORATOR    SUCTION    CONDENSER

SHOWN IN COOLING POSITION

**Figure 15-54.** Cutaway of Reversing Valve (*Courtesy of Ranco, Inc.*)

from the thermostat or the defrost control.

The refrigerant gas path is schematically diagrammed through the main valve showing the sliding port at a position of rest over two tube openings as it transfers both coils from the opening phases of cooling, defrosting, and heating. (See Figure 15-54.) The solenoid coil is not energized in a normal cooling cycle.

When adding refrigerant to these units or making any type of adjustment, the manufacturer's specifications must be followed. The trial and error method is too time-consuming and the desired conditions may never be reached because of the numerous overlaps and interlocks incorporated into these units. Therefore, for comfort and economy's sake follow the proper specifications.

## FURNACES

Furnaces are generally considered as being the heating part of an air conditioning system. They also serve as the air handling and filtering section of a complete system. Furnaces may be designed to use the various types of gases, fuel oil, or electrical resistance elements as the heat source. The type of heat source will also govern the type of heat exchanger construction.

**GAS FURNACES:** Gas furnaces are manufactured in three designs: upflow, downflow, and horizontal. This terminology refers to the direction of air flow. Each furnace is specifically designed for a particular application. (See Figure 15-55.) Notice that the burners and flue pipe connection are in a fixed location and cannot be changed.

In the downflow furnace, the fan is located on top and the air is blown down through the heat exchanger (see Figure 15-56). The duct work which distributes the air to the separate areas is connected to the bottom of the furnace.

The horizontal furnace is generally installed in the attic or a crawl space under the building. The air is blown in a horizontal direction. (See Figure 15-57.)

# PARTS ARRANGEMENT

FLUE PIPE

BURNER ACCESS PANEL

FAN/LIMIT CONTROL

UNIT RATING PLATE

GAS VALVE

PILOT

BURNERS

WIRING MAKE-UP BOX

HAMMOCK FILTER
(Slab Filter On Some Models)

BLOWER ACCESS PANEL

LINERS

HEAT EXCHANGER

BLOWER MOTOR

BLOWER

**Figure 15-55.** Cutaway of Upflow Furnace *(Courtesy of Lennox Industries, Inc.)*

Generally, these furnaces may be operated with either natural or LP gases. Minor adjustments may be needed along with changing the orifice and in some cases the burner assembly to the proper size. The manufacturer's recommendations should be followed regarding these items. Natural gas furnaces require a gas pressure regulator. LP furnaces do not require a gas pressure regulator. LP gas is regulated at the storage tank.

**OIL FURNACES:** These furnaces are designed to use domestic fuel oil as the heat source. They also are available in upflow, horizontal, and counterflow. (See Figures 15-58, 15-59, and 15-60.) The horizontal units are designed to be installed on combustible flooring in an attic space, on a slab in a crawl space, or they can be suspended in a basement, utility room, furnace room, or crawl space. (See Figure 15-61.) The counterflow oil furnace is designed for installation in either the counterflow or horizontal position. (See Figure 15-62.)

**ELECTRIC FURNACES:** Electric furnaces can be installed in the upflow, downflow, and horizontal positions. (See Figure 15-63.) Note that these furnaces do not have burners or a combustion type heat exchanger as do the gas and oil-fired furnaces. (See Figure 15-64.) Instead electric heating elements are used as the heat source. There are no vent pipe or gas pipe connections—only electrical connections. It is mainly for this reason that these furnaces are multipositional.

All furnaces should be installed with the proper clearances on the rear, sides, and front. These clearances are specified in the correct installation instructions for the particular furnace being installed.

**HEAT RISE:** The heat rise through a furnace can be defined as the temperature increase of the air as it passes through the heating device. That is, the temperature of the supply air minus the return air temperature.

**Example:** Air is entering the heating unit at 70°F (21.1°C) and leaving at 145°F (63°C). The heat rise is 145°F − 70°F = 75°F (41.7°C).

The heat rise through a unit is determined by the amount of heat input to the furnace and the volume of air flowing through the furnace. Gas and oil furnaces are rated from 40°F (22.4°C) to 100°F (56°C) heat rise. Electric furnaces are generally rated from 40°F (22.4°C) to 60°F (33.6°C) heat rise.

1. Burners
2. Heat Exchanger
3. Blower
4. Controls
5. Draft Diverter
6. Limit Control
7. Fan Control
8. Cabinet

**Figure 15-56.** Cutaway of Counterflow Furnace (*Courtesy of Southwest Manufacturing Company*)

1. Burners
2. Heat Exchanger
3. Blower
4. Controls
5. Draft Diverter
6. Limit Control
7. Fan Control
8. Cabinet

**Figure 15-57.** Horizontal Furnace (*Courtesy of Southwest Manufacturing Company*)

FLUE OUTLET

HEAT EXCHANGER

BURNER ACCESS PANEL

IGNITION TRANSFORMER

FAN/LIMIT CONTROL

OIL PUMP

PRIMARY SAFETY CONTROL

COMBUSTION AIR
BLOWER HOUSING

COMBUSTION AIR
BLOWER MOTOR

BLOWER/FILTER
ACCESS PANEL

BLOWER

FILTER

BLOWER MOTOR
(Belt Drive Shown)

**Figure 15-58.** Cutaway of Upflow Oil Furnace (*Courtesy of Lennox Industries, Inc.*)

**Figure 15-59.** Cutaway of Horizontal Oil Furnace (*Courtesy of Lennox Industries, Inc.*)

Crawl Space Horizontal Installation

Suspended Horizontal Installation

**Figure 15-60.** Cutaway of Counterflow Oil Furnace (*Courtesy of Lennox Industries, Inc.*)

**Figure 15-61.** Horizontal Furnace Installations (*Courtesy of Lennox Industries, Inc.*)

**Figure 15-62.** Counterflow Installation (*Courtesy of Lennox Industries, Inc.*)

## OIL BURNERS

Fuel oils, like natural and LP gases, are excellent heating fuels. However, the lighting and burning of fuel oil requires special equipment. Fuel oil in its liquid form will not burn; it must be either atomized or vaporized. From this, we can define a fuel oil burner as a mechanical device that prepares oil for combustion. The actual burning of the fuel takes place in the firebox, with the atomizing method being the most popular. The atomizing type of burner prepares the oil for burning by breaking it up into a fog-like vapor. This is accomplished in three

Horizontal installation with
cooling coil and electronic air cleaner

Up-Flo installation
with cooling coil.

Down-Flo installation
with cooling coil.

Horizontal installation in closet.

**Figure 15-63.** Electric Furnace Installations (*Courtesy of Lennox Industries, Inc.*)

**Figure 15-64.** Cutaway of Electric Furnace (*Courtesy of Lennox Industries, Inc.*)

**Figure 15-65.** Low Pressure Oil Burner (*Courtesy of The Carlin Company*)

ways: (1) by forcing the oil under pressure through a nozzle (air or steam may be forced through with the oil, or the oil may be alone), (2) by allowing the oil to flow out the end of small tubes that are rapidly rotated on a distributor, or (3) by forcing the oil off the edge of a rapidly rotating cup that may be mounted either vertically or horizontally.

Fuel oil burners are divided into three types depending on the operating principle. They are the high pressure, low pressure, and rotary type burners. (See Figure 15-65.)

**HIGH PRESSURE OIL BURNER:** A high pressure oil burner is made up of an electric motor blower, fuel unit, and an ignition transformer. The motor is mounted on the motor shaft and the fuel unit is directly connected to the end of the motor shaft. The fuel unit consists of a strainer, a pump, and a pressure-regulating valve.

High pressure oil burners do not use primary and secondary air as such. They are designed to atomize the fuel oil and mix this

atomized oil with the air before ignition. (See Figure 15-66.)

The fuel oil is forced through the burner nozzle at approximately 100 psig (7.03 $kg/cm^2$). The atomized oil has enough velocity at the nozzle to cause a low pressure area into which the combustion air is drawn. The combustion air is also under force from the fan which creates a turbulence in the air. All of these together are used to create a complete mixing of the fuel oil and the air. This combustible mixture is then ignited at the end of the burner.

In operation, when the thermostat demands heat for the building, the fuel oil is

Oil at 100 psi (7.03 $kg/cm^2$)

Air

Air

**Figure 15-66.** High Pressure Oil Burner Operation

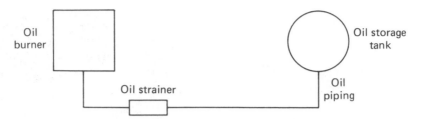

**Figure 15-67.** Basic Fuel Oil System

drawn from the storage tank. It passes through a strainer, where the solid particles are removed, to the pump which forces it to the pressure-regulating valve. At this point, the oil hesitates until the pressure has reached 100 psig (7.03 kg/cm²). Then the pressure-regulating valve opens and allows the oil to proceed to the burner nozzle where it is atomized. After the fuel leaves the nozzle, it is mixed with the proper amount of air and ignited by a high voltage transformer —approximately 10,000 volts and 25 milliamperes. (See Figure 15-67.)

When sufficient heat has been supplied to the structure, the thermostat will signal the burner motor to stop. As it reduces its speed, the oil pressure from the fuel pump is also reduced. As the oil pressure is reduced below 90 psig (6.33 kg/cm²), the pressure-regulating valve stops the flow of oil to the nozzle and the flame is extinguished because of the lack of fuel. The oil burner will remain at rest until the building requires additional heat, at which time the sequence is started again.

**LOW PRESSURE OIL BURNER:** The low pressure oil burner is similar in appearance to the high pressure burner. It also operates much the same as the high pressure burner except for the method used to atomize the oil. In the low pressure burner, the oil is atomized by compressed air, similar to a paint spray gun. The primary air and fuel

are forced through the nozzle at the same time with 1–15 psig (.0703 to 1.05 kg/cm²) pressure. The secondary air is supplied in the same manner as the air used by the high pressure burner (see Figure 15-68).

Low pressure oil burners operate quite satisfactorily, but each manufacturer uses different mechanical means of mixing the air and the oil. This requires special service knowledge and techniques for each burner. Because simplicity of service is almost essential, low pressure oil burners are not a common item.

**ROTARY OIL BURNER:** Rotary oil burners incorporate a completely different principle from either the high pressure or the low pressure burners. The rotary burner employs a rotating cup to atomize the oil. The fuel oil is forced to the edge of the cup in small droplets because of the high speed of the cup. As the droplets leave the edge of the cup, they are dispersed into a current of rapidly moving air which is provided by a centrifugal blower. (See Figure 15-69.) These droplets are further atomized by this air and

**Figure 15-68.** Low Pressure Oil Burner Operation

**Figure 15-69.** Rotary Oil Burner Operation (*Courtesy of Ray Burner Co.*)

the foglike fuel is projected parallel to the axis of the cup. The atomized oil is then ignited in much the same manner as any other type of oil burner. In some rotary burners the cup may be replaced with small tubes extending to the face of the rotating member. The oil is forced to the ends of the tubes where centrifugal force distributes the oil into the air stream.

## BOILERS

Hydronic heating maintains comfort in the home by circulating hot water to the secondary heat exchanger. This system basically consists of a boiler, a pump, a secondary heat exchanger, and a series of piping (see Figure 15-70). The principle of hydronics dates back to ancient times when the Romans used it by circulating warm water through the walls and floors. Today, however, hydronics bears no resemblance to the

**Figure 15-70.** Basic Hot Water Heating System

ancient methods. Modern systems incorporate many refinements, such as zoning, which allows the temperature to vary in different areas of the building.

**HOT WATER BOILERS:** A hot water boiler maintains water temperatures between 120°F and 210°F (49°C and 99°C). This water is pumped through piping to the secondary heat exchanger. The term boiler may bring to mind large heating plants such as those found in schools or other large buildings. These are, however, commercial type boilers. Actually, a modern cast-iron boiler which is used in residential heating is very compact. Most units are the size of an automatic washing machine, and some are as small as a suitcase and even can be hung on the wall. (See Figure 15-71.)

**STEAM BOILERS:** The widespread use of steam for space heating today points up a long recognized fact that steam as a heating medium has numerous basic characteristics which can be advantageously employed. Some of the most important advantages are as follows.

**Steam's Ability to Give Off Heat:** Properties of saturated steam may be found in steam tables which give much information regarding the temperature and heat contained in 1 lb (453.6 g) of steam for any pressure. (See Table 15-5.) For example, to change 1 lb of

**Figure 15-71.** Water Boiler (*Courtesy of Weil-McLain Co., Inc., Hydronic Division*)

water to steam at 212°F (100°C) at atmospheric pressure requires a heat content of 1150.4 Btus (289.9 kcal). This is made up of 180.1 Btus (45.4 kcal) of sensible heat (the heat required to raise 1 lb of water from 32°F to 212°F (0.°C to 100°C) and 970.3 Btus (244.5 kcal) of latent heat. Latent heat is the heat added to change the 1 lb of water at 212°F into steam. This stored-up heat is required to transform the water into steam, and it reappears as heat when the process is reversed to condense the steam to water.

Because of this fact, the high latent heat of vaporization of 1 lb (453.6 g) of steam permits a large quantity of heat to be transmitted efficiently from the boiler to the heating unit with little change in temperature.

**Steam Promotes Its Own Circulation:** For example, steam will flow naturally from a higher pressure (as generated in the boiler) to a lower pressure (existing in the steam lines). Circulation or flow is caused by the lowering of the steam pressure along the steam supply mains and in the heating units due to the pipe friction. Also, the condensation process of the steam as it gives up heat to the space being heated aids in circulation. (See Figure 15-72.) Because of this fact, the natural flow of steam does not require a pump such as that needed for hot water heating, or a fan as used in warm air heating.

**Steam Heats More Readily:** Steam circulates through a heating system faster than other fluid media. This can be important where fast pick-up of the space temperature is desired. It will also cool down more readily when circulation is stopped. This is an important consideration in spring and fall when comfort conditions can be adversely affected by long heating-up or slow cooling-down periods. (See Figure 15-73.)

## Table 15-5

### Properties of saturated steam (approx.)

| Absolute pressure | Gage reading at sea level | Temp. ° F | Heat in water Btu per lb. | Latent heat in steam (Vaporization) Btu per lb. | Volume of 1 lb steam ft$^3$ | Wgt. of water lb. per ft$^3$ |
|---|---|---|---|---|---|---|
| 0.18 | 29.7 | 32 | 0.0 | 1076 | 3306 | 62.4 |
| 0.50 | 28.4 | 59 | 27.0 | 1061 | 1248 | 62.3 |
| 1.0 | 28.9 | 79 | 47.0 | 1049 | 653 | 62.2 |
| 2.0 | 28 | 101 | 69 | 1037 | 341 | 62.0 |
| 4.0 | 26 | 125 | 93 | 1023 | 179 | 61.7 |
| 6.0 | 24 | 141 | 109 | 1014 | 120 | 61.4 |
| 8.0 | 22 | 152 | 120 | 1007 | 93 | 61.1 |
| 10.0 | 20 | 161 | 129 | 1002 | 75 | 60.9 |
| 12.0 | 18 | 169 | 137 | 997 | 63 | 60.8 |
| 14.0 | 16 | 176 | 144 | 993 | 55 | 60.6 |
| 16.0 | 14 | 182 | 150 | 969 | 48 | 60.5 |
| 18.0 | 12 | 187 | 155 | 986 | 43 | 60.4 |
| 20.0 | 10 | 192 | 160 | 983 | 39 | 60.3 |
| 22.0 | 8 | 197 | 165 | 980 | 36 | 60.2 |
| 24.0 | 6 | 201 | 169 | 977 | 33 | 60.1 |
| 26.0 | 4 | 205 | 173 | 975 | 31 | 60.0 |
| 28.0 | 2 | 209 | 177 | 972 | 29 | 59.9 |
| 29.0 | 1 | 210 | 178 | 971 | 28 | 59.9 |
| 30.0 | 0 | 212 | 180 | 970 | 27 | 59.8 |
| 14.7 | 0 | 212 | 180 | 970 | 27 | 59.8 |
| 15.7 | 1 | 216 | 184 | 968 | 25 | 59.8 |
| 16.7 | 2 | 219 | 187 | 966 | 24 | 59.7 |
| 17.7 | 3 | 222 | 190 | 964 | 22 | 59.6 |
| 18.7 | 4 | 225 | 193 | 962 | 21 | 59.5 |
| 19.7 | 5 | 227 | 195 | 960 | 20 | 59.4 |
| 20.7 | 6 | 230 | 196 | 958 | 19 | 59.4 |
| 21.7 | 7 | 232 | 200 | 957 | 19 | 59.3 |
| 22.7 | 8 | 235 | 203 | 955 | 18 | 59.2 |
| 23.7 | 9 | 237 | 205 | 954 | 17 | 59.2 |
| 25 | 10 | 240 | 208 | 952 | 16 | 59.2 |
| 30 | 15 | 250 | 219 | 945 | 14 | 58.8 |
| 35 | 20 | 259 | 228 | 939 | 12 | 58.5 |
| 40 | 25 | 267 | 236 | 934 | 10 | 58.3 |
| 45 | 30 | 274 | 243 | 929 | 9 | 58.1 |
| 50 | 35 | 281 | 250 | 924 | 8 | 57.9 |
| 55 | 40 | 287 | 256 | 920 | 8 | 57.7 |
| 60 | 45 | 293 | 262 | 915 | 7 | 57.5 |
| 65 | 50 | 298 | 268 | 912 | 7 | 57.4 |
| 70 | 55 | 303 | 273 | 908 | 6 | 57.2 |
| 75 | 60 | 308 | 277 | 905 | 6 | 57.0 |
| 85 | 70 | 316 | 286 | 898 | 5 | 56.8 |
| 95 | 80 | 324 | 294 | 892 | 5 | 56.5 |
| 105 | 90 | 332 | 302 | 886 | 4 | 56.3 |
| 115 | 100 | 338 | 309 | 881 | 4 | 56.0 |
| 140 | 125 | 353 | 325 | 868 | 3 | 55.5 |

(Courtesy ITT Hoffman Specialty)

**Figure 15-72.** Steam and Condensate Flow-in Lines (*Courtesy of ITT Hoffman Specialty*)

(a) Steam Boiler
Water at Boiler Water Level Only

(b) Hot Water Boiler
System Completely Filled With Water

**Figure 15-73.** Boiler Water Comparison (*Courtesy of ITT Hoffman Specialty*)

## PIPING ARRANGEMENTS:

Steam heating systems fall into two basic classifications —one-pipe and two-pipe. These names are descriptive of the piping arrangement used to supply steam to the heating unit and to return condensate from the unit. A one-pipe system consists of a heating unit with a single pipe connection through which steam flows to it and condensate returns from it using the same line at the same time. In a like manner, a two-pipe system consists of a heating unit with two separate pipe connections—one used for the steam supply and the other for condensate return.

**One-Pipe System:** Modern, automatically-fired boilers promote rapid steaming and assure quick pick-up of the space temperature from a cold start. The natural circulation of steam in the system, in combination with the simplicity of piping and air venting, makes this type of system a desirable method of heating as well as the most economical.

A one-pipe system, properly designed for gravity return of the condensate to the boiler with open air vents on the ends of the mains on each heating unit, requires a minimum of mechanical equipment (see Figure 15-74). The result is a low initial cost for a very dependable system.

**Two-Pipe System:** Although a two-pipe system has fundamental differences from a one-pipe system, many components and piping installation practices are common to both systems. Also, the two-pipe system employs many of the advantages of using steam as a heating medium. They are applicable in a variety of structures from small residences to large commercial buildings, office buildings, apartment buildings, and industrial complexes.

Two-pipe systems are designed to operate at pressures ranging from subatmospheric (vacuum) to high pressure. Although they may use many practical piping arrangements to provide upflow or downflow systems, they are conveniently classified by the method of condensate return to the boiler—by gravity or by use of any one of several mechanical means.

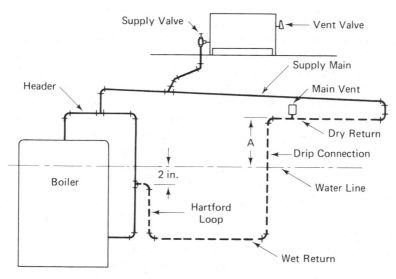

**Figure 15-74.** Basic One-Pipe Upfeed System (*Courtesy of ITT Hoffman Specialty*)

## HUMIDIFICATION

One of the most important aspects of air conditioning, humidification, is unfortunately one of the least understood. This is undoubtedly because it is intangible; you can't see it, you can't touch it, it has no odor, no color, and no sound.

To be sure that we are starting out on common ground, let's define some of the words that we are going to use in the discussion of humidity and humidification.

**Humidity:** The water vapor within a given space.

**Absolute Humidity:** The weight of water vapor per unit volume.

**Percentage Humidity:** The ratio of the weight of water vapor per pound of dry air to the weight of water vapor per pound of dry air saturated at the same temperature.

**Relative Humidity:** The amount of water vapor (percent) actually in the air compared to the maximum amount that the air could hold under the same conditions. The warmer the air, the more moisture it can hold. Air in a building heated to 70°F (21.1°C) can hold about 8 grains (.52 g) of moisture per cubic foot (.0283 m³). This is 100% relative humidity. If there are only 2 grains (.13 g) per cubic foot (.0283 m³) in the building, this is ¼ of the amount of moisture that the air can hold. Therefore, the relative humidity is ¼ or 25%; the air could hold 4 times as much water.

The important thing to remember is that when air is heated it can hold more moisture. This reduction of relative humidity is taking place in every unhumidified or under-humidified building where heating is used.

To solve this problem, we add moisture so there is more water available for the air to absorb. We humidify because there are benefits that are as important as heating to overall indoor comfort and well-being during the heating season.

These benefits can be grouped into three general classifications.

1. Comfort.

2. Preservation.

3. Health.

**COMFORT:** Have you ever stepped out of the shower and noticed how warm the bathroom is? It is probably about 75°F (23.9°C) in the bathroom, and the relative humidity is probably about 70–80% because of the water vapor that was added to the air while showering. Now, if the phone rings, and you step out into the hall to answer it, you nearly freeze. Yet the temperature there is probably about 70°F (21.1°C)—just 5°F (2.78°C) cooler than in the bathroom and you are shivering. This shivering is because you just became an evaporative cooler. The air out in the hall is dry, the relative humidity is possibly about 10–15%. You are wet and this thirsty air absorbs moisture from your skin. The water evaporates, and as it does, your skin is cooled. This same type of thing happens day after day, every winter, in millions of buildings. People are turning thermostats up to 75°F and higher in order to feel warm because it feels drafty and chilly when the evaporative cooling process is going on. The proper relative humidity level makes 70°F feel more like 75°F.

The chilling effect is not the only discomfort caused by too dry air. Static electricity is usually an indication of low relative humidity and a condition that is consistently annoying. Proper relative humidity will alleviate this discomfort.

**PRESERVATION:** The addition or reduction of moisture drastically affects the qualities, the dimensions, and the weight of hygroscopic materials. Wood, leather, paper, and cloth, even though they feel dry to the touch, contain water—not a fixed amount, but an amount that will vary greatly with the relative humidity level of the surrounding air. For example, 1 ft$^3$ (.0283 m$^3$) of wood with a bone-dry weight of 30 lbs (13.6 kg) at 60% relative humidity will hold over 3 pints (1.41 1) of water. If the relative humidity is lowered to 10%, the water held by the wood will not even fill a one-pint (.473-1) bottle. We have, in effect, withdrawn 2½ pints (1.2 1) of water from the wood by lowering the relative humidity from 60% to 10%.

This type of action goes on, not only with wood, but with every single material in the building that has the capability of absorbing and releasing moisture. Paper, plaster, fibers, leather, glue, hair, skin, and so on will shrink as they lose water and swell as they take on water. If the water loss is rapid, warping or cracking takes place. As the relative humidity changes, the condition and dimensions of the materials change as constantly as the weather. This is why proper relative humidity is important.

Too much moisture also can be damaging. Everyone has seen windows fog during the winter, maybe a little fog on the lower corners, maybe a whole window fogging or completely frosting over. This latter condition is an indication that the indoor relative humidity is too high.

The formation of condensation is due to the effect of vapor pressure. Dalton's Law explains vapor pressure. It states, "In a gaseous mixture, the molecules of each gas are evenly dispersed throughout the entire volume of the mixture." Taking the building as the volume involved, water vapor molecules move throughout the entire building. Because of the tendency of these molecules to disperse evenly, or to mix, the moisture in the humidified air moves toward drier air. In other words, in a building the moist indoor air attempts to reach the drier outside air. It moves toward the windows, where there is a lower temperature, and therefore, an increase in relative humidity to a point at which the water vapor will condense out of the air and onto the cold surface of the window. This is the dew point and it occurs at various temperatures, depending upon the type of windows in the building.

Usually, condensation on inside windows is a type of measurement of the allowable relative humidity inside a building. We can further assume that if this condensation activity is taking place on windows, it also is taking place within the walls if there is no vapor barrier (see Figure 15-75).

The typical outside wall has dry wall (or

**Figure 15-75.** Outside Wall Construction (*Courtesy of Research Products Corp.*)

plaster), a vapor barrier (on the warm side of the insulation), the insulation, air space, sheathing, building paper, and siding. Given an indoor temperature of 70°F (21.1°C) and a relative humidity of 35%, and an outside temperature of 0°F (−17.8°C), the temperature through the wall drops to about 63°F (17.2°C) at the vapor barrier, down to 17°F (−8.3°C) at the sheathing, and on down to 0°F (−17.8°C) outside. If a psychrometric chart was checked, we would discover that 70°F indoor air at 35% relative humidity has a dew point of 41°F (5°C). That temperature occurs right in the middle of the insulation. This is where condensation occurs—and where the trouble is —without a vapor barrier and a humidifier that can be controlled.

Properly controlled relative humidity, is the important factor in avoiding damaging effects of too dry air and too high relative humidity.

**HEALTH:** In the struggle between the nose and the air conditioning equipment, sometimes the heater wins and sometimes the cooler; but seldom does the nose win. The nasal mucus contains some 96 percent water. It is more viscous than mucus elsewhere in the body and even slight drying increases the viscosity enough to interfere with the work of the cilia. Demands on the nasal glands are great even under excellent conditions and they cannot cope with extreme dryness indoors in winter.

Experience has shown that with approaching winter, the first wave of dry nose patients appear in the doctor's office when the relative humidity indoors falls to 25%. It would seem, therefore, that 35% would be regarded as a passing grade, but 40% is something to try for. It boils down to this: a pint of water (.473 l) is a lot of water for a small nose to turn out. In disease or

old age, it simply does not deliver enough moisture; drainage stops and germs take over.

Dr. Joseph Lubart, an expert on the common cold, states in the New York State *Journal of Medicine:*

> Prevention of the common cold at present is our nearest approach to a cure. The most important prevention measure would appear to be proper regulation of the humidity especially during the heating season with its distressing drying of the indoor air and the creation of an environment favorable to the cold bug.

**TYPES OF HUMIDIFIERS:** There are many types of humidifiers available. They vary in price, capacity, and in principle of operation. For classification purposes, it is simpler, and more logical, to consider humidifiers in three general types:

1. Pan type.

2. Atomizing type.

3. Wetted element type.

The pan type is the simplest type of humidifier. It has a low capacity. On a hot radiator, it may evaporate .07 lbs (.317 kg) of water per hour. In the warm air plenum of a heating furnace, it would evaporate approximately 1.5 lbs (.680 kg) of water per hour.* To increase the capacity, the air-to-water surfaces can be increased by placing water wicking plates in the pan. The capacity goes up as the air temperature in the furnace plenum increases. Greater capacity is also possible through the use of a steam, hot water, or electric heating element immersed in the water (see Figure 15-76). A

* A plenum is a pressurized chamber with one or more distributing ducts connected to it.

**Figure 15-76.** Pan Type Humidifier (Courtesy of Research Products Corp.)

**Figure 15-77.** Atomizing Type Humidifier (Courtesy of Research Products Corp.)

**Figure 15-78.** Wetted Element Type Humidifier (Courtesy of Research Products Corp.)

1200-watt heating element, for example, in a container with water supplied by a float valve could produce 4 lbs (1.8 kg) of moisture per hour.

The second type of humidifier is the atomizing type. This device atomizes the water by throwing it from the surface of a rapidly revolving disc. It is generally a portable or console unit. However, it also can be installed so that the water particles will be directed into a ducted central system. (See Figure 15-77.)

The third type is the wetted element type. In its simplest form, it operates in the manner of an evaporative cooler. Here air is either pushed or pulled through a wetted element or filter and evaporative cooling takes place. By increasing the air flow or by supplying additional heat, the evaporation rate of the humidifier can be increased. The heat source for evaporation can be from an increase in water temperature or an increase in air temperature. (See Figure 15.78.)

The air for evaporation can be taken from the heated air of the furnace plenum and directed through the humidifier by the humidifier fan (see Figure 15-79). It also can be drawn through the wetted element by the air pressure differential of the furnace blower system (see Figure 15-80).

Furnace-mounted humidifiers, usually of the wetted element type, can be constructed so that they produce 10 or more pounds (4.536 kg) of moisture per hour. Because of their high capacity, this type usually has a humidistat or control which will actuate a relay or water valve and start a fan that

**Figure 15-79.** Humidifier Installation (*Courtesy of Research Products Corp.*)

**Figure 15-80.** Bypass Type Humidifier (*Courtesy of Research Products Corp.*)

operates until the control is satisfied. Normally more water is supplied to the unit than is evaporated, and this flushing action washes a large portion of the hardness salts from the evaporative element to the floor drain, eliminating them from the humidifying system.

## BASIC HEATING CONTROL

Certain controls are required in order for a heating system to provide safe, economical, and automatic heating for a building. Other more sophisticated controls of the system may be added to these basic controls.

**THERMOSTAT:** The purpose of a thermostat is to signal the equipment when to operate. During the heating season, when the temperature inside the conditioned space drops below the thermostat set point, an electrical circuit will be completed and the furnace will provide the required amount of heat. When sufficient heat has been added to the building, the thermostat will interrupt the electrical circuit and the heating equipment will stop the heating process.

Thermostats are available to provide almost any operation desired. The majority of thermostats are single-stage heating and single-stage cooling (see Figure 15-81). However, some equipment operation will demand that other staging be used, such as double-stage heating and double-stage cooling (see Figure 15-82). These thermostats are capable of controlling heating equipment in two stages, or inputs. The cooling can also be controlled in two parts. When less conditioning is used, only that much of the equipment is operated. However, if the lower capacity is not capable of maintaining the desired temperature the thermostat will cause the equipment to operate at full capacity.

**Figure 15-81.** Single-Stage Thermostat (*Courtesy of Honeywell Inc.*)

**Figure 15-82.** Double-Stage Heating, Double-Stage Cooling Thermostat (*Courtesy of Honeywell Inc.*)

**Figure 15-83.** Main Gas Valve (*Courtesy of White-Rodgers Division, Emerson Electric Co.*)

**Heat Anticipator:** Thermostats are provided with heat anticipators which cause the thermostat to stop the heating equipment just a little bit before the space is actually up to temperature. This is to prevent overheating of the space. The heat anticipator is a resistor wired in series with the temperature control circuit and provides heat inside the thermostat. This heat is what causes the thermostat to signal the heating equipment early. The heat anticipator is usually adjustable and should be set to match the current draw of the temperature control circuit. (See Figure 15-82.)

**MAIN GAS VALVE:** The purpose of the main gas valve is to allow gas to flow to the main burners or stop this flow on demand from the thermostat. The valve is mounted physically in the gas line to the heating equipment. Electrically it is placed in the temperature control circuit. (See Figure 15-83.) These valves usually contain all the elements of a gas manifold. These components are: solenoid gas stop valve, pilot

safety control, gas pressure regulator, pilot burner control, and main line gas shut-off.

**PILOT SAFETY:** The purpose of a pilot safety device is to interrupt the electrical circuit, or interrupt both the gas supply and the electrical control circuit when the pilot burner will not safely light the main burner. These are safety devices and should not be bypassed to allow the furnace to operate. They are termed 90 percent safety and 100 percent safety. A 90 percent safe system will allow pilot gas to flow at all times unless it is manually shut-off, while a 100 percent safe system will not allow any gas to flow when the pilot is not safe. (See Figure 15.84.) Most pilot safety devices are incorporated in the main gas valve and are 100 percent safe systems. When LP gas is being used in heating equipment the 100 percent safe system

**Figure 15-84.** One-Hundred Percent Pilot Safety Shut-Off (*Courtesy of Honeywell Inc.*)

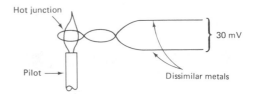

**Figure 15-85.** Thermocouple Principle

contacts which are operated by a bimetal sensing element. The element is placed inside the furnace heat exchanger. This element may be of the helical type, blade type as shown in Figure 15-86, or disc type (see Figure 15-87).These controls are not usually adjustable and are set to open the contacts at approximately 200°F and to close the contacts at approximately 170°F (77°C). These controls are sometimes manufactured

must be used. However, 90 percent safe systems may be used on natural gas units.

**THERMOCOUPLE:** The thermocouple provides the electrical power necessary to operate the pilot safety device. These devices are made of dissimilar metals so that when heated one metal will repel electrons while the other will attract them. Thermocouples are mounted in the pilot flame on one end and connected to the pilot safety device on the other end. When the pilot gas is lit and the flame encases the thermocouple, a small electrical current of approximately 25 to 35 millivolts is generated (see Figure 15-85). This small electric current operates an electromagnetic coil inside the pilot safety device.

**Figure 15-86.** Blade Type Limit Control (*Courtesy of White-Rodgers Division, Emerson Electric Co.*)

(a)             (b)

**Figure 15-87.** Disc Type Limit Control: (a) Front View; (b) Reverse View (*Courtesy of White-Rodgers Division, Emerson Electric Co.*)

**LIMIT CONTROL:** The limit control is a safety device that interrupts the temperature control circuit if temperatures inside the furnace heat exchanger should reach 200°F (93°C). It has one set of normally closed in the same housing as the fan control, but their function is completely different.

**FAN CONTROL:** The purpose of the fan control is to stop and start the fan in response to the temperature inside the furnace heat exchanger. In operation, when the heat exchanger temperature reaches approximately 150°F (66°C) the normally open contacts close and start the fan motor. When the temperature drops to approximately 100°F (37.8°C) the contacts open and stop the fan motor. This is done to prevent the circulation of cold air within the conditioned space. These controls also are operated by a bimetal sensing element (see Figure 15-88). In some instances it is desirable to have a fixed time before the fan will come on. This is accomplished by use of an electrically operated fan control. (See Figure 15-89.) The heater element in this control receives electrical power from the control circuit at the same time the gas valve is energized. After a preset time, the fan is started by the contacts closing. (See Figure 15-90.)

**Figure 15-89.** Fan Timer Switch (*Courtesy of Honeywell Inc.*)

**Figure 15-90.** Fan Timer Switch Wiring Diagram

**Figure 15-88.** Fan Control (*Courtesy of Honeywell Inc.*)

**FIRE STAT:** The fire stat is a safety device mounted in the fan section of the furnace that senses the return air temperature. It has a set of normally closed contacts which open and stop fan operation when the return air temperature reaches approximately 160°F (71°C). This is done to prevent the fan

**Figure 15-91.** Fire Stat Wiring Diagram

from agitating a fire in the building (see Figure 15-91).

**TRANSFORMER:** The transformer is a device used to provide the low voltage for the temperature control system. Line voltage is connected to the primary side and 24 volts is provided by the secondary side of the transformer. (See Figure 15-92.) Transformers are rated by their voltampere (VA) capacity. Usually a 40-VA transformer is used on heating and cooling applications.

**BOILER PRESSURE CONTROL:** This device is used to control the pressure inside a steam-heating boiler. A bellows is connected to the inside of the boiler. When the steam pressure reaches the set point, the main gas valve is closed and steam is no longer generated. (See Figure 15-93.) These controls also may be used as boiler pressure limit controls. The control setting is merely set at a higher pressure.

**IMMERSION CONTROL:** This is a temperature control which is used inside a hot water boiler. The sensing element is inserted directly into the hot water. The action of

**Figure 15-92.** Twenty-four Volt Transformer top, foot mounted; bottom, plate mounted (*Courtesy of Honeywell Inc.*)

the main gas valve is controlled by this device in response to water temperature. (See Figure 15-94.) This control may also be used as a high limit control.

**PROTECTORELAY CONTROL:** This control is used for the automatic recycling of an intermittent ignition oil burner heating

**Figure 15-93.** Boiler Pressure Control *(Courtesy of Honeywell Inc.)*

① MODELS WITH FIXED DIFFERENTIALS DO NOT INCLUDE ADJUSTING WHEEL.

② ON MODELS WITH VERTICAL MOUNTING OF IMMERSION WELL, ELEMENT IS ATTACHED TO BOTTOM OF THE CASE.

**Figure 15-94.** Immersion Control *(Courtesy of Honeywell Inc.)*

INTERNAL VIEW OF R8185E.

**Figure 15-95.** Protectorelay Control *(Courtesy of Honeywell Inc.)*

system. (See Figure 15-95.) The control senses the flue gas temperature of an oil burner system. If the flue gas temperature has not reached the designated temperature within a predetermined period of time, the control will stop the oil burner. After a given period of time the control will recycle the oil burner in an attempt to provide heat to the building. A schematic of a protectorelay control is shown in Figure 15-96.

**BASIC WARM AIR SYSTEM WIRING DIAGRAM:** There are three different circuits and voltages in a modern 24-volt temperature control system. (See Figure 15-97.) The circuits are: (1) fan or circulator circuit, (2) pilot safety circuit, and (3) temperature control circuit. The fan or circula-

**Figure 15-96.** Wiring Diagram of Protectorelay Control (*Courtesy of Honeywell Inc.*)

**Figure 15-97.** Basic Warm Air System Wiring Diagram

tor circuit is line voltage. The pilot safety circuit is millivolts. The temperature control circuit is 24 volts (low voltage).

**WARM AIR FURNACE SEQUENCE OF OPERATION:** In operation, when the thermostat demands heat, the main gas valve opens and allows gas to flow to the main

burners. When the temperature inside the furnace heat exchanger reaches approximately 150°F (66°C), the fan control starts the fan motor which circulates the air inside the conditioned space. The furnace is now in the heating cycle. When the thermostat is satisfied, the electrical circuit to the main gas valve is interrupted and the valve closes. The flame inside the furnace is extinguished and the temperature inside the furnace starts to fall. When this temperature falls to approximately 100°F (37.8°C), the fan control stops the fan motor and the air circulation stops. The furnace now awaits a command from the thermostat to start a heating cycle.

## SAFETY PROCEDURES

Heating equipment is probably the most dangerous of all the equipment used in the industry. This is because of the combustion process which provides heat for the building. Therefore, a service technician must practice safety procedures to prevent loss of life, property, and equipment.

1. Do not tamper with the safety controls. If a problem arises with one, replace it.

2. Avoid lighting the pilot until the fire box has been cleared of any unburned fuel.

3. Follow the manufacturer's directions carefully. He has spent much time and money developing these directions and they should be heeded.

4. Always follow the building and safety codes which are effective in your locality.

5. Be sure the unit is properly grounded electrically.

6. Do not make adjustments without knowing what the results will be.

7. Do not alter the venting system in any way.

8. Be certain that proper ventilation air is supplied in the equipment room.

9. Never allow a unit to be operated when a bad heat exchanger is suspected.

10. Never leave a furnace in operation that is not functioning properly.

## SUMMARY

• A heat source is any substance that produces heat while going through a given process.

• The main sources of heat are solids, liquids, gases, electricity, and solar energy.

• Coal is the main solid used for heating purposes.

• Fuel oil is the most common liquid used for heating purposes.

• Natural and LP gases are the most common gases used for heating purposes.

• LP gas is a mixture of butane and propane.

• Butane has a boiling point of 31.1°F (−.6°C) and a heat content of 3267 Btus/ft³ (23.3 kcal/m³).

• Propane has a boiling point of −43.8°F (−42°C) and a heat content of 2521 Btus/ft³ (635.29 kcal/m³).

• Electricity has a heating value of 3412 Btus/kW (859 kcal/kW).

• The types of electrical heating elements are: (1) open wire, (2) open ribbon, and (3) tubular cased.

• The major methods of electric heating are: resistance, heat pump, and a combination of the two.

• The heat pump is the most efficient electric heating method.

• Auxiliary heat is usually required when heat pumps are used in cold climates.

• The three basic requirements for combustion are: (1) fuel, (2) oxygen, and (3) heat.

• The limits of flammability should be observed when making adjustments on gas burners.

• Ten cubic feet (.28 m³) of air are required to burn just 1 cubic foot (m³) of natural gas.

• Excess air is supplied at the rate of 50 percent of the required air for combustion.

• The products of combustion produced when 1 ft³ (m³) of natural gas is completely burned are: 8 ft³ (.23 m³) of nitrogen, 1 ft³ (.0283 m³) of carbon dioxide, and 2 ft³ (.057 m³) of water vapor.

• The orifice is the opening through which gas is admitted to the main burner.

• The orifice and the burner must be matched in Btu (cal) ratings.

• A gas burner is defined as a device that provides for the mixing of gas and air in the proper ratio to ensure satisfactory combustion.

• Primary air is admitted to the main burner through the face of the burner.

• Forced draft burners are used on rooftop units.

• Power burners have a very wide range of capacities.

• The automatic pilot is probably the most important safety device used on modern gas heating equipment.

• The millivolt type pilot burner can be divided into two types: thermocouple and thermopile.

• The bimetal pilot uses a sensing element made of dissimilar metals.

• The liquid-filled pilot functions from the force exerted when a liquid is heated and a vapor is formed.

• There are basically two types of flames: yellow and blue.

• The heat exchanger is the heart of the heating system.

• The two general classifications of primary heat exchangers are: (1) barrel and (2) sectionalized.

• Hydronic heating uses both primary and secondary heat exchangers.

• The building requires additional insulation when electric heating is used.

• Electrical resistance heating involves: resistance elements, baseboard elements, and duct heaters.

• The minimum air velocity required for electric resistance heating is determined on the basis of entering air temperature and watts/ft² of the cross-sectional duct area.

• A heat pump can be defined as an air conditioning system that moves heat both to and from the conditioned area.

• Heat pumps are classified according to their heat source and are air-to-air, water-to-air, water-to-water, and ground-to-air.

• During the cooling season the heat pump operates as a normal cooling system.

• During the heating season the refrigerant cycle is reversed by a reversing valve.

• During the heating season the outdoor coil must be defrosted to keep the efficiency up on a heat pump.

• A heat pump should never be used for reheat purposes.

• In cold climates heat pump units are

equipped with supplemental electric heat to provide the added capacity needed.

- Furnaces are manufactured in upflow, downflow, and horizontal designs.

- The heat rise through a furnace can be defined as the temperature increase of the air as it passes through the heating device.

- There are basically three types of oil burners: (1) high pressure, (2) low pressure, and (3) rotary.

- In the high pressure oil burner the oil is forced through the burner nozzle at approximately 100 psig (7.03 kg/cm²).

- The low pressure oil burner forces the oil through the nozzle at from 1 to 15 psig (.0703 to 1.05 kg/cm²).

- The rotary oil burner uses a cup rotating at high speed to atomize the oil.

- A hot water boiler maintains water temperature between 120°F and 210°F (49°C and 99°C).

- Steam heating has numerous basic characteristics which can be advantageously employed.

- Steam-heating systems can be divided into two classifications: one-pipe and two-pipe.

- Humidification is one of the most important aspects of air conditioning.

- Relative humidity is the amount of moisture actually in the air as compared to the amount that it can hold.

- When air is heated it can hold more moisture.

- The three reasons for controlling humidity are: (1) comfort, (2) preservation, and (3) health.

- The three types of humidifiers are:

(1) pan type, (2) atomizing type, and (3) wetted element type.

- The purpose of the thermostat is to signal the equipment when to operate.

- The purpose of the main gas valve is to allow gas to flow to the main burners or stop this flow on demand from the thermostat.

- The purpose of the pilot safety device is to interrupt the electrical circuit, or interrupt both the gas supply and the electrical circuit when the pilot burner will not safely ignite the main burner.

- The thermocouple provides the electrical power necessary to operate the pilot safety device.

- The limit control is a safety device which interrupts the temperature control when the temperature inside the furnace heat exchanger reaches approximately 200°F (93°C).

- The purpose of the fan control is to stop and start the fan in repsonse to the temperature inside the furnace heat exchanger.

- The fire stat is a safety device installed in the fan compartment that stops the fan when the return air temperature reaches approximately 160°F (71°C).

- A transformer is used to reduce line voltage to 24 volts for use in the temperature control circuit.

- Boiler pressure controls are used to control the steam pressure inside a steam-heating boiler.

- Immersion controls are used to control the water temperature inside a hot water boiler.

- Protectorelay controls are used for the

automatic recycling of intermittent ignition oil burner heating systems.

## REVIEW EXERCISES

1. What are the four basic sources of heat used in modern heating units?

2. What is LP gas made of?

3. What is the boiling point of propane?

4. What is the ignition point of natural gas?

5. Name the components of natural gas and their percentages.

6. What is a definite advantage of LP gas?

7. Name the major methods of using electricity for heating purposes.

8. Why is a heat pump less efficient at low outdoor temperatures?

9. Define the combustion process.

10. What are the basic requirements for combustion?

11. What is excess oxygen?

12. Define a gas orifice.

13. Define a main gas burner.

14. Name the four different types of main gas burners.

15. Where is the primary air introduced into the main burner?

16. Where are forced draft burners popular?

17. What is probably the most important safety device used on modern gas heating equipment?

18. Name the types of pilot burners.

19. What are the two basic types of flames?

20. What controls the flow of combustion and excess air through a heat ex-changer?

21. In what types of heating equipment are secondary heat exchangers used?

22. Why is electrical heating beneficial from an ecological standpoint?

23. How is the air flow determined in electric heaters?

24. Define a heat pump.

25. How does a heat pump move heat to the inside of the conditioned area?

26. Why is the defrost cycle needed on a heat pump?

27. Is a heat pump designed to be used for reheat purposes?

28. What is the purpose of the reversing valve on a heat pump?

29. Define heat rise through a furnace.

30. Name the different types of oil burners.

31. What is the temperature range desired on a hot water boiler?

32. Name the piping arrangements used in steam-heating systems.

33. Which can hold the most moisture, cold air or warm air?

34. Name the three classifications of benefits of controlling humidity.

35. Name the three types of humidifiers.

36. What is the purpose of the thermostat?

37. What is the purpose of the pilot safety?

38. What is the limit control?

39. What provides the low voltage for the temperature control system on a modern heating furnace?

40. What device provides automatic recycling of intermittent ignition oil burners?

# Air Conditioning (Cooling)

Most people have little appreciation of the basic principles of air conditioning, probably because the public in general only became conscious of air conditioning in about 1920. It was at this time that the large scale use of air conditioning on trains and in theaters began. It was these systems that first exposed large numbers of people to the comfort and advantages of summer cooling and at the same time caused the erroneous impression that air conditioning is synonymous with cooling.

The cooling process involves air circulation, cooling, dehumidifying, and cleaning. The other processes are accomplished during the heating process.

## DEFINITION

The accepted definition of the term air conditioning is: the simultaneous mechanical control of temperature, humidity, air purity, and air motion. Unless all of these conditions are controlled, the term air conditioning cannot be properly applied to any system or equipment. It should be noted that the control of temperature can mean either cooling or heating. The control of humidity can mean either humidifying or dehumidifying. Thus an industrial system which provides an indoor condition of 150°F (66°C) db at 75% rh can just as well be called air conditioning as a system designed to provide indoor conditions of 80°F (26.7°C) db and 50% rh. Similarly a system which only cools a space without regard to the relative humidity or air purity or motion cannot be properly called a true air conditioning system. An air conditioning system can maintain any atmospheric condition regardless of variations in the outdoor atmosphere.

## HUMAN COMFORT

The two primary reasons for using air conditioning are: to improve the control of an industrial process and to maintain human comfort. The conditions to be maintained in an industrial process are dictated by the very nature of the process or the materials being handled. In a comfort system however, the conditions to be maintained are determined by the requirements of the human body. Therefore, an understanding of the essential body functions is basic to an understanding of air conditioning.

Human comfort is dependent upon how fast the body loses heat. The human body might be compared to a heating unit that uses food as its fuel. Food is composed of carbon and hydrogen. The energy contained in the fuel, food in our case, is released by oxidation. The oxygen used in the process comes from the air, and the principal products of combustion are carbon dioxide and water vapor. Doctors call this process metabolism.

The human body is basically a constant temperature machine. The internal temperature of the human body is 98.6°F (37°C) which is maintained by a delicate temperature regulating mechanism. Because the body always produces more heat than it needs, heat rejection is a constant process. The main object of air conditioning is to assist the body in controlling the cooling rate. This is true for both the heating and cooling seasons. In summer, the job is to increase the cooling rate; in winter, it is to decrease the cooling rate.

In the air conditioning process, there are three ways that the body gives off heat: (1) convection, (2) radiation, and (3) evaporation. In most cases the body uses all three methods at the same time.

**CONVECTION:** In the convection process, the air close to the body becomes warmer than the air farther away from the skin. Because the warm air is lighter than the cool air, the warm air rises. This warm air is replaced by the cooler air and the cooling by convection is a continuous process. As this

air becomes warm, it also floats upward. Even though the deep body temperature remains at 98.6°F (37°C), the human skin temperature will vary. The skin temperature may vary from 40°F to 105°F (4.4°C to 40°C) in relation to the temperature, humidity, and velocity of the surrounding air. If the temperature of the surounding air drops, the temperature of the skin also will drop.

**RADIATION:** Heat radiates directly from the body to any cooler surface just as the rays of the sun travel through space and warm the surface of the earth. Heat may flow from the skin to any surface or object that is cooler than the body. This process is independent of the convection process. The temperature of the air between the person and the cooler surface has no effect on the radiation process. The same principle applies when a person is warmed by a camp fire. The side next to the fire gets warm while the other side is cool. The air temperature between the person and the fire is the same as the air temperature on the person's other side.

**EVAPORATION:** Evaporation heat regulation is the body process that maintains life outside of an air conditioncd space. In this process moisture, or perspiration, is given off through the pores of the skin. When this moisture evaporates, it absorbs heat from the body and cools it. The effect of evaporation can be felt more easily when alcohol is put on the skin because the alcohol vaporizes more readily and absorbs heat faster. This evaporation turns the moisture into low pressure steam or vapor and is a continuous process. When drops of sweat appear on the skin it means that the body is producing more heat than it can reject at the normal rate.

## CONDITIONS THAT AFFECT BODY COMFORT

There appears to be no set rule as to the best conditions for all people. In the same atmospheric conditions, the young, healthy person may be slightly warm while an elderly one may be cool.

The three conditions that affect the ability of the body to give off heat are: (1) temperature, (2) relative humidity, and (3) air motion. A change in any one of these conditions will either speed up or slow down the cooling process.

**AIR TEMPERATURE:** Air at a temperature lower than the skin will speed up the convection process. The cooler the air, the more heat the body will lose through convection. Heat always flows from a warm place to a cooler place. The greater the temperature difference, the faster the heat will flow. If this difference in temperature is too great, the body will lose heat more rapidly than it should and it will become uncomfortable.

If the air temperature is higher than the skin temperature, the convection process will be reversed. The body heat will be increased. It can be seen that the air temperature has a very important effect on human comfort. Experience shows that air temperatures ranging from 72°F to 80°F (22.2°C to 26.7°C) feel comfortable to most people.

The temperature of any surrounding surfaces is also important because this temperature affects the rate of radiation from the body. The lower the surface temperature, the more heat is given off by the body through radiation. As the temperature difference between the surface and the body is decreased, the rate of radiation is decreased. The radiation process will be reversed if the surrounding surface temperature is higher than the body temperature. When this hap-

pens, the body must give off more heat through the convection and evaporation processes.

**RELATIVE HUMIDITY:** Relative humidity regulates the amount of heat that the body can reject through evaporation. Relative humidity is a measure of the amount of moisture in the air. It is an indication of the ability of the air to absorb moisture. Relative humidity is basic to the air conditioning process.

As an example, let us consider one cubic foot (m³) of air at 70°F (21.1°C) which contains four grains (.26 g) of water vapor. (See Figure 16-1.) A grain is a very small amount of water. Actually, it takes 7000 grains (453.6 g) to make one pound.

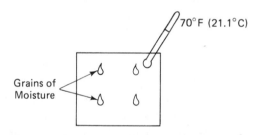

**Figure 16-1.** One Cubic Foot of Air at 70°F (21.1°C) with 4 Grains (.26 g) of Water Vapor

The relative humidity can be determined in the following manner. There are only four grains (.26 g) of water vapor in the cubic foot (m³) of air. If that cubic foot of air held all the moisture it could possibly hold at that temperature, it would actually hold eight grains (.52 g) of water vapor, (see Figure 16-2) and would be said to be saturated.

To find the relative humidity, divide the moisture actually present in the air by the amount that the air could hold at a saturated condition at the same temperature. (See

**Figure 16-2.** Cubic Foot of Saturated Air

Figure 16-3.) This process tells us that the relative humidity is 50%. Relative humidity is an indication of the amount of moisture in the air compared to the amount of moisture that could be present at that same temperature. The relative humidity will change with a change in temperature.

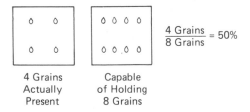

4 Grains                    Capable
Actually                    of Holding
Present                     8 Grains

**Figure 16-3.** Determination of Relative Humidity

For example: the temperature of the air is raised to 92°F (33.3°C) without adding any moisture. When the humidity tables are checked, we can see that one cubic foot (m³) of air at 92°F will hold 16 grains (1.04 g) of water vapor when saturated. The relative humidity in this example is 4 grains (.26 g) divided by 16 grains or 25%. (See Figure 16-4.)

If the air which surrounds the body has a low relative humidity, the body will give off more heat through evaporation. If the air

**Figure 16-4.** Relative Humidity at a Higher Temperature

surrounding the body has a high relative humidity, the body will give off less heat through evaporation. A conditioned air temperature at 80°F (26.7°C) and 50% relative humidity will be reasonably comfortable.

**AIR MOVEMENT:** An increase in the rate of evaporation of perspiration from the body is largely a result of air movement. Evaporation is dependent upon the ability of air to absorb moisture. As the air moves across the skin, the moisture-laden air is replaced by drier air which allows more moisture to evaporate from the skin.

If the air is allowed to remain still (static), the air next to the skin will absorb moisture until the saturation point is reached. As the saturation point is reached, the evaporation process is slowed down. The moisture will evaporate more slowly and will eventually stop when the saturation point is reached and the person will feel uncomfortable.

The movement of air also speeds up the convection process. This is possible because the warm air next to the skin is replaced by cooler air and heat is given up from the body to the air.

Air movement also removes heat from other substances such as walls, ceilings, and other objects surrounding the body, thus, tending to speed up the radiation process. It must be remembered that air motion is one of the conditions which affects the comfort of human beings.

**AIR DISTRIBUTION**

The air introduced into a conditioned space should be distributed so that there will be only minor temperature differences between the floor and ceiling—from floor level to 6 ft (1.8m) above the floor—and between the inside and outside walls. The proper quantity of air can be delivered to the different areas of the room by a careful consideration of the cooling and heating requirements. However, these requirements must be met without drafts. Generally, an air velocity of 15 to 25 fpm (4.6 to 7.6 mpm) is considered still air while air moving with a velocity of 65 fpm (19.8 mpm) is considered drafty to most people.

**AIR DISTRIBUTION REQUIREMENTS:**
A good air distribution system must do the following:

1.    Mix the conditioned air with enough room air so that upon reaching the occupied zone the air stream will not be so cold as to be objectionable.

2.    Allow the air velocity to be reduced sufficiently before reaching the occupied zone so as to prevent drafts.

3.    Provide a turbulent, underflowing air motion within the occupied zone.

4.    Keep any air noise from the supply and return grills to an unobjectionable level.

The requirements may seem difficult to meet but it should be remembered that an air conditioning unit may have the very best of components and still be, from the owner's viewpoint, unsatisfactory if the air distribution does not provide comfort for the occupants.

**FAN REQUIREMENTS:** The fans used in modern air conditioning systems fall into two general classes depending on the direction of the air flow through them. The first class is the axial flow type through which the air flows in the direction of and parallel to the fan shaft. This type of fan is commonly known as a *propeller fan.* (See Figure 16-5.) It consists essentially of two or more blades which extend from the shaft.

**Figure 16-5.   Axial Flow Type Fan**

Each blade is bent or twisted to give the desired pitch. The blades are often made from formed sheet metal but are sometimes made from castings and molded from plastic. This type of fan is best suited for applications where it is necessary to move large volumes of air against low resistances, such as air-cooled condensers, humidifiers, and cooling tower fans. (See Figure 16-6.)

**Figure 16-6.   Air-Cooled Condenser (Courtesy of Dearborn Stove Co.)**

The second class of fan is the radial flow type through which the air flow is outward from the fan shaft. This outward air flow is produced by centrifugal force. Therefore, this type is commonly known as a *centrifugal fan* or blower. It consists of a series of blades mounted around the circumference of a circle with the shaft in the center. The blades themselves are parallel to the shaft and may be either forward curved or backward curved (see Figure 16-7). The fan using forward curved blades will generally run

Figure 16-7.  Centrifugal fans

**Figure 16-8.**  Centrifugal fan scroll

at a lower speed than a backward curved fan for the same air delivery conditions. However, the fan that uses backward curved blades will tend to be nonoverloading while a forward curved fan will tend to overload as the pressure is reduced and the air volume is increased. Therefore, unless a very careful analysis is made as to the exact operating pressure, there is considerably more danger of overloading and burning out the motor with forward curved fans.

The fan wheel of a centrifugal blower is not sufficient by itself to provide a usable flow of air. A casing or housing is always necessary to collect and direct the air flow as it leaves the fan blades. This housing, or scroll, usually starts with a very small area just after the discharge outlet and progressively increases in size around the circumference of the shell until the maximum size is reached at the discharge opening (see Figure 16-8). This shape is known as a *volute*. The air inlet to the wheel usually takes the form of a circular opening in the casing in one or both sides. The diameter of this opening is as large as possible without exceeding the inside diameter of the blower wheel.

Centrifugal fans are capable of delivering large volumes of air against considerable resistance. It is for this reason that they are the first choice in air conditioning and ventilating systems where it is necessary to overcome the resistance of the air filters, cooling coils, heat exchangers, ducts, and outlets.

**FAN NOISES:** Fans and blowers can be designed and operated at such speeds as to provide almost any desired velocity and air volume. However, the difficulty in air conditioning work is the noise produced by the fans and blowers. The noise is caused mainly by the successive air waves which are caused by the fan blades and by the turbulence caused by the resistance of the air being forced through the system. A small amount of noise may be caused by the vibrating fan blades. This noise is relatively minor, however, when compared with the noise that may be created by a high velocity fan. The humming noise produced by a fan is in effect a miniature siren. The loudness and the pitch of the sound depend upon the tip speed of the fan, the general shape of the blades, and the number and angle of the blades. Thus, it can be seen that there is a definite relation between the amount of noise and the quantity of air being delivered. To be suitable, an air conditioning system must deliver the proper quantity of air with a minimum of fan noise.

**HEATING CHAMBER:** In most air conditioning systems the next device the air encounters is the heating chamber (see Figure 16-9). This device can be either a gas-fired furnace heat exchanger, an electric heating element, or a hot water or steam coil, depending on the system design. It is here that the air is heated during the heating season. The heat gained by the air is termed the temperature rise. During the cooling season this apparatus does not generally supply heat.

1. Heating Chamber
2. Burners
3. Controls
4. Insulation
5. Cabinet
6. Blower

**Figure 16-9.** Heating Chamber (*Courtesy of Dearborn Stove Co.*)

**COOLING CHAMBER:** Upon leaving the heating chamber, the air flows through the cooling chamber. The cooling process is generally done by a refrigeration evaporator (see Figure 16-10). However, the cooling process may also be accomplished by a chilled water coil. During the cooling process the temperature of the air is lowered. When there is excessive moisture in the air some of it is removed to lower the humidity. There must be a drain line connected to drain away this condensate. The moisture or humidity level must be kept low to produce effective cooling. The cooling chamber is not generally used during the heating process even though air will still flow through it.

**FEATURES:**

1. HEADER PLATES – Heavy gauge steel or aluminum.
2. EXPANSION VALVE — thermostatic non-adjustable type. Mounted at the factory.
3. CONDENSATE PAN — heavily undercoated. 7/8" O.D.S. copper tubing drain connection.
4. SUCTION AND LIQUID LINE — stubs for sweat connections.
5. COIL CIRCUITS — are properly manifolded and circuited for maximum performance.
6. COOLING COIL — aluminum finned, copper tube "A" type, designed for minimum resistance and maximum heat transfer.

**Figure 16-10.** Cooling Coil (*Courtesy of Southwest Manufacturing Company*)

**HUMIDIFIER:** As the air leaves the cooling chamber it enters the area where a humidifier may be located (see Figure 16-11). Not all systems include a humidifier. The humidifier is used to add moisture to the air during the heating season. As the air leaves the heating chamber the relative humidity is lower. This warm, dry air should be humidified for comfort and health reasons. The desired moisture is added to the air by evaporation, or by spraying moisture directly into the air stream, depending on the type of humidifier used. These units are not generally used during the cooling season because the removal of moisture is part of the cooling process. To add moisture at this time would be expensive and undesirable.

**Figure 16-11.** Humidifier location

**SUPPLY DUCTS:** The air is then forced through a series of pipes known as air ducts which direct the air to the desired area. This duct system must be properly designed if satisfactory operation is to be obtained. The air outlet location must also receive a great deal of consideration in order to maintain proper air distribution within the conditioned space. These air ducts are insulated to prevent the escape of heat during the heating season and the absorption of heat during the

cooling season. This insulation must include a vapor barrier to prevent the condensation of moisture on the colder surfaces during the cooling process. (See Figure 16-12.) The duct system is used during both the heating and cooling processes.

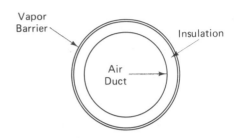

**Figure 16-12.** Air duct with insulation and vapor barrier

**SUPPLY OUTLETS:** Supply outlets are devices placed on the end of the supply duct in the conditioned space (see Figure 16-13). They help to distribute the air properly in the room. Some outlets are used to fan the air, while others are used to direct the air in a jet stream. Other types are designed to do both. They are used to control the direction of air flow. The directional control, along with the location and number of outlets in the room, contribute tremendously to the satisfactory operation and comforting effect of the air pattern.

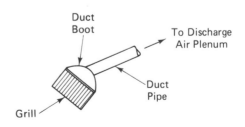

**Figure 16-13.** Supply outlet

**CONDITIONED SPACE:** The conditioned space is one of the most important parts of the air distribution system. If an enclosed space did not exist, the air could not be reclaimed and the circulation of conditioned air would not be possible. It is because of this that the conditioned space is so important. The material and workmanship used to build the space are also important because they help to control the heat loss or heat gain inside the space. (See Figure 16-14.)

**RETURN DUCTS:** Return ducts are used to connect the conditioned space and the air-handling equipment together (see Figure 16-16). The return ducts are generally designed for less resistance to air flow than the supply ducts. This is done to ensure sufficient air return to the fan. Return ducts are sometimes, but not always, insulated. If the duct is to be run through a hot or cold space, it should be insulated to prevent heat loss or heat gain into the air inside the duct.

**Figure 16-14.  Conditioned space**

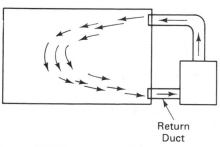

**Figure 16-16.  Return duct**

**RETURN OUTLETS:** Return outlets are the openings through which the air from the conditioned space is allowed to enter the return duct system. They are generally located opposite the supply outlet (see Figure 16-15). This is not always possible nor feasible but should be done when possible. They should be properly sized to allow air to leave the room with as little resistance as possible.

**FILTERS:** Filters are located at the inlet of the fan to prevent dust particles from entering the equipment. (See Figure 16-17.) Their only purpose is to clean the air. They should always be properly located to protect the fan, heating chamber, and cooling chamber from dust particles. If this dust is allowed to enter the equipment it will gradually collect on the surfaces and will reduce the air

**Figure 16-15.  Location of return outlets**

**Figure 16-17.  Location of filter**

flow and adversely affect system operation. Filters are made from many materials such as spun glass and composition plastic. More efficient types of filters operate on the electronic principle.

## AIR CONDITIONING TERMINOLOGY

Before a study of air conditioning can be undertaken, an understanding of the terms used is necessary. The following definitions explain the most common terms.

**DRY AIR:** Dry air is air that contains no water vapor. In nature, air contains some water vapor.

**ABSOLUTE HUMIDITY:** Absolute humidity is the actual quantity or weight of moisture present in the air in the form of water vapor. It is expressed as the weight of water vapor in grains per pound of air. There are 7000 grains of moisture to one pound of moisture. A rise or fall in temperature without a change in the quantity of moisture in the air does not affect the absolute humidity.

**RELATIVE HUMIDITY (RH):** Relative humidity is the ratio between the moisture content of a given quantity of air at any temperature and pressure compared to the maximum amount of moisture which that same quantity of air is capable of holding at the same temperature and pressure. Relative humidity is expressed as a percentage. Note particularly the difference between relative humidity and absolute humidity. The relative humidity can be found by dividing the amount of moisture actually present by the maximum moisture-holding capacity of the air at that same temperature and pressure.

**Example:** If one cubic foot ($m^3$) of air contains 4 grains (.26 grams) of moisture and is capable of holding 8 grains (.52 grams) of moisture, what is the relative humidity?

$$rh = \frac{\text{Absolute Humidity}}{\text{Maximum Amount It Can Hold}}$$

$$= \frac{4}{8} = 50\%$$

A rise in air temperature will cause a decrease in the ratio or relative humidity. A fall in air temperature will cause a rise in the ratio or relative humidity, unless the saturation point (100% rh) has been reached. Beyond this point, condensation will take place as the temperature falls. When air contains all of the moisture it can hold at any given temperature and pressure, it is said to be saturated. When air is saturated, the relative humidity is 100%.

**HUMIDIFICATION:** The process of adding moisture to the air is called humidification.

**DEHUMIDIFICATION:** The process of removing moisture from the air is called dehumidification.

**SATURATED AIR:** Saturated air is air which contains all the water vapor it is capable of holding at that particular pressure and temperature. When air is saturated the dry bulb, the wet bulb, and the dew point temperatures are the same.

**DRY BULB TEMPERATURE:** The temperature of the air as indicated by an ordinary thermometer is known as the dry bulb temperature and is a measure of the sensible heat content of the air.

**WET BULB TEMPERATURE:** The wet

bulb temperature is the temperature of the air as measured by an ordinary thermometer with the bulb covered with wet cloth or gauze (see Figure 16-18). The temperature is taken after the thermometer has been exposed to a rapidly moving air stream. The temperature indicated on a wet bulb thermometer will be depressed lower than a dry bulb thermometer reading because of the evaporation of the moisture in the wick. The heat necessary for the evaporation of the water from the wick is removed from the air while the air is passing over the thermometer. The thermometer has been exposed to the air stream for a sufficient length of time when the temperature falls to a point of equilibrium which is determined by the rate of moisture evaporation from the wick and the quantity of sensible heat in the air.

The point of equilibrium is the temperature at which the rate of sensible heat transfer from the air to the water on the wet bulb is equal to the rate at which heat is transferred from the wet bulb to the air by evaporation of the water. For any given dry bulb temperature and any given relative humidity, the point of equilibrium will be the same. Since the rate of evaporation from the wick is determined by the quantity of moisture in the air, as well as the sensible heat, it can be seen that the wet bulb temperature is indicative of the total heat contained in the air.

**WET BULB DEPRESSION:** The difference between the dry bulb temperature and the wet bulb temperature for any given condition is called the wet bulb depression except at the saturation point when the two temperatures are the same.

**DEW POINT TEMPERATURE:** The dew point temperature is the temperature at which water vapor will start to condense out of the air. The quantity of water vapor in the air is always the same at any given dew point temperature. Therefore, the quantity of moisture in the air can be measured by the dew point temperature. When air is at the dew point temperature, it is holding all the moisture it can hold at that temperature. As long as there is no removal or addition of moisture, the dew point temperature will remain constant. No latent heat is given to or released from a mixture of air and water vapor unless some moisture is added to or removed from the air.

**SATURATION TEMPERATURE:** When air is at the dew point temperature, it is also at the saturation temperature and is considered to be saturated with moisture. The relative humidity is 100% at the saturation temperature. That is, it contains all the moisture it can hold at that temperature.

**TOTAL HEAT:** When the sensible heat and the latent heat are added together, the sum is known as the total heat. In air conditioning, 0°F (−17.8°C) is usually taken as the point from which heat content is measured. Total heat is also indicated by a wet bulb thermometer.

**Figure 16-18.** Wet bulb thermometer

**SATURATED GAS (VAPOR):** When a gas or water vapor is at the temperature of the boiling point which corresponds to its pressure, it is also at its saturation temperature. This saturated state is the condition under which a gas exists above its liquid in a closed container, as in a refrigerant drum or an evaporator. A saturated gas contains no superheat.

**POUND OF AIR:** When this term is used it means a pound of *dry* air. The following terms are examples of the use of the term pound of air: total heat per pound of air, moisture per pound of air, or latent heat per pound of air. In each case this means a pound of dry air and does not refer to a pound of the mixture.

**CUBIC FOOT OF AIR:** This term merely indicates a cubic foot of air and is used in expressing the quantity of moisture present in one cubic foot of dry air when it is saturated.

**VENTILATION:** The process of supplying or removing air from a space is known as ventilation whether it is conditioned air, used air, or fresh air.

**EFFECTIVE TEMPERATURE:** Effective temperature is a temperature which is determined by experiment. It is not a temperature measured on a thermometer, but is a measure of personal comfort as felt by an individual. Effective temperature is produced by the correct combination of dry bulb temperature, relative humidity, and air motion.

**COMFORT ZONE:** This is the range of temperature and humidity in which most people feel comfortable. The outer limits of the comfort zone are not clearly defined due to dependence upon outside conditions.

**OVERALL COEFFICIENT OF HEAT TRANSMISSION:** The quantity of Btus (cal) which are transmitted each hour through one square foot (m²) of any material, or combination of materials, for each degree of temperature difference between the two sides of a partition is known as the overall coefficient of heat transmission. Note that the difference between the two sides of the material is the difference in temperature of the air on both sides, and not the difference in temperature between both surfaces of the material.

**STATIC PRESSURE:** Static pressure is the pressure which a gas or air exerts at right angles against the walls of an enclosure-duct. Static head or static pressure of air is usually measured in inches (mm) of water column by use of manometers and pitot tubes. The frictional loss in air ducts is known as pressure drop or loss of static head.

**VELOCITY PRESSURE:** This is the pressure due to the air movement. Velocity pressure is created by the energy of motion or kinetic energy in the moving air. It is measured in inches (mm) of water column.

**TOTAL PRESSURE:** Total pressure is the sum of the static pressure and velocity pressure and is a measure of the total energy of the air.

**STACK EFFECT:** This is an upward flow or draft created by the tendency of heated air to rise in a vertical or inclined duct. Warm air rises because it expands and becomes lighter when heated. This principle is of special importance in the heating of tall buildings, where the entire building acts as a chimney (stack) due to the warmer air inside of it. The warm air rises and escapes out of openings or cracks on the upper floors while cold air is drawn in through doors, windows, cracks, etc. on the lower floors to replace the warm, rising air. As a result, extra radiation must be used on the lower floors to offset the chilling effect of the incoming cold air.

**PLENUM CHAMBER:** The plenum chamber is an equalizing chamber or air supply compartment to which the various distributing ducts of an air conditioning or ventilating system are connected. The plenum chamber is maintained under pressure which causes the air to flow into the different distribution ducts.

## PSYCHROMETRICS

The science which deals with the relationships that exist within a mixture of air and water vapor is known as *psychrometrics*. In air conditioning, psychrometrics involves the measuring and determining of the different properties of air both inside and outside of the conditioned space. Psychrometrics also can be used to establish the conditions of air which will provide the most comfortable conditions in a given air conditioning application.

## PSYCHROMETRIC CHART

A psychrometric chart is a diagram which represents the various relationships that exist between the heat and moisture content of air and water vapor. Different air conditioning manufacturers have slightly different versions of the psychrometric chart, locating the various properties in different places on the chart. The factors shown on a complete psychrometric chart are: dry bulb temperature, wet bulb temperature, dew point temperature, relative humidity, total heat, vapor pressure, and the actual moisture content of the air.

**IDENTIFICATION OF LINES AND SCALES ON A PSYCHROMETRIC CHART:** The following illustrations will help to locate the different lines and scales on a psychrometric chart. Consider the psychrometric chart as being a shoe with the toe pointing to the left and the heel on the right (see Figure 16-19).

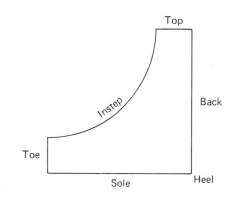

**Figure 16-19.** Psychometric chart outline

**Dry Bulb Temperature Lines:** The dry bulb temperature scale is located on the "sole" of the psychrometric chart (see Figure 16-20). The dry bulb lines extend vertically upward from the sole. There is one line for each degree of temperature.

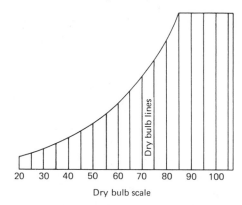

**Figure 16-20.** Dry bulb lines (*Courtesy of Research Products Corp.*)

**Wet Bulb Temperature Lines:** The wet bulb temperature scale is found along the "instep" of the chart extending from the toe to the top (see Figure 16-21). The wet bulb lines extend from the instep diagonally downward to the right. There is one line for each degree of temperature.

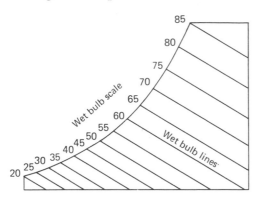

**Figure 16-21.** Wet bulb lines (*Courtesy of Research Products Corp.*)

**Relative Humidity Lines:** On a complete psychrometric chart, the relative humidity lines are the only curved lines on it (see Figure 16-22). The various relative humidities are indicated on the lines themselves. There is no coordinate scale as with the other air properties.

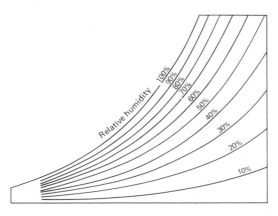

**Figure 16-22.** Relative humidity lines (*Courtesy of Research Products Corp.*)

**Absolute Humidity Lines:** The scale for absolute humidity is a vertical scale on the right side of the psychrometric chart (see Figure 16-23). The absolute humidity lines run horizontally to the left from this scale.

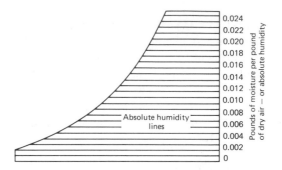

**Figure 16-23.** Absolute humidity lines (*Courtesy of Research Products Corp.*)

**Dew Point Temperature Lines:** The scale for the dew point temperature is identical to the scale for the wet bulb temperature (see Figure 16-24). The dew point temperature lines run horizontally to the right, however, not diagonally as is the case with the wet bulb temperature lines.

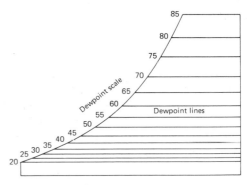

**Figure 16-24.** Dew point lines (*Courtesy of Research Products Corp.*)

**Specific Volume Lines:** The specific volume scale is located along the sole of the chart from 12.5 ft³ to 14.5 ft³ (see Figure 16-25). The cubic feet lines extend diagonally up-

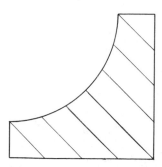

**Figure 16-25.** Cubic feet lines

ward to the left from the sole to the instep of the chart. The specific volume lines on the chart are identified in terms of cubic feet per pound of air.

**Enthalpy Lines:** The enthalpy scale is located along the instep of the chart (see Figure 16-26). The enthalpy lines are the same as the wet bulb lines on the psychrometric chart. Enthalpy is total heat content. It can be used with the psychrometric chart to measure any heat change that takes place in any given psychrometric process. Both sensible and latent heat can be measured by use of enthalpy. It is a quick means of finding either of these factors.

When we put together the seven charts we have just covered, we will have a complete psychrometric chart (see Figure 16-27).

**USE OF THE PSYCHROMETRIC CHART:** Any two of the foregoing properties which are located on the psychrometric chart will determine definitely the condition of the air. If we choose any dry bulb temperature and any wet bulb temperature, the point at which these two lines intersect on the chart is the point which indicates the condition of the air at those given temperatures. The condition of the air at this point is definitely fixed. Similarly, the condition of the air at any other point on the psychrometric chart is fixed by the particular dry bulb and wet bulb temperatures. Because the possible combinations of any two temperatures are unlimited, there are an infinite number of possible conditions of air, and an equally infinite number of possible points that may be plotted on the chart.

When a fixed air condition has been located at a point on the chart, all of the other properties of that sample of air can be determined from the chart. Similarly, by use of the psychrometric chart, any two properties of the air and water vapor combination are sufficient to determine a condition of the air and all its other properties.

**Example 1:** Given a dry bulb temperature of 95°F (35°C) and a dew point temperature of 54°F (12.2°C), find the wet bulb temperature (see Figure 16-28).

Solution:

Step 1. Locate 95°F (35°C) on the dry bulb scale.

2. Draw a line straight upward to the instep.

3. Follow the instep down until 54°F (12.2°C) is found.

4. Extend this point horizontally to the right until the 95°F (35°C) dry bulb line is crossed.

**Figure 16-26.** Enthalpy lines

Figure 16-27.

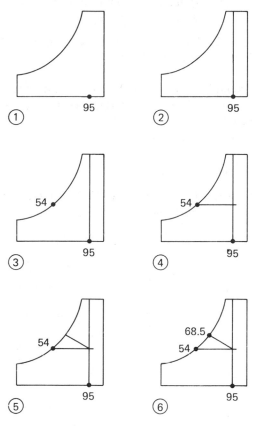

**Figure 16-28.** Finding wet bulb temperature

5. Extend the point upward to the left to the wet bulb scale on the instep.

6. Read the wet bulb temperature of 68.5°F (20.3°C).

**Example 2:** Given a wet bulb temperature of 74°F (23.3°C) and a dew point temperature of 70°F (21.1°C), find the dry bulb temperature (see Figure 16-29).

Solution:

Step 1. Locate 74°F (23.3°C) on the wet bulb scale.

2. Draw a line diagonally down the right to the back of the chart.

3. Locate 70°F (21.1°C) on the dew point scale.

4. Extend this point horizontally to the right until the wet bulb line is crossed.

5. Extend this point vertically downward to the dry bulb scale.

6. Read the dry bulb temperature of 84°F (28.9°C).

**Example 3:** Given a wet bulb temperature of 73°F (22.8°C) and a dry bulb temperature

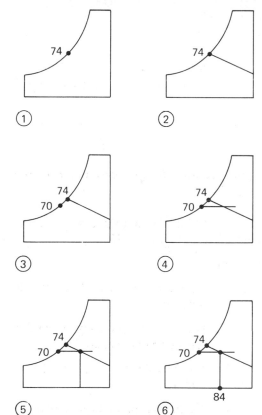

**Figure 16-29.** Finding dry bulb temperature

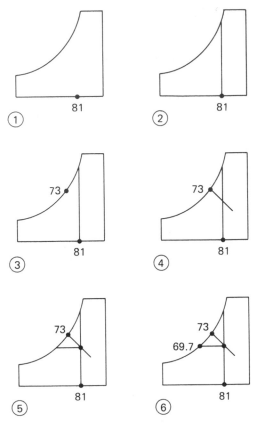

**Figure 16-30.** Finding dew point temperature

of 81°F (27.2°C), find the dew point temperature (see Figure 16-30).

Solution:

Step 1. Locate 81°F (27.2°C) on the dry bulb scale.

   2. Draw a line vertically upward to the instep of the chart.

   3. Locate 73°F (22.8°C) on the wet bulb scale.

   4. Extend this point diagonally downward until the dry bulb line is crossed.

   5. Extend this point horizontally to the left to the instep.

   6. Read a dew point temperature of 69.7°F (20.9°C).

**Example 4:** Given a dry bulb temperature of 95°F (35°C) and a dew point temperature of 54° (12.2°C), find the relative humidity (see Figure 16-31).

Solution:

Step 1. Locate 95°F (35°C) on the dry bulb scale.

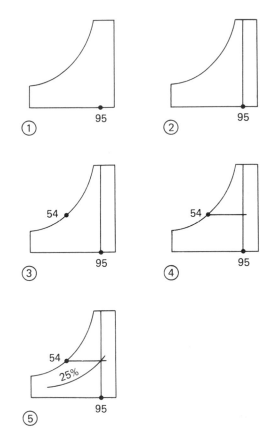

**Figure 16-31.** Finding relative humidity

2. Draw a line vertically upward to the instep of the chart.

3. Locate 54°F (12.2°C) on the dew point scale.

4. Extend this point horizontally to the right until the dry bulb line is crossed.

5. Read the relative humidity of 25% at this point.

**Example 5:** Given a wet bulb temperature of 74°F (23.3°C) and a dew point temperature of 70°F (21.1°C), find the total heat content of the air (see Figure 16-32).

Solution:

Step 1. Locate 74°F (23.3°C) on the wet bulb scale.

2. Draw a line diagonally upward to the left to the total heat scale.

3. Read the total heat to be 36.91 Btus (9.3 kcal) per pound (.4536 kg) of air.

**Example 6:** Given a dry bulb temperature of 95°F (35°C) and a dew point temperature of 54°F (12.2°C), find the total heat content of the air (see Figure 16-33).

Solution:

Step 1. Locate 95°F (35°C) on the dry bulb scale.

2. Draw a line vertically upward to the instep of the scale.

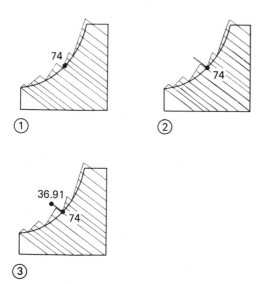

**Figure 16-32.** Finding total heat content using wet bulb temperature

**Figure 16-33.** Finding total heat content using dry bulb temperature

3.  Locate 54°F (12.2°C) on the dew point scale.

4.  Extend this point horizontally to the right until the dry bulb line is crossed.

5.  Extend this point diagonally upward to the left to the total heat scale.

6.  Find the total heat to be 32.32 Btus (8.14 kcal) per pound (.4536 kg) of air.

This total heat is found by interpolation.

Total heat at 69°F (20.6°C)
wet bulb = 32.71 Btu (8.24 kcal)

Total heat at 68°F (20°C)
wet bulb =31.92 Btu (8.04 kcal)

Difference =     .79 Btu (.199 kcal)

68.5° − 68° = .5°
.5° × .79  = .395

therefore

Total heat at 68.5°F (20.3°C)
wet bulb = 31.92 + .395
= 32.315 or 32.32 Btus
(8.14 kcal).

**CHANGE IN TOTAL HEAT CONTENT:**
During the cooling season, we are primarily interested in the amount of heat that must be removed so that the air is cooled enough to fulfill the inside design conditions. In the heating season, heat is added to the air to meet the inside design conditions. Suppose for instance that the outside design condition is 75°F (23.9°C) wet bulb and it is desired to maintain a 67°F (19.4°C) wet bulb temperature inside the conditioned space. The amount of total heat that must

be removed per pound (453.6 g) of dry air is found by the following method:

Total heat at 75°F (23.9°C)
wet bulb = 37.81 Btu/lb
or 20.79 kcal/kg of air

Total heat at 67°F (19.4°C)
wet bulb = 31.15 Btu/lb
or 17.13 kcal/kg of air

Difference = 6.66 Btu/lb
or 3.66 kcal/kg of air

Therefore, the total heat that must be removed in cooling air from 75°F (23.9°C) wet bulb down to 67°F (19.4°C) wet bulb is 6.66 Btus (1.68 kcal) per pound (453.6 g) of dry air (see Figure 16-34).

**Figure 16-34.**   Change in total heat

**SENSIBLE HEAT:** Sensible heat is heat that can be added to or removed from a substance without a change of state. If that substance is air, changing the sensible heat will result only in a change in temperature. The sensible heat content is indicated by the dry bulb temperature. A change in dry bulb temperature will result in a change in sensible heat only because there is no change of state.

**Example 7:** During the heating season air is to be heated from a 65°F (18.3°C)) dry bulb temperature and a 50°F (10°C) wet bulb temperature to an 88°F (31.1°C) dry bulb temperature and a 60°F (15.6°C) wet bulb temperature. Find the sensible heat that must be added per pound of dry air (see Figure 16-35).

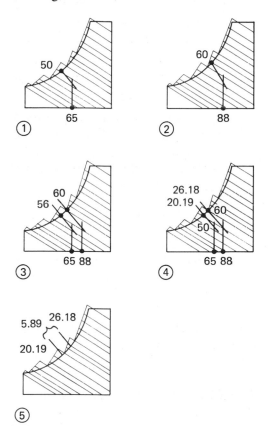

**Figure 16-35.** Sensible heat change on heating air

Solution:

Step 1. Locate 65°F (18.3°C) dry bulb and 50°F (10°C) wet bulb.

2. Locate 88°F (31.1°C) dry bulb and 60°F (15.6°C) wet bulb.

3. Extend these points diagonally upward to the left to the total heat scale.

4. Read 20.19 and 26.18 Btus (11.1 kcal/kg and 14.4 kcal/kg)

5. Obtain the difference:

Total heat at 60°F (15.6°C) wet bulb = 26.18 Btu/lb. (14.4 kcal/kg)

Total heat at 50°F (10°C) wet bulb = 20.19 Btu/lb (11.1 kcal/kg)

Sensible heat added = 5.99 Btu/lb (3.3 kcal/kg of dry air)

This heat change is sensible heat only because there was no change in the moisture content of the air.

**LATENT HEAT:** The dew point temperature is always an indication of the moisture content of the air. Any change in the dew point temperature will result in a change in moisture content. The moisture content can only be changed by changing the dew point temperature. Therefore, it should be noted that as long as the dew point temperature remains constant there will be no change in the moisture content or water vapor which is present in the air.

The moisture content scale used on most charts indicates grains of moisture per pound (453.6 g) of dry air. The dew point temperature is read on the same scale as the wet bulb temperature.

The latent heat of vaporization is the amount of heat in Btu (cal) required to change a liquid to a gas at a constant temperature. If we now have a given number of grains of moisture per pound (453.6 g) of dry air, there must have been a certain amount of heat required to vaporize this moisture into the air. This is the latent heat content of the air and water vapor mixture.

**Example 8:** Air at 75°F (23.9°C) dry bulb and 57°F (13.9°C) wet bulb is to be conditioned to obtain a 75°F (23.9°C) dry bulb temperature and a 70°F (21.1°C) wet bulb temperature. Find the amount of latent heat added and the grains (453.6 g) of moisture added. (See Figure 16-36.)

Solution:

Step 1.  Locate 75°F (23.9°C) dry bulb and 57°F (13.9°C) wet bulb.

2.  Extend this point to the total heat scale and read 24.5 Btus (6.2 kcal).

3.  Locate 75°F (23.9°C) dry bulb and 70°F (21.1°C) wet bulb.

4.  Extend this point to the total heat scale and read 34 Btus (8.6 kcal).

5.  The heat added to this air is 34 − 24.5 = 9.5 Btus (2.4 kcal) per pound (453.6 g) of dry air.

6.  Extend these points to the right to the grains of moisture scale and read 44 grains (2.85 g) at the beginning of the process and 102 grains (6.61 g) of moisture per pound (453.6 g) of dry air at the end of the process.

7.  The grains of moisture added per pound (453.6 g) of dry air is 102 − 44 = 58 grains of moisture added.

This heat change is latent heat only because there was no change in the dry bulb temperature of the air.

**SENSIBLE HEAT RATIO:** Sensible heat ratio is the sensible heat removed in Btu (cal) divided by the total heat removed in Btu (cal). The sensible heat ratio indicates the sensible heat percentage of total heat removed. The sensible heat ratio is above 50 percent in most comfort air conditioning. This is because most comfort air conditioning systems remove more sensible heat than latent heat.

The ratio will vary with the type of installation. A residence may have a .7 or .8

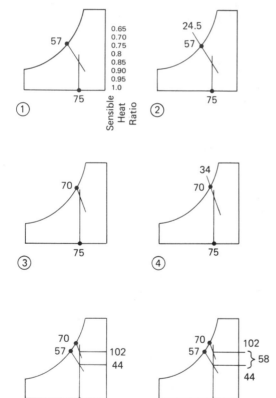

**Figure 16-36.** Latent heat change and grains of moisture added

sensible heat factor, while a restaurant may have a .5 or .6 sensible heat ratio. This indicates that 70 or 80 percent of the total change in heat removed from the air is sensible heat.

The sensible heat ratio scale is located on the right hand side of the psychrometric chart. When the conditions of air are plotted on the chart and a line is drawn through them and extended to the sensible heat ratio scale the ratio is read directly. (See Figure 16-37.) When the line is at approximately a 45° angle, the sensible heat ratio is 50 percent or .50. This indicates that half of the heat removed is sensible and half is latent.

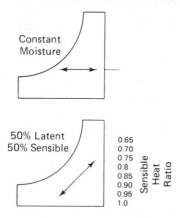

**Figure 16-37.** Sensible heat ratio line

**Example 9:** The desired room conditions are 80°F (26.7°C) dry bulb and 67°F (19.4°C) wet bulb. The supply air conditions are 60°F (15.6°C) dry bulb and 58°F (14.4°C) wet bulb. Find the sensible heat factor (see Figure 16-38).

Solution:

Step 1. Plot the given conditions on the psychrometric chart.

2. Draw a straight line between the two points.

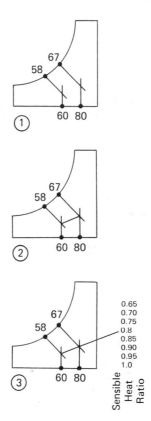

**Figure 16-38.** Finding sensible heat ratio (line method)

3. Extend the line to the sensible heat ratio scale.

4. Read .8 on the sensible heat ratio scale.

Another method of finding the sensible heat ratio is to use a ratio of the total heat removed to the sensible heat removed from the air. If we use the same conditions as stated in Example 9, we can figure the sensible heat ratio as follows (see Figure 16-39).

Solution:

Step 1. Plot the given conditions on the psychrometric chart.

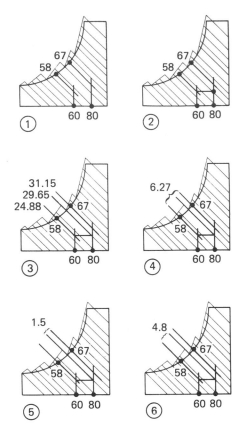

**Figure 16-39.** Calculated sensible heat ratio

2. Extend the final air condition [60°F (15.6°C) dry bulb and 58°F (14.4°C) wet bulb] point horizontally to the right until the 80°F (26.7°C) dry bulb line is crossed.

3. Extend the 80°F (26.7°C) dry bulb line vertically downward to the sole of the chart. This forms a reverse "L."

4. Extend the room air point, the supply air point, and the intersection of the two sides of the "L" to the total heat scale.

5. Read the total heat for each line extended; room conditions 31.15 Btus (7.85 kcal); supply conditions 24.88 Btus (6.3 kcal); and intersection of "L" 29.65 Btus (7.5 kcal).

6. Calculate total heat removed 31.15 − 24.88 = 6.27 Btus (1.5 kcal).

7. Calculate latent heat removed 31.15 − 29.65 = 1.5 Btus (.38 kcal).

8. Calculate sensible heat removed 29.65 − 24.88 = 4.8 Btus (1.2 kcal).

9. Calculate sensible heat ratio
$$\frac{4.8}{6.27} = .76\%$$

This indicates that 76 percent of the heat removed is sensible heat.

**MIXING TWO QUANTITIES OF AIR AT DIFFERENT CONDITIONS:** The science of air conditioning involves the process of mixing air, such as when the conditioned air supply is mixed with the room air or when it is mixed with bypass air. Another example is when fresh air is supplied in some proportion for ventilation purposes. When the initial condition and the quantities of these two air sources are known, the condition of the final mixture can easily be found by use of the psychrometric chart.

**Example 10:** Outdoor air at 95°F (35°C) dry bulb and 75°F (23.9°C) wet bulb, point *A*, is to be mixed with return air at 70°F (21.1°C) dry bulb and 10% relative humidity, point *B*. The mixture is to consist of 25 percent outdoor air and 75 percent return air. Find the resulting dry bulb and wet bulb temperatures of the mixture (see Figure 16-40).

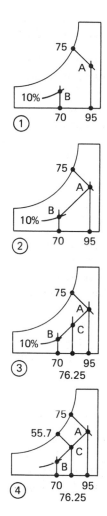

**Figure 16-40.** Mixing two quantities of air

Solution:

Step 1. Plot the two points, *A* and *B,* on the chart.

2. Draw a line between the two points.

3. Locate the dry bulb temperature by adding the percentages of each dry bulb temperature, that is

25% of 95° = 23.75 (−4°C)

75% of 70° = 52.5 (−11.4°C)

Resulting dry bulb temperature of mixture = 76.25 (24.4°C)

4. Locate 76.25°F (24.4°C) dry bulb on the scale and extend this point to the mixture line, point *C*.

5. Extend point *C* to the wet bulb scale and read 55.7°F (13°C).

It is very important to note that it is permissible to find the resulting dry bulb temperature of a mixture of air by the percentage method, but it is not permissible to use this method for determining the resulting wet bulb temperature. For example, if the percentage method was used in the preceding example the results would be as follows:

25% of 75° = 18.75
75% of 47° = 32.25
_____
51.00  *incorrect* resulting wet bulb temperature

The percentage method can be used to find the resulting wet bulb temperature by indirect application. In order to do this it is necessary first to find the total heat for each condition of the air and then to apply the percentage method to find the resulting total heat of the air mixture. From the total heat of the mixture it is then possible to locate the corresponding correct wet bulb temperature. The preceding example would then be:

**Example 11:** The total heat at 75°F (23.9°C) is 37.81 Btus (9.53 kcal). The total heat at 47°F (8.3°C) is 18.60 Btu (4.7 kcal). Find the corresponding wet bulb temperature.

Solution:

    25% of 37.81 =   9.45 Btus (2.38 kcal)

    75% of 18.60 = 13.95 Btus (3.52 kcal)

    Total heat of moisture

    = 23.40 Btus (5.89 kcal)

The corresponding wet bulb temperature is 55.6°F (13.1°C).

## HEAT LOAD CALCULATION

Everything in nature reacts against the air conditioning process. During the summer, the outdoor elements tend to make the cooling equipment work harder to remove the unwanted heat from within the space. During the winter, the outdoor elements tend to make the heating equipment work hard to supply warmth to the conditioned space.

The process of estimating the size of air conditioning equipment required to serve the conditioned space is much like the process used in refrigeration calculations. The chief difference is that there are more sources of heat involved in air conditioning estimating.

An accurate load estimate will indicate four things:

1.   The true cooling or heating load requirements.

2.   The most economical equipment selection.

3.   The possibilities for least cost with the greatest load reduction.

4.   The most efficient air distribution system.

**HEAT SOURCES:** Heat sources are listed under two general classifications:

1.   Those which result in an internal heat load on the conditioned air space.

2.   Those which result in an external heat load on the conditioned space.

The external heat load requires extra evaporator capacity but it does not affect the conditioned air after it has been delivered to the conditioned space. The following list is a classification of these sources.

1.   Heat sources which result in an internal heat load:

    a.   Heat conduction through the walls, roof, windows, etc.
    b.   Solar heat gain.
    c.   Duct heat gain (when outside the conditioned space).
    d.   Occupant heat load.
    e.   Lighting load.
    f.   Electric equipment and appliances.
    g.   Outside air infiltration.

2.   Heat sources which result in an external heat load:

    a.   Ventilation air.
    b.   Any heat added to the air after it leaves the conditioned space.

The heat that enters a conditioned space may come from any or all of the above listed sources. An accurate survey must be made on every estimate so that the heat load can be calculated correctly. The purpose of cooling equipment is to remove unwanted heat and maintain the space at the desired conditions. Therefore, caution must be exercised to make sure that the total heat load used is the load for the peak period. Many of the heat gains are at their peak at different times. For example, a restaurant may have a peak sun load at noon, while the peak occupant load may not be until seven o'clock in the evening. These two peak loads should not be added together. Another example would be when the lighting heat load is larger than the solar load from the sun and would not occur at the same time. It can be seen that the heat load on an air conditioning system is subject to great variation. Because of this, the heat load is calculated on the basis of one hour and not 24 hours. The cooling system selected must be able to handle this *peak load hour*. It must be remembered that the peak load hour of the day is when the sum of the heat loads is at a peak and not a sum of the heat loads.

**CONDUCTED HEAT LOAD:** Heat always flows from a warm temperature to a cooler temperature. Therefore, when the temperature inside a conditioned space is lowered, heat will be conducted through the walls, doors, windows, etc., from the outside to the inside. (See Figure 16-41.) The amount of heat which will flow through a wall whose temperatures differ on each side will depend on three factors:

1. The area of the wall.

2. The heat-conducting characteristics of the wall, called the *overall coefficient of heat transmission,* commonly designated as the *U* factor.

U = Heat Transfer

**Figure 16-41. Heat conducted through a wall**

3. The difference in the temperature between the spaces which are separated by the wall.

These three factors also are applied to all forms of construction such as doors, windows, and roofs. The formula used to calculate heat flow is:

$$Q = A \times U \times (T_1 - T_2)$$

where:

$Q$ = heat transferred per hour in Btu (cal)

$A$ = net area of the wall, window, door, etc.

$U$ = the number of Btu (cal) which will flow through one square foot ($m^2$) of the material, in one hour, with a temperature difference of one degree F (.56°C) between the two sides.

$T_1$ = outside temperature

$T^2$ = inside temperature

The temperature differential, $T_1 - T_2$, is often given as TD. When this designation is used the formula becomes:

$$Q = A \times U \times TD.$$

The *U* factors for most materials have been tabulated and placed in tables for convenience. (See Table 16-1.) A more complete listing of *U* factors can be found in the American Society of Heating, Refrigerating, and Air Conditioning Engineers' Society *Guide and Data Book.* The guide provides the most accurate information available on this subject.

When a heat load is to be calculated, each type of construction must be considered and calculated separately.

## Table 16-1

### Partial List of Heat Transfer Coefficient *U* Factors

#### EXTERIOR WALLS

| | |
|---|---|
| Wood siding, 2″ by 4″ studs, plastered | .25 |
| Wood siding, 2″ by 4″ suds, plastered ½″ insulation | .17 |
| Brick veneer, 2″ by 4″ studs, plastered | .27 |
| Brick veneer, 2″ by 4″ studs, plastered ½″ insulation | .18 |
| Stucco, 2″ by 4″ studs, plastered | .30 |
| Brick wall, 8″ thick, no inside finish | .50 |
| Brick wall, 8″ thick, furred, plastered on wood lath | .30 |
| Brick wall, 12″ thick, furred and plastered on wood lath | .24 |
| Hollow tile, 10″ thick | .39 |
| 4″ brick hollow tile, 12″ thick, furred, plastered wood lath | .20 |
| 4″ brick faced concrete, 10″ thick | .48 |

#### INTERIOR WALLS OR PARTITIONS

| | |
|---|---|
| Studding, wood lath and plaster both sides | .34 |
| Studding, wood lath and plaster both sides, ½″ insulation | .21 |
| 4″ hollow tile, plastered both sides | .40 |
| 4″ common brick, plastered both sides | .43 |

#### FLOORS, CEILINGS, AND ROOFS

| | |
|---|---|
| Plastered ceiling without flooring above | .62 |
| Plastered ceiling with 1″ flooring | .28 |
| Plastered ceiling on 4″ concrete | .59 |
| Suspended plastered ceiling under 4″ concrete | .37 |
| Concrete 4″ (floor or ceiling) | .65 |
| Concrete 8″ (floor or ceiling) | .53 |
| Average wood floor | .35 |
| Average concrete roof | .25 |
| Average wood roof | .30 |

#### GLASS AND DOORS

| | |
|---|---|
| Window glass, single | 1.13 |
| Double glass windows | .45 |
| Doors, thin panel | 1.13 |
| Doors, heavy panel (1¼″) | .59 |
| Doors, heavy panel (1¼″) with glass storm door | .38 |

**Example 12:** Consider a wall 80 ft (24.32 m) long and 8 ft (2.43 m) high and constructed of 8-inch (20.32 cm) thick brick, furred and plastered with wood lath. There is one thin panel door, 6 ft by 2½ ft (1.8 m by .76 m). There will be a temperature difference (TD) of 15°F (8.4°C). The $U$ factors found in Table 16-1 for this type of construction are .30 for the wall and 1.13 for the door.

Solution:

$$\text{Gross area of wall} = 640 \text{ ft}^2 \ (59.10 \text{ m}^2)$$

$$\text{Area of door} = \underline{15 \text{ ft}^2} \ (1.37 \text{ m}^2)$$

$$625 \text{ ft}^2 \ (57.73 \text{ m}^2)$$

Substitute the values in the formula $Q = U \times A \times TD$

Heat gain through wall $= 625 \times \ .30 \times 15 = 2812.5$ Btu (708.7 kcal)

Heat gain through door $= \ 15 \times 1.13 \times 15 = \ \underline{254.3}$ Btu (64 kcal)

$$\text{Total Heat Gain} = 3066.8 \text{ Btu} \ (772.8 \text{ kcal})$$

**Example 13—Part 1:** This example is somewhat more complicated than Example 12 and will use Figure 16-42 to complete the computations, which will be completed in three separate steps. The exterior walls will be considered first. The wall is 10 ft. long and 8 ft. high (3.04 m × 2.43 m) and constructed of 8″ (20.32 cm) thick brick, furred and plastered on wood lath.

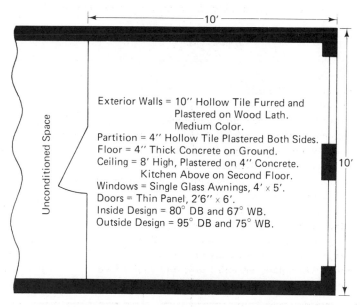

Exterior Walls = 10″ Hollow Tile Furred and Plastered on Wood Lath. Medium Color.
Partition = 4″ Hollow Tile Plastered Both Sides.
Floor = 4″ Thick Concrete on Ground.
Ceiling = 8′ High, Plastered on 4″ Concrete. Kitchen Above on Second Floor.
Windows = Single Glass Awnings, 4′ x 5′.
Doors = Thin Panel, 2′6″ x 6′.
Inside Design = 80° DB and 67° WB.
Outside Design = 95° DB and 75° WB.

Unconditioned Space

**Figure 16-42.** Sketch for example 13

Solution:          Gross area of exterior wall =

3 (10 ft × 8 ft) (3.04 m × 2.43 m) = 240 ft² (22.16 m²)

Area of exterior glass =

2 (4 ft × 5 ft) (1.2 m × 1.5 m) =  40 ft² (3.6 m²)

Net exterior wall = 200 ft² (18.56 m²)

From Table 16-1 $U$ factor for wall = .30
$U$ factor for glass = 1.13

TD = 95°F − 80°F (35°C − 26.7°C) = 15°F (8.3°C)

Heat gain through wall = 200 × .30 × 15 = 900 Btu/hr (226.8 kcal)

Heat gain through glass = 40 × 1.13 × 15 = 678 Btu/hr (170.85 kcal)

Table 16-2

Temperature Differential for Abnormal Conditions
(Based on 15° normal differential for exterior walls)

| STRUCTURE | TD |
| --- | --- |
| Walls—exterior | 15 |
| Glass in exterior walls | 15 |
| Glass in partitions | 10 |
| Partitions next to unconditioned rooms | 10 |
| Partitions next to laundries, kitchens, boiler rooms, or store show windows having a large lighting load | 30 |
| Floors above unconditioned rooms | 10 |
| Floors on ground or over unvented spaces or basements | 0 |
| Floors above laundries, kitchens, or boiler rooms | 40 |
| Floors above vented spaces | 15 |
| Ceilings with unconditioned rooms above | 10 |
| Ceilings with laundries, kitchens, etc., above | 25 |
| Ceilings with roof directly above (no attic) | 15 |
| Ceilings with totally enclosed attic and roof above | 20 |
| Ceilings with cross-vented attic and roof above | 15 |

Note: If normal differential is 20°, add 5° to above values

This heat loss is all sensible heat because it was calculated on a difference in dry bulb temperatures.

**TEMPERATURE DIFFERENCES (TD):** In practice, the 15°F (8.34°C) does not remain constant for all surfaces when calculating the conduction heat load. For example, if the conditioned space is located on the opposite side of a partition from a space which has a high temperature, like a boiler room, it is obvious that a 15°F temperature differential is not adequate. In cases where a 15°F TD for the exterior walls would be normal, the TD across the partition would be higher than 15°F. Table 16-2 gives the TD in degrees which should be used if the normal differential for the exterior walls is 15°F. If the normal TD is to be maintained is 20°F (11.12°C) for the exterior walls then 5°F (2.8°C) must be added to the TD given in the table. For example, for ceilings with laundries or kitchens located above the table shows an indicated TD of 25°F (14°C) when the normal TD is 15°F. If the normal TD is 20°F then 5°F must be added to the 25°F which indicates a TD of 30°F (16.8°C) for this particular ceiling.

**Example 13**—Part 2: To calculate the heat gain through the partition, floor, and ceiling we will use Table 16-2.

Solution:

$$\text{Gross area of partition} = 10 \text{ ft} \times 8 \text{ ft } (3.04 \text{ m} \times 2.4 \text{ m})$$
$$= 80 \text{ ft}^2 \ (24.3 \text{ m}^2)$$

$$\text{Gross area of door} = 6 \text{ ft} \times 2\frac{1}{2} \text{ ft } (1.8 \text{ m} \times .76 \text{ m})$$
$$= 15 \text{ ft}^2 \ (4.56 \text{ m}^2)$$

$$\text{Net area of partition} = 65 \text{ ft}^2 \ (19.76 \text{ m}^2)$$

Table 16-1 indicates: $U$ factor for partition $= .40$

$U$ factor for thin panel door $= 1.13$

Table 16-2 indicates that partitions which are located next to unconditioned spaces have a TD of 10°F (5.6°C) when the normal TD of 15°F (8.4°C) is used. This TD also applies to the door.

Heat gain through the partition $= 65 \times .40 \times 10 = 260$ Btu/hr (65.52 kcal)

Heat gain through the door $= 15 \times 1.13 \times 10 = 169.5$ Btu/hr (42.7 kcal)

The area of the floor $= 10 \text{ ft} \times 10 \text{ ft } (3.04 \text{ m} \times 3.04 \text{ m}) = 100 \text{ ft}^2 \ (30.4 \text{ m}^2)$

The area of the ceiling $= 10 \text{ ft} \times 10 \text{ ft } (3.04 \text{ m} \times 3.04 \text{ m}) = 100 \text{ ft}^2 \ (30.4 \text{ m}^2)$

Table 16-1 indicates: $U$ factor for the floor $= .65$

$U$ factor for the ceiling $= .59$

Table 16-2 indicates: TD for the floor $= 0°F \ (-17.8°C)$

TD for the ceiling $= 25°F \ (-3.9°C)$

Heat gain through the ceiling $= 100 \times .59 \times 25 = 1475$ Btu/hr (371.7 kcal)

Heat gain through the floor $= 100 \times .65 \times 0 = 0$ Btu/hr (0 kcal)

The total heat gain by conduction can now be found by adding together all the calculated values:

| | | | |
|---|---|---|---|
| Exterior walls | = | 780.0 Btu/hr | (196.56 kcal) |
| Exterior glass | = | 678.0 Btu/hr | (170.85 kcal) |
| Partition | = | 260.0 Btu/hr | ( 65.52 kcal) |
| Door in partition | = | 169.5 Btu/hr | ( 42.7  kcal) |
| Ceiling | = | 1475.0 Btu/hr | (371.7  kcal) |
| Floor | = | 0  Btu/hr | ( 0   kcal) |
| Total heat gain | = | 3362.5 Btu/hr | (847.35 kcal) |

The air conditioning unit designed for this building must remove 3362.5 Btu (847.35 kcal) each hour for the heat which was gained by conduction. This is the amount of heat that will be conducted through the exterior walls and surfaces. The other sources of heat which also must be removed by the air conditioning unit will be taken up in the following examples.

**DESIGN TEMPERATURES:** The various parts of the country undergo a considerable variation in temperatures daily. For example, a city like Minneapolis which is located in the northern part of the country may have very high temperatures during parts of the summer season. However, on an average the temperatures in Minneapolis are lower than the average temperatures in Dallas. When calculating the conducted heat load for an air conditioning system, it is necessary to choose the proper design temperature. The Weather Bureau records indicate that extremely high temperatures occur during less than 10 percent of the time. Therefore, the design temperature for air conditioning is not based on the highest temperature occurring in that locality, but on its average maximum temperature. It would

not be economically feasible to install an air conditioning unit which would provide the cooling required for the highest temperatures, since they would occur for only a small percentage of the time.

The design dry bulb temperatures for the various cities throughout the United States have been placed in tables for convenience. (See Table 16-3.) The wet bulb temperature and the summer wind velocity in miles per hour are also shown. For example, the correct outside design conditions for Detroit would be 95°F (35°C) dry bulb, 75°F (23.9°C) wet bulb, and a wind velocity of 10.3 mph (16.58 km/hr).

**SOLAR HEAT LOAD:** When a roof, ceiling, wall, or window is exposed to the direct rays of the sun, the surface will warm up very quickly, thus giving the effect of a greater TD between the inside and outside surfaces. The factors which affect the quantity of solar heat gained are:

1.   The time of day.

2.   The direction in which the exposed surfaces face.

3.   The color of the exposed surface.

## Table 16-3
### Design Wet Bulb and Dry Bulb Summer Temperatures and Wind Velocities

| CITY & STATE | db | wb | mph | CITY & STATE | db | wb | mph |
|---|---|---|---|---|---|---|---|
| ALABAMA | | | | NEBRASKA | | | |
| Birmingham | 95 | 78 | 5.2 | Lincoln | 95 | 75 | 9.3 |
| Mobile | 95 | 80 | 8.6 | NEVADA | | | |
| ARIZONA | | | | Reno | 95 | 65 | 7.4 |
| Phoenix | 105 | 76 | 6.0 | NEW HAMPSHIRE | | | |
| ARKANSAS | | | | Manchester | 90 | 73 | 5.6 |
| Little Rock | 95 | 78 | 7.0 | NEW JERSEY | | | |
| CALIFORNIA | | | | Trenton | 95 | 78 | 10.0 |
| Los Angeles | 90 | 70 | 6.0 | NEW YORK | | | |
| San Francisco | 90 | 65 | 11.0 | Buffalo | 93 | 75 | 12.2 |
| COLORADO | | | | New York | 95 | 75 | 12.9 |
| Denver | 95 | 64 | 6.8 | NEW MEXICO | | | |
| CONNECTICUT | | | | Santa Fe | 90 | 65 | 6.5 |
| New Haven | 95 | 75 | 7.3 | NORTH CAROLINA | | | |
| DELAWARE | | | | Asheville | 90 | 75 | 5.6 |
| Wilmington | 95 | 78 | 9.7 | Wilmington | 95 | 79 | 7.8 |
| DIST. COLUMBIA | | | | NORTH DAKOTA | | | |
| Washington | 95 | 78 | 6.2 | Bismarck | 95 | 73 | 8.8 |
| FLORIDA | | | | OHIO | | | |
| Jacksonville | 95 | 78 | 8.7 | Cincinnati | 95 | 78 | 6.6 |
| GEORGIA | | | | Cleveland | 95 | 75 | 9.9 |
| Atlanta | 95 | 76 | 7.3 | OKLAHOMA | | | |
| IDAHO | | | | Oklahoma City | 101 | 76 | 10.1 |
| Boise | 95 | 65 | 5.8 | OREGON | | | |
| ILLINOIS | | | | Portland | 90 | 65 | 6.6 |
| Chicago | 95 | 75 | 10.2 | PENNSYLVANIA | | | |
| INDIANA | | | | Philadelphia | 95 | 78 | 9.7 |
| Indianapolis | 95 | 76 | 9.0 | RHODE ISLAND | | | |
| IOWA | | | | Providence | 93 | 75 | 10.0 |
| Des Moines | 95 | 77 | 6.6 | SOUTH CAROLINA | | | |
| KANSAS | | | | Charleston | 95 | 80 | 9.9 |
| Wichita | 100 | 75 | 11.0 | SOUTH DAKOTA | | | |
| KENTUCKY | | | | Sioux Falls | 95 | 75 | 7.6 |
| Louisville | 95 | 76 | 8.0 | TENNESSEE | | | |
| LOUISIANA | | | | Memphis | 95 | 78 | 7.5 |
| New Orleans | 95 | 79 | 7.0 | TEXAS | | | |
| MAINE | | | | Dallas | 100 | 78 | 9.4 |
| Portland | 90 | 73 | 7.3 | El Paso | 100 | 69 | 6.9 |
| MARYLAND | | | | Houston | 95 | 78 | 7.7 |
| Baltimore | 95 | 78 | 6.9 | UTAH | | | |
| MASSACHUSETTS | | | | Salt Lake City | 92 | 63 | 8.2 |
| Boston | 92 | 75 | 9.2 | VERMONT | | | |
| MICHIGAN | | | | Burlington | 90 | 73 | 8.9 |
| Detroit | 95 | 75 | 10.3 | | | | |

Table 16-3   (Cont'd)

| CITY & STATE | db | wb | mph | CITY & STATE | db | wb | mph |
|---|---|---|---|---|---|---|---|
| MINNESOTA | | | | VIRGINIA | | | |
| Minneapolis | 95 | 75 | 8.4 | Richmond | 95 | 78 | 6.2 |
| MISSISSIPPI | | | | WASHINGTON | | | |
| Vicksburg | 95 | 78 | 6.2 | Seattle | 85 | 65 | 7.9 |
| MISSOURI | | | | WEST VIRGINIA | | | |
| Kansas City | 100 | 76 | 9.5 | Parkersburg | 95 | 75 | 5.3 |
| St. Louis | 95 | 78 | 9.4 | WISCONSIN | | | |
| MONTANA | | | | Milwaukee | 95 | 75 | 10.4 |
| Helena | 95 | 67 | 7.3 | WYOMING | | | |
| | | | | Cheyenne | 95 | 65 | 9.2 |

4.   The type of surface, whether it is smooth or rough.

5.   The latitude.

The time of day is very important because the solar effect will vary greatly from one hour to another. There are tables and charts designed to provide the excess differential, due to the solar effect, for any wall or roof for any hour of the day. (See Figure 16-43.) Notice that the solar effect on the east wall is the greatest at 10:00 A.M. It is the greatest on the south walls and the roof at 2:00 P.M., and on the west walls at 6:00 P.M. When the rays of the sun are perpendicular to the surface, the rate of heat absorption is the greatest. However, due to the time lag in the heat passing through the wall, it is not noticed in the conditioned space until sometime later. This time lag will vary depending on the construction of the surface. For instance, a light frame construction may have a lag of less than one hour while a heavy masonry type construction may have a lag of several hours. The average type construction will usually have a time lag of two hours. For example, at noon the rays of the sun are perpendicular to the roof, but the peak effect is not felt until 2:00 P.M.

There is no time lag through glass because the sun's rays pass directly through the glass and release heat when they strike objects in the room. The solar heat gain through glass is, therefore, maximum during the time when the sun shines through the window with the greatest intensity. When a building is to be air conditioned, the windows and glass are usually protected from the direct rays of the sun.

Walls and roofs constructed with dark surfaces absorb a great deal more heat than light-colored surfaces. Because of this, roofs are usually coated with a light color or a reflective type paint. The solar heat may be omitted when there is a ventilated air space, attic, or room above the conditioned space, or when the roof is shaded by adjacent buildings, etc., except for a very short period at noon. When roofs are partly shaded, calculate the heat gain for the exposed part only. When the roof is sprayed with water to cool the surface, the solar heat load need not be calculated for the roof.

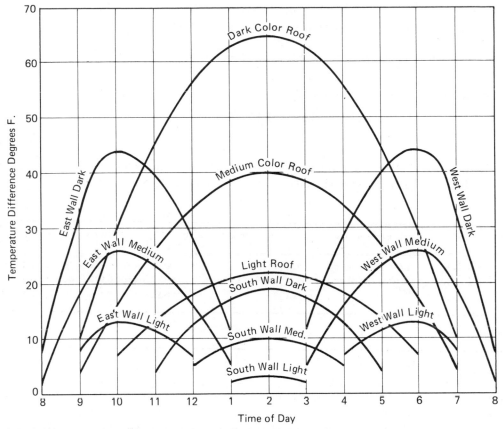

**Figure 16-43.** Solar chart

The peak solar heat load in a one-story building will usually occur when the sun's effect on the roof is at a peak, except when steps have been taken to reduce the solar load. Ordinarily, the areas of the east and west walls and glass, and to some extent the south walls and glass, are the factors which govern the heat load. Considerable care must be exercised in calculating the solar load. Good common sense used when inspecting and determining the peak heat load is an absolute necessity.

When calculating the solar heat load, remember that it is calculated separately from the conduction heat load and is in addition to the conduction heat load.

The solar heat load is calculated exactly like the conduction heat load; that is, the area is found and multipied by the same $U$ factor which is a characteristic of the wall. This factor is then multiplied by the solar temperature difference found from Figure 16-43.

**Example 13**—Part 3: Referring back to Figure 16-42, we can now calculate the solar load on the east wall and glass.

Solution:

The gross area of the wall = 10 ft × 8 ft (3.04 m × 2.43 m) = 80 ft² (7.39 m²)

Area of the glass = 2 × 4 ft × 5 ft (21.2 m × 1.5 m) = *40* ft²
(3.6 m²)

As before, the U factor for the wall = .30

U factor for the glass = 1.13

Figure 16-43 indicates that the peak solar load on a medium-colored east wall would occur at 10:00 A.M. The combined solar heat load will be calculated as follows:

| A | 8:00 A.M. U × TD | 9:00 A.M. U × TD | 10:00 A.M. U × TD |
|---|---|---|---|
| East wall | 40 ft² × .30 × 2°F = 24 | 40 ft² × .30 × 18°F = 216 | 40 ft² × .30 × 26°F = 312 |
| East glass | 40 ft² × 1.13 × 52°F = 2350.4 | 40 ft² × 1.13 × 46°F = 2079.2 | 40 ft² × 1.13 × 32°F = 1446.4 |
| Total Btu | 2374.4 | 2295.2 | 1758.4 |

| A | 8:00 A.M. U × TD | 9:00 A.M. U × TD | 10:00 A.M. U × TD |
|---|---|---|---|
| East wall | 3.7 m² × 1.46 × 1.12°C = 6.05 kcal | 3.7 m² × 1.46 × 10°C = 54.02 kcal | 3.7 m² × 1.46 × 14.56°C = 78.6 kcal |
| East glass | 3.7 m² × 5.51 × 29.12°C = 593.6 kcal | 3.7 m² × 5.51 × 25.76°C = 525.17 kcal | 3.7 m² × 5.51 × 17.9°C = 364.9 kcal |
| Total kcal | 599.65 | 579.19 | 443.5 |

Notice especially that the combination heat gain may not come at a time when either of the separate loads is at a peak. This calculation shows that the peak solar load on this east wall and glass will be 2371.2 Btu (597.54 kcal) per hour at 8:00 A.M. Remember that this is in addition to the conducted heat load and would occur only while the sun is shining. The conducted heat load would be effective even when the sun is not shining.

**OCCUPANT LOAD:** The heat gain for people has also been calculated and placed in tables. (See Table 16-4.) Notice that this is the first item that has been divided into sensible and latent heat. The use of this table simply involves finding out how many people fall into each classification and multiplying this number by the proper value from the table.

The amount of heat given off by a person is dependent upon how active the person is. For example, people who are seated and at rest, as in the home, in restaurants, or in theaters, give off approximately 350 Btus (88.2 kcal) per hour of total heat. In the case of medium activity the total heat may be increased to 850 Btus (214.2 kcal) per hour.

Table 16-4

Body Heat of Occupants

|  | SENSIBLE | | LATENT | | TOTAL | |
| --- | --- | --- | --- | --- | --- | --- |
|  | Btu/hr | (kcal) | Btu/hr | (kcal) | Btu/hr | (kcal) |
| Theater | 195 | ( 49.1) | 155 | ( 39 ) | 350 | ( 88.2) |
| Office | 200 | ( 50.4) | 250 | ( 63 ) | 450 | (113.4) |
| Dancing | 245 | ( 61.7) | 605 | (152.5) | 850 | (214.2) |
| Bowling | 465 | (117.2) | 985 | (248.2) | 1450 | (365.4) |

**Example 14:** In a restaurant having five employees and a capacity of 30 customers, the occupant heat load can be calculated as follows:

Solution:

$$\text{Sensible heat customers} = 30 \times 195 = 5850 \text{ Btu/hr } (1474.2 \text{ kcal/hr})$$

$$\text{Sensible heat employees} = 5 \times 245 = 1225 \text{ Btu/hr } (308.7 \text{ kcal/hr})$$

$$\text{Latent heat customers} = 30 \times 155 = 4650 \text{ Btu/hr } (1171.8 \text{ kcal/hr})$$

$$\text{Latent heat employees} = 5 \times 605 = 3025 \text{ Btu/hr } (762.3 \text{ kcal/hr})$$

$$\text{Total body heat} = 5850 + 1225 + 4650 + 3025 = 14750 \text{ Btu/hr } (3717 \text{ kcal/hr})$$

**INFILTRATION:** Infiltration is a source of external heat load for the air conditioning apparatus. Infiltration will occur in almost every building because it is virtually impossible to build an airtight room. Air will gain entrance into the conditioned space through the cracks around doors, windows, etc., unless the space is kept under a positive pressure. In a residence or small office where there are no excessive tobacco smoke or other objectionable odors, natural infiltration will supply enough air to satisfy the ventilation air requirements for the space. To satisfy the ventilation air requirements the opening must be in excess of, or at least equal to, 5 percent of the flow space. Also, the space must be large enough to allow 50 ft² (15.2 m²) of space and a minimum of 500 ft³ (14.15 m³) of volume for each oc-

cupant. When these conditions are fulfilled and sufficient infiltration is obtained for the fresh air requirements, forced ventilation will not be necessary unless it is required by local codes and ordinances. (See Table 16-5.) The better designed installations do not depend on air infiltration to satisfy the ventilation air requirements.

Tables have been constructed so that air infiltration through doors and windows can be more easily determined by the designer. (See Table 16-6.) When there are two exposed walls, use the wall which has the greater air leakage. If more than two walls are exposed, use either the one wall that has the greatest amount of leakage or one half of the total of all of the walls, whichever has the larger quantity of infiltration air.

## Table 16-5

### Minimum Outdoor Air Required for Ventilation
(Subject to local code regulations)

| APPLICATION | CFM PER PERSON |
| --- | --- |
| Apartment or residence | 10 to 15 |
| Auditorium | 5 to 7½ |
| Barber shop | 10 to 15 |
| Bank or beauty parlor | 7½ to 10 |
| Broker's board room | 25 to 40 |
| Church | 5 to 7½ |
| Cocktail lounge | 20 to 30 |
| Department store | 5 to 7½ |
| Drug store | 7½ to 10 |
| Funeral parlor | 7½ to 10 |
| General office space | 10 to 15 |
| Hospital rooms (Private) | 15 to 25 |
| Hospital rooms (Wards) | 10 to 15 |
| Hotel room | 20 to 30 |
| Night clubs and taverns | 15 to 20 |
| Private office | 15 to 25 |
| Restaurant | 12 to 15 |
| Retail shop | 7½ to 10 |
| Theater (Smoking permitted) | 10 to 15 |
| Theater (Smoking NOT permitted) | 5 to 7½ |

**Example 15:** Determine the air infiltration into a room which has three exposed walls. One wall has three average fit weather-stripped double hung windows measuring 2 ft × 5 ft (.6 m × 1.5 m). The opposite wall has one industrial type steel sash window measuring 3 ft × 3 ft (.9 m × .9 m). The third wall has a poorly fitted weather-stripped door measuring 3 ft × 6 ft (.9 m × 1.8 m). The average wind velocity is 5 miles (8 km) per hr during the cooling season. Determine the total crack length around the windows.

Solution:

The cracks at the top, center and bottom of

$$\text{the windows} = 2 \text{ ft} \times 3 \text{ ft} (.6 \text{ m} \times .9 \text{ m}) = 6 \text{ ft} (1.8 \text{ m})$$

$$\text{The vertical cracks} = 5 \text{ ft} \times 2 \text{ ft} (1.5 \text{ m} \times 6 \text{ m}) = 10 \text{ ft} (3 \text{ m})$$

$$\text{Total crack length for one window} = 16 \text{ ft} (4.86 \text{ m})$$

$$\text{Total crack length for 3 windows} = 3 \times 16 = 48 \text{ ft} (14.6 \text{ m})$$

Table 16-6

Infiltration Per Foot of Crack Per Hour in Cubic Feet

| TYPE OF OPENING | CONDITION | WIND VELOCITY, MILES PER HOUR | | | | | |
|---|---|---|---|---|---|---|---|
| | | 5 | 10 | 15 | 20 | 25 | 30 |
| Double hung wood window | Average fit, not weatherstripped | 6 | 20 | 40 | 60 | 80 | 100 |
| Double hung wood window | Average fit, weatherstripped | 5 | 15 | 25 | 35 | 50 | 65 |
| Double hung wood window | Poor fit, weatherstripped | 6 | 20 | 35 | 50 | 70 | 90 |
| Double hung metal window | Not weather-stripped | 20 | 45 | 70 | 100 | 135 | 170 |
| Double hung metal window | Weatherstripped | 6 | 18 | 30 | 44 | 58 | 75 |
| Steel Sash | Casement, good fit | 6 | 18 | 30 | 44 | 58 | 75 |
| Steel Sash | Casement, average fit | 12 | 30 | 50 | 75 | 100 | 125 |
| Steel Sash | Industrial type | 50 | 110 | 175 | 240 | 300 | 375 |
| Doors | Good fit, not weatherstripped | 30 | 70 | 110 | 155 | 200 | 250 |
| Doors | Good fit, weatherstripped | 15 | 35 | 55 | 75 | 100 | 125 |
| Doors | Poor fit, not weatherstripped | 55 | 140 | 225 | 310 | 400 | 500 |
| Doors | Poor fit, weatherstripped | 30 | 70 | 110 | 155 | 200 | 250 |

NOTE: Use of storm sash permits a 50% infiltration reduction on poorly fitting windows only.

A wind velocity of five miles (8 km) per hour for an average fit weather-stripped double hung window allows an infiltration of 5 ft$^3$ (.14 m$^3$) per hour per foot (.304 m) of crack.

Infiltration air per hour through first wall = 48 ft (14.6 m) × 5 = 240 ft$^3$ (6.8 m$^3$)

The steel sash window has a total crack length of 3 ft × 4 ft (.9 m × 1.2 m) = 12 ft (3.6 m)

Infiltration air for an industrial steel sash at a 5 mile (8 km) per hr wind velocity is 50 ft$^3$ (1.4 m$^3$) per hr.

Infiltration per hour in second wall = 12 ft × 50 = 600 ft$^3$ (16.98 m$^3$).

Infiltration through the door in the third     wall $= 2(6$ ft $+ 3$ ft$) = 18$ ft $= 2(1.8$ m $+$ .9 m$) = 5.4$ m

A poorly fitted weather-stripped door admits 30 ft$^3$ (.85 m$^3$) of infiltration air per hour with a 5 mile (8 km) per hour wind velocity.

$18 \times 30 = 540$ ft$^3$ (15.28 m$^3$) of infiltration air per hr in the third wall.

Total infiltration through the three walls $=$     240 ft$^3$ (  6.8  m$^3$) first wall

                                                 600 ft$^3$ (16.98 m$^3$) second wall
                                                 540 ft$^3$ (15.28 m$^3$) third wall
Total infiltration $= 1\overline{380}$ ft$^3$ (    39 m$^3$)

One-half of the total infiltration $= 1380 \div 2 = 690$ ft$^3$ (27.16 m$^3$)

The quantity is greater than the infiltration through any individual wall. This should be the quantity used in determining whether or not there is sufficient fresh air being admitted to the space by infiltration. Because the air conditioning unit must cool all of the air which enters by infiltration, it must be added to the load on the unit. However, the quantity of air entering by infiltration is not sufficient to provide the minimum fresh air requirements. Fresh air must be taken in by the cooling unit to provide the requirements. (Refer to Table 16-5.) This extra quantity of fresh air will prevent natural infiltration. This extra amount of air being introduced into the space will build up an internal pressure which will cause the air to leak out rather than in. This is the superior arrangement because all of the air that enters the conditioned space must first pass through the air conditioning equipment.

**INFILTRATION HEAT LOAD:** From the foregoing explanation and examples, it should be evident that infiltration air will be an internal heat load on the space. Therefore, it must be divided into its sensible and latent heat components. This is not the same for ventilation air which is a heat load on the cooling unit but not on the room. When ventilation air is supplied to the conditioned space, it is only necessary to figure the total heat load of the air.

**Example 16:** The total infiltration air which was figured in Example 15 was 690 ft$^3$ (19.53 m$^3$) per hr. Since the basis of the psychrometric chart which we use to find the properties of air is one pound (.4536 g) of dry air, the cubic feet (m$^3$) per hour must be changed to pounds (g) per hour. If there is an outside air temperature of 95°F (35°C) dry bulb and 75°F (23.9°C) wet bulb, which are the design conditions for New York City (refer to Table 16-3) and if the inside design conditions are 80°F (26.7°C) dry bulb and 67°F (19.4°C) wet bulb, the psychrometric chart indicates that dry air at 95°F dry bulb weighs .0715 lbs per ft$^3$ (.0011 g/cm$^3$).

Total air per hour = 690 × .0715 = 49.335 lbs (22.38 kg) of dry air per hour

Total heat at 75°F (23.9°C) wet bulb = 37.81 Btu/lb (20.79 kcal/kg) of dry air.
Total heat at 67°F (19.4°C) wet bulb = 31.15 Btu/lb (17.13 kcal/kg)
Total heat removed = 6.66 Btu/lb ( 3.66 kcal/kg)

Total heat from infiltration = 49.335 lbs air (22.38 kg) × 6.66 Btu (3.66 kcal/kg) = 328.57 Btu/hr (82.80 kcal).

This total heat must now be divided into the sensible and latent heat loads.

Sensible heat content at 95°F (35°C) dry bulb = 22.95 Btu/lb (12.62 kcal/kg) of dry air.
Sensible heat content at 80°F (26.7°C) dry bulb = 19.32 Btu/lb (10.6 kcal/kg of dry air.
Total sensible heat removed = 3.63 Btu/lb (1.99 kcal/kg) of dry air

Total sensible heat from infiltration = 49.335 lbs air (22.38 kg) × 3.63 Btu/lb (1.99 kcal/kg) = 179.09 Btu/lb (98.5 kcal/kg) of dry air.

The latent heat to be removed is total heat less sensible heat. Therefore, the latent heat removed = 328.57 Btu (180.7 kcal/kg) − 179.09 Btu (98.5 kcal/kg) = 149.48 Btu (82.2 kcal/kg)

**VENTILATION HEAT LOAD:** The amount of infiltration air should be compared with the minimum fresh air requirements and if it is found to be insufficient, ventilation air must be brought in through the air conditioning equipment. The ventilation heat load and the infiltration heat load should not be used on the same job. In Example 15, it was found that 690 ft³ (19.53 m³) of air entered the room by infiltration. If the room is an office which has two occupants, Table 16-5 indicates that 1200 ft³ (33.96 m³) of air per hour are required because each occupant requires 10 ft³ (.28 m³) per min, i.e.: 2 occupants × 10 cfm (.28 m³) × 60 min per hr = 1200 ft³ (33.96 m³) of air per hour.

Therefore, the amount of infiltration air would not be sufficient and the heat load due to the introduction of 1200 ft³ (33.96 m³) of air would be added to the conduction heat load. It would not, however, be necessary to divide the total ventilation heat load into the sensible and latent heat loads. The ventilation heat load would be calculated exactly like the infiltration heat load.

**ELECTRIC APPLIANCE LOAD:** Only 5 percent or less of the electricity used by an electric light is converted into light. The other 95 percent is given off in the form of heat. Because of this, the lighting load on a conditioned space must be calculated on the basis of the total amount of electricity consumed by the light. The figure used in calculating this heat load is 3.41 Btus (859.3 cal) per watt. Only the lights used during the peak load period should be included in the load calculation. A business located on a lower floor of a large building probably uses lights all the time, while it would receive very little, if any, solar effect. These conditions require that the lighting load be included in the estimate. If a unit is installed where the sun will shine into the windows and additional lighting is also required, both loads must be included in the calculation. When fluorescent lighting is used, there must be an addition of 25 percent of the lighting load included in the estimate because of the heat generated by the ballast.

Appliances such as toasters, hair driers, etc., are figured on the same basis as electric lighting; that is, 3.41 Btus (859.3 cal) per watt. If all the heat generated by these appliances is dissipated into the conditioned space, the total input wattage is multiplied by 3.41. Consideration should be given to the amount of time and the period that these appliances are operated. A toaster in a restaurant, for example, may operate continuously during the breakfast period, but not be used at all during the peak load period. It, therefore, should not be included in the estimate.

Varying amounts of heat are given off by electric motors. Larger motors are more efficient than the smaller horsepower motors and, therefore, generate less heat to be dissipated into the conditioned space. Equivalent values for motors have been placed in tables for convenience. (See Table 16-7.) It

should be noted that the values for motors are given on the basis of Btu (cal) per horsepower per hour.

**GAS AND STEAM APPLIANCES:** These appliances include good cooking and warming devices, such as steam tables, coffee urns, etc. (See Table 16-8.) Each pound of steam condensed by the coil in a steam table must be added to the heat load. This is equal to 960 Btus per pound (528 kcal/kg) of steam condensed. Because of the moisture given off by the heated food, this heat must be considered as half sensible heat and half latent heat. Coffee urns heated by gas emit the total amount of heat generated by the burning gas. This heat also is considered as being half sensible and half latent. The amount of heat given off for coffee urns is calculated by figuring one cubic foot (.0283 m³) of natural gas or two cubic feet (.057 m³) of manufactured gas per gallon of rated capacity of the urn. If exhaust fans are placed over the urns, the sensible and latent heat loads may be reduced by 50 percent.

**DUCT HEAT GAIN:** When ducts are to be installed in hot places such as attics or boiler rooms, there is an additional load placed on the cooling unit. Any heat which is added to the air from the time it leaves

Table 16-7

Heat from Electric Appliances

| ITEM | AMOUNT OF HEAT |
|------|----------------|
| Lights | 3.41 Btu per watt (.86 kcal) |
| Appliances (Toasters, Hair driers, etc.) | 3.41 Btu per watt (.86 kcal) |
| Motors—⅛ to ½ hp | 4250 Btu per hp per hr (1071 kcal) |
| —¾ to 3 hp | 3700 Btu per hp per hr ( 932.4 kcal) |
| —5 to 20 hp | 3000 Btu per hp per hr ( 756 kcal) |

Table 16-8

Heat from Gas and Steam Appliances

| ITEM | BTU PER HR (KCAL) | | |
|---|---|---|---|
| | SENSIBLE | LATENT | TOTAL |
| Steam tables, per ft² top surface | 1000 ( 252    kcal) | 1000 ( 252    kcal) | 2000 ( 504    kcal) |
| Restaurant coffee urns | 5000 (1260   kcal) | 5000 (1260   kcal) | 10000 (2520   kcal) |
| Natural gas per ft³ (m³) | 500 ( 126    kcal) | 500 ( 126    kcal) | 1000 ( 252    kcal) |
| Manufactured gas per ft³ (m³) | 275 (   69.3 kcal) | 275 (   69.3 kcal) | 550 ( 138.6 kcal) |
| Steam condensed in warming coils, per pound (.4536 kg) | 480 ( 120.9 kcal) | 480 ( 120.9 kcal) | 960 ( 241.92 kcal) |

NOTE: 1. If there are exhaust hoods over steam tables and coffee urns, include only 50% of the heat load from these sources.

2. Coffee urns may also be figured on the basis of consuming approximately 1 ft³ of natural gas or approximately 2 ft³ of manufactured gas per hour per gallon rated capacity.

the cooling coil until it gets back to the coil must be removed by the air conditioning unit. Heat transfer factors have been placed in tables for convenience. (See Table 16-9.) It should be noted that when the entire supply duct is located inside the conditioned space, no duct heat gain needs to be calculated. Any heat gain will come from the conditioned space. This becomes a credit on the amount of cooling required in the exact amount of heat gain to the ducts.

Since the amount of air which must be supplied to the conditioned space is determined from the amount of sensible heat that must be removed, it follows that the duct sizes are indirectly related to the sensible heat gain of the space. Therefore, it is reasonable that the duct heat gain be calculated as a percentage of the sensible heat gain. This percentage will vary from 0% to a maximum of 5%. The length of the duct run through an unconditioned space, as well as the surrounding air temperature, must be given consideration. The above mentioned percentages are based on the assumption that all the ducts installed in unconditioned spaces will be insulated.

The procedure for figuring duct heat gain is the same which is used for calculating the heat gain for any other heat flow. (Refer

Table 16-9

Heat Transfer Factors for Ducts

| DUCT | BTU PER FT² PER HR PER 1°F DIFFERENCE |
|---|---|
| Sheet metal, not insulated | 1.13 (.28 kcal) |
| Average insulation, ½ in. thick (12.7 mm) | .41 (.10 kcal) |

to Table 16-9.) The area of the exposed duct is multiplied by the *U* factor found in the table. This total is then multiplied by the dry bulb temperature difference between the air inside the duct and the air outside the duct.

**FAN HEAT GAIN:** To maintain the proper air circulation, a fan is used to push the air through the system. The fan depends upon the amount of air circulated, which is also dependent upon the total sensible heat calculated. The fan power allowance also may be taken as a percentage of the sensible heat. This percentage will usually be between 3% and 4% of the total sensible heat load. It is, however, customary to allow 5% as a combined factor for the duct heat gain and the fan heat gain.

**SAFETY FACTOR:** Because of the many variables involved, heat load calculation is not an exact science like mathematics, chemistry, physics, etc. For example, heat transfer coefficients are calculated with great accuracy for most types of construction, but the design engineer has no positive assurance that the construction under consideration is identical in all respects to the test panels from which the coefficients were established. Also, internal loads such as people, lighting, and equipment, as well as shading factors, color of the exterior surfaces, etc. may vary from those upon which the design was based. Because of this, it is advisable to apply a factor of safety to the calculated heat load. This safety factor should vary from a maximum—on very small installations where any variation may seriously affect the performance of the system—to a minimum—on very large installations where there is a greater chance that the possible variations may cancel out one another. For the average installation, a safety factor of 10% is usually adequate. It must be noted that the safety factor must be applied to both the sensible and latent heat loads so that the relation between the two are not upset.

**EQUIPMENT SELECTION:** The Btu (cal) capacity of the equipment must be as close as possible to the calculated value. However, it must always be equal to or greater than the calculated heat gain. The equipment size should always be slightly larger than that required; never in great excess. If the calculated size is not available, only the next size larger should be chosen.

**APPARATUS DEW POINT AND AIR QUANTITY:** On the psychrometric chart, the point where the sensible heat ratio line and the saturation curve intersect is known as the apparatus dew point (ADP). This point represents the lowest temperature at which air can be supplied to the conditioned space and still pick up the required amount of sensible and latent heat. If air at a higher ADP was supplied to the conditioned space, the quantity of air could be adjusted so that the required amount of sensible heat would be picked up. However, this air quantity would not pick up a sufficient amount of latent heat. The resultant air would then permit a higher relative humidity inside the conditioned space than would be desired. If the ADP was too low, the relative humidity also would be too low. However, if air at any condition which falls on the sensible heat ratio line between condition (Point *A*) and ADP (Point *B*) is supplied, the quantity of air can be adjusted so that the exact amount of both sensible and latent heat and the design conditions can be met. (See Figure 16-44.)

The ADP will actually very closely approach the cooling coil surface temperature.

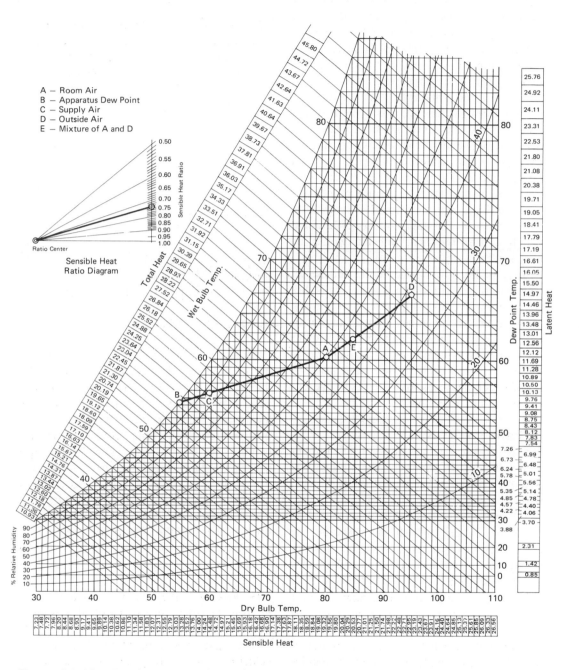

**Figure 16-44.** Psychrometric chart showing ADP

A — Room Air
B — Apparatus Dew Point
C — Supply Air
D — Outside Air
E — Mixture of A and D

Sensible Heat Ratio Diagram

Ratio Center

Sensible Heat Ratio

Total Heat

Wet Bulb Temp.

% Relative Humidity

Dry Bulb Temp.

Sensible Heat

Dew Point Temp.

Latent Heat

Therefore, a very large amount of coil surface would be needed if the conditioned air was to be supplied at the ADP. The air entering the cooling coil will be at its maximum temperature and the maximum air temperature differential will exist between the air and the coil surface. As the air progresses through the coil, it is cooled so that the temperature differential progressively decreases. The first row of tubing will do the most work and each following row will do less work. Therefore, to cool the air to the ADP would take an infinite number of tubes and the last rows would do very little work because the temperature differential which causes the heat transfer would be very small.

In actuality, the choice of the number of rows of tubing in depth usually falls between a minimum of three rows and a maximum of eight rows. When a large amount of sensible heat is to be removed, a smaller number of rows can be used. Conversely, as the latent heat load increases, more rows of tubing become necessary from both the engineering and economic standpoints.

**BYPASS FACTOR:** The bypass factor refers to that percentage of the total amount of air which passes through the cooling coil without coming into contact with the coil surfaces and, therefore, is not conditioned. The bypass factor is dependent upon the velocity of the air through the coil and the actual design of the coil itself. The number of rows of tubing in the direction of air flow also is a determining factor. Therefore, since coils vary in design from one manufacturer to another, the bypass factor also will be different. However, the average bypass factors of coils which are presently manufactured and used with normal air velocities are shown in Table 16-10.

It can be seen from the table that if a 4-row coil were used, 20 percent of the total

Table 16-10

Bypass Factor for Different Depths of Coil

| NUMBER OF ROWS | AVERAGE BYPASS FACTOR |
|---|---|
| 3 | 35% |
| 4 | 20% |
| 5 | 10% |
| 6 | 5% |
| 7 | 2.5% |
| 8 | 0.0% |

air passing through the coil would not be treated. This amount of air would not be useful in cooling the conditioned space.

## DUCT SYSTEM DESIGN

When a new air conditioning system is being considered, the type and design of the duct system is as critical as the load calculation. The location of the outlet, the air velocity, and the cubic feet (.0283 m³) of air per minute are of prime importance in calculating the duct system. The design engineer may make drawings where each pipe, fitting, duct, or other piece of the equipment is shown in detail as it actually appears.

After the heat load has been calculated for the space and the individual rooms within the conditioned space, the duct work can be calculated and sized.

**AIR VELOCITIES:** Air velocity is the rate of speed at which the air travels through the ducts or openings. There are suggested velocities that should be maintained. Higher velocities will result in increased air noise and electrical power consumption. (See Table 16-11.) Note that the maximum duct velocities are given in parentheses.

Table 16-11

Recommended and Maximum Duct Velocities

| LOCATION | RESIDENCES | SCHOOLS AND PUBLIC BUILDINGS | INDUSTRIAL BUILDINGS |
|---|---|---|---|
| Main Ducts | 700– 900 (1000) | 1000–1300 (1400) | 1200–1800 (2000) |
| Branch Ducts | 600 (700) | 600– 900 (1000) | 800–1000 (1200) |
| Branch Risers | 500 (650) | 600– 700 (900) | 800 (1000) |
| Outside Air Intakes | 700 (800) | 800 (900) | 1000 (1200) |
| Filters | 250 (300) | 300 (350) | 350 (350) |
| Heating Coils | 450 (500) | 500 (600) | 600 (700) |
| Air Washers | 500 | 500 | 500 |
| Suction Connections | 700 (900) | 800 (1000) | 1000 (1400) |
| Fan Outlets | 1000–1600 (1700) | 1300–2000 (2200) | 1600–2400 (2800) |

**DUCT DESIGN:** Ducts made of metal are preferred because of the smooth surfaces which do not add resistance to the air flow. Metal ducts also can be formed into compact sizes and shapes, and placed in difficult locations.

Pressure losses which occur in duct systems are caused by the velocity of the air flow, number of fittings, and friction of the air against the walls of the duct. The use of coils, air filters, dampers, and deflectors all add to the resistance to the air flow.

When a more extensive system is to be designed, a good practice is to gradually lower the velocity in both the main duct and the remote branches. This type of design has the following advantages:

1. It enables the air to be distributed in a more uniform manner.

2. It decreases the air friction in the smaller ducts where it would otherwise be the greatest.

3. When the velocity is reduced there is a recovery of velocity pressure which compensates for the duct friction.

Dampers and deflectors should be placed at all points necessary to assure the proper balance of the system. It is difficult to design a duct system which will provide the correct amount of air at each outlet. It is therefore necessary to use dampers and deflectors so that the air supply may be directly proportioned after the system is placed in operation.

**DUCT CALCULATIONS:** The required duct sizes and velocities can be figured by use of the following general formula:

$$Q = A \times V$$

where:

$Q$ = Discharge of air in cfm $(m^3/m)$

$A$ = Cross-sectional area of the opening in ft² (m²)

V = Velocity of air in feet per minute (mpm)

**Example 17:** Given 1500 cfm (42.45 m³/m) of air to be moved at a velocity of 1000 feet (304 m) per minute, find the area of duct needed.

Solution:

$$Q = A \times V$$

$$1500 = A \times 1000$$

$$A = \frac{1500}{1000} = 1.5 \text{ ft}^2 \text{ (.1395 m}^2\text{)}$$

To change these terms to square inches (cm²), multiply $A$ times 144 (929 cm²) as follows:

$$1.5 \times 144 = 216 \text{ in.}^2 \text{ (.1395 m}^2\text{)}$$

This gives the area of the duct or opening. This duct can be made in any shape to suit the available space.

The areas of circumferences of circles and the sides of squares of the same area are given in tables such as Table 16-12. To find the equivalent diameter of a circular duct or the side of a square duct note that 216 in.² falls between 213.8 and 226.9 in.² Using the largest value, 226.9, find the diameter of the circular duct to be 15 in.

If the space available for the duct is 18 in. (45.72 cm), then a 12-in. × 18-in. (30.48 cm × 45.72 cm) duct could be used:

$$\frac{216}{18} = 12 \text{ (30.48 cm)}.$$

**FRICTION LOSS CALCULATION:** In order to determine the friction loss of a duct system, it is necessary to know the velocity and size of the duct. The friction loss per hundred feet (30.4 m) of a duct of a given size may be found in Table 16-13.

Allowances also must be made for elbows and bends. These allowances may be calculated by adding to the total length of duct. Add 8 diameters for a round elbow and 5 diameters for a square or rectangular elbow which is equipped with turning vanes.

**Example 18:** Given a blower moving air through an inlet duct 42 inches (106.68 cm) in diameter and 100 feet (30.4 m) in length, and supplying air through a system of ducts at 4 points, figure the pressure loss. (See Figure 16-45 and Table 16-13.)

Solution: Section A—100 ft (30.4) of straight duct 36 in. (91.44 cm) in diameter, an air velocity of 1200 ft (364.8 m) per minute and a friction loss of .069 in. (.175 cm).

Section B—80 ft (24.32 m) of straight duct, 30 in. (76.2 cm) in diameter, an air velocity of 1000 ft (304 m) per minute. For a run of 100 ft (30.4 m) it is found that the loss is .058 in. (.147 cm). Because this run is for 80 ft (24.32 m) take $\frac{8}{10}$ of .058 or .046.

Section C—One 90° bend, 2 ft (.608 m), the center line radius equal to 1½ times the diameter. The friction loss is equal to 8 diameters or 16 ft (4.86 m) of straight duct. This length must be added to the straight run of 50 ft (15.2 m), making a total of 66 ft (20.06 m) at a velocity of 900 ft/min (273.6 pm). Therefore, take $\frac{66}{100}$ of .058 in. (.147 cm) or .037 in. (.099 cm).

Section D—In this section there is one 90° bend. The duct is 18 in. (45.72 cm) in diameter. The friction loss is equal to 8 diameters or 12 ft (3.65 m) of straight duct. Add this to the straight run of 40 ft (12.16 m) for a total of 52 ft (15.8 m). Therefore, take $\frac{52}{100}$ of .060 in. (.15 cm) and the result is .031 in. (.08 cm).

**Figure 16-45.** Figuring friction loss of a system

Section E—This duct is 100 ft (30.4 m) long, and 42 in. (106.68 cm) in diameter. Air travels through this duct at a velocity of 1000 ft (304 m) per minute. It is found that the pressure loss is .041 in. (.104 cm).

To find the total friction losses, add the loss in each section which amounts to a total of .224 inch (.569 cm). In this case the blower selected must be capable of delivering the required volume at ¼ inch (6.5 mm) static pressure.

The example was based on round ducts. If the ducts are to be square or rectangular, however, it would only be necessary to reduce the rectangular ducts to round ducts of equivalent area and proceed with the calculations just described.

## AIR CONDITIONING EQUIPMENT

The refrigeration equpiment used in air conditioning work is much like that used in refrigeration work. The major difference is the duct work. The fact that higher temperatures are encountered necessitates some different design characteristics. Otherwise, the refrigeration cycle is the same as that used in refrigeration applications.

**COMPRESSORS:** The compressors used in air conditioning units are basically the same as those discussed in Chapter 4. They may be either the reciprocating, rotary, or centrifugal type. However, the reciprocating type is most commonly used. (See Figure 16-46.) When used on air-cooled condenser systems, it is mounted outside of the condensing unit. This moves the compressor noise outside and allows more air to flow over it to aid in cooling. (See Figure 16-47.)

**CONDENSERS:** The function of an air conditioning condenser is to remove heat from the refrigerant and cause the vapor to change to a liquid. Most air conditioning condensers are of the air-cooled type (see Figure 16-48) but there also are water-cooled and evaporative condensers in use.

# Table 16-12

## Table of Areas and Circumferences of Circles and the Sides of Squares of the Same Area

| Diameter in Inches | Area of Circle in Square Feet | Area of Circle in Square Inches | Sides of Squares of Same Area in Inches | Circumference of Circle in Inches | Diameter in Inches | Area of Circle in Square Feet | Area of Circle in Square Inches | Sides of Squares of Same Area in Inches | Circumference of Circle in Inches | Diameter in Inches |
|---|---|---|---|---|---|---|---|---|---|---|
| 1 | .005 | .8 | .9 | 3.1 | 16½ | 1.48 | 213.8 | 14.6 | 51.8 | 32 |
| 1½ | .012 | 1.7 | 1.3 | 4.7 | 17 | 1.58 | 226.9 | 15.0 | 53.4 | 32½ |
| 2 | .022 | 3.1 | 1.7 | 6.2 | 17½ | 1.67 | 240.5 | 15.5 | 54.9 | 33 |
| 2½ | .034 | 4.9 | 2.2 | 7.8 | 18 | 1.77 | 254.4 | 15.9 | 56.5 | 33½ |
| 3 | .049 | 7.0 | 2.6 | 9.4 | 18½ | 1.87 | 268.8 | 16.4 | 58.1 | 34 |
| 3½ | .067 | 9.6 | 3.1 | 10.9 | 19 | 1.97 | 283.5 | 16.8 | 59.6 | 34½ |
| 4 | .087 | 12.5 | 3.5 | 12.5 | 19½ | 2.07 | 298.6 | 17.2 | 61.2 | 35 |
| 4½ | .110 | 15.9 | 3.9 | 14.1 | 20 | 2.18 | 314.1 | 17.7 | 62.8 | 35½ |
| 5 | .136 | 19.6 | 4.4 | 15.7 | 20½ | 2.29 | 330.0 | 18.1 | 64.4 | 36 |
| 5½ | .165 | 23.7 | 4.8 | 17.2 | 21 | 2.41 | 346.3 | 18.6 | 65.9 | 36½ |
| 6 | .196 | 28.2 | 5.3 | 18.8 | 21½ | 2.52 | 363.0 | 19.0 | 67.5 | 37 |
| 6½ | .230 | 33.1 | 5.7 | 20.4 | 22 | 2.64 | 380.1 | 19.5 | 69.1 | 37½ |
| 7 | .268 | 38.4 | 6.2 | 21.9 | 22½ | 2.76 | 397.6 | 19.9 | 70.6 | 38 |
| 7½ | .306 | 44.1 | 6.6 | 23.5 | 23 | 2.84 | 415.4 | 20.3 | 72.2 | 38½ |
| 8 | .350 | 50.2 | 7.0 | 25.1 | 23½ | 3.01 | 433.7 | 20.8 | 73.8 | 39 |
| 8½ | .393 | 56.7 | 7.5 | 26.7 | 24 | 3.14 | 452.3 | 21.2 | 75.3 | 39½ |
| 9 | .442 | 63.6 | 7.9 | 28.2 | 24½ | 3.28 | 471.4 | 21.7 | 76.9 | 40 |
| 9½ | .493 | 70.8 | 8.4 | 29.8 | 25 | 3.41 | 490.8 | 22.1 | 78.5 | 40½ |
| 10 | .545 | 78.5 | 8.8 | 31.4 | 25½ | 3.54 | 510.7 | 22.6 | 80.1 | 41 |
| 10½ | .600 | 86.5 | 9.3 | 32.9 | 26 | 3.69 | 530.9 | 23.0 | 81.6 | 41½ |
| 11 | .660 | 95.0 | 9.7 | 34.5 | 26½ | 3.83 | 551.5 | 23.4 | 83.2 | 42 |
| 11½ | .722 | 103.8 | 10.1 | 36.1 | 27 | 3.98 | 572.5 | 23.9 | 84.8 | 42½ |
| 12 | .785 | 113.1 | 10.6 | 37.6 | 27½ | 4.13 | 393.9 | 24.3 | 86.3 | 43 |
| 12½ | .852 | 122.7 | 11.0 | 39.2 | 28 | 4.28 | 615.7 | 24.8 | 87.9 | 43½ |
| 13 | .922 | 132.7 | 11.5 | 40.8 | 28½ | 4.42 | 637.9 | 25.2 | 89.5 | 44 |
| 13½ | .994 | 143.1 | 11.9 | 42.4 | 29 | 4.59 | 660.5 | 25.7 | 91.1 | 44½ |
| 14 | 1.07 | 153.9 | 12.4 | 43.9 | 29½ | 4.74 | 683.4 | 26.1 | 92.6 | 45 |
| 14½ | 1.15 | 165.1 | 12.8 | 45.5 | 30 | 4.92 | 706.8 | 26.5 | 94.2 | 45½ |
| 15 | 1.23 | 176.7 | 13.2 | 47.1 | 30½ | 5.07 | 730.6 | 27.0 | 95.8 | 46 |
| 15½ | 1.31 | 188.6 | 13.7 | 48.6 | 31 | 5.23 | 754.7 | 27.4 | 97.3 | 46½ |
| 16 | 1.40 | 201.0 | 14.1 | 50.2 | 31½ | 5.42 | 779.3 | 27.9 | 98.9 | 47 |

Table 16-12   (cont'd)

| Area of Circle in Square Feet | Area of Circle in Square Inches | Sides of Square of Same Area in Inches | Circumference of Circle in Inches | Diameter in Inches | Area of Circle in Square Feet | Area of Circle in Square Inches | Sides of Squares of Same Area in inches | Circumference of Circle in Inches |
|---|---|---|---|---|---|---|---|---|
| 5.75 | 804.2 | 28.3 | 100.5 | 47½ | 12.31 | 1772.0 | 42.1 | 149.2 |
| 5.58 | 829.5 | 28.8 | 102.1 | 48 | 12.55 | 1809.5 | 42.5 | 150.7 |
| 5.93 | 855.3 | 29.2 | 103.6 | 48½ | 12.85 | 1847.4 | 42.9 | 152.3 |
| 6.11 | 881.4 | 29.6 | 105.2 | 49 | 13.10 | 1885.7 | 43.4 | 153.9 |
| 6.31 | 907.9 | 30.1 | 106.8 | 49½ | 13.40 | 1924.4 | 43.8 | 155.5 |
| 6.47 | 934.8 | 30.5 | 108.3 | 50 | 13.65 | 1963.5 | 44.3 | 157.0 |
| 6.67 | 962.1 | 31.0 | 109.9 | 50½ | 13.95 | 2002.9 | 44.7 | 158.6 |
| 6.87 | 989.8 | 31.4 | 111.5 | 51 | 14.20 | 2042.8 | 45.2 | 160.2 |
| 7.08 | 1017.8 | 31.9 | 113.0 | 51½ | 14.45 | 2083.0 | 45.6 | 161.7 |
| 7.28 | 1046.3 | 32.3 | 114.6 | 52 | 14.75 | 2123.7 | 46.0 | 163.3 |
| 7.46 | 1075.2 | 32.7 | 116.2 | 52½ | 15.05 | 2164.7 | 46.5 | 164.9 |
| 7.69 | 1104.4 | 33.2 | 117.8 | 53 | 15.35 | 2206.1 | 46.9 | 166.5 |
| 7.88 | 1134.1 | 33.6 | 119.3 | 53½ | 15.62 | 2248.0 | 47.4 | 168.0 |
| 8.10 | 1164.1 | 34.1 | 120.9 | 54 | 15.92 | 2290.2 | 47.8 | 169.6 |
| 8.30 | 1194.5 | 34.5 | 122.5 | 54½ | 16.20 | 2332.8 | 48.3 | 171.2 |
| 8.52 | 1225.4 | 35.0 | 124.0 | 55 | 16.45 | 2375.8 | 48.7 | 172.7 |
| 8.72 | 1256.6 | 35.4 | 125.6 | 55½ | 16.80 | 2419.2 | 49.1 | 174.3 |
| 8.95 | 1288.2 | 35.8 | 127.2 | 56 | 17.10 | 2463.0 | 49.6 | 175.9 |
| 9.17 | 1320.2 | 36.3 | 128.8 | 56½ | 17.45 | 2507.1 | 50.0 | 177.5 |
| 9.40 | 1352.6 | 36.8 | 130.3 | 57 | 17.72 | 2551.7 | 50.5 | 179.0 |
| 9.64 | 1385.4 | 37.2 | 131.9 | 57½ | 18.00 | 2596.7 | 50.9 | 180.6 |
| 9.85 | 1418.6 | 37.6 | 133.5 | 58 | 18.35 | 2642.0 | 51.4 | 182.2 |
| 10.10 | 1452.2 | 38.1 | 135.0 | 58½ | 18.65 | 2687.8 | 51.8 | 183.7 |
| 10.31 | 1486.1 | 38.5 | 136.6 | 59 | 18.95 | 2733.9 | 52.2 | 185.3 |
| 10.55 | 1520.5 | 38.9 | 138.2 | 59½ | 19.35 | 2780.5 | 52.7 | 186.9 |
| 10.80 | 1555.2 | 39.4 | 139.8 | 60 | 19.65 | 2827.7 | 53.1 | 188.4 |
| 11.05 | 1590.4 | 39.8 | 141.3 | 60½ | 19.95 | 2874.7 | 53.6 | 190.0 |
| 11.30 | 1625.9 | 40.3 | 142.9 | 61 | 20.39 | 2922.5 | 54.0 | 191.6 |
| 11.55 | 1661.9 | 40.7 | 144.5 | 61½ | 20.62 | 2970.6 | 54.5 | 193.2 |
| 11.80 | 1698.2 | 41.2 | 146.0 | 62 | 20.95 | 3019.1 | 54.9 | 194.8 |
| 12.10 | 1734.9 | 41.6 | 147.6 | 62½ | 21.32 | 3068.0 | 55.4 | 196.4 |

## Table 16-13

### Friction Loss Table

PRESSURE LOSS IN INCHES OF WATER, DUE TO FRICTION OF AIR, PER 100 FT. OF PIPE

| Diam. of pipe in inches | Area in sq. inches | VELOCITY OF AIR IN FEET PER MINUTE | | | | | | | | | | | | | | |
|---|---|---|---|---|---|---|---|---|---|---|---|---|---|---|---|---|
| | | 200 | 400 | 600 | 800 | 900 | 1000 | 1200 | 1400 | 1600 | 1800 | 2000 | 2400 | 3000 | 4500 | 6000 |
| 3 | 7.0 | .023 | .092 | .208 | .370 | .468 | .581 | .832 | 1.14 | 1.48 | 1.87 | 2.32 | 3.33 | 5.20 | 11.70 | 20.80 |
| 4 | 12.5 | .017 | .070 | .157 | .280 | .350 | .440 | .624 | .862 | 1.12 | 1.40 | 1.76 | 2.50 | 3.94 | 8.85 | 15.76 |
| 5 | 19.6 | .014 | .056 | .125 | .226 | .281 | .348 | .498 | .721 | .902 | 1.12 | 1.39 | 1.99 | 3.12 | 7.03 | 12.48 |
| 6 | 28.2 | .011 | .046 | .104 | .185 | .234 | .290 | .417 | .568 | .741 | .935 | 1.16 | 1.67 | 2.60 | 5.84 | 10.40 |
| 7 | 38.4 | .010 | .040 | .089 | .158 | .201 | .249 | .357 | .487 | .633 | .805 | 1.00 | 1.43 | 2.22 | 5.01 | 8.88 |
| 8 | 50.2 | .009 | .034 | .078 | .138 | .176 | .218 | .312 | .427 | .554 | .703 | .872 | 1.25 | 1.95 | 4.38 | 7.80 |
| 9 | 63.6 | .008 | .031 | .069 | .123 | .156 | .194 | .277 | .379 | .493 | .625 | .776 | 1.11 | 1.73 | 3.89 | 6.92 |
| 10 | 78.5 | .007 | .028 | .063 | .111 | .141 | .175 | .250 | .342 | .445 | .562 | .700 | 1.00 | 1.56 | 3.51 | 6.24 |
| 12 | 113.1 | .006 | .023 | .052 | .092 | .117 | .145 | .208 | .284 | .369 | .467 | .580 | .832 | 1.30 | 2.92 | 5.20 |
| 14 | 153.9 | .005 | .021 | .049 | .086 | .109 | .136 | .194 | .266 | .345 | .437 | .544 | .776 | 1.21 | 2.74 | 4.84 |
| 16 | 201.0 | .004 | .017 | .039 | .070 | .088 | .109 | .156 | .214 | .278 | .352 | .436 | .624 | .977 | 2.20 | 3.91 |
| 18 | 254.4 | .004 | .015 | .035 | .060 | .078 | .097 | .138 | .190 | .242 | .312 | .387 | .552 | .867 | 1.95 | 3.47 |
| 20 | 314.1 | .004 | .014 | .031 | .055 | .070 | .087 | .124 | .170 | .221 | .280 | .348 | .496 | .778 | 1.75 | 3.11 |
| 22 | 380.1 | .003 | .013 | .028 | .050 | .064 | .079 | .113 | .156 | .202 | .256 | .317 | .452 | .711 | 1.60 | 2.84 |
| 24 | 452.3 | .003 | .012 | .026 | .046 | .058 | .073 | .104 | .142 | .164 | .234 | .290 | .416 | .650 | 1.46 | 2.60 |
| 28 | 615.7 | .003 | .010 | .022 | .040 | .052 | .062 | .089 | .122 | .159 | .202 | .249 | .356 | .558 | 1.26 | 2.23 |
| 30 | 706.8 | .003 | .009 | .020 | .037 | .047 | .058 | .083 | .114 | .148 | .187 | .232 | .332 | .520 | 1.17 | 2.08 |
| 36 | 1017 | .002 | .008 | .017 | .031 | .039 | .048 | .069 | .095 | .123 | .156 | .193 | .276 | .433 | .976 | 1.73 |
| 42 | 1385 | .002 | .007 | .015 | .026 | .033 | .041 | .059 | .081 | .106 | .134 | .166 | .236 | .371 | .836 | 1.48 |
| 48 | 1809 | .002 | .006 | .013 | .023 | .029 | .036 | .052 | .071 | .092 | .117 | .144 | .208 | .325 | .731 | 1.30 |
| 54 | 2290 | .001 | .005 | .012 | .020 | .026 | .032 | .046 | .063 | .082 | .104 | .129 | .184 | .288 | .650 | 1.15 |
| 60 | 2827 | .001 | .005 | .010 | .018 | .023 | .029 | .042 | .057 | .074 | .094 | .116 | .168 | .260 | .585 | 1.04 |

Note: For other lengths of pipe the loss is directly proportional to the length.

**Figure 16-46.** Cutaway of hermetic compressor (*Courtesy of Sundstrand Corp.*)

The efficiency of the condenser determines to a great extent, the energy efficiency ratio (EER) of the unit. The EER is determined by dividing the Btu (cal) rating of the unit by the wattage drawn by the unit. The higher the EER, the more efficient the unit. There are several units on the market now with an EER of 8.5. The EER is certified by the Air Conditioning and Refrigeration Institute (ARI), and is known as the ARI Rating.

**EVAPORATORS:** Air conditioning evaporators are manufactured in three types: (1) upflow, (2) downflow, and (3) horizontal. (See Figure 16-49.) This terminology refers to the direction of air flow through the coil. The purpose of the evaporator is to remove heat and moisture from the air which is delivered to the conditioned space. A drain pan is incorporated in the manufacture of the coil unit to collect and dispose of the condensation due to cooling the air below the dew point temperature. The drain connection must be connected to a drain to dispose of the condensate properly.

**FLOW CONTROL DEVICE:** The flow control device used on air conditioning installations may be either the capillary tube or a thermostatic expansion valve. Capillary tubes are used on systems up to approximately 3 tons in capacity. The larger units require a thermostatic expansion valve to control the flow of refrigerant properly.

**AIR HANDLING UNIT:** The air handling unit contains the blower and either the heating equipment, the cooling equipment, or both the heating and cooling equipment. In most residences this is the heating furnace. In any case, the blower must be capable of delivering the required amount of air at the proper velocity with the design static pressure of the duct system.

**TYPES OF AIR CONDITIONING SYSTEMS:** Air conditioning systems are generally classified as being either self-contained or remote systems. Both of these types can be installed as a central unit.

**Self-Contained Systems:** Self-contained systems have all of the components installed on one base (see Figure 16-50), with the possible exception of the condenser. Water-cooled systems have a water tower placed at some location outside the building with the water piped to the condenser. Air-cooled systems may be installed on an outside wall with the necessary provisions made to allow the condenser air to circulate outdoors freely. (See Figure 16-51.)

Self-contained units may be larger commercial units known as roof-mount units

**Figure 16-47.** Air-cooled condensing unit *(Courtesy of Dearborn Stove Co.)*

**Figure 16-48.** Air-cooled condenser *(Courtesy of Dearborn Stove Co.)*

**Figure 16-49.** Air conditioning evaporators (*Courtesy of Dearborn Stove Co.*)

(see Figure 16-52), window units, or the smaller commercial units in Figure 16-50. Duct work may be connected to the supply and return air openings and ducted to and from the space.

**Remote Split Systems:** These systems are in more common use on residential and small commercial installations. They have the evaporator, blower, and furnace all located in a closet in an equipment room with the

**FEATURES:**

1. FRONT RETURN — Air grille may be interchanged with lower back panel for rear return.

2. AMPLE SPACE — Provided above evaporator coil for field installation of steam, or hot water coil.

3. BLOWER — Full size for maximum air movement at low noise level.

4. MOTOR — has integral overload protection. Variable speed sheave and V-belt drive.

5. CABINET — Heavy gauge, electrolytic zinc coated steel, finished in attractive two-tone acrylic enamel.

6. INSULATION — Thermal and acoustical, spun glass insulation.

7. EXPANSION VALVE — Thermostatic non-adjustable type. Mounted at the factory.

8. COIL — Aluminum finned, copper tubing, manifolded and circuited for maximum heat transfer.

9. FILTERS — Chemically treated, disposable type, easy to change.

**Figure 16-50.** Cutaway of self-contained unit (*Courtesy of Southwest Manufacturing Company*)

**Figure 16-51.** Self-contained condenser connections

condensing unit. (See Figure 16-53.) The refrigerant is piped from the condensing unit to the indoor coil through copper tubing. The air is ducted from the air conditioning equipment to the space to be conditioned.

Most residential units have the indoor section centrally located within the space which eliminates the need for a return air duct system.

In order to obtain the proper and most

**Figure 16-52.** Roof-mount self-contained unit cutaway (*Courtesy of Lennox Industries*)

**Figure 16-53.** Remote split system *(Courtesy of Southwest Manufacturing Company)*

efficient operation of a system, all of the pieces must be matched as far as Btu (cal) ratings. Most manufacturers build equipment to fit almost any need and the designated combination of fan, furnace, coil, and condensing unit must be used. After the heat load has been calculated, tables can be consulted to determine the desired combination. Care should be exercised in choosing the equipment to obtain the most satisfactory operation.

## SAFETY PROCEDURES

When working on air conditioning units, it is important to remember that the conditioned air must contain enough oxygen to sustain life. Working on a complete air conditioning system involves many of the dangers involved in refrigeration systems and heating systems.

1. Always wear goggles when working with the refrigerant in these systems.

2. Always wear gloves around duct work to prevent cutting the hands and fingers.

3. Provide protective shields around operating motors and fans.

4. Be sure that the electrical power is off before servicing electrical components.

5. Keep hands, feet, and clothing clear of belts, fans, and rotating parts.

6. Use only safe ladders when working on the roofs of buildings.

7. Never carry equipment, tools, or supplies up or down a ladder. Transport these items with a rope or a hoist.

8. Avoid contact with electrical circuits when the area around the equipment is wet.

9. Do not leave loose, light objects on a roof during high winds.

10. Be sure that refrigerant being charged into a system is the same as what the unit is rated to use.

## SUMMARY

• Air conditioning is the simultaneous mechanical control of temperature, humidity, air purity, and air motion.

• The control of humidity can mean either humidifying or dehumidifying.

• The two primary reasons for using air conditioning are: (1) to improve the control of an industrial process and (2) to maintain human comfort.

• Human comfort depends upon how fast the body loses heat.

• The human body is basically a constant temperature machine.

• The human body gives off heat in three ways: (1) convection, (2) radiation, and (3) evaporation.

• The three conditions that affect the

ability of the body to give off heat are: (1) temperature, (2) relative humidity, and (3) air motion.

• The air should be distributed into the room to provide only minor temperature differences within the occupied zone.

• The two general types of fans used on air conditioning installations are: propeller fan and centrifugal blower.

• The air noise caused by air movement is the result of successive air waves which are caused by the fan blades.

• The heating chamber is the place where the heat may be added to the air during the heating season.

• The cooling chamber is the place where the cooling and dehumidifying of the air occurs.

• The humidifier is the device which adds moisture to the air during the heating season.

• The supply air ducts direct the conditioned air to the conditioned space.

• Supply air outlets are the grills on the end of the supply air ducts which distribute the air within the conditioned space.

• The conditioned space is one of the most important parts of the air distribution system.

• The return outlets are the openings through which the air from the conditioned space is allowed to enter the return duct system.

• The return ducts are used to connect the conditioned space and the air handling equipment together.

• Filters are located on the air inlet side of the fan to prevent the circulation of dust particles through the system.

• The science which deals with the relationships that exist within a mixture of air and water vapor is known as psychrometrics.

• A psychrometric chart is a diagram which represents the various relationships that exist between the heat and moisture content of air and water vapor.

• Any two of the air properties which are located on the psychrometric chart will definitely determine the condition of the air.

• Sensible heat is the heat that can be added to or removed from a substance without a change of state.

• The sensible heat content of air is indicated by the dry bulb temperature.

• The dew point temperature is always an indication of the moisture content of the air.

• The moisture content scale used on most psychrometric charts indicates grains of moisture per pound (453.6 g) of dry air.

• Sensible heat ratio is the sensible heat removed in Btu (cal) divided by the total heat removed in Btu (cal), usually about 50 percent in most comfort air conditioning installations.

• The calculation of the heat load is similar to the calculation of a refrigeration heat load, with the major difference that more sources of heat are involved in air conditioning calculations.

• Heat sources are listed under two general classifications: (1) internal heat load and (2) external heat load.

• The amount of heat which will flow through one square foot ($m^2$) per degree temperature difference through a substance is known as the $U$ factor.

• The design temperatures are the average high temperatures for a given locality and not the highest temperatures in that locality.

• A solar heat load is gained from the direct rays of the sun, therefore, solar heat gain occurs only while the sun is shining.

• The peak heat load may not occur until hours after the sun is at its peak because of the resistance to heat flow of the construction materials.

• People who are active create a greater heat load for the cooling equipment than people who are at rest.

• Infiltration air enters the conditioned space through windows, doors, and cracks and is a source of internal heat load for the cooling equipment.

• The outdoor wind velocity affects the infiltration of air into the conditioned space.

• Ventilation air is air which is introduced into the equipment before it enters the conditioned space and is considered to be an external heat load.

• Electric appliances and lighting introduce a heat load of 3.41 Btus (859.3 cal) per watt into the conditioned space.

• Electric motors also are a source of heat in varying amounts depending upon the motor horsepower.

• Steam appliances supply a heat load of 960 Btu per pound (528 kcal/kg) of steam condensed.

• Appliances heated by gas emit the total amount of heat generated by the burning gas.

• When ducts are installed in hot places, an additional heat load is placed upon the cooling equipment.

• The motor used to drive the fan also emits heat which must be removed by the cooling equipment. The amount of heat transferred is in direct proportion to the amount of work done.

• It is advisable to apply a safety factor of about 10% to the calculated heat load because of the many variables involved.

• The capacity of the chosen equipment must be as close as possible to the calculated value but never less than the calculated value.

• On the psychrometric chart, the point where the sensible heat ratio line and the saturation curve intersect is known as the apparatus dew point (ADP).

• The ADP is the lowest temperature at which air supplied to the conditioned space can still pick up the required amount of both sensible and latent heat.

• The bypass factor refers to that percentage of the total amount of air which passes through the cooling coil without coming into contact with the coil surfaces and being conditioned.

• The duct system design is as critical as the heat load calculation.

• Air velocity is the rate of speed that the air travels through the ducts or openings.

• Air ducts made of metal are preferred because of the smooth surfaces which do not add resistance to the air flow.

• In order to determine the friction loss of a duct system, it is necessary to know the velocity and size of duct.

• The major differences in the equipment used in air conditioning work and that used in refrigeration work are the design characteristics used to compensate for the higher temperatures encountered.

• Air conditioning equipment is generally classified as being either a self-contained or a remote split system.

## REVIEW EXERCISES

1. What are the four basic functions of a true air conditioning system?

2. What are the two primary reasons for using air conditioning?

3. Name the three ways a human body gives off heat.

4. What are the three conditions that affect the ability of the human body to give off heat?

5. Why should the conditioned air be distributed into the space?

6. Name the general types of fans used in air conditioning work.

7. What is the cause of air noise in an air conditioning system?

8. What is the purpose of a humidifier?

9. Where is the heat added to the air during the heating season?

10. What device carries out the dehumidification process in a cooling unit?

11. What part of the air conditioning system directs the air to the conditioned space?

12. Why is the conditioned space one of the most important parts of the air distribution system?

13. What is the purpose of an air filter?

14. What is psychrometrics?

15. What device is used to determine the various relationships that exist between the heat and moisture content of air and water vapor?

16. How many of the air properties must be known before the rest can be determined?

17. What does the dry bulb temperature indicate?

18. What indicates the moisture content of the air?

19. How is the moisture content indicated on a psychrometric chart?

20. What is the sensible heat ratio in most comfort conditioning installations?

21. How are the heat sources on an air conditioning unit classified?

22. What are design temperatures?

23. From where does solar heat gain come?

24. Does the peak heat load occur at the same time the sun is the hottest?

25. What is infiltration air?

26. Do people who are active create more or less heat load than people who are at rest?

27. Where is ventilation air introduced into the system?

28. Is ventilation air an internal or external heat load?

29. What is the Btu (cal) rating of electric appliances?

30. Why is an additional heat load placed on the equipment when the ducts are located in a hot place?

31. What safety factor percentage is generally used on an average job?

32. Should the capacity of the chosen equipment be less than the calculated capacity?

33. What is the apparatus dew point?

34. To what does the bypass factor refer?

35. What are the two classifications of air conditioning equipment?

# Automotive Air Conditioning

Automotive air conditioning systems use the same basic refrigeration system as residential and commercial air conditioning systems. The major differences between an automotive and any other air conditioning system are: (1) the manner used to combine the different components, (2) the manner used to drive the compressor, (3) the fast temperature change inside the automobile, and (4) the complex system of air control dampers which is used on an automobile.

Automotive heating is accomplished in much the same manner as heating with a hot water boiler. In the automobile, the hot water is taken from the radiator and passed through a heating coil where air is blown over it and heated.

Automotive air conditioning units use doors and dampers to direct the air through the desired function. They also are used to control the volume of air that passes through the desired coil and the amount that bypasses the coil. This bypass is done to aid in temperature control. Each manufacturer will use a different arrangement and method of control for these dampers.

## OPERATION

The automotive air conditioning system functions much the same as any other type of air conditioning system. Its purposes are to remove heat and humidity from inside the automobile and deposit this heat and humidity in the outside air, and to clean the air.

The principles of heat transfer used are the same as in other systems. The refrigeration system is divided into a high pressure side and a low pressure side. (See Figure 17-1.) The high pressure side contains the following components: (1) discharge side of the compressor and the high side service valve, (2) condenser, (3) receiver-drier, (4) inlet side of the expansion valve, and (5) connecting hoses to all highside components.

The low side contains the following components: (1) outlet side of the expansion valve, (2) evaporator, (3) suction side of

EXPANSION VALVE

COLD AIR 34°

EVAPORATOR

28°

SUCTION LINE 16°

WARM AIR 85°

A

COMPRESSOR

HIGH PRESSURE LINE (DISCHARGE) 130°

LIQUID LINE 110°

CONDENSER

HOT AIR

SIGHT GLASS

RECEIVER/DRYER

PRIMARY LIQUID LINE

RAM AIR 100°

HIGH PRESSURE **GAS**    HIGH PRESSURE **LIQUID**    LOW PRESSURE    LOW PRESSURE **GAS**

**Figure 17-1.** Basic refrigeration system

the compressor and its low side service valve, and (4) connecting hoses to the low side components.

## BASIC COMPONENTS

The refrigeration system used in automotive air conditioning consists of the following components: (1) compressor, (2) magnetic clutch, (3) condenser, (4) receiver-drier, (5) expansion valve, (6) evaporator, (7) suction throttling valve, and (8) high pressure relief valve. The refrigerant used in automotive air conditioning is R–12, which is presently the standard type of refrigerant. In most cases the evaporator and expansion valve are installed inside the automobile while the other components are located under the hood. The following is a description of the various components.

**COMPRESSOR:** The compressors used in automotive air conditioning units are of the reciprocating type. There are three styles of compressors in use: the two-cylinder, the six-cylinder, and the five-cylinder. (See Figure 17-2.) They are all driven by a belt, or belts, from the engine crankshaft (see Figure 17-3). The functions of the compressor are: to create the pressure required for condensing the refrigerant, to circulate the refrigerant through the system, and to create the low pressure required for evaporation of the refrigerant in the evaporator.

Automotive air conditioning compressors are of the open type. That is, they use a shaft seal to prevent the leakage of refrigerant and oil around the shaft where it extends through the compressor body. Since these seals require oil to prevent leakage, they must be operated periodically so that oil will keep the seal faces wet.

There are two service valves located on the compressor for use in servicing the re-frigeration system. One is the discharge service valve and the other is the suction service valve. Some service valves are equipped with schrader valves which allow pressure to pass when the service hose is connected to it, much like the valve stem in an automobile tire. (See Figure 17-4.) Others use a stem-operated type service valve to aid in servicing. The easiest way to determine which is which is by checking the refrigerant lines. The discharge line goes to the condenser, while the suction line goes to the evaporator.

**MAGNETIC CLUTCH:** All automotive air conditioning compressors in use today use a magnetic clutch to engage and disengage the compressor on demand from the thermostat inside the automobile. A magnetic clutch also is used when a defrost cycle is needed and is manually turned off when the unit is not in use.

All clutches operate basically on the same principle of electromagnetism. They are made in two general types: the stationary field and the rotating field.

**Figure 17-2. (a)** Sankyo compressor (*Courtesy of Sankyo International* [USA], *Inc.*)

## sankyo compressor cutaway

| | | | |
|---|---|---|---|
| 1 | ELECTROMAGNETIC CLUTCH | 9 | PISTON |
| 2 | SHAFT SEAL | 10 | CYLINDER HEAD GASKET & VALVE PLATE GASKET |
| 3 | CLUTCH BEARING | 11 | VALVE PLATE ASSEMBLY |
| 4 | NEEDLE THRUST BEARINGS | 12 | CYLINDER HEAD |
| 5 | FRONT HOUSING & O-RING | 13 | SERVICE PORT |
| 6 | CAM ROTOR | 14 | SYSTEM HOSE CONNECTION |
| 7 | ANTIROTATION GEAR | 15 | OIL FILLER PLUG |
| 8 | CYLINDER BLOCK | 16 | PLANET PLATE |

**Figure 17-2. (b)** Cutaway of Sankyo compressor (*Courtesy of Sankyo International [USA], Inc.*)

**Figure 17-3.** Compressor drive pulley arrangement

**Stationary Field Clutch:** The stationary field clutch is probably the most desirable clutch because it has fewer parts to wear out. (See Figure 17-5.) The magnetic field is mounted on the compressor body by some mechanical means depending on the type of field and the type of compressor. The clutch rotor is held on the armature by a bearing and snap rings. The armature is mounted on the com- pressor crankshaft.

When there is no electrical current being applied to the clutch field, there is no magnetic field and the rotor is free to rotate on the armature. The armature does not rotate when electrical current is not applied.

As the temperature inside the automobile rises, the thermostat contacts close and electrical current is applied to the clutch field. A

**Figure 17-4.** Schrader value in-service valve

**Figure 17-5.** Stationary field magnetic clutch

magnetic attraction is then set up between the field and the armature which attracts the armature to the rotor. This magnetic attraction causes the armature and rotor to act as a single piece and the complete assembly turns while the field remains stationary. The compressor crankshaft begins turning and the refrigeration process is started.

When the temperature inside the automobile is satisfied the thermostat contacts open, interrupting the electrical current to the clutch. The magnetic field fades away and the armature snaps away from the rotor and stops. The rotor continues to turn, but the compressor stops until electric current is again applied to the field.

**Rotating Field Clutch:** The rotating field clutch operates the same as the stationary field clutch except for the difference in placement of the field. The field is incorporated into the rotor and turns with it. The electric current is fed to the field through brushes which are mounted on the compressor.

The electric current is fed through the brushes and sets up a magnetic field which attracts the armature to the rotor. The complete assembly turns, including the field, which causes the compressor to turn.

Slots are machined in both the armature

and rotor, in both types of clutches, which concentrate the magnetic field and increase the magnetic attraction between them.

Because the clutch is engaged and disengaged at high speeds as required for proper temperature control inside the automobile, considerable scoring of both the armature and rotor will occur. This scoring is normal and should not cause undue alarm.

**CONDENSER:** The purpose of the condenser is to receive the hot-high pressure refrigerant vapor from the compressor and cool the vapor until it condenses into a liquid. This is accomplished because heat always flows from a warm substance to a cooler substance. The air passes through the condenser coil and removes the heat.

The condenser is usually mounted in front of the radiator, and, in fact, resembles it. (See Figure 17-6.) Air is forced through the condenser in two ways: (1) it is pulled through the condenser by the radiator fan and (2) it is forced through the condenser as the car moves along the road. This is known as *ram air*.

The compressor increases the pressure over the refrigerant vapor in the condenser. The heat intensity is actually concentrated into a smaller space, thus raising the refrigerant temperature higher than the temperature of the ambient air. Conditions such as a dirty condenser, loose belts, and improper condenser location will result in poor operation of the condenser and a decrease in overall system efficiency and must be avoided.

**RECEIVER-DRIER:** Automotive air conditioning units are more susceptible to leaks than other units because of the open type compressor and the greater amount of vibration to which they are exposed. Over a period of time small leaks will occur requiring the addition of refrigerant. Also, the evaporator refrigerant requirements vary be-

**Figure 17-6.** Automotive air-conditioning system (*Courtesy of Chrysler Corporation*)

cause of the changing heat load, the efficiency of the condenser, and the compressor speed.

A small receiver tank is used in the system to compensate for these variables. Refrigerant is stored in the receiver until it is needed by the evaporator. When a receiver tank is added to the system approximately 1 to 1½ lbs (453.6 to 680.4 g) of additional refrigerant is needed. (See Figure 17-7.)

A drying agent, termed a desiccant, such as silica-gel or molecular seive is placed inside the receiver during the manufacturing process. This is the only point in the entire refrigeration system where damaging moisture and acid can be removed. (See Figure 17-8.) If the desiccant reaches its saturation point, that is, when it has absorbed all the moisture it can hold, moisture and acid will be released into the refrigerant circuit.

This moisture will cause the expansion valve to freeze up and cause compressor damage. If the expansion valve freezes up, the flow

**Figure 17-7.** Receiver-drier (*Courtesy of Refrigeration and Air-Conditioning Division, Parker-Hannifin*)

**Figure 17-8.** Cutaway of receiver-drier

of refrigerant will stop and a decrease in cooling capacity will result. Because of this, the receiver-drier should be replaced each time the system is opened for repair or service.

When a refrigeration system is opened for repairs or service, foreign particles can enter the system. Moisture and air also will enter and will cause the refrigerant to deteriorate. This deteriorating action will cause a hydrolizing action resulting in corrosion inside the system. This corrosion is released into the system in the form of solid particles which will eventually stop the flow of refrigerant through the expansion valve. A filter screen is placed inside the receiver-drier to catch these particles. If this screen catches enough of these foreign particles to reduce the flow of refrigerant, the cooling will be reduced inside the automobile.

A sight glass is usually incorporated into the receiver-drier to aid in servicing the unit. (See Figure 17-9.) A quick check of the sight glass will reveal several conditions to the service technician as follows:

1.  A clear sight glass (indicated by a solid stream of refrigerant) indicates that the system has a sufficient amount of refrigerant or possibly an overcharge. An overcharge will be indicated by higher than normal gauge readings.

2.  Foam or a steady stream of bubbles indicate that the system is short on refrigerant.

3.  Oil streaks on the sight glass indicate that the system is completely out of refrigerant.

4.  Occasional bubbles indicate that the system is slightly low on refrigerant or that the receiver-drier is saturated and is releasing moisture into the system.

5.  A clouded sight glass indicates that

**Figure 17-9.** Sight glass in receiver-drier

the desiccant is breaking down and is circulating through the system.

**EXPANSION VALVE:** The thermostatic expansion valve is the most popular flow control device used on automotive air conditioning systems. This device reduces the high pressure refrigerant liquid to a low pressure liquid refrigerant. (See Figure 17-10.) The orifice in the expansion valve is the point where the pressure reduction takes place. The liquid refrigerant has the lowest temperature upon leaving the expansion valve and entering the evaporator. The expansion valve is controlled by the sensing bulb which is fastened to the tubing at the evaporator outlet. The expansion and contraction of the charge inside the bulb, in response to the evaporator outlet temperature, causes the valve to open and close as needed.

The thermostatic expansion valve has three functions. It controls the flow of refrigerant into the evaporator, it provides a quick refrigerant pressure pulldown on start-up, and it overcomes high discharge pressures during idle and low speed operation.

There are two types of thermostatic expansion used on automotive air conditioning systems: (1) internally equalized, which is the most common, and (2) externally equalized, which is used where special control is needed. (See Figure 17-11.)

**Expansion Tube:** Some General Motors units use an expansion tube in place of the expansion valve (see Figure 17-12). The flow rate through the expansion tube depends primarily on the amount of subcooling that takes place in the condenser. Subcooling is the cooling of the liquid refrigerant after it has been condensed to a liquid. The expansion tube is located at the inlet of the evaporator like the expansion valve (see Figure 17-13).

**EVAPORATOR:** The evaporator is the point in the system that absorbs the heat from the interior of the automobile. It is the forced air finned type of coil.

It is in the evaporator where the refrigerant, which is under low pressure, evaporates and absorbs heat. The heat is absorbed from the air as the blower moves the air through the coil. (See Figure 17-14.) This flow of

CAPILLARY TUBE

DIAPHRAGM

SPRING

INTERNAL EQUALIZER PORTS

PIN

OUTLET TO EVAPORATOR

REMOTE BULB

BODY

INLET FROM RECEIVER

**Figure 17-10.** Internal view of a thermostatic expansion valve

Externally Equalized

Internally Equalized

**Figure 17-11.** Typical thermostatic expansion valve

TUBE—EXPANSION

Contains a fixed orifice through which liquid refrigerant passes from the high pressure liquid line into the evaporator. Located in the inlet tube of the evaporator core.

**Figure 17-12.** Expansion tube (*Courtesy of Chevrolet Division, General Motors Corp.*)

**Figure 17-13.** Expansion tube location (*Courtesy of Chevrolet Division, General Motors Corp.*)

air is essential to the evaporation of the liquid refrigerant. The amount of heat exchanged depends on the temperature difference of the air and the refrigerant. The greater the differential, the greater the amount of heat that will be transferred from the air to the refrigerant. A high heat load will allow for the rapid transfer of heat to the refrigerant. When the blower is on high speed it will deliver the greatest volume of air through the evaporator and produce rapid evaporation of the refrigerant. As the inside of the automobile is cooled, a temperature will be reached where very little extra cooling will be experienced if the fan is allowed

**Figure 17-14.** Automotive air-conditioning evaporator (*Courtesy of Chevrolet Division, General Motors Corp.*)

to continue operating on high speed. A decrease in air volume will result when the fan speed is reduced. However, the lower volume of air will allow the air to remain in contact with the evaporator coil longer. More heat will be transferred to the refrigerant, resulting in colder air being discharged inside the automobile.

The changing of the state of the refrigerant inside the evaporator is as important to the overall efficiency of the unit as the air flow over the coil. The liquid refrigerant which is admitted to the evaporator by the expansion valve changes to a vapor as it absorbs heat from the air. Some liquid must be present throughout the entire length of the evaporator in order to obtain full capacity of the unit.

A flooded evaporator is the result of too much refrigerant being fed into the evaporator by the expansion valve. When this condition occurs, the excess of refrigerant is allowed to pass through the evaporator and into the suction line and eventually into the compressor. This condition can result in serious damage to the compressor. There also will be insufficient cooling inside the automobile.

When too little refrigerant is fed into the evaporator a starved evaporator is the result. When this condition occurs, liquid refrigerant does not reach through the entire length of the coil. Therefore, heat transfer is not accomplished through the complete length of the evaporator resulting in reduced capacity.

**SUCTION THROTTLING VALVES:** Any device which is used to regulate the flow of refrigerant from the evaporator to the compressor is termed a suction throttling valve or regulator. (See Figure 17-15.) These devices are located in the suction line between the tail end of the evaporator and the compressor suction service valve.

There are several different types of suction throttling valves and it may be difficult to determine exactly which type is used on a particular unit. However, on inspection of the unit, if no valves are found in the suction line it may be assumed that a thermostat is used to operate the clutch when an icing condition ocurs.

The different types of suction throttling valves are:

**1. POA (pilot operated absolute) Valve:** These valves were used on General Motors units during the period of 1966–1973. Ford Motor Corporation is presently using them on their systems, but call them STV valves.

**Figure 17-15.** Suction throttling valve cutaway (*Courtesy of Chevrolet Division, General Motors Corp.*)

They are a unique type of valve because of the "straight through" refrigerant flow and the sealed housing. They are basically spring-loaded valves which are controlled by an evacuated bellows and needle-valve assembly which are located within the valve body. Operation does not depend upon atmospheric pressure and, therefore, is not affected by any changes in altitude. (See Figure 17-16.) If the evaporator pressure is above approximately 28.5 psi (2 kg/cm²) with the compressor running, the pressure differential between the valve inlet and valve outlet pressures will cause the bellows to be compressed which will open the valve and permit a free flow of refrigerant to the compressor. If the evaporator pressure drops below 28.5 psi, however, the bellows will

expand and close the needle valve. This bellows action tends to equalize the pressure on each side of the valve and permit the valve spring to move the valve needle toward the closed position. When the flow of refrigerant is slowed down by the valve closing, the evaporator outlet pressure will increase and prevent freezing of the evaporator coil. When the evaporator pressure increases above 28.5 psi, the valve will open again permitting the full flow of refrigerant to the compressor. This cycling of the valve is constantly repeated throughout the operation of the unit.

**2. Evaporator Pressure Regulator (EPR):** These valves are used on 1961 and later Chrysler Corporation automobiles except

Suction Throttling Valve
Refrigerant Flow

Liquid Bleed Line
Connector (From Regulated
Evaporator Pressure)

Protection
Cap

Expansion Valve
Equalizer Line Connection
(To Compressor Suction Pressure)

Regulated Evaporator
Pressure Test Connector
(Schrader Valve)

From Regulated
Evaporator Pressure

To Compressor
Suction Pressure

To Compressor
Suction Pressure

Pressure Differential Valve
(Starts to Open @ 10 psi
Fully Open @ 20 psi)

Regulated Refrigerant
Pressure (Evaporator)

Non-Regulated Refrigerant
(Compressor Suction)

**Figure 17-16.** POA valve (*Courtesy of Ford Motor Corp.*)

Dart, Valiant, and automobiles which are equipped with hang-on type air conditioning units. EPR valves are suction throttling valves which are located at the compressor (see Figure 17-17). These valves are designed to prevent evaporator icing by controlling the suction pressure and keeping it above the freezing temperature. They operate the same as the earlier external type EPR valve with the exception that the valve piston is controlled by a gas-filled bellows. The EPR valve maintains the evaporator pressure between 22 psi and 26 psi (1.55 kg/cm$^2$ and 1.83 kg/cm$^2$). (See Figure 17-18.)

This valve cannot be adjusted like the older type. A replacement of the complete valve is required when a malfunction occurs.

The compressor and EPR are both equipped with an oil passage which allows oil to pass from the suction line into the compressor crankcase. This oil passage has two functions: (1) it allows any oil which has left the compressor to return to it and (2) it is used to pressurize the compressor crankcase and prevent it from falling below atmospheric pressure. This is necessary because air and moisture could be drawn into the crankcase through the shaft seal if the crankcase pressure was allowed to fall below atmospheric pressure.

**3. Evaporator Temperature Regulator (ETR):** In the Chrysler Corporation Auto-Temp System of 1967 and later, the EPR valve has been replaced with the ETR. (See Figure 17-19.) The EPR and the ETR valves are designed to accomplish the same thing. They control the flow of refrigerant

**Figure 17-17.** EPR valve location (*Courtesy of Chrysler Corporation*)

from the evaporator to prevent freezing of the evaporator. The major difference between the two valves is that the EPR valve is designed to modulate the flow of refrigerant while the ETR is controlled by a solenoid which is built into the valve. Therefore, it is either fully open or fully closed. It is a normally open valve. (See Figure 17-20.)

An ETR switch is a temperature sensitive control which actuates the solenoid. It is located in the rear of the evaporator case and has a sensing capillary tube which extends through the fins on the evaporator coil to sense the evaporator temperature. When the evaporator falls to a point at which the evaporator might freeze up, the ETR switch contacts close, thus closing the ETR valve. When the evaporator temperature increases sufficiently to prevent freezing, the switch contacts open and allow the full flow of refrigerant to pass to the compressor.

**Figure 17-18.** EPR valve (*Courtesy of Chrysler Corporation*)

**Figure 17-19.** ETR valve location (*Courtesy of Chrysler Corporation*)

**Figure 17-20.** ETR valve *(Courtesy of Chrysler Corporation)*

**VALVES IN RECEIVER (VIR):** With the 1973 model, all General Motors automobiles were equipped with a valves-in-receiver (VIR) assembly. The VIR assembly combines the thermostatic expansion valve, the POA valve, the receiver-drier, and the sight glass into a single unit. The VIR is mounted next to the evaporator and has eliminated the external equalizer connection between the expansion valve and the POA valve. This was accomplished by drilling a hole in the divider between the POA valve and the thermostatic expansion valve cavities in the valve body. (See Figure 17-21.)

This valve was redesigned in 1975 and called the evaporator equalized valves in receiver (EEVIR). This new design reduced the fuel consumption of the automobile.

These new valves, the EEVIR, are identified by their gold finish as opposed to the silver finish of the VIR assembly.

The valve housing and the expansion valve are two components of the EEVIR that must not be interchanged with the components of the older VIR assembly.

In operation, the liquid refrigerant flows from the condenser into the receiver-drier assembly when it is dehydrated by the desiccant. It then flows from the receiver section through the filter screen at the inlet of the pick-up tube, through the pick-up tube to the expansion valve cavity.

The expansion valve controls the flow of refrigerant into the evaporator by sensing both the temperature and pressure of the re-

**Figure 17-21.** Cutaway of VIR assembly *(Courtesy of Chrysler Corporation)*

frigerant as it flows through the VIR assembly. This control is accomplished by the power diaphragm of the thermostatic expansion valve. The diaphragm pressure is affected by the temperature of the refrigerant from the evaporator on its way through the VIR inlet to the POA valve. The end of the power diaphragm is now in the refrigerant vapor stream from the evaporator and any increase in temperature of the refrigerant vapor will cause the power diaphragm to expand. This expansion of the diaphragm will move the needle away from its seat allowing more refrigerant to flow. A drop in temperature of the refrigerant vapor will reverse the action of the power diaphragm and allow the valve to decrease the flow of refrigerant.

The pressure of the refrigerant vapor from the evaporator also will affect the operation of the power diaphragm. The pressure is sensed through the equalizer port between the power diaphragm section of the expansion valve cavity and the POA valve cavity, thereby eliminating the need for an external equalizer line.

The POA valve controls the flow of refrigerant to maintain an evaporator pressure of 30 psi (2.1 kg/cm$^2$). A temperature of 32°F (0°C) is maintained with this pressure. This pressure and temperature combination prevents freezing up of the evaporator, which would block the flow of air through the coil.

In operation, the evaporator pressure is applied to the inlet end of the valve piston. The refrigerant then flows through the piston screen and a hole which is drilled through the piston to the bellows cavity. When the pressure from the evaporator increases, the force of the POA valve piston spring and the bellows cavity pressure is overcome allowing the piston to move and open the main valve port. The refrigerant then flows through the main valve port back to the

compressor suction. The opening of the valve allows the refrigerant to flow to the compressor. The evaporator pressure is lowered to 30 psi (2.1 kg/cm$^2$). The force of the piston spring and the bellows cavity pressure overcomes the evaporator pressure and the valve blocks the valve port and reduces the flow of refrigerant to the compressor.

When the evaporator pressure is greater than the pressure in the bellows cavity, refrigerant flows through an orifice in the piston to maintain the control pressure in the bellows cavity. (See Figure 17-22.) This control pressure affects the piston valve movement and the resulting movement of the valve needle. Refrigerant is allowed to pass from the piston spring area of the bellows cavity through two small holes in the bellows retainer in the cavity surrounding the bellows. With an increase in pres-

**Figure 17-22.** Refrigerant flow through EEVIR (*Courtesy of Chevrolet Division, General Motors Corp.*)

sure around the evacuated bellows, it will contract, raising the needle from the seat. The refrigerant will then flow from the bellows cavity through the needle seat orifice and into the suction line. When the flow of refrigerant through the needle seat orifice is greater than the refrigerant flow through the piston orifice into the bellows cavity, the bellows cavity pressure will decrease. The bellows will expand and move toward the seat, closing off the orifice. The balance of forces acting upon the valve and the controlled flow of refrigerant tend to keep the POA valve piston and needle valve modulating the flow of refrigerant to the compressor.

These POA valves are set at the factory and cannot be adjusted or repaired in the field. If the POA valve is not functioning properly, the complete assembly must be replaced.

**EEVIR EQUALIZER FUNCTIONS:** The equalizer port is designed to speed up the opening of the expansion valve when certain conditions exist. When the POA valve is modulating toward the closed position and slowing down the cooling, the first part of the evaporator coil would eventually become warm enough to open the expansion valve, because of the pressure of the R–22 refrigerant vapor in the power diaphragm. When the automobile is moving at high speed, this time lag would be too great to allow an even outlet air temperature because of the time required to cool the evaporator coil after the expansion valve has opened. This fluctuation in temperature has been eliminated by use of the equalizer feature which helps the expansion valve power diaphragm to open the valve without waiting for the expansion valve to warm up.

A pressure drop in the equalizer port and the resulting drop in pressure under the diaphragm would cause the valve to operate as if the expansion valve sensing element had been heated and pressure exerted on the top of the diaphragm. Thus, the expansion valve will open and allow refrigerant to flow to the evaporator. This is possible because of the reduced pressure under the diaphragm, not the increase in pressure on top of the diaphragm.

## COMPRESSOR CONTROLS

The controls that operate the compressor are electrical switches. They are installed in series with the magnetic clutch so that the desired function can be obtained by any one of the controls. They are designed to regulate the operation of the compressor. These controls are as follows:

**THERMOSTAT SWITCH:** This control is placed inside the automobile and senses the evaporator temperature. It is connected into the compressor clutch electrical circuit and when the evaporator approaches the freezing temperature, the thermostat switch will interrupt the electrical circuit to the clutch and stop operation of the compressor. When the evaporator temperature rises to a predetermined temperature, the electrical circuit is completed, again energizing the magnetic clutch and compressor operation is resumed. Some of these controls have a knob which allows the operator to change the cut-in and cut-out points of the control. Others have an adjustment screw which is used to regulate the switch operation. There are two types of thermostatic switches in use. They are: (1) bimetal and (2) bellows. (See Figure 17-23.)

**Bimetal Thermostatic Switch:** The actuating device on these controls is a bimetal arm which expands and contracts with a change

**Figure 17-23.** Thermostatic switches (*Courtesy of Chevrolet Division, General Motors Corp.*)

in temperature. This control is mounted inside the evaporator housing which allows the discharge air from the evaporator to flow directly over the bimetal arm. This cold air causes the bimetal arm to contract and open the switch contacts. The evaporator discharge air becomes warmer and causes the bimetal arm to expand and close the switch contacts, starting operation of the compressor. Cooling is resumed until the discharge air temperature drops to the switch cut-out point and the cycle is repeated.

**Bellows Thermostatic Switch:** These switches consist of a gas-filled bellows and capillary tube assembly. The capillary tube is the sensing device and is usually inserted into the fins of the evaporator to obtain a close temperature control of the evaporator. An increase in evaporator temperature will cause an increase in pressure inside the bellows assembly. This increase in pressure will cause the switch contacts to close, completing the electrical circuit to the compressor clutch which starts operation of the compressor. As the evaporator temperature is lowered, the pressure within the bellows assembly will drop causing the bellows to contract. Thus, the electrical circuit to the clutch is interrupted and the cooling process is stopped. There is a definite differential between the

cut-in and the cut-out temperatures of the switch. This differential allows the evaporator to warm before the compressor is started again. An adjustment screw is preset at the factory to maintain a difference of about 12°F (6.67°C). However, field adjustments may be made when necessary. A control knob is provided so that the operator can adjust the switch to obtain the desired temperature.

**AMBIENT SWITCH:** Compressor operation when air conditioning is not required may result in compressor damage. Ambient switches are designed to prevent operation of the compressor when the outside air temperature is too low. The switch will normally close at approximately 32°F (0°C). However, some manufacturers may require a different setting. When this switch is closed, the electrical circuit to the compressor clutch will be completed which will allow operation of the compressor. Temperatures below this setting will interrupt the electrical circuit to the compressor clutch and prevent operation of the compressor.

**SUPERHEAT SWITCH:** These switches include a diaphragm and sensing tube assembly which is charged with R–114 refrigerant. The sensing tube is fitted into the suction cavity of the rear compressor head and senses the temperature of the suction gas. The suction gas temperature influences the internal pressure of the diaphragm and sensing tube assembly. The suction pressure exerts an external influence on the diaphragm. (See Figure 17-24.)

The switch is designed so that only a low-pressure high-temperature condition inside the compressor will cause an electrical contact which is welded to the diaphragm to touch the terminal pin, which may cause a fuse to blow. Refrigerant conditions of low-pressure low-temperature or high-pressure high-temperature will not cause the contacts to close.

**Figure 17-24.** Superheat switch (*Courtesy of Chevrolet Division, General Motors Corp.*)

**THERMAL FUSE:** The thermal fuse consists of a temperature sensitive fuse link, a wire-wound resistor, and the necessary electrical terminals. These are all mounted in a plastic casing. (See Figure 17-25.)

**Figure 17-25.** Thermal fuse (*Courtesy of Chevrolet Division, General Motors Corp.*)

The thermal fuse allows a time delay before the fuse link is blown. Conditions such as a low refrigerant charge, complete loss of refrigerant charge, a defective POA valve, a defective expansion valve, or an improperly located thermal fuse can cause the thermal fuse to blow and interrupt the electrical circuit to the compressor clutch. The thermal fuse prevents possible damage to the compressor due to these conditions.

**LOW REFRIGERANT CHARGE PROTECTION SYSTEM:** Many of the later model General Motors automobiles include a low refrigerant charge protection system. This system consists of the superheat switch, the thermal fuse, the air conditioning control switch, and the ambient switch. During operation, electric current flows through the air conditioning control switch, the ambient switch, and the thermal fuse to actuate the compressor clutch coil. (See Figure 17-26.) If a partial or total refrigerant charge is lost the superheat switch, mounted in the rear compressor head, will sense the high suction

**Figure 17-26.** Low refrigerant charge protection system (*Courtesy of Chevrolet Division, General Motors Corp.*)

gas temperature and its contacts will close. When the superheat switch contacts close, electric current flows through the resistance heater in the thermal fuse. The heater provides heat to the fuse link which warms it to its specific melting point. When the fuse opens, the electrical current is interrupted to the compressor clutch. The clutch will disengage, preventing damage to the compressor due to a loss of refrigerant. The cause for the shortage of refrigerant must be found and repaired and the unit recharged with refrigerant or the replacement fuse will be blown.

## AUTOMOTIVE TEMPERATURE CONTROL SYSTEMS

During recent years there has been an increased demand for passenger comfort in automobiles. This demand has led to a greater sophistication of automotive air conditioning systems. Many manufacturers have combined the control of heating and cooling systems into one temperature control package.

Most of the current control systems are similar in their theory of operation. The reheat principle is used almost exclusively. This system directs the air through the evaporator coil where it is cooled and dehumidified, then routed either through or around the heater core and into the air distribution system. (See Figure 17-27.)

**MANUAL CONTROL SYSTEMS:** The basic difference between the systems being used is the method of moving the various components. Some control systems use mechanical levers and cables, some use vacuum-actuated components, while other systems use a combination of both of these

**Figure 17-27.** Typical automotive air-conditioning air flow schematic (*Courtesy of Chevrolet Division, General Motors Corp.*)

methods. The following is a brief discussion of the various components which are involved in the control of an automotive air conditioning system.

**Control Assembly:** This is the first component that is used in an automobile air conditioning system. (See Figure 17-28.) This assembly consists of levers or a series of switches which are located on the control panel. The levers or buttons are used to control the mode of operation. The passenger can control whether the system is heating, cooling, or defrosting by changing the control settings on the control panel.

The system selection switch controls the source and the flow of air. For example, when the selector is placed in a position which calls for cooling, an electrical circuit is completed to the compressor clutch through the control panel switch and the ambient switch. If the ambient air temperature is above the setting of the ambient switch, the switch contacts will be closed and the compressor will operate. If the control switch or lever were placed in any of the other positions, the compressor would not usually operate.

The system selector switches or levers also determine the direction of air flow. Changing the mode of operation changes the position of a rotary vacuum valve, or switch, which is located on the control panel. (See Figure 17-29.) The vacuum valve, or switch, controls the position of the air doors located in the duct system. These doors determine the routing of the flow of air. In a push button type control system, the buttons act as a switch valve which directs the vacuum to one side or the other of a vacuum-actuator diaphragm. The actuator diaphragm is connected to an air door by a rod and linkage. Movement of the diaphragm will either open or close the air door. (See Figure 17-30.) The same function could be accomplished by use of levers and cables without the vacuum actuator.

The vacuum in these systems is created in the engine manifold. A vacuum storage tank is sometimes used which tends to eliminate fluctuations in the control system. The vacuum is usually routed to the various vacuum motors by use of a vacuum valve, or switch, which is usually located on the rear of the control panel.

**Figure 17-28.** Control assembly (*Courtesy of Chevrolet Division, General Motors Corp.*)

**Temperature Control:** The temperature lever normally determines the temperature of the discharge air by positioning an air door which is located in the air ducts. This is called an air temperature door and its position is usually controlled by use of a bowden cable which connects the temperature lever to the air temperature door. In some later models, the temperature lever is connected to a vacuum valve. This vacuum valve directs the vacuum to an air inlet diaphragm through the system selector rotary vacuum valve. (See Figure 17-31.) This control system will determine the air tem-

**Figure 17-29.** Auto air-conditioning vacuum control system (*Courtesy of Chevrolet Division, General Motors Corp.*)

perature which is being recirculated inside the passenger compartment. Fast recirculation of the air inside will give a fast cool down on hot, humid days.

**Water Valve:** The water valve is used to control the flow of engine coolant through the heater core. When more coolant is allowed to flow through the heater core, more

heat is provided in the passenger compartment. When the valve is closed, less coolant is allowed to flow, which reduces the amount of heat provided for the passenger compartment.

**Heater Core:** The heater core is where the heat is transferred from the engine coolant to the air. The coolant flow through the core

**Figure 17-30.** Typical vacuum-actuated system (*Courtesy of Chevrolet Division, General Motors Corp.*)

begins at the front of the engine intake manifold and flows to the inlet port of the heater core, through the heater core, out of the outlet port of the heater core, and to the suction of the engine water pump.

**Blower Speed:** Normally a blower switch is used to control the air flow by regulating the speed of the blower motor. Several different blower speeds may be available with the control numbers varying with the make and model of the particular automobile.

In checking and repairing automotive air conditioning systems, many malfunctions are traced to conditions which exist in the vacuum system. Therefore, it is necessary that all vacuum connections, vacuum hoses, and check valves be checked if a vacuum problem is suspected. An ohmmeter and a circuit diagram are usually all that is needed to find electrical shorts, grounds, and loose connections.

**AUTOMATIC TEMPERATURE CONTROL SYSTEMS:** Automatic temperature control systems control both the heating and

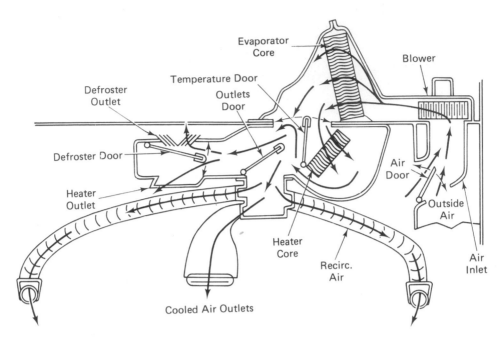

**Figure 17-31.** Typical air flow schematic (*Courtesy of Chevrolet Division, General Motors Corp.*)

cooling modes. Therefore, the amount of each is automatically supplied to produce the desired temperature in the passenger compartment. The temperature is selected by use of a temperature dial similar to a room thermostat. This control may be set to maintain a passenger compartment temperature between 65°F and 85°F (18.3°C and 29.4°C). A sliding control is usually used to select the specific mode of operation. When the temperature selection has been made, the system operates automatically regardless of any changes in weather conditions. (See Figure 17-32.)

Individual control systems may vary in design, but they will all have the following four basic components:

1.   Sensor string

2.   Control panel

3.   Transducer

4.   Power servo

**Sensor String:** The sensor is a thermistor type of resistor. The resistance value of a thermistor varies inversely as the temperature changes. That is, a rise in temperature causes its resistance value to decrease. Conversely, a fall in temperature of a thermistor causes its resistance value to increase. A sensor string is a series of thermistors placed at strategic points throughout the system to perform a particular function. (See Figure 17-33.)

The most popular locations for these sensors are:

1.   The in-car sensor is located on the control panel to sense the air temperature within the passenger compartment.

**Figure 17-32.** Typical automatic temperature control system (*Courtesy of Chevrolet Division, General Motors Corp.*)

**Figure 17-33.** Typical thermistor sensor (*Courtesy of Chevrolet Division, General Motors Corp.*)

2.   The ambient sensor is generally located in the fresh air duct so that it can sense the incoming air temperature. This sensor also may sense the sunlight effect on the passenger compartment.

3.   The duct sensor is located in the discharge air duct plenum to sense the discharge air temperature.

These sensors are connected in series with the temperature selector and with each other. This is done to provide an electrical signal which will reflect the various temperatures of the sensors and temperature selector. This electrical signal is fed to the system amplifier.

**Control Panel:** The control panel is located

AUTOMOTIVE AIR CONDITIONING

on the instrument panel. It contains the amplifier and the temperature dial. The amplifier provides an output voltage from the voltage which is received from the sensors and the selector switch. The amplifier can be mounted on the control panel or it may be mounted separately. (See Figure 17-34.)

**Figure 17-34.** Typical control panel (Courtesy of Chevrolet Division, General Motors Corp.)

**Transducer:** The transducer is an electro-mechanical device that changes the electrical signal from the amplifier into a vacuum signal which controls operation of the power servo (see Figure 17-35).

**Power Servo:** The power servo is the control component of the system. It operates in response to the signal from the sensor to program for the desired operation. It is connected to the air temperature door and positions the door to provide either heating or cooling to the passenger compartment. It also regulates the vacuum control and the blower motor speeds to maintain passenger compartment temperature in response to the temperature dial setting (See Figure 17-36.)

**Figure 17-35.** Typical transducer (Courtesy of Chevrolet Division, General Motors Corp.)

## UNIVERSAL AUTOMATIC TEMPERATURE CONTROL ANALYZER II

The analyzer is basically a device used to substitute resistances when connected into the wiring harness of the automobile. When properly connected it bypasses the sensors and the temperature selector and allows for

**Figure 17-36.** Typical power servo (Courtesy of Chevrolet Division, General Motors Corp.)

manual control of the system. By switching in preset resistance values, the complete system operation can be checked. (See Figure 17-37.)

The tester vacuum gauge samples the vacuum at the blower master switch to indicate changes in the transfer valve and mode changes. The voltmeter is used to check operation of the amplifier and servo as well as operation of the sensors and the selector dial potentiometer.

## INSTALLATION PROCEDURES

These procedures do not refer to any particular make or model of automobile air conditioner. They are merely a guide for use in supplementing the schematic drawings which are included in each manufacturer's mount and drive kit.

Universal type auto air conditioning units are fairly easy to install. All of the components are prefabricated by the manufacturer and are accompanied by detailed instructions and schematics. The automotive air conditioning unit consists of a belt-driven compressor which is mounted on the engine, a condenser which is mounted between the radiator and grill, a receiver-drier which is mounted on the fender well, and an evaporator case assembly which is mounted under the dash panel.

The following steps are suggested by the manufacturer to save time, money, and labor. The directions which are given (right and left) are given from the driver's seat.

*Important:* All refrigerant lines and fittings should remain capped or plugged until connected. This is to eliminate the possibility of dirt and moisture contamination of the system. Use one or two drops of clean refrigeration oil on all flare fittings to insure leakproof connections. Do not use excessive torque when tightening the fittings. Be sure

to disconnect the battery negative cable before starting the following procedure.

1. Remove the radiator from the automobile if the crankshaft pulley is not readily accessible. To do this, first drain the radiator, then disconnect the radiator hoses and transmission cooler lines, remove the radiator shroud, remove the support bolts, and lift the radiator out carefully.

2. The compressor mount and drive assembly should be mounted according to the instructions included in each kit. In some cases the compressor mounting bolts may not be accessible. In such cases the compressor should be installed on the bracket before installation of the bracket on the engine.

3. Install the clutch on the compressor crankshaft. Follow the instructions in the clutch box. Be sure to check the alignment of the Woodruff key on the compressor crankshaft with the key way in the clutch hub.* After the clutch field coil has been installed on the compressor, place the clutch rotor-pulley assembly on the shaft. Rotate the pulley manually to be certain that there is no interference between the clutch field and the rotor. Secure the rotor assembly to the compressor crankshaft with the washer and cap screw which are provided. Tighten the cap screw again after the installation is completed. A loose cap screw on the clutch may ruin the Woodruff key, the clutch hub, or the tapered compressor crankshaft.

4. Install the condenser between the radiator and grill with the connections on the

* A woodruff key is a crescent-shaped piece of metal that fits in a slot on the compressor shaft and a matching slot in the clutch hub to prevent rotation of the hub.

1. Tee Fitting Assortment
2. Vacuum Hose Assortment
3. Temperature Dial Calibrator
4. Volt & Ohm Probe & Clip
5. Clip to Clip Test Lead
6. 1970 Chevrolet Adapter
7. 1967-1968 Oldsmobile Adapter
8. 1968-1969 Chevrolet Adapter
9. 1968-1970 Buick Adapter
10. 1969-1970 Cadillac Adapter
11. GM Programmer Vacuum Plug Adapter
12. GM Programmer Service Stand
13. GM Programmer Adjustment Tool
14. GM Programmer Door Turning Tool
15. Vacuum Pump Hose

**Figure 17-37.** Universal automatic temperature analyzer II (*Courtesy of Chevrolet Division, General Motors Corp.*)

same side of the automobile as the compressor. Locate the condenser about ½ inch to 1½ inch (12.7 mm to 38.1 mm) away from the radiator and as high as possible. Use the sheet metal screws and the brackets provided to secure the condenser to the radiator yoke. Do not crossthread or twist the condenser fittings.

5. If necessary, cut two holes in the upper part of the radiator yoke on the same side as the compressor. Use a 1¼-inch (31.9 mm) hole saw to make the holes. Run the refrigerant lines into the engine compartment. Put the provided grommets (line protectors) in the holes to prevent damage to the refrigerant lines. It may be necessary to relocate some of the parts to make room for the condenser.

6. The receiver-drier should be mounted as nearly as possible upright on the fender well. Use the sheet metal screws and clamp which are provided. The inlet fitting is connected to the ⅜-inch (9.525-mm) fitting on the condenser.

7. The evaporator is to be mounted underneath the dash. Attach the hanger brackets to each end of the evaporator using the screws and washers which are provided. Leave all bolts and nuts loose so that the evaporator can be properly positioned under the dash.

8. Hold the evaporator case under the dash to determine the best location. Be sure that the brake and accelerator are not crowded. With the evaporator parallel to the underside of the dash, mark the hole locations for the hanger brackets, drain holes in the transmission tunnel, and the locations on the firewall for routing the refrigerant lines. Be sure that the hole locations in the firewall clear the engine and its components. Drill the holes as marked. Do not attempt to cut the carpet with drills or hole saws.

9. Route the refrigerant lines from the engine compartment, through the holes in the firewall and secure them to the proper fittings on the evaporator assembly. Be sure the provided grommets are installed in the holes in the firewall. Wrap the exposed portion of the suction line connection on the evaporator with insulating tape.

10. Secure the evaporator assembly underneath the dash. If a long firewall brace is used, extreme care must be used when drilling into the end of the evaporator case or blower housing. Cut one end of the drain hose at an angle to prevent its butting against the transmission beneath the floorboard. Push the drain hose through the hole in the floor, then slip the other end over the aluminum drain tube in back of the evaporator drain pan.

11. Complete the electrical wiring according to the provided diagram. Be sure that the wire with an in-line fuse holder is connected to the ignition switch accessory terminal.

12. Pull the two refrigerant lines through the firewall toward the compressor to remove any slack from the passenger compartment. Connect the suction hose to the suction service valve on the compressor, the liquid line to the outlet fitting on the receiver-drier, and the discharge line to the discharge service valve on the compressor. All refrigerant hoses should be secured away from exhaust manifolds and pipes, hot sections of the engine, and any sharp edges.

## PROCEDURE FOR EVACUATING AND CHARGING AN AUTOMOBILE AIR CONDITIONER

It is necessary to evacuate an air conditioning system after installation or any time that the system has been serviced so that it is purged of all refrigerant. Evacuation is neccessary to rid the system of all air and moisture that has entered the system during the installation or servicing operations. As the pressure is lowered in the system, the temperature of any moisture present in the system also will be lowered. The moisture will boil and it is possible to remove it in the form of vapor.

**PROCEDURE:** Following is an outline of the procedure used to connect the vacuum pump to the system to be evacuated.

1. Connect the manifold gauge set to the system (see Figure 17-38).

2. Close the gauge manifold hand valves.

3. Remove the protective caps from the inlet and exhaust of the vacuum pump. Be sure that the port cap is removed from the exhaust port to avoid damage to the vacuum pump.

4. Connect the center gauge manifold line to the inlet of the vacuum pump. Connect the high pressure manifold hose to the discharge service valve on the compressor. The compressor head will be marked with the letter D or DISC. Connect the low pressure manifold hose to the suction service valve on the compressor. The compressor will be marked with the letter S or SUCTION.

COMPOUND      PRESSURE

GAGE SET

**Figure 17-38.** System hook-up for evacuation

**SYSTEM EVACUATION:** The following steps outline the procedure for evacuating an air conditioning system after the proper connections have been made as outlined under the above procedure.

1.  Start the vacuum pump.

2.  Open the low pressure manifold hand valve and observe the compound gauge needle. It should be indicating a vacuum inside the system. After the vacuum pump has operated about 5 minutes, the compound gauge should indicate at least a 20-inch (25.4-cm) vacuum. The high pressure gauge should be slightly below zero.

3.  If the high pressure gauge needle does not drop below zero, a system blockage is indicated. If the system is blocked, discontinue evacuation. Remove the obstruction. Continue with the evacuation procedure.

4.  Operate the vacuum pump for approximately 15 minutes and observe the gauges. The compound gauge should indicate a reading of 24 to 26 inches (15.2 cm to 10.2 cm) of vacuum if there is no leakage in the system. If the desired vacuum has not been reached, close the low pressure hand valve and observe the compound gauge. If the compound gauge indicates an increase in pressure, indicating a loss of vacuum, there is a leak which must be repaired before continuing the evacuation procedure.

5.  Operate the vacuum pump for at least 30 minutes. After the evacuation process is completed, close the low side pressure manifold hand valve. Shut off the vacuum pump and disconnect the center manifold hose from the vacuum pump. Replace the protective caps on the inlet ports of the vacuum pump.

6.  Note the compound gauge. It should indicate 26 to 29 inches (10.2 cm to 2.54 cm) of vacuum. The compound gauge should remain at this reading. If the system fails to meet this requirement, a partial charge must be put in the system and the system leak checked. After the leak has been found and repaired, the system must be purged of refrigerant and completely re-evacuated. If the system holds the specified vacuum (see Figure 17-39) it may be charged with freon.

**SYSTEM CHARGING PROCEDURE USING ONE-POUND CANS OF R–12:** Following is an outline of the charging procedure used on automotive air conditioning units:

1.  The gauge manifold should still be connected for the evacuation procedure, but the center hose should be loose.

2.  Attach the freon dispenser valve to the freon can. Connect the center manifold hose to the dispenser valve and pierce the can by turning the shut-off valve in a clockwise direction. After piercing the can, back the shut-off valve out in a clockwise direction. The center hose is now charged with

**Figure 17-39.** Gauge readings with system under a vacuum

both refrigerant and air. Loosen the center hose connector at the manifold until a hiss can be heard. Allow the refrigerant to escape for a few seconds and then retighten the connector. The center hose is now completely free of air. (See Figure 17-40.)

COMPOUND                    PRESSURE

R – 12

**Figure 17-40.** System hook-up for charging

3.    Open the high pressure gauge manifold hand valve and allow as much refrigerant to enter the system as will flow in freely. This will vary from ½ to 1 lb (226.8 to 453.6 g) of refrigerant. After this partial charging, both gauges should indicate approximately the same reading. Close the high pressure manifold hand valve.

4.    Start the engine and adjust the speed to maintain approximately 1250 rpm. Turn the blower switch to the high speed position and set the thermostat to the coldest setting.

Open the low pressure manifold hand valve to allow refrigerant to enter the system. Add refrigerant until the sight glass in the top of the receiver-drier is clear and no bubbles are visible.

5.    Turn the low pressure manifold hand valve off and check the readings of both gauges. The proper readings should be approximately 10 to 1; that is, the high pressure gauge should indicate approximately 10 times the reading of the compound gauge. However, these readings will vary with the ambient temperature. The gauge readings must be taken with the engine running. (See Table 17-1.) If the gauge readings are approximately 10 to 1, and the sight glass is clear, the unit will be properly charged and producing cold air.

**Table 17-1**
**Gauge Readngs**
**on a Properly Charged System**

| PRESSURE—TEMPERATURE RELATIONSHIP | | |
|---|---|---|
| AMBIENT TEMPERATURE F° | LOW PRESSURE GAUGE READING | HIGH PRESSURE GAUGE READING |
| 80 | 16 | 150 - 170 |
| 85 | 18 | 165 - 185 |
| 90 | 20 | 175 - 195 |
| 95 | 22 | 185 - 205 |
| 100 | 24 | 210 - 230 |
| 105 | 26 | 230 - 250 |
| 110 | 28 | 250 - 270 |
| 115 | 30 | 265 - 285 |
| Engine operating 1500 to 1750 RPM & Correct Refrigerant Charge. (Equivalent to 30 M.P.H.) | | |

A typical refrigerant piping schematic of an underdash unit is shown in Figure 17-41. A typical 8-cylinder mount and drive pulley installation is shown in Figure 17-42.

When malfunctions occur in an auto air conditioning system, there are several standard service procedures which can be followed. The guide on pages 629-31 should be helpful in solving any problems.

A. CONDENSER
B. COMPRESSOR
C. DISCHARGE HOSE (COMPRESSOR TO CONDENSER)
D. LIQUID HOSE (CONDENSER TO RECEIVER-DRIER)
E. CONDENSER MOUNTING BRACKETS (4)
F. RECEIVER-DRIER
G. LIQUID HOSE (RECEIVER-DRIER TO EXPANSION VALVE)
H. SUCTION HOSE (EVAPORATOR TO COMPRESSOR)
I. EVAPORATOR HANGER BRACKETS (1 LEFT - 1 RIGHT)
J. EVAPORATOR
K. DRAIN HOSE (2)
L. CLAMP FOR RECEIVER-DRIER

**Figure 17-41.** Typical underdash air-conditioner layout

# 1970 thru 1977 DODGE PICK-UP & VAN V-8
## 318, & 360 CU. IN. DISPLACEMENT
### WITH OR WITHOUT POWER STEERING
## COMPRESSOR MOUNT & DRIVE PULLEY INSTALLATION

25 TY-RAP PLASTIC STRAP

26 5700023 DRIVE BELT

CLUTCH

IDLER

FAN

P/S

ALT.

D/P

4 9555160 DRIVE PULLEY

20 (4)

21 (6)

12 5/16-18 x 7/8 BOLT (6)

5 5013060 PILOT ADAPTER

Reuse O.E.M. Bolt

13 3/8-16 x 7/8 BOLT

13 3/8-16 x 7/8 BOLT

3 7834206 COMPRESSOR STRAP

3

22

22

22   23

13 3/8-16 x 7/8 BOLT

22

2 7834205 IDLER BRACE

19 3/8-16 NUT

18 1/2-20 x 1 BOLT

14 3/8-16 x 1 BOLT

22

24

1 7734201 COMPRESSOR MOUNT

9

6 9522050 IDLER PULLEY

10

11 6477008 SPACER

7 8240115 ECCENTRIC

8 5112410 SHOULDER BOLT

22 (4)

22

16 3/8-16 x 1-1/2 BOLT

17 3/8-16 x 5-1/2 BOLT (2)

15 3/8-16 x 1-1/4 BOLT (4)

Mapco USA   10-1-76

**Figure 17-42.**   Typical 8-cylinder mount and drive pulley installation

# AUTO AIR CONDITIONING
# DIAGNOSIS AND ANSWER GUIDE

When you find conditions of no cooling; or insufficient cooling; First: make sure that you connect all gauges properly. At this time find out what is causing the EXISTING TROUBLE. Check the POSSIBLE CAUSES and correct with the ANSWER.

## A. REFRIGERATION PROBLEMS

### 1. LOW HEAD PRESSURE (Existing Trouble)

| POSSIBLE CAUSE | ANSWER |
|---|---|
| a. Leak in system | Repair Leak |
| b. Defective expansion valve | Replace valve |
| c. Suction valve closed | Open valve |
| d. Freon shortage | Add freon |
| e. Plugged receiver drier | Replace receiver drier |
| f. Compressor suction-valve leaking | Replace valve |
| g. Bad reed valves in compressor | Replace reed valve |

### 2. HIGH HEAD PRESSURE (Existing Trouble)

| POSSIBLE CAUSE | ANSWER |
|---|---|
| a. Air in system | Recharge system |
| b. Clogged condenser | Clean condenser |
| c. Discharge valve closed | Open valve |
| d. Overcharged system | Remove some refrigerant |
| e. Insufficient condenser air | Install large fan |
| f. Loose fan belt | Tighten fan belt |
| g. Condenser not centered on fan or too close to radiator | Center and check distance from radiator |

### 3. LOW SUCTION PRESSURE (Existing Trouble)

| POSSIBLE CAUSE | ANSWER |
|---|---|
| a. Refrigerant shortage | Add refrigerant |
| b. Worn compressor piston | Replace compressor |
| c. Compressor head gasket leaking | Replace head gasket |
| d. Kinked or flattened hoses | Replace hose |
| e. Compressor suction valve leaking | Change valve plate |
| f. Moisture in system | Replace drier |
| g. Trash in expansion valve screen | Replace drier |

### 4. HIGH SUCTION PRESSURE (Existing Trouble)

| POSSIBLE CAUSE | ANSWER |
|---|---|
| a. Loose expansion bulb | Tighten bulb clamp |
| b. Overcharged system | Remove some refrigerant |
| c. Expansion valve stuck open | Replace valve |
| d. Compressor reed valves | Replace valves in Compressor |
| e. Leaking head gasket on-compressor | Replace head gasket |

### 5. COMPRESSOR NOT WORKING (Existing Trouble)

| POSSIBLE CAUSE | ANSWER |
|---|---|
| a. Broken drive belt | Replace belt |
| b. Broken clutch wire | Repair wire |
| c. Broken compressor piston | Replace compressor |
| d. Bad thermostat switch | Replace thermostat |
| e. Bad clutch coil | Replace clutch coil |

## 6. ENGINE OVERHEATING (Existing Trouble)

| POSSIBLE CAUSE | ANSWER |
|---|---|
| a. Fan belt slipping | Tighten belt |
| b. Engine out of time | Tune engine |
| c. Leaky radiator cap | Replace cap |
| d. Radiator water low | Fill radiator |
| e. Clogged condenser coil fins | Clean fins |
| f. Engine cooling system clogged | Flush radiator and engine block |
| g. Not enough air over radiator | Install large fan |

## 7. EVAPORATOR NOT COOLING (Existing Trouble)

| POSSIBLE CAUSE | ANSWER |
|---|---|
| a. Frozen coil, switch set too high | Defrost coil by turning thermostat down |
| b. Faulty clutch | Check clutch wire and thermostat |
| c. Drive belt slipping | Tighten belt |
| d. Hot air leaks into car | Close heater vents or air vents |
| e. Plugged receiver drier | Replace receiver drier |
| f. Capillary tube broken | Replace expansion valve |
| g. Shortage of freon | Add freon |
| h. High head pressure | See #2 problem |
| i. Low suction pressure | See #3 problem |
| j. High suction pressure | See #4 problem |
| k. Frozen expansion valve | Evacuate system and replace receiver drier |
| l. Defective expansion valve | Replace valve |

## 8. FROZEN EVAPORATOR COIL (Existing Trouble)

| POSSIBLE CAUSE | ANSWER |
|---|---|
| a. Faulty thermostat | Replace thermostat |
| b. Thermostat not set properly | Set to driving conditions |
| c. Insufficient evaporator air | Turn switch to higher setting |

# B. MECHANICAL PROBLEMS

## 9. BELT TROUBLE (Existing Trouble)

| POSSIBLE CAUSE | ANSWER |
|---|---|
| a. Pulley's not in line | Aline pulley's |
| b. Belt too tight or too loose | Adjust correctly |
| c. Wrong belt | Replace belt |
| d. Overcharged system or excessive head pressure | Discharge some freon |
| e. Bad bearing in idler pulley | Replace bearing |

## 10. EXCESSIVE VIBRATION OF COMPRESSOR AND MOUNT (Existing Trouble)

| POSSIBLE CAUSE | ANSWER |
|---|---|
| a. Head pressure too high | Check procedure #2 |
| b. Loose or broken bolts in mount | Replace or tighten bolt |
| c. No lock washers or bolts | Install lock washers |
| d. Crankshaft pulley not on straight and tight | Tighten all bolts |
| e. Clutch not tight on compressor | Tighten bolts |
| f. Too much freon | Discharge some freon |
| g. Worn or frozen bearings in idler pulley | Replace bearings |
| h. Loose or defective belt | Tighten or replace |

## 11. NOISY CLUTCH (Existing Trouble)

| POSSIBLE CAUSE | ANSWER |
|---|---|
| a. Be sure coil properly installed | Center clutch and coil |
| b. Be sure pulley is tightly bolted to compressor shaft | Tighten bolt |
| c. Check key on shaft of compressor | Line key with clutch |

# MECHANICAL PROBLEM

## 12. CLUTCH DOES NOT WORK (Existing Trouble)

| POSSIBLE CAUSE | ANSWER |
|---|---|
| a. Check fuse | Replace if bad |
| b. Check for broken or loose wires to clutch | Replace or repair wire |
| c. Check for short in clutch coil | Replace coil |
| d. Check voltage at clutch | Check connections and insulation on wires |
| e. Check thermostat | Replace if bad |
| f. Check blower switch at all positions | Replace if inner brass race is burned |
| g. Defective compressor (Frozen) | Replace compressor |

## 13. BLOWER DOES NOT WORK (Existing Trouble)

| POSSIBLE CAUSE | ANSWER |
|---|---|
| a. Check fuse | Replace fuse |
| b. Check for broken or loose connections | Repair connection |
| c. Check control switch | Replace if defective |
| d. Check fan motor and be sure it is not locked or dragging | Center motor and adjust blower clearance |
| e. Check voltage at motor | Check wires and replace motor if defective |
| f. Check ground wire | Correct problem |

## 14. BLOWER RUNS TOO SLOWLY (Existing Trouble)

| POSSIBLE CAUSE | ANSWER |
|---|---|
| a. Check for loose wires or shorts | Correct problem |
| b. Check for binding shaft | Center motor and blower |
| c. Check for burned out resistors in controls | Replace switch |
| d. Check voltage at motor | |
| e. Check Allen Set screw on blower wheel | Tighten set screw |

## 15. LEAKING COMPRESSOR SEAL OR GASKETS (Existing Trouble)

| POSSIBLE CAUSE | ANSWER |
|---|---|
| a. Check for low freon charge and leaks around compressor seal with leak detector | |
| b. If oil is on compressor body, clutch and underside of car hood above compressor | Replace seal, or gasket |
| | Check oil level |
| | Replace drier if moisture present |

## 16. LEAKING HOSES OR FITTINGS (Existing Trouble)

| POSSIBLE CAUSE | ANSWER |
|---|---|
| a. Check distance from exhaust manifold | Moves hoses |
| b. Check grommets in fire wall | Install grommets |
| c. Check wiper cables and arms | Move hose |
| d. Check flare seats | Replace hose or fittings |
| e. Check hose to fitting connection | Replace hose |

## SAFETY PROCEDURES

Because the engine must be running during the servicing operations on an automotive air conditioning unit, there are several important safety precautions which must be observed.

1. Be sure that the exhaust fumes are properly channeled to the outside of the building.

2. Do not touch the exhaust manifold or serious burns may result.

3. Do not leave loose tools lying on the radiator or other areas where they may fall into the fan.

4. Keep hands, feet, and loose clothing away from fan belts and fans.

5. Keep tools and hands away from the spark plugs because of the high voltage.

6. Do not attempt to remove the radiator cap from a warm radiator.

7. Keep the gauge service lines clear of pulleys, belts, and fans.

8. Be sure to block the wheels of the car when the engine is running.

9. Be sure to wear goggles when servicing the system with refrigerant.

10. Use a protective shield over the fan blade.

## SUMMARY

• The major differences between automotive air conditioning and any other air conditioning system are: (1) the manner used to drive the compressor, (2) the manner used to combine the different components, (3) the fast temperature change inside the automobile, and (4) the complex system of air control dampers used in an automobile.

• Automotive heating is accomplished in much the same manner as heating a home with a hot water boiler.

• The refrigeration system used on automotive air conditioning units is the same as that used on any air conditioning system.

• The automotive air conditioning system consists of the following components: (1) compressor, (2) magnetic clutch, (3) condenser, (4) receiver-drier, (5) expansion valve, (6) evaporator, (7) suction throttling valve, and (8) high pressure relief valve.

• The compressors used on automotive air conditioning units are open type.

• The magnetic clutch is used to engage and disengage the compressor on demand from the thermostat.

• The two types of magnetic clutches are: (1) stationary field and (2) rotating field.

• The condenser is usually mounted in front of the automobile radiator.

• Automotive air conditioning units are more susceptible to leaks than other types of installations because of the vibration encountered.

• The thermostatic expansion valve is the most popular type of flow control device used on automotive air conditioning systems.

• The thermostatic expansion valve has three functions: (1) it controls the flow of refrigerant into the evaporator, (2) it provides a quick refrigerant pressure pulldown on start-up, and (3) it overcomes high discharge pressures during idle and low speed operation.

- Some General Motors units use an expansion tube in place of an expansion valve which operates on the amount of subcooling provided by the condenser.

- POA and STV valves are basically spring-loaded valves which are controlled by an evacuated bellows and needle-valve assembly which are located within the valve body.

- EPR valves are suction throttling valves which are located at the compressor. They are designed to prevent icing of the evaporator by controlling the suction pressure.

- ETR valves are controlled by a solenoid which is built into the valve. It is either fully open or fully closed.

- The ETR valve is controlled by the ETR switch.

- The VIR assembly combines the thermostatic expansion valve, the POA valve, the receiver-drier, and the sight glass into a single unit.

- The VIR assembly was redesigned and called the EEVIR (evaporator equalized valves in receiver) and reduced the fuel consumption of the automobile.

- The equalizer port on the EEVIR is designed to speed up the opening of the expansion valve under certain conditions.

- The thermostat switch is placed inside the automobile and senses the evaporator temperature.

- The bellows thermostatic switch has its capillary tube inserted into the fins of the evaporator to maintain close temperature control of the evaporator.

- Ambient switches are designed to prevent compressor operation when the outside air temperature is too low.

- Superheat switches are fitted into the compressor cavity on the rear compressor head and sense the temperature of the suction gas and cause a fuse to the compressor clutch to be blown.

- The thermal fuse allows a time delay before the fuse link is blown when overheating conditions occur inside the compressor.

- The low refrigerant charge protection system consists of the superheat switch, the thermal fuse, the air conditioning control switch, and the ambient switch.

- The control assembly is mounted inside the passenger compartment and is designed to select the mode of operation.

- The temperature lever normally determines the temperature of the discharge air by positioning an air door which is located in the air ducts.

- Normally a blower switch is used to control the air flow by regulating the speed of the blower motor.

- Automatic temperature control systems control both the heating and cooling modes.

- The sensor string consists of thermistors connected in series with each other and in series with the temperature selection.

- The amplifier provides an output voltage from the voltage which is received from the sensors and the selector switch.

- The transducer is an electromechanical device that changes the electrical signal from the amplifier into a vacuum signal which controls operation of the power servo.

- The power servo is the control component of the system. It operates in response to the signal from the sensor to program for the desired operation.

• The universal automatic temperature control analyzer is basically a device used to substitute resistances when connected into the wiring harness of the automobile.

## REVIEW EXERCISES

1. List the major differences between automotive air conditioning and other types of air conditioning units.

2. What is the purpose of the doors and dampers on automotive air conditioning units?

3. What type of compressors are used on automotive air conditioning units?

4. Name the types of magnetic clutches which are used on automotive air conditioning units.

5. What causes the scoring of a magnetic clutch?

6. How is the condenser cooled on an automotive air conditioning unit?

7. How much additional refrigerant is needed in a system when a receiver-drier is used?

8. What is the purpose of the desiccant in a receiver-drier?

9. What is the most popular flow control device used on automotive air conditioning units?

10. Why does the air become colder when the volume of air is reduced?

11. What is meant by the term "flooded evaporator"?

12. What is a suction throttling valve?

13. What operates the ETR valve?

14. What components are incorporated into the VIR assembly?

15. How is an EEVIR valve identified?

16. What pressure does the POA valve maintain in the evaporator?

17. What must be done if the POA valve in an EEVIR assembly is not working properly?

18. What is the function of the EEVIR equalizer?

19. What two types of thermostatic switches are most commonly used in automotive air conditioning units?

20. What is the normal operating temperature of an ambient switch?

21. What is the purpose of a superheat switch?

22. What conditions will cause the fuse to blow in a thermal fuse?

23. What component will cause the low refrigerant charge protection system to function?

24. What are the three methods of controlling the various components on an automotive air conditioning system?

25. What is a sensor?

26. What is the function of the transducer?

27. What does the power servo control?

28. What components does the automatic temperature control analyzer replace?

29. What vacuum should be obtained when evacuating an air conditioning system?

30. What means can be used to determine whether or not a system has a sufficient refrigerant charge?

# Solar Energy

The average efficiency of use of coal and oil to the present time has reached no more than 20 percent of the demand. The demand for energy has only been about 20 percent of what is potentially recoverable by using these fuels.

Wood, peat, and other vegetable matter provide approximately 15 percent of present day fuel requirements. Hydroelectric and nuclear energy provide approximately 2 percent of the total demand.

In the United States the average energy demand per person is approximately 18,000 kWh/year/person (killowatt hours per person per year). The future energy demand is estimated at 20,000 kWh/year/person. When this is combined with the estimate of total population in a steady state to be 10 billion, we can arrive at an ultimate energy demand of approximately 200 trillion kWh/year. This is approximately 16 times the present level of usage. This increase arises from an estimated increase in population of three times the present population and an increase in average energy demand per person by more than five times the present demand.

Fuel shortages and the increase in their costs have rekindled man's interest in harnessing the sun's energy. In the past few years, millions of dollars have been spent in learning about the properties and potentials of the star nearest the Earth.

## BRIEF HISTORY OF SOLAR ENERGY

The use of the sun's energy to heat buildings and water is not new. It has been traced back several thousand years to when the ancient Greeks used it. A large majority of the principles of solar heat were worked out in the early days of this century. The solar technology developed then has undergone very few changes except in the materials used. Solar heating has not been developed into wide use because of several factors. As long as the conventional fuels were apparently abundant and cheap, there was no incentive for the development of solar heating on a large scale. Solar heating is now coming into the competition of the marketplace because the conventional fuels are in short supply.

Every year the amount of solar energy which falls in this country is almost one thousand times greater than the total yearly energy consumption of the country. When this is added to the facts that solar energy is nonpolluting, inexhaustible, and could significantly reduce the dependence of the United States on imported fuels, it is easy to understand the interest in this source of energy.

## BASIC SOLAR THEORY

The Earth is 93 million miles from the sun. The sun, sometimes called a solar fusion furnace, provides 442 Btus (111.4 kcal) per hour for every square foot (m²) of upper atmosphere. By the time this heat reaches the Earth, however, the average heat value has dropped to between 250 and 300 Btus (63 kcal to 75.6 kcal) per hour per square foot. This is equal to approximately one horsepower per square yard (m²). This is one reason why we are not likely ever to have solar-powered automobiles.

There is, however, a variation in the availability of solar energy in different geographic areas and seasons of the year. (See Table 18-1.) The values are expressed in Btu/ft²/hr as measured by a device called a pyroheliometer. (See Figure 18-1.)

There are a few areas in the United States which receive less than 50 percent of the normal sunshine during the months of January and February, the coldest part of the year. (See Table 18-2.) This will be more than sufficient for home heating and domestic hot water heating.

In practical solar engineering, there are

**Figure 18-1.** Pyroheliometer (*Courtesy of Rho Sigma, Inc.*)

two basic approaches to collecting thermal energy from the sun: concentrating collectors, which focus the energy to a point for collection (see Figure 18-2), and using flat plate collectors, which capture the solar output directly as it falls on the surface (see Figure 18-3). A basic example of a concentrating collector is the use of a magnifying glass to set fire to a piece of paper. Either a lens or a curved mirror can be used to concentrate the energy into a smaller area. The major advantage of the concentrating collector is its ability to generate very high temperatures, much higher than the flat plate collector. The disadvantage of the concentrating collector is that as the sun moves, so does the spot where the energy is focused. Therefore, the collector must move constantly to follow the sun. This method requires expensive equipment which has eliminated it as a practical means of home heating.

The flat plate collector has been accepted almost universally as the best device for gathering solar energy for low temperature applications, such as home heating. (See

## Table 18-1
## Daily Insolation on a South-Facing Roof*

### 24° N. LATITUDE
BTU/Hr.-Sq. Ft. on South Facing Surface
Angle with Horizontal

| Date | 0° | 14° | 24° | 34° | 54° | 90° |
|---|---|---|---|---|---|---|
| Sept. 21 | 2194 | 2342 | 2366 | 2322 | 2212 | 992 |
| Oct. 21 | 1928 | 2198 | 2314 | 2364 | 2346 | 1442 |
| Nov. 21 | 1610 | 1962 | 2146 | 2268 | 2324 | 1730 |
| Dec. 21 | 1474 | 1852 | 2058 | 2204 | 2286 | 1808 |
| Jan. 21 | 1622 | 1984 | 2174 | 2300 | 2360 | 1766 |
| Feb. 21 | 1998 | 2276 | 2396 | 2446 | 2424 | 1476 |
| Mar. 21 | 2270 | 2428 | 2456 | 2412 | 2298 | 1022 |
| Apr. 21 | 2454 | 2458 | 2374 | 2228 | 2016 | 488 |
| May 21 | 2556 | 2447 | 2286 | 2072 | 1800 | 246 |
| June 21 | 2574 | 2422 | 2230 | 1992 | 1700 | 204 |
| July 21 | 2526 | 2412 | 2250 | 2036 | 1766 | 246 |
| Aug. 21 | 2408 | 2402 | 2316 | 2168 | 1958 | 470 |

### 32° N. LATITUDE

| Date | 0° | 22° | 32° | 42° | 52° | 90° |
|---|---|---|---|---|---|---|
| Sept. 21 | 2014 | 2288 | 2308 | 2264 | 2154 | 1226 |
| Oct. 21 | 1654 | 2100 | 2208 | 2252 | 2232 | 1588 |
| Nov. 21 | 1280 | 1816 | 1980 | 2084 | 2130 | 1742 |
| Dec. 21 | 1136 | 1704 | 1888 | 2016 | 2086 | 1794 |
| Jan. 21 | 1288 | 1839 | 2008 | 2118 | 2166 | 1779 |
| Feb. 21 | 1724 | 2188 | 2300 | 2345 | 2322 | 1644 |
| Mar. 21 | 2084 | 2378 | 2403 | 2358 | 2246 | 1276 |
| Apr. 21 | 2390 | 2444 | 2356 | 2206 | 1994 | 764 |
| May 21 | 2582 | 2454 | 2284 | 2064 | 1788 | 469 |
| June 21 | 2634 | 2436 | 2234 | 1990 | 1690 | 370 |
| July 21 | 2558 | 2422 | 2250 | 2030 | 1754 | 458 |
| Aug. 21 | 2352 | 2388 | 2296 | 2144 | 1934 | 736 |

### 40° N. LATITUDE

| Date | 0° | 30° | 40° | 50° | 60° | 90° |
|---|---|---|---|---|---|---|
| Sept. 21 | 1788 | 2210 | 2228 | 2182 | 2074 | 1416 |
| Oct. 21 | 1348 | 1962 | 2060 | 2098 | 2074 | 1654 |
| Nov. 21 | 942 | 1636 | 1778 | 1870 | 1908 | 1686 |
| Dec. 21 | 782 | 1480 | 1634 | 1740 | 1796 | 1646 |
| Jan. 21 | 948 | 1660 | 1810 | 1906 | 1944 | 1726 |
| Feb. 21 | 1414 | 2060 | 2162 | 2202 | 2176 | 1730 |
| Mar. 21 | 1852 | 2308 | 2330 | 2284 | 2174 | 1484 |
| Apr. 21 | 2274 | 2412 | 2320 | 2168 | 1956 | 1022 |
| May 21 | 2552 | 2442 | 2264 | 2040 | 1760 | 724 |
| June 21 | 2648 | 2434 | 2224 | 1974 | 1670 | 610 |
| July 21 | 2534 | 2409 | 2230 | 2006 | 1728 | 702 |
| Aug. 21 | 2244 | 2354 | 2258 | 2104 | 1894 | 978 |

### 48° N. LATITUDE

| Date | 0° | 38° | 48° | 58° | 68° | 90° |
|---|---|---|---|---|---|---|
| Sept. 21 | 1522 | 2102 | 2118 | 2070 | 1966 | 1546 |
| Oct. 21 | 1022 | 1774 | 1860 | 1890 | 1866 | 1626 |
| Nov. 21 | 596 | 1336 | 1448 | 1518 | 1544 | 1442 |
| Dec. 21 | 446 | 1136 | 1250 | 1326 | 1364 | 1304 |
| Jan. 21 | 596 | 1360 | 1478 | 1550 | 1578 | 1478 |
| Feb. 21 | 1080 | 1880 | 1972 | 2024 | 1978 | 1720 |
| Mar. 21 | 1578 | 2208 | 2228 | 2182 | 2074 | 1632 |
| Apr. 21 | 2106 | 2358 | 2266 | 2114 | 1902 | 1262 |
| May 21 | 2482 | 2418 | 2234 | 2010 | 1728 | 982 |
| June 21 | 2626 | 2420 | 2204 | 1950 | 1644 | 874 |
| July 21 | 2474 | 2386 | 2200 | 1974 | 1694 | 956 |
| Aug. 21 | 2086 | 2300 | 2200 | 2046 | 1836 | 1208 |

Courtesy of Sol-R-Tech

# Table 18-2

## Mean Percentage of Possible Sunshine for Selected Locations

| State and Station | Years | Jan. | Feb. | Mar. | Apr. | May | June | July | Aug. | Sept. | Oct. | Nov. | Dec. | Annual |
|---|---|---|---|---|---|---|---|---|---|---|---|---|---|---|
| Ala. Birmingham | 56 | 43 | 49 | 56 | 63 | 66 | 67 | 62 | 65 | 66 | 67 | 58 | 44 | 59 |
| Montgomery | 49 | 51 | 53 | 61 | 69 | 73 | 72 | 66 | 69 | 69 | 71 | 64 | 48 | 64 |
| Alaska Anchorage | 19 | 39 | 46 | 56 | 58 | 50 | 51 | 45 | 39 | 35 | 32 | 33 | 29 | 45 |
| Fairbanks | 20 | 34 | 50 | 61 | 68 | 55 | 53 | 45 | 35 | 31 | 28 | 38 | 29 | 44 |
| Juneau | 14 | 30 | 32 | 39 | 37 | 34 | 35 | 28 | 30 | 25 | 18 | 21 | 18 | 30 |
| Nome | 29 | 44 | 46 | 48 | 53 | 51 | 48 | 32 | 26 | 34 | 35 | 36 | 30 | 41 |
| Ariz. Phoenix | 64 | 76 | 79 | 83 | 88 | 93 | 94 | 84 | 84 | 89 | 88 | 84 | 77 | 85 |
| Yuma | 52 | 83 | 87 | 91 | 94 | 97 | 98 | 92 | 91 | 93 | 93 | 90 | 83 | 91 |
| Ark. Little Rock | 66 | 44 | 53 | 57 | 62 | 67 | 72 | 71 | 73 | 71 | 74 | 58 | 47 | 62 |
| Calif. Eureka | 49 | 40 | 44 | 50 | 53 | 54 | 56 | 51 | 46 | 52 | 48 | 42 | 39 | 49 |
| Fresno | 55 | 46 | 63 | 72 | 83 | 89 | 94 | 97 | 97 | 93 | 87 | 73 | 47 | 78 |
| Los Angeles | 63 | 70 | 69 | 70 | 67 | 68 | 69 | 80 | 81 | 80 | 76 | 79 | 72 | 73 |
| Red Bluff | 39 | 50 | 60 | 65 | 75 | 79 | 86 | 95 | 94 | 89 | 77 | 64 | 50 | 75 |
| Sacramento | 48 | 44 | 57 | 67 | 76 | 82 | 90 | 96 | 95 | 92 | 82 | 65 | 44 | 77 |
| San Diego | 68 | 68 | 67 | 66 | 66 | 60 | 60 | 67 | 70 | 70 | 70 | 76 | 71 | 68 |
| San Francisco | 64 | 53 | 57 | 63 | 69 | 70 | 75 | 68 | 63 | 70 | 70 | 62 | 54 | 68 |
| Colo. Denver | 64 | 67 | 67 | 65 | 63 | 61 | 69 | 68 | 68 | 71 | 71 | 67 | 65 | 67 |
| Grand Junction | 57 | 58 | 62 | 64 | 67 | 71 | 79 | 76 | 72 | 77 | 74 | 67 | 58 | 69 |
| Conn. Hartford | 48 | 46 | 55 | 56 | 54 | 57 | 60 | 62 | 60 | 57 | 55 | 46 | 46 | 56 |
| D.C. Washington | 66 | 46 | 53 | 56 | 57 | 61 | 64 | 64 | 62 | 62 | 61 | 54 | 47 | 58 |
| Fla. Apalachicola | 26 | 59 | 62 | 62 | 71 | 77 | 70 | 64 | 63 | 62 | 74 | 66 | 53 | 65 |
| Jacksonville | 60 | 58 | 59 | 66 | 71 | 71 | 63 | 62 | 63 | 58 | 58 | 61 | 53 | 62 |
| Key West | 45 | 68 | 75 | 78 | 78 | 76 | 70 | 69 | 71 | 65 | 65 | 69 | 66 | 71 |
| Miami Beach | 48 | 66 | 72 | 73 | 73 | 68 | 62 | 65 | 67 | 62 | 62 | 65 | 65 | 67 |
| Tampa | 63 | 63 | 67 | 71 | 74 | 75 | 66 | 61 | 64 | 64 | 67 | 67 | 61 | 68 |
| Ga. Atlanta | 65 | 48 | 53 | 57 | 65 | 68 | 68 | 62 | 63 | 65 | 67 | 60 | 47 | 60 |
| Hawaii. Hilo | 9 | 48 | 42 | 41 | 34 | 31 | 41 | 44 | 38 | 42 | 41 | 34 | 36 | 39 |
| Honolulu | 53 | 62 | 64 | 60 | 62 | 64 | 66 | 67 | 70 | 70 | 68 | 63 | 60 | 65 |
| Lihue | 9 | 48 | 48 | 48 | 46 | 51 | 60 | 58 | 59 | 67 | 58 | 51 | 45 | 54 |
| Idaho Boise | 20 | 40 | 48 | 59 | 67 | 68 | 75 | 89 | 86 | 81 | 66 | 46 | 37 | 66 |
| Pocatello | 21 | 37 | 47 | 58 | 64 | 66 | 72 | 82 | 81 | 78 | 66 | 48 | 36 | 64 |
| Ill. Cairo | 30 | 46 | 53 | 59 | 65 | 71 | 77 | 82 | 79 | 75 | 73 | 56 | 46 | 65 |
| Chicago | 66 | 44 | 49 | 53 | 56 | 63 | 69 | 73 | 70 | 65 | 61 | 47 | 41 | 59 |
| Springfield | 59 | 47 | 51 | 54 | 58 | 64 | 69 | 76 | 72 | 73 | 64 | 53 | 45 | 60 |
| Ind. Evansville | 48 | 42 | 49 | 55 | 61 | 67 | 73 | 78 | 76 | 73 | 67 | 52 | 42 | 64 |
| Ft. Wayne | 48 | 38 | 44 | 51 | 55 | 62 | 69 | 74 | 69 | 64 | 58 | 41 | 38 | 57 |
| Indianapolis | 63 | 41 | 47 | 49 | 55 | 62 | 68 | 74 | 70 | 68 | 64 | 48 | 39 | 59 |
| Iowa Des Moines | 66 | 56 | 56 | 56 | 59 | 62 | 66 | 75 | 70 | 64 | 64 | 53 | 48 | 62 |
| Dubuque | 54 | 48 | 52 | 52 | 58 | 60 | 63 | 73 | 67 | 61 | 55 | 44 | 40 | 57 |
| Sioux City | 52 | 55 | 58 | 58 | 59 | 63 | 67 | 75 | 72 | 67 | 65 | 53 | 50 | 63 |
| Kans. Concordia | 52 | 60 | 60 | 62 | 63 | 65 | 73 | 79 | 76 | 72 | 70 | 64 | 58 | 67 |
| Dodge City | 70 | 67 | 66 | 68 | 68 | 68 | 74 | 78 | 78 | 76 | 76 | 67 | 61 | 71 |
| Wichita | 46 | 61 | 63 | 64 | 64 | 66 | 73 | 80 | 77 | 73 | 69 | 67 | 59 | 69 |
| Ky. Louisville | 59 | 41 | 47 | 52 | 57 | 64 | 68 | 72 | 69 | 68 | 64 | 51 | 39 | 59 |
| La. New Orleans | 89 | 49 | 50 | 57 | 63 | 66 | 64 | 58 | 60 | 64 | 70 | 60 | 46 | 59 |
| Shreveport | 18 | 48 | 54 | 58 | 60 | 69 | 78 | 79 | 80 | 79 | 77 | 65 | 60 | 69 |
| Maine Eastport | 58 | 45 | 51 | 52 | 52 | 51 | 53 | 55 | 57 | 54 | 50 | 37 | 40 | 50 |
| Mass. Boston | 67 | 47 | 56 | 57 | 56 | 59 | 62 | 64 | 63 | 61 | 58 | 48 | 44 | 61 |
| Mich. Alpena | 45 | 29 | 43 | 52 | 56 | 59 | 64 | 70 | 64 | 52 | 44 | 24 | 22 | 51 |
| Detroit | 69 | 34 | 42 | 48 | 52 | 56 | 64 | 69 | 66 | 61 | 54 | 35 | 29 | 52 |
| Grand Rapids | 56 | 26 | 37 | 48 | 54 | 60 | 66 | 72 | 67 | 58 | 50 | 31 | 22 | 49 |
| Marquette | 55 | 31 | 40 | 47 | 52 | 53 | 56 | 63 | 57 | 47 | 38 | 24 | 24 | 47 |
| S. Ste. Marie | 60 | 28 | 44 | 50 | 54 | 54 | 59 | 63 | 59 | 46 | 36 | 21 | 22 | 47 |
| Minn. Duluth | 49 | 47 | 55 | 60 | 58 | 58 | 60 | 68 | 63 | 53 | 47 | 36 | 40 | 55 |
| Minneapolis | 45 | 49 | 54 | 55 | 57 | 60 | 64 | 70 | 66 | 56 | 51 | 38 | 40 | 56 |
| Miss. Vicksburg | 66 | 46 | 50 | 57 | 64 | 69 | 73 | 69 | 72 | 74 | 71 | 60 | 45 | 64 |
| Mo. Kansas City | 69 | 55 | 57 | 59 | 60 | 64 | 70 | 76 | 73 | 70 | 67 | 59 | 52 | 65 |
| St. Louis | 68 | 48 | 49 | 56 | 59 | 64 | 68 | 72 | 68 | 72 | 64 | 56 | 48 | 61 |
| Springfield | 45 | 48 | 54 | 57 | 60 | 63 | 69 | 77 | 72 | 71 | 65 | 58 | 48 | 63 |
| Mont. Havre | 55 | 49 | 58 | 61 | 63 | 63 | 65 | 78 | 75 | 64 | 57 | 48 | 46 | 62 |
| Helena | 65 | 46 | 55 | 58 | 60 | 59 | 63 | 77 | 74 | 63 | 57 | 48 | 43 | 60 |
| Kalispell | 50 | 28 | 40 | 49 | 57 | 58 | 60 | 77 | 73 | 61 | 50 | 28 | 20 | 53 |
| Nebr. Lincoln | 55 | 57 | 59 | 60 | 60 | 63 | 69 | 76 | 71 | 67 | 66 | 59 | 55 | 64 |
| North Platte | 53 | 63 | 63 | 64 | 62 | 64 | 72 | 78 | 74 | 72 | 70 | 62 | 58 | 68 |
| Nev. Ely | 21 | 61 | 64 | 68 | 65 | 67 | 79 | 79 | 81 | 81 | 73 | 67 | 62 | 72 |
| Las Vegas | 19 | 74 | 77 | 78 | 81 | 85 | 91 | 84 | 86 | 92 | 84 | 83 | 75 | 82 |
| Reno | 51 | 59 | 64 | 69 | 75 | 77 | 82 | 90 | 89 | 86 | 76 | 68 | 56 | 76 |
| Winnemucca | 53 | 52 | 60 | 64 | 70 | 76 | 83 | 90 | 90 | 86 | 75 | 62 | 53 | 76 |
| N.H. Concord | 44 | 48 | 53 | 55 | 53 | 51 | 56 | 57 | 58 | 55 | 50 | 43 | 43 | 52 |
| N.J. Atlantic City | 62 | 51 | 57 | 58 | 59 | 62 | 65 | 67 | 66 | 65 | 54 | 58 | 52 | 60 |

| State and Station | Years | Jan. | Feb. | Mar. | Apr. | May | June | July | Aug. | Sept. | Oct. | Nov. | Dec. | Annual |
|---|---|---|---|---|---|---|---|---|---|---|---|---|---|---|
| N. Mex. Albuquerque | 28 | 70 | 72 | 72 | 76 | 79 | 84 | 76 | 75 | 81 | 80 | 79 | 70 | 76 |
| Roswell | 47 | 69 | 72 | 75 | 77 | 76 | 80 | 76 | 75 | 74 | 74 | 74 | 69 | 74 |
| N.Y. Albany | 63 | 43 | 51 | 53 | 53 | 57 | 62 | 63 | 61 | 58 | 54 | 39 | 38 | 53 |
| Binghamton | 63 | 31 | 39 | 41 | 44 | 50 | 56 | 54 | 51 | 57 | 43 | 29 | 26 | 44 |
| Buffalo | 49 | 32 | 41 | 49 | 51 | 59 | 67 | 70 | 67 | 60 | 51 | 31 | 28 | 53 |
| Canton | 43 | 37 | 47 | 50 | 48 | 54 | 61 | 63 | 61 | 54 | 45 | 30 | 31 | 49 |
| New York | 83 | 49 | 56 | 57 | 69 | 62 | 65 | 66 | 64 | 64 | 61 | 53 | 50 | 59 |
| Syracuse | 49 | 31 | 38 | 45 | 50 | 58 | 64 | 67 | 63 | 56 | 47 | 29 | 26 | 50 |
| N.C. Asheville | 57 | 48 | 53 | 56 | 61 | 64 | 63 | 59 | 59 | 62 | 64 | 59 | 48 | 58 |
| Raleigh | 61 | 50 | 56 | 57 | 64 | 67 | 65 | 62 | 62 | 63 | 64 | 62 | 52 | 61 |
| N. Dak. Bismarck | 65 | 52 | 58 | 66 | 57 | 68 | 61 | 73 | 69 | 62 | 59 | 49 | 48 | 60 |
| Devils Lake | 55 | 53 | 60 | 59 | 60 | 59 | 62 | 71 | 67 | 59 | 58 | 44 | 45 | 58 |
| Fargo | 39 | 47 | 55 | 56 | 58 | 62 | 63 | 73 | 69 | 60 | 57 | 39 | 46 | 59 |
| Williston | 43 | 51 | 59 | 60 | 63 | 66 | 66 | 78 | 75 | 65 | 60 | 48 | 48 | 63 |
| Ohio Cincinnati | 44 | 41 | 46 | 52 | 56 | 62 | 69 | 72 | 68 | 68 | 60 | 46 | 39 | 57 |
| Cleveland | 65 | 29 | 36 | 45 | 52 | 61 | 67 | 71 | 68 | 62 | 54 | 32 | 25 | 50 |
| Columbus | 65 | 36 | 44 | 49 | 54 | 63 | 68 | 71 | 68 | 66 | 60 | 44 | 36 | 55 |
| Okla. Oklahoma City | 62 | 57 | 60 | 63 | 64 | 65 | 74 | 78 | 78 | 74 | 68 | 64 | 57 | 67 |
| Oreg. Baker | 46 | 41 | 49 | 56 | 61 | 63 | 67 | 83 | 81 | 74 | 62 | 46 | 37 | 60 |
| Portland | 69 | 27 | 34 | 41 | 49 | 52 | 55 | 70 | 65 | 55 | 42 | 28 | 23 | 48 |
| Roseburg | 29 | 24 | 32 | 40 | 51 | 57 | 59 | 79 | 77 | 68 | 42 | 28 | 18 | 51 |
| Pa. Harrisburg | 60 | 43 | 52 | 55 | 57 | 61 | 65 | 68 | 63 | 82 | 58 | 47 | 43 | 57 |
| Philadelphia | 66 | 45 | 56 | 57 | 58 | 61 | 62 | 64 | 61 | 62 | 61 | 53 | 49 | 57 |
| Pittsburgh | 63 | 32 | 39 | 45 | 50 | 57 | 62 | 64 | 61 | 60 | 54 | 39 | 30 | 51 |
| R.I. Block Island | 48 | 45 | 54 | 47 | 56 | 58 | 60 | 62 | 62 | 60 | 59 | 50 | 44 | 56 |
| S. C. Charleston | 61 | 58 | 60 | 65 | 72 | 73 | 70 | 66 | 66 | 67 | 68 | 68 | 57 | 66 |
| Columbia | 55 | 53 | 57 | 62 | 68 | 69 | 68 | 63 | 65 | 64 | 68 | 64 | 51 | 63 |
| S. Dak. Huron | 62 | 66 | 62 | 60 | 62 | 65 | 68 | 76 | 72 | 66 | 61 | 52 | 49 | 63 |
| Rapid City | 53 | 58 | 62 | 63 | 62 | 61 | 66 | 73 | 73 | 69 | 66 | 58 | 54 | 63 |
| Tenn. Knoxville | 62 | 42 | 49 | 53 | 59 | 64 | 64 | 64 | 59 | 64 | 64 | 53 | 41 | 57 |
| Memphis | 55 | 44 | 51 | 57 | 64 | 68 | 74 | 73 | 74 | 70 | 69 | 58 | 45 | 64 |
| Nashville | 63 | 42 | 47 | 54 | 60 | 65 | 69 | 69 | 68 | 69 | 65 | 53 | 41 | 60 |
| Tex. Abilene | 14 | 64 | 68 | 73 | 66 | 73 | 86 | 83 | 85 | 73 | 71 | 72 | 66 | 73 |
| Amarillo | 54 | 71 | 71 | 75 | 75 | 75 | 82 | 81 | 81 | 79 | 76 | 70 | 70 | 76 |
| Austin | 33 | 46 | 50 | 57 | 60 | 62 | 72 | 76 | 79 | 70 | 70 | 57 | 49 | 63 |
| Brownsville | 37 | 44 | 49 | 51 | 57 | 65 | 73 | 78 | 78 | 67 | 70 | 54 | 44 | 61 |
| Del Rio | 36 | 53 | 55 | 61 | 63 | 66 | 75 | 80 | 69 | 66 | 68 | 65 | 54 | 65 |
| El Paso | 53 | 74 | 77 | 81 | 85 | 87 | 87 | 78 | 78 | 80 | 82 | 80 | 73 | 80 |
| Ft. Worth | 33 | 56 | 57 | 66 | 67 | 75 | 78 | 74 | 74 | 70 | 63 | 58 | | 68 |
| Galveston | 66 | 50 | 50 | 55 | 61 | 69 | 76 | 72 | 71 | 70 | 74 | 62 | 49 | 63 |
| San Antonio | 57 | 48 | 51 | 56 | 58 | 60 | 69 | 74 | 75 | 69 | 67 | 55 | 49 | 62 |
| Utah Salt Lake City | 22 | 48 | 53 | 54 | 62 | 68 | 80 | 80 | 65 | 69 | 46 | 44 | | 66 |
| Vt. Burlington | 54 | 34 | 43 | 48 | 47 | 53 | 59 | 62 | 59 | 51 | 43 | 25 | 24 | 46 |
| Va. Norfolk | 60 | 50 | 57 | 60 | 63 | 67 | 66 | 66 | 66 | 63 | 64 | 60 | 51 | 61 |
| Richmond | 56 | 49 | 55 | 59 | 63 | 67 | 66 | 65 | 62 | 63 | 64 | 58 | 50 | 61 |
| Wash. North Head | 44 | 28 | 37 | 42 | 48 | 48 | 48 | 50 | 46 | 48 | 41 | 31 | 27 | 41 |
| Seattle | 26 | 27 | 34 | 42 | 48 | 48 | 45 | 48 | 52 | 56 | 53 | 36 | 28 | 45 |
| Spokane | 62 | 26 | 41 | 53 | 63 | 64 | 68 | 82 | 79 | 68 | 53 | 28 | 22 | 58 |
| Tatoosh Island | 49 | 26 | 36 | 39 | 45 | 47 | 46 | 48 | 47 | 58 | 38 | 26 | 23 | 40 |
| Walla Walla | 44 | 24 | 35 | 51 | 63 | 67 | 72 | 86 | 84 | 72 | 59 | 33 | 20 | 60 |
| Yakima | 18 | 34 | 49 | 62 | 70 | 72 | 74 | 86 | 88 | 74 | 61 | 38 | 29 | 63 |
| W. Va. Elkins | 55 | 33 | 37 | 42 | 47 | 55 | 55 | 56 | 53 | 55 | 51 | 41 | 33 | 48 |
| Parkersburg | 62 | 30 | 36 | 42 | 49 | 60 | 60 | 60 | 53 | 37 | 29 | | | 48 |
| Wis. Green Bay | 57 | 44 | 51 | 55 | 56 | 58 | 64 | 70 | 66 | 56 | 47 | 38 | 36 | 56 |
| Madison | 59 | 44 | 49 | 52 | 56 | 60 | 63 | 73 | 67 | 62 | 56 | 41 | 38 | 56 |
| Milwaukee | 59 | 44 | 48 | 53 | 56 | 60 | 65 | 73 | 67 | 62 | 56 | 44 | 39 | 57 |
| Wyo. Cheyenne | 63 | 65 | 66 | 64 | 61 | 59 | 68 | 70 | 68 | 69 | 69 | 65 | 63 | 66 |
| Lander | 57 | 66 | 70 | 71 | 66 | 67 | 65 | 74 | 76 | 75 | 72 | 67 | 61 | 69 |
| Sheridan | 52 | 56 | 61 | 62 | 61 | 61 | 67 | 76 | 74 | 67 | 60 | 53 | 52 | 64 |
| Yellowstone Park | 35 | 39 | 51 | 55 | 57 | 62 | 65 | 67 | 61 | 63 | 45 | 38 | | 56 |
| P.R. San Juan | 57 | 64 | 69 | 71 | 66 | 59 | 62 | 65 | 67 | 61 | 63 | 65 | 65 | 65 |

* Courtesy of Sol-R-Tech
Based on period of record through December 1959, except in a few instances. These charts and tabulation derived from "Normals, Means, and Extremes" table in U.S. Weather Bureau publication *Local Climatological Data*.

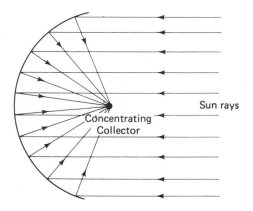

**Figure 18-2.** Concentrating collector (*Courtesy of Sol-R-Tech*)

Figure 18-4.) The basic principle of operation is very simple. The flat plate collector consists of a flat metal plate which is coated with a surface that absorbs sunlight and is heated by the rays from the sun. There are two covers over the flat plate to keep heat from being transferred to the ambient air. A working fluid is used to carry the heat from the plate to the place where it is to be used for heating the building or hot water tank.

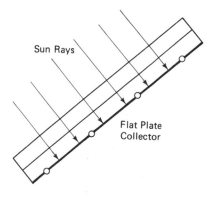

**Figure 18-3.** Flat plate collector (*Courtesy of Sol-R-Tech*)

There are two types of flat plate collectors. One type uses air and the other liquid as the working fluid. Although systems using air can be made more simple than systems which use a liquid, air does not have the ability to absorb and carry the heat that water and many other fluids do. Although both types of systems are in use, the liquid systems outnumber the air type by a large margin.

The material used for the absorber plate must conduct heat readily. Copper and aluminum are usually the first choices. Even though copper has definite advantages as far as corrosion resistance, the substantially lower cost of aluminum will usually make it the choice of the two.

The coating used for the absorber plate is of great importance. It should, ideally, absorb all the energy which falls on it, while re-emitting none of it. It is an unfortunate fact of nature that most objects emit radiation most easily at the same frequencies that they absorb them best. Because of this, the best approach to the ideal surface is one which absorbs the visible sunlight, but does not emit the very low frequency radiation which is associated with heat. These selective surfaces are available, but their present cost is very high, and their durability is doubtful.

In actual outdoor tests, it was found that the best coating is one of the simplest, that is, a layer of flat black paint. Almost all collectors presently on the market use this type of surface.

The latest major component of the flat plate collector is the cover plate. Glass is an excellent material optically because it is transparent to visible light but opaque to infrared. This is the reason a greenhouse retains heat. However, it is heavy in weight and subject to breakage. There are a number of plastic materials available that offer

**Figure 18-4.** Exploded view of flat plate collector (*Courtesy of Sol-R-Tech*)

the optical properties of glass without the disadvantages.

The number of cover plates is important to the performance of the collector. If too few are used, sufficient insulation from the cold amibent air will not be provided. Too many will reduce the amount of sunlight reaching the absorber plate. Two cover plates have been found to be the optimum number for most areas.

It is not enough to collect the required amount of energy while the sun is shining. Some of the energy must be stored for use at night and during cloudy periods. The material which is used for the heat storage me-

dium is crucial to the performance of the system. Presently, the main choices are water or rock in the form of crushed stone or concrete.

The specific heat and heat capacity of common materials which are used for thermal storage in solar heating systems vary. (See Table 18-3.) Specific heat is a measure of the amount of heat a pound of a material can hold for every degree its temperature is raised. Water has a specific heat of one because the Btu (cal) is defined as the amount of heat required to raise the temperature of one pound of water one degree.

As seen in Table 18-3, water will store

Table 18-3

Specific Heat and Heat Capacity of Common Storage
Materials Used in Solar Heating Systems

| MATERIAL | SPECIFIC HEAT BTU/LB DEG. F | DENSITY LB/FT$^3$ | HEAT CAPACITY BTU/FT$^3$-DEG. |
|---|---|---|---|
| Water | 1.00 | 62.5 | 62.5 |
| Stone | 0.21 | 180 | 36 |
| Iron ore (magnetite) | 0.165 | 320 | 53 |
| Concrete | 0.27 | 140 | 38 |

five times as much heat as stone and four times as much heat as concrete. There are no readily available, inexpensive materials which come close to the heat storage ability of water. Eutectic salts are being developed that store heat by melting rather than by getting hotter. However, these heat fusion materials are still in the experimental stage and have not been proven.

## DOMESTIC HOT WATER

This is probably the most simple and most easily installed system of all solar heating systems. It is generally recommended for existing homes and to be used to decrease or eliminate utility bills. The required equipment consists of a solar collector array on the roof, a heat exchanger, a water storage tank, circulator pumps, controls, and an auxiliary water heater, which could be the one in use.

When sunlight strikes a collector panel, several things take place.

1. Solar radiation passes through the outer, fiberglass-reinforced polyester skin. This layer acts as a weather conduction, and infrared barrier.

2. The solar radiation then passes through a ½ inch (12.7 mm) desiccated air space which eliminates condensation and acts as a conductive barrier.

3. Next it passes through an interior layer of transparent temperature resistant polyvinyl fluoride "Tedlar" which acts as a thermal and infrared barrier.

4. It then passes through a second ½ inch (12.7 mm) air space which acts as a convective barrier.

5. Finally, the solar radiation is absorbed by the blackened tube-in-sheet collector plate of 1100 aluminum alloy.

6. The working fluid within the collector, inhibited ethylene glycol, captures the heat which is generated by the absorbed radiation and transfers it to the material being heated (either domestic hot water or space heating) by means of the appropriate plumbing.

7. Conductive losses are kept to a minimum by incorporating 1 inch (25.4 mm) of semirigid fiberglass insulation on back of the panel.

8. All layers are then compressed and sealed by a rigid aluminum $T$ edge to ensure panel stability and aesthetic appearance.

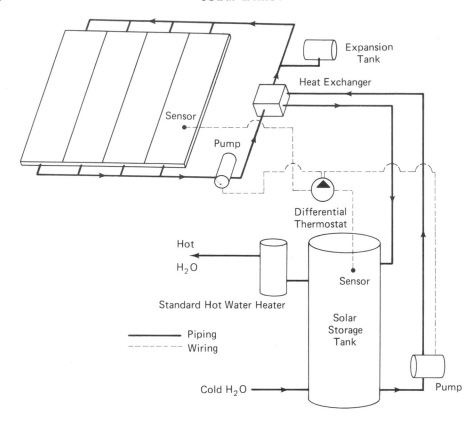

**Figure 18-5.** Residential Solar Hot Water System *(Courtesy of Sol-R-Tech)*

## CHANGING SOLAR ENERGY INTO HOT WATER

The heat transfer fluid (ethylene glycol anti-freeze) is pumped through a heat exchanger and gives up its energy to the water being pumped from the solar hot water supply tank. (See Figure 18-5.) The electrical consumption of these two pumps is minimal. Even in the highest electrical rate areas, the pumps will not cost more than about $1.50 per month to run. When there is a demand for hot water, the replacement water is already solar heated, so the thermostat will not signal for more gas, oil, or electricity to raise its temperature. When cloudy days or extreme cold interfere with solar hot water production, the standard hot water heater will take over the load. With a properly designed system, 75 to 95 percent of the domestic hot water will be free from utility rates. The pumps are controlled by a solar differential thermostat which turns them on when the temperature of the collector exceeds the solar hot water supply tank temperature. The thermostat also turns off the water pumps when the temperature of the solar collector falls below the solar hot water supply tank temperature. Everything operates automatically. When there has not been enough sunlight to keep the water heated, the auxiliary heater will take over and provide hot water automatically.

## SOLAR COLLECTOR EFFICIENCY

The amount of heat captured by a solar collector is dependent upon five variables, the first three of which are: materials used in fabrication, ambient temperature, and solar intensity. (See Figure 18-6.) All are inherently related. As the outside temperature falls, the collector efficiency will decrease. Clouds, haze, smog, etc. also decrease the efficiency as well as dirty plates or shaded panels.

The fourth factor is concerned with the orientation of the collector. South-facing roofs are best for the collection of solar energy. Southeast and southwest are also acceptable. Even east-west alignment will work for solar hot water if the appropriate zoning valves are used. For example, a home at 40° North latitude with a 40° roof pitch and with collectors facing east and west will collect (if plumbed with a differential valve so that only one side operates at a time) 80 percent of the energy of south-facing panels

in the month of January and 130 percent of that collected during the month of July.

The fifth factor which affects solar collector performance is the roof pitch on the building. For the greatest efficiency, the sun should be perpendicular to the collector. For North America, a roof pitch of 15° plus the local latitude is ideal. For example, the city of Washington, D.C. is located at 39° North latitude. A solar heated house should have a roof pitch of 54° although anything between 39° and 54° would be adequate for heating purposes. However, a flatter roof would have a tendency to produce more hot water and reduce hot water bills.

## COLLECTOR SURFACE NEEDED

Because most people are interested in how they can save money with solar water heating, all the variables have been taken into account, such as: ambient temperautre, efficiency, roof orientation, solar radiation, and cloud cover for 80 American cities. (See Table 18-4.)

**Example:** A family with four members in Columbia, Ohio would require 4 times 0.7 or 2.8 collectors in the summertime and 4 times 3.1 or 12.4 collectors in the wintertime to furnish 90 to 100 percent of their hot water load.

A similar family in Newark, New Jersey, would use the panel requirements for Seabrook because Newark is not listed.

When determining how many collectors are needed, always round off the panel requirement. For example, if 3.2 panels are required, then order 3. If 4.5 panels are needed, order 5. Also, remember, the number of panels which deliver 90 to 100 percent of the midwinter demand will also supply 90 to 100 percent of the yearly hot water supply.

**Figure 18-6.** Solar collector efficiency (*Courtesy of Sol-R-Tech*)

## Table 18-4

### Panel Requirements per Person

| CITY | APPROXIMATE LATITUDE | SUMMER PANEL REQUIREMENT (per person) | WINTER PANEL REQUIREMENT |
|------|---------------------|--------------------------------------|--------------------------|
| Albuquerque, New Mexico | (35°) | 0.5 | 1.4 |
| Annette Island, Alaska | (55°) | 1.2 | 4.0 |
| Apalachicola, Florida | (30°) | 0.7 | 1.2 |
| Astoria, Oregon | (46°) | 0.9 | 2.8 |
| Atlanta, Georgia | (33°) | 0.7 | 1.8 |
| Barrow, Alaska | (71°) | 1.3 | 175.4 |
| Bethel, Alaska | (60°) | 1.2 | 1.12 |
| Bismarck, North Dakota | (47°) | 0.7 | 2.9 |
| Blue Hill, Massachusetts | (42°) | 0.9 | 2.9 |
| Boise, Idaho | (43°) | 0.6 | 2.9 |
| Boston, Massachusetts | (42°) | 0.8 | 2.9 |
| Brownsville, Texas | (26°) | 0.6 | 1.2 |
| Caribou, Maine | (47°) | 0.9 | 4.0 |
| Charleston, South Carolina | (33°) | 0.7 | 1.5 |
| Cleveland, Ohio | (41°) | 0.8 | 3.2 |
| Columbia, Missouri | (39°) | 0.7 | 2.3 |
| Columbus, Ohio | (40°) | 0.7 | 3.1 |
| Davis, California | (38°) | 0.6 | 2.1 |
| Dodge City, Kansas | (38°) | 0.6 | 1.6 |
| East Lansing, Michigan | (42°) | 0.8 | 3.7 |
| East Wareham, Massachusetts | (42°) | 0.9 | 2.4 |
| Edmonton, Alberta | (53°) | 0.9 | 4.6 |
| El Paso, Texas | (32°) | 0.5 | 1.3 |
| Ely, Nevada | (39°) | 0.6 | 1.8 |
| Fairbanks, Alaska | (65°) | 0.9 | 83.3 |
| Fort Worth, Texas | (33°) | 0.6 | 1.4 |
| Fresno, California | (37°) | 0.6 | 1.9 |
| Gainesville, Florida | (29°) | 0.7 | 1.2 |
| Glasgow, Montana | (48°) | 0.6 | 3.4 |
| Grand Junction, Colorado | (39°) | 0.6 | 1.8 |
| Grand Lake, Colorado | (40°) | 0.8 | 2.6 |
| Great Falls, Montana | (47°) | 0.7 | 2.8 |
| Greensboro, N.C. | (36°) | 0.7 | 1.9 |
| Griffin, Georgia | (33°) | 0.7 | 1.7 |
| Hatteras, North Carolina | (35°) | 0.6 | 1.5 |
| Indianapolis, Indiana | (40°) | 0.7 | 2.9 |
| Inyokern, California | (35°) | 0.5 | 1.1 |
| Ithaca, New York | (42°) | 0.8 | 3.6 |
| Lake Charles, Louisiana | (30°) | 0.7 | 1.4 |
| Lander, Wyoming | (30°) | 0.7 | 2.2 |
| Las Vegas, Nevada | (36°) | 0.5 | 1.2 |
| Lemont, Illinois | (41°) | 0.8 | 2.6 |
| Lexington, Kentucky | (38 ) | 0.7 | 2.0 |
| Lincoln, Nebraska | (41°) | 0.8 | 2.1 |
| Little Rock, Arkansas | (34°) | 0.7 | 1.8 |
| Los Angeles, California | (34°) | 0.7 | 1.2 |
| Madison, Wisconsin | (43°) | 0.8 | 2.8 |
| Matanuska, Alaska | (61°) | 1.1 | 14.2 |
| Medford, Oregon | (42°) | 0.6 | 3.1 |
| Miami, Florida | (26°) | 0.7 | 0.9 |
| Midland, Texas | (32°) | 0.6 | 1.6 |
| Nashville, Tennessee | (36°) | 0.7 | 2.1 |
| Newport, Rhode Island | (41°) | 0.8 | 2.4 |
| New York, New York | (41°) | 0.8 | 2.5 |
| Oak Ridge, Tennessee | (36°) | 0.7 | 2.1 |
| Oklahoma City, Oklahoma | (35°) | 0.6 | 1.4 |
| Ottawa, Ontario | (45°) | 0.8 | 3.6 |
| Phoenix, Arizona | (33°) | 0.5 | 1.2 |
| Portland, Maine | (43°) | 0.8 | 2.3 |
| Rapid City, South Dakota | (44°) | 0.7 | 2.2 |
| Riverside, California | (34°) | 0.6 | 1.2 |
| St. Cloud, Minnesota | (45°) | 0.7 | 3.4 |
| Salt Lake City, Utah | (41°) | no data | 2.3 |
| San Antonio, Texas | (29°) | 0.6 | 1.3 |
| Santa Maria, California | (35°) | 0.7 | 1.4 |
| Sault Ste. Marie, Michigan | (46°) | 0.8 | 3.8 |
| Sayville, New York | (40°) | 0.8 | 2.4 |
| Schenectady, New York | (43°) | 0.9 | 3.7 |
| Seattle, Washington | (47°) | 0.9 | 3.7 |
| Seabrook, New Jersey | (39°) | 0.8 | 2.4 |
| Spokane, Washington | (47°) | 0.6 | 4.0 |
| State College, Pennsylvania | (41°) | 0.8 | 3.1 |
| Stillwater, Oklahoma | (36°) | 0.7 | 1.6 |
| Tampa, Florida | (28°) | 0.7 | 1.0 |
| Toronto, Ontario | (43°) | 0.8 | 3.6 |
| Tucson, Arizona | (32°) | 0.6 | 1.1 |
| Upton, New York | (41°) | 0.8 | 2.2 |
| Washington, D.C. | (39°) | 0.7 | 2.1 |
| Winnipeg, Manitoba | (50°) | 0.8 | 5.0 |

(Courtesy of Sol-R-Tech)

## SPACE HEATING WITH SOLAR ENERGY

There are several important criteria relating to the basic structure of the house. Obviously, to heat with the sun, sunlight must be available. The ideal location of the collectors should be a clear southern exposure, with no objects around to shade the collector panels. Local conditions, such as frequent fogginess, can present problems.

The trees should be trimmed to provide a clear path for the winter sun. If the system is designed for space heating only, deciduous trees can usually be ignored, because they will have lost their leaves by the time the heat is needed.

The ideal roof orientation which will carry the collectors is due south. (See Figure 18-7.) The only stipulation to the south-facing collector is that the sky remain clear throughout the day. In many areas, however, there is a tendency for clouds to develop at about the same time each day during some periods of the year. For example, in Vermont the winter days often start out sunny, then become cloudy in the afternoon. Because of this the solar collectors in this geographical area are faced about 15° to 25° to the east of south in order to make more use of the morning sun. Since there is presently no source of weather data which would show this kind of trend, personal judgment and observation will be needed to determine if a correction of this sort is needed, and in which direction to make it.

The best roof pitch for any space heating application corresponds to the latitude (measured in degrees) plus 15°. It is possible to compensate for much lower or higher angles by adding more panels. However, this is not recommended if it can be avoided.

In general the roof should have as many square feet (m²) of solar panels as equal ⅓ to ½ the floor space in the home.

**Figure 18-7.** Roof orientation (*Courtesy of Sol-R-Tech*)

**STORAGE LOCATION:** The heat storage tank should be located under a heated portion of the structure. Any heat lost from the tank will then go toward heating the house. In a new home this is best accomplished by burying the tank under the cellar or ground floor. In an existing home, the basement or garage is usually the only available location.

**GENERAL CONSIDERATIONS:** It should go without saying that when a solar heating system is being designed, the designer should first be sure that the home is well insulated and is free of major air leaks. Insulation should be the heaviest available, normally R–13 in the walls and R–22 in the attic. The windows should be double-glazed (thermopane) or have storm windows. A tight, well-insulated house is important with any heating system, and an absolute necessity for the best performance with solar heat.

## SOLAR HEAT OPERATION

When a system is to be used for solar space heating the heat exchanger can be eliminated. (See Figure 18-8.) Since no one will be drinking the water in the heat storage tanks, it is possible to add a corrosion inhibitor that allows the water to be circulated directly through the solar panels. With this one change, the collector and storage parts of the system work in much the same way as in the domestic water heating system. A heat pump is added to distribute the heat to the house.

The use of a heat pump in a solar heating system allows the collectors to operate at a much lower temperature than would otherwise be possible, thereby doubling the efficiency of the collector. Also, during the summer season the refrigeration cycle of the heat pump can be reversed to provide central air conditioning with no extra equipment.

There are several types of solar-assisted heat pumps presently manufactured. Probably the most simple type uses a hot water coil between the outdoor coil and the outdoor fan. (See Figure 18-9.) The hot wa-

**Figure 18-8.** Solar system without heat exchanger (*Courtesy of Sol-R-Tech*)

**Figure 18-9.** Solar-assisted heat pump

ter from the heat storage tank is pumped through the water coil to provide extra heat when the ambient temperature is low and the efficiency of the heat pump has dropped. When this type is used, a conventional type heat pump with the added coil will operate satisfactorily.

The balanced flow pump is operated on an internal pressure control to maintain water pressure in the lines connected to it. A solenoid valve is wired to each heat pump to shut off the flow of water when the heat pump is not running. (See Figure 18-10.) This will cause the water pump to shut off automatically when no water is being used.

**HEAT PUMP CUT-OFF:** Some heat pump units must be equipped with a low temperature cut-off control to prevent operation of the compressor when the water temperature is below 40°F (4.4°C). This is accomplished by wiring a thermostat in series with the compressor contactor coil. (See Figure 18-11.)

**HEAT PUMP SUPPLY PUMP:** This is usually the smallest available balanced flow pump unit. It must be equipped with an internal pressure control to turn the pump off

**Figure 18-10.** Balanced flow pump for solar-assisted heat pump (*Courtesy of Sol-R-Tech*)

Contactor

Coil

Contactor

Block

"Klixon"
Schematic

"Klixon"
Top View

Note: On DW-40H
Contactor Leads are
Top and Bottom

Water
Inlet

Tape in Place
Warp with Insulating
Loam Tape (Arma-Flex)

**Figure 18-11.** Heat pump freeze protector (*Courtesy of Sol-R-Tech*)

when no water is being used. A good example of this type of pump is one which is used for shallow well installations. There are many ways to provide this assistance, but this is the most simple and economical method.

## HEAT PUMP VERSUS FIN-TUBE RADIATION

In areas where solar radiation is scarce, the heat pump is the best system to choose because it can still extract heat from the solar storage tank when the temperature is at 40°F (4.4°C).

Fin-tube coils can be installed in sunnier and warmer localities. In a fin-tube system the solar heated water is pumped directly through the heat exchanger tubes which are located inside the system duct work. (See Figure 18-12.) Air is blown across the tubes, warmed, and forced into the rooms requiring heating. Obviously, the storage tank temperature must be higher for such a system to operate efficiently.

Heated Air
to Space

Heated Air
to Space

Water
Coil

Pump

Water
Storage

**Figure 18-12.** Direct solar heating system

## WATER TREATMENT

Even though the getter column is used, it is not possible to circulate untreated water through the solar panels. To do so would cause severe corrosion problems. The following steps are recommended for water treatment:

1. When filling the water storage tank pass the water through a water softener which is designed to remove the minerals present in the water supply in that locality. Contact a local water treatment company for advice.

2. Add 50 parts per million (ppm) of a sodium chromate corrosion inhibitor, such as Barclay Chemical *Inhibitor C*. An inexpensive color-test kit should be purchased to allow for the measurement of concentration of the inhibitor.

3. The water is now ready for use.

## HEAT STORAGE

A typical heat storage tank is usually very simple. (See Figure 18-13.) The dimensions can be varied to fit the space without affecting the performance of the system as long as the thickness of insulation and relative positions of the pipes are maintained.

The most simple method of waterproofing the heat storage tank is with *Thoro-Seal*, a cement waterproofing compound which is available at most local building supply stores. Apply the Thoro-Seal to the inside of the tank according to the directions on the package.

As has already been mentioned, the heat storage tank should be located under the heated portion of the home. If this is not possible, the tank insulation should be increased to 4 inches (10.16 cm) on the sides and 6 inches (15.24 cm) on the top.

## DISTRIBUTION OF HEAT

Any new home being designed with solar heat in mind should be planned around forced warm air heating. This type of system

**Figure 18-13.** Storage tank piping arrangement and panel pipe sensor (*Courtesy of Sol-R-Tech*)

will accommodate both heat pumps and direct heat exchanger systems. Existing homes with this type of heating system can readily be adapted to solar heating.

Houses that use steam or hot water heating present a more difficult problem, especially if baseboard heaters or radiators are used. Unfortunately, the average temperature required in a hot water system is 180°F to 210°F (82°C to 99°C), while the temperature that can be reached by the solar panels is usually approximately 150°F (66°C) during the winter season. This could mean that almost twice as much baseboard heating would be needed to heat the building, making such a system uneconomical.

## PANEL WATER SUPPLY PUMP

Any ⅓-hp pump can be used to supply the solar collector array, as long as it is self-priming. If self-priming pumps are not available a foot valve must be installed in the suction line to aid in keeping the pump primed. A check valve should also be installed just beyond the pump discharge.

## OPTIONAL DIRECT HEAT EXCHANGER

By adding a hot water coil and providing a three-way solenoid valve to divert the flow of water from the heat pump to the coil, it is possible to provide direct solar heating during some seasons of the year. (See Figure 18-14.) The three-way solenoid should

**Figure 18-14.** Optional direct heat exchanger solar heating system (*Courtesy of Sol-R-Tech*)

be actuated by a thermostat in the storage tank so that when the water temperature exceeds 100°F (37.8°C), the direct heat exchanger coil is used. The coil uses the blower already present in the heat pump, preventing the needless duplication of equipment. (See Figure 18-15.)

## SPACE HEATING EXISTING HOMES

The equipping of already existing homes with solar heat is a topic of interest to many who wish to remain in their present home while still enjoying the benefits of solar heating. However, the addition of solar heat to an existing home is a difficult and expensive operation. Many of the installation procedures that are easy during construction later become almost impossible.

The optimum time for adding solar heating to a home is during remodeling, or when putting on an addition to the house. Additions are especially helpful, since they provide an area where a storage tank can be buried under the house, and the roof can be oriented for optimum collector performance.

Some of the questions that should be answered when determining how feasible it would be to add solar heating to an already existing home are:

1. What kind of heating system is presently used? If it is warm air, fine. If any other type is used, would it be possible to install a warm air system? If not, then solar heating should not be considered. However, domestic hot water from solar energy can still be used.

2. Is there a large, south-facing roof area at an angle of 45° to 60°?

Note: Field-Supplied Water Solenoid Valve for Heating and Cooling

**Figure 18-15.** Typical wiring diagram for field-supplied water solenoid (*Courtesy of Sol-R-Tech*)

3. Is there an area, preferably underneath a heated space, where a storage tank could be buried?

4. Is the building well-insulated?

If all these questions were answered yes, then the home is a good candidate for solar heating. If the answer to Question 1, 2, or 3 was no, then the addition of solar heating would probably not be feasible. If the answer to Question 4 was no then add to the insulation and proceed with the installation.

## STEPS IN DESIGNING A SOLAR HEATING SYSTEM

The recommended major steps that should be used in designing a solar heating system may be summarized as follows:

1. Calculate the total net heat loss of the house as follows:

 a. Select the design outdoor weather conditions, such as: temperature, wind direction, and wind velocity.

 b. Select the indoor air temperature that is to be maintained in each room during the coldest weather.

 c. Estimate the temperatures in any adjacent unheated spaces.

 d. Select or compute the heat transfer coefficients for outside walls and glass, and for inside walls adjacent to unheated spaces.

 e. Determine the net wall, roof, ceiling, glass, etc., areas.

 f. Compute the heat transfer loss for each surface.

 g. Compute the heat equivalent of the infiltration losses and the total heat required for positive ventilation, if used.

 h. Sum up the heat losses.

 i. Determine the internal heat generated (range, lights, etc.). This item should be subtracted from the sum of the heat losses.

 j. Additional heating capacity may be required for intermittently heated buildings.

2. Heating requirement for the design period: A decision must be made concerning the maximum number of hours during which the required heat will be supplied either by the storage system or by the auxiliary heating system. Assume, for instance, that 48 hours is chosen. By using the sum obtained in section 1h above, compute the 48-hour requirement, $Q_D$ (design load). Part of this heat will be supplied by the auxiliary heater.

The capacities of the solar collector, heat storage unit, and the auxiliary heater are interrelated. In the cold climate zone a full capacity auxiliary heating unit must be provided to supplement the solar heating system. Otherwise, cost of the collector and storage system will be too high. The auxiliary heating unit will seldom be called upon for heat delivery over an extended period; but the capacity is nevertheless required. The cost of an additional furnace capacity beyond average needs is rather low.

3. Solar collector size: The component selection procedure for the solar heating of a residential home may be somewhat simplified. In fact, the size of the solar collector is usually limited by the maximum roof area available facing the southern direction. Therefore, the approximate solar collector surface area is known. This area, however, should be large enough so that the collector will be capable of providing the major portion of the design load ($Q_D$).

4. Selection of design parameters: Choose the type of solar collector, flow rate of the heat transfer fluid (water), and assume a reasonable collector efficiency ($\eta$).

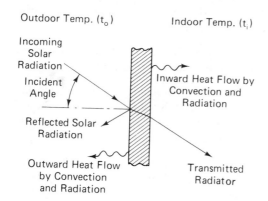

Outdoor Temp. $(t_o)$            Indoor Temp. $(t_i)$

Incoming
Solar
Radiation

Incident
Angle                          Inward Heat Flow by
                               Convection and
                               Radiation

Reflected Solar
Radiation

Outward Heat Flow             Transmitted
by Convection                 Radiator
and Radiation

**Figure 18-16.** Heat balance for sunlit glazing material

Determine the design daily total radiation $(Q_H)$ which is incident on the collector. (See Figure 18-16.) The total useful heat supplied by the collector will be given by the relation:

$$Q_U = \eta Q_H$$

where

$Q_U$ = Btu (cal) per day heat loss
$\eta$ = Collector efficiency
$Q_H$ = Total collector incident radiation

The difference between $Q_U$ and $Q_D$ should be supplied by the auxiliary heater. (See Figure 18-17.)

5. Design temperatures of the collector and fluid: By using the following equations, compute the average collector temperature $(T)$, and the water temperature rise $(T_f)$ in the collector. If the water temperature at the collector inlet and the exit are $T_1$ and $T_2$, respectively, then:

$$T = 0.5(T_1 + T_2)$$

$$T_f = T_2 - T_1.$$

Solar
Radiation, $Q_H$          Collector

Optical Losses
Thermal Losses

Angle of
Collector Tilt

South                Heated
                     Fluid

$Q_u$

                     Heat Losses
Solar                              Auxiliary
Heat       Hot Water              Heat
Storage    Preheat

           Space Heat

**Figure 18-17.** Residential solar heating system

Thus, $T_1$ and $T_2$ are determined. These are the daily average water temperatures. In reality, these temperatures are changing with time.

6. Head storage capacity: In its simplest form, it may be assumed that the temperature rise in the storage is equal to the temperature rise in the collector ($T_C$). Then the heat capacity ($C$) of the unit is found from:

$$Q_U = C(T_f)$$

7. Auxiliary heater capacity: The auxiliary heating unit capacity should be adequate to supply at least the difference between $Q_D$ and $Q_U$ during the assumed design period.

8. Air flow rate: If air is used, for instance, to remove heat from the storage to the rooms, compute the required air flow rate and the blower capacity.

*NOTE:* Since many of these procedures were covered in Chapter 16, it may be helpful to review that chapter before actually attempting to estimate these calculations.

## SAFETY PROCEDURES

The use of solar energy is comparatively new in the refrigeration and air conditioning industry. However, the safety practices used in the other areas also apply in solar use.

1. Obtain help when moving heavy objects.

2. When cleaning collectors wear sunglasses to protect the eyes from the rays of the sun.

3. Do not touch the reflective plate with the bare hands, since personal burns and etching of the collector surface may result.

4. Always use a safe ladder when working on roof tops.

5. Be careful not to cause separation of the tube and the collector plate.

6. Do not allow the flow of working fluid to stop with some of it still in the collector when the sun is shining on it. Excessive temperatures may result.

7. Be sure that the electrical power is off before attempting work on the solution pumps.

8. Do not tamper with the controls for the tracking system or the efficiency of the collector may be decreased.

9. Do not set the working fluid temperature control too high because excessive pressure may result and rupture the storage tank.

10. Be sure that the working fluid drains from the collector when the pump is off or that there is sufficient antifreeze to prevent freezing.

11. When making line connections, use two wrenches to prevent twisting of the tubing.

## SUMMARY

• Solar energy has been traced back to the ancient Greeks.

• Solar energy has not been developed into wide usage because of economic factors.

• The annual solar energy falling in this country is almost 1000 times greater than the total yearly consumption of the country.

• The sun provides between 250 and 300 Btus (63 kcal to 75.6 kcal) per hour per square foot ($m^2$) of the earth's surface.

• The two types of practical solar collectors are: concentrating collector and flat plate collector.

• The major advantage of the concentrating collector is its ability to generate very high temperatures.

• Concentrating collectors must move with the sun.

• The flat collector is considered the best device for gathering solar energy for low temperature applications, such as home heating.

• A "working fluid" is used to carry the heat from the collector to the storage area.

• The coating used for the absorber plate is of great importance and is usually a flat black paint.

• If too few cover plates are used, an insufficient amount of insulation from the cold ambient air will be provided.

• If too many cover plates are used, the amount of sunlight reaching the absorber plate will be reduced.

• Water will store five times as much heat as stone, and four times as much as concrete.

• The required equipment for solar heating of domestic hot water is: a solar collector array, a heat exchanger, a water storage tank, circulator pumps, controls, and an auxiliary water heater.

• With a properly designed system 75 to 95 percent of the domestic hot water will be free of utility rates when heated with solar energy.

• The amount of heat collected by a solar collector is dependent upon: materials used in fabrication, ambient temperature, solar intensity, orientation of the collector, and the roof pitch.

• The ideal orientation of the roof that will carry the collectors is due south.

• The best roof pitch for a space heating application corresponds to the latitude (measured in degrees) plus 15°.

• When considering solar heating, the home must be well-insulated and be free of any major air leaks.

• Solar heating can be applied to heat pump systems to increase their efficiency even in low ambient temperatures.

• On heat pump systems solar heated water is pumped through an auxiliary coil which adds heat to the refrigerant during the heating season.

• The water circulating through the collector plates should be treated to prevent corrosion.

• A solar space heating system should be planned around a forced warm air system.

• Baseboard heaters and radiators cannot be used with solar heating.

• Heat pump systems can use an optional direct heat exchanger to supply the total heat when the storage water is above 100°F (37.8°C).

• The addition of solar heating to an existing home is difficult and expensive.

## REVIEW EXERCISES

1. Why is solar energy not in wide use today?

2. How much solar heat strikes the earth's surface per hour?

3. What are the two major types of solar collectors?

4. Which type of collector will provide the hottest water?

5. Which type of collector is most popular?

6. What type of protective coating is best suited for solar collectors?

7. At present, what is the best heat storage medium for solar heat storage?

8. Why is glass not used for collector cover plates?

9. What is generally used as the heat transfer fluid?

10. What are the five variables that determine how much heat can be collected by a solar collector?

11. What is the ideal orientation of a solar collector?

12. What is the best roof pitch for a roof that is to hold the collectors?

13. In general, how many collector plates are needed for space heating?

14. What type of insulation is recommended for the walls and attic of a home to be heated by solar energy.

15. At what storage temperature should some heat pump compressors be shut off?

16. What type of heat distribution system should be used with solar heating?

17. What are two major considerations that must be given a home when designing a solar heating system?

18. How is the efficiency of a heat pump increased when assisted by solar heating?

19. Are baseboard heaters and radiators suitable for use with solar heating?

20. At what storage temperature can an optional direct heat exchanger be used?

# Glossary

**Absolute humidity:** The weight in grains of water vapor actually contained in one ft³ (.0283m³) of the air and moisture mixture.

**Absolute pressure:** The sum of gauge pressure plus atmospheric pressure.

**Absolute temperature:** The temperature at which all molecular motion of a substance stops. At this temperature the substance theoretically contains no heat.

**Absorbent:** A substance that has the ability to absorb another substance.

**Absorptivity:** The ratio of radiant energy absorbed by an actual surface at a given temperature to that absorbed by a black body at the same temperature.

**Accessible hermetic:** A single unit containing the motor and compressor; this unit may be field serviceable.

**Accumulator:** A shell placed in the suction line to prevent liquid refrigerant from entering the compressor by vaporizing the liquid.

**Acid condition:** A condition in which the refrigerant and/or oil in the system is contaminated with other fluids that are acidic in nature.

**ACR tubing:** Tubing used in the refrigeration and air conditioning industry. The ends are sealed and the tubing is dehydrated.

**Activated alumina:** A chemical made up of aluminum oxide which is used as a desiccant for refrigeration driers.

**Activated carbon:** A specially processed carbon commonly used to clean air.

**Actuator:** That part of a regulating valve that changes fluid or thermal or electrical energy into mechanical motion to open or close valves, dampers, etc.

**Adiabatic compression:** The compression of a gas without the addition or removal of heat.

**Aeration:** The combining of a substance with air.

**Agitator:** A device used to induce motion in a confined fluid.

**Air:** An elastic gas. Air is a mechanical mixture of oxygen and nitrogen and slight traces of other gases; it also may contain moisture known as humidity. Dry air weighs 0.075 lb/ft³ (.0012 g/m³). One Btu (252 calories) will raise the temperature of 55 ft³ (1.56 m³) of air 1°F (.56°C).

**Air change:** The number of times in an hour the air in a room is changed either by mechanical means or by the infiltration of outside air leaking into the room through cracks around doors, windows, and other openings.

**Air cleaner:** A device designed to remove airborne impurities such as dust, fumes, and smoke. Air cleaners include air washers and air filters.

**Air coil:** The coils on refrigeration systems and air conditioning units through which air is blown.

**Air conditioning:** The simultaneous control of the temperature, humidity, air motion, and air distribution within an enclosure. Where human comfort and health are involved, a reasonable air purity with regard to dust, bacteria, and odors is also included. The primary requirement of a good air conditioning system is a good heating system.

**Air conditioning unit:** A device designed for the treatment of air; it consists of a means for ventilation, air circulation, air cleaning, and heat transfer with a control to maintain the temperature within the prescribed limits.

**Air diffuser:** An air distribution element or outlet designed to direct the flow of air in the desired patterns.

**Air handler:** A unit consisting of the fan or blower, heat transfer element, filter, and housing components of an air distribution system.

**Air infiltration:** The leakage of air into a house through cracks and crevices and through doors, windows, and other openings by wind pressure and/or temperature difference.

**Air-sensing thermostat:** A thermostat that operates as a result of air temperature. The sensing bulb is located in the air stream.

**Air, standard:** Air whose temperature is 68°F (20°C), whose relative humidity is 36%, and whose pressure is 14.7 psi (1.03 kg/cm²).

**Algae:** A low form of plant life found in water. It is especially troublesome in water towers.

**Alternating current:** Electrical current in which the direction of current reverses periodically, usually 60 times per second. Also called 60 cycle current.

**Altitude correction:** The change in atmospheric pressure at higher altitudes has the same effect on a vapor pressure bellows as does a barometric change. The effects of altitude are greater and of a permanent nature so that adjustments may be made in the control settings to correct for the effect of altitude.

**Ambient temperature:** The temperature of the air that surrounds an object.

**Ammeter:** An electric meter calibrated in amperes that is used to measure current flow in a circuit.

**Amperage:** The unit of electrical current equivalent to a flow of one coulomb per second.

**Amplifier:** An electrical device that increases the electron flow in a circuit.

**Analyzer:** A device used to check the condition of electrical components.

**Anemometer:** An instrument used to measure the velocity of air motion.

**Anneal:** To heat treat metal in order to obtain the desired softness and ductility.

**Anode:** The positive terminal of an electrolytic cell or battery.

**Apparatus dew point:** The dew point of the air leaving the air conditioning coil.

**Armature:** Part of an electric motor or generator which is moved by magnetism.

**ASTM standards:** Standards set by the American Society of Testing Materials.

**Atmospheric pressure:** The pressure exerted by the atmosphere in all directions, as indicated by a barometer. Standard atmospheric pressure at sea level is considered to be 14.7 lb/in² (1.03 kg/cm²).

**Atom:** The smallest part of a substance that can exist alone or in combination with other elements.

**Atomize:** To change a liquid to small particles or fine spray.

**Automatic control:** The control of the various functions of a piece of equipment without manual adjustment.

**Automatic defrost:** A method of removing ice and frost from the evaporator automatically.

**Automatic expansion valve (AEV):** A pressure-controlled valve that reduces the high-pressure liquid refrigerant to a low-pressure liquid and vapor refrigerant.

**Automatic ice maker:** A refrigeration unit designed to produce a quantity of ice automatically.

**Back pressure:** The pressure in the low pressure side of the refrigeration system.

**Back seating:** The wide open position of a service valve which usually closes off the gauge connection fitting.

**Baffle:** A plate or wall for deflecting gases or fluids.

**Ball check valve:** A valve assembly that allows the flow of fluid in only one direction.

**Barometer:** An instrument that measures atmospheric pressure.

**Battery:** An electricity-producing device that uses the interaction of metals and chemicals to create a flow of electrical current.

**Bearing:** A low-friction device used for supporting and aligning moving parts.

**Bellows:** A corrugated cylindrical container that expands and contracts as the pressure inside changes to provide a seal during the movement of parts.

**Bellows seal:** A means of preventing leakage of a fluid by rotating parts. The sealing material expands and contracts with a change in pressure.

**Bending spring:** A spring put on the inside or the outside of a tube to prevent collapse during the bending process.

**Bimetal strip:** A temperature-regulating device that works on the principle that two dissimilar metals have unequal expansion rates upon a change in temperature.

**Bleed:** To release pressure slowly from a system or cylinder by opening a valve slightly.

**Bleed valve:** A valve with a small opening through which pressure is permitted to escape at a slow rate when the valve is closed.

**Boiler:** A closed vessel in which steam is generated or in which water is heated.

**Boiler heating surface:** The area of the heat transfer surfaces in contact with the water (or steam) in the boiler on one side and the fire or hot gases on the other.

**Boiling point:** The boiling temperature of a liquid under a pressure of 14.7 psi (1.03 kg/cm$^2$).

**Bourdon tube:** A thin-walled tube of elastic metal used in pressure gauges. It is flattened and bent into a circular shape and tends to straighten as the pressure inside is increased.

**Boyle's Law:** A law of physics that deals with the volume of gases as the pressure varies. If the temperature remains constant, the volume varies. If the pressure is increased, the volume is decreased; if the pressure is reduced, the volume is increased.

**Braze:** To join metals with a nonferrous filler using heat between the temperature of 800°F (427°C) and the melting point of the base metal.

**Breaker strip:** A strip of plastic used to cover the opening between the case and the liner of a refrigerator or freezer.

**British thermal unit (Btu):** The quantity of heat required to raise the temperature of 1 lb (453.6 grams) of water 1°F (.56°C). (This measurement is approximate but is sufficiently accurate for any work discussed in this book.)

**Burner:** A device in which the mixing of gas and air occurs.

**Butane:** A liquid hydrocarbon commonly used as a heating fuel.

**Bypass:** A pipe or duct usually controlled by a valve or a damper for short circuiting the flow of a fluid.

**Calibrate:** To position the pointers; also, needles used to determine accurate measurements.

**Calorie:** The calorie, as used in engineering, is a large heat unit, and is equal to the amount of heat required to raise 1 kilogram of water 1°C.

**Capacitance:** The property of an electric current that permits the storage of electrical energy in an electrostatic field and the release of that energy at a later time.

**Capacitor:** An electrical device capable of storing electrical energy.

**Capacitor-start motor:** A motor that uses a capacitor in the starting circuit to increase the starting torque.

**Capacity:** The capacity of a refrigeration unit is the heat-absorbing capacity per unit of time. Usually measured in Btu per hour.

**Capillary tube:** A tube of small diameter used to regulate the flow of liquid refrigerant into the evaporator.

**Carbon dioxide:** A compound made up of carbon and oxygen which is sometimes used as a refrigerant. Its refrigerant number is R–744.

**Carbon monoxide:** A colorless, odorless, and poisonous gas produced when hydrocarbon fuels are burned with too little air.

**Carbon tetrachloride:** A colorless, nonflammable, and very toxic liquid sometimes used as a solvent. It should never be allowed to touch the skin and the fumes should not be inhaled.

**Cathode:** The negative terminal of an electrical device. The electrons leave the device through this terminal.

**Cavitation:** A gaseous condition found in low pressure places of a liquid stream.

**Centigrade:** A type of thermometer on which the freezing temperature is 0° and the boiling temperature is 100° when water is the medium tested.

**Centimeter:** A metric unit of linear measurement equal to .3937 in.

**Central fan system:** A mechanical indirect system of heating, ventilating, or air conditioning consisting of a central plant where the air is conditioned and then circulated by fans or blowers through a system of distributing ducts.

**Centrifugal compressor:** A pump that uses centrifugal force to compress gaseous refrigerants.

**Change of state:** The changing of a substance from one state to another, such as from a liquid to a vapor or a solid to a liquid.

**Charge:** The amount of refrigerant in a system. Also, to put the refrigerant into a system.

**Charging board:** A specially designed panel fitted with gauges, valves, and refrigerant cylinders which is used for evacuating and charging refrigerant and oil into a system.

**Charles' Law:** States that the volume of a given mass of gas at a constant pressure varies according to its temperature.

**Check valve:** A device that allows the flow of a fluid in only one direction.

**Chill factor:** A calculated figure based on the dry bulb temperature and wind velocity.

**Chimney:** A vertical shaft that carries vent gases from gas heating equipment to the atmosphere.

**Chimney effect:** The tendency in a duct or other vertical air passage for air to rise when heated due to its decrease in density.

**Circuit:** The tubing, piping, or electrical wiring that permits flow from the energy source through the circuit and back to the energy source.

**Circuit breaker:** A safety device used to protect an electrical circuit from overload conditions.

**Circuit, parallel:** The arrangement of electrical devices so that the current is equally divided between each circuit.

**Circuit, pilot:** A secondary circuit used to control or signal a device in the main circuit. Usually of a different voltage.

**Circuit, series:** The arrangement of electrical devices in such a way that the current passes through each of the devices one after the other.

**Clearance pocket:** The space in the cylinder above the piston at the end of the compression stroke.

**Closed circuit:** A completed electrical circuit through which the electrons are flowing.

**Closed cycle:** Any cycle in which the refrigerant is used over and over again.

**Clutch, magnetic:** A clutch used on an automotive compressor to start and stop the compressor on demand from the thermostat.

**$CO_2$ Indicator:** An instrument used to indicate the percentage of carbon dioxide present in the flue gases.

**Coefficient of Heat Transmission (U):** The amount of heat transmitted from air to air in 1 hr/ft² of the wall, floor, roof, or ceiling for a difference in temperature of 1°F (.56°C) between the air inside and the air outside of the wall, floor, roof, or ceiling.

**Coefficient of Performance (COP):** The ratio of work performed to energy used.

**Cold:** The absence of heat.

**Cold junction:** The part of a thermocouple that absorbs heat from a flame or ambient air.

**Cold storage:** The process of preserving perishables on a large scale by means of refrigeration.

**Cold wall:** A type of refrigerator construction in which the inner liner is used as the cooling surface.

**Combustible liquid:** A liquid whose flash point is at or above 140°F (60°C); usually considered to be a Class 3 liquid.

**Comfort cooling:** Refrigeration for comfort as opposed to refrigeration for food storage.

**Comfort zone:** The range of effective temperatures over which the majority of adults feel comfortable.

**Commercial refrigeration:** A reach-in or service refrigerator of commercial size with or without the means of cooling.

**Commutator:** The part of a rotor in an electric motor that conveys the electric current to the motor windings.

**Compound gauge:** An instrument used to indicate the pressure both above and below atmospheric pressure. Commonly used to measure low side pressures.

**Compression ratio:** The ratio of the volume of the clearance space to the total volume of the cylinder. It is also used as the ratio of the absolute suction pressure to the absolute discharge pressure.

**Compression system:** A refrigeration system in which the pressure-imposing element is mechanically operated. Distinguished from an absorption system which uses no compressor.

**Compressor:** The device used for increasing the pressure on the refrigerant.

**Compressor displacement:** The volume in inches (mm) represented by the area of the piston head multiplied by the length of the piston stroke.

**Compressor seal:** The device that prevents leakage between the crankshaft and the compressor body on an open type compressor.

**Condensate:** The moisture resulting from the removal of heat from a vapor to bring it below the dew point temperature.

**Condenser:** The vessel or arrangement of pipes in which the vaporized refrigerant is liquified by the removal of heat.

**Condenser fan:** The fan that forces air over the condenser coil.

**Condensing pressure:** The pressure inside the condenser at which the refrigerant vapor gives up latent heat of vaporization and changes to a liquid. This pressure varies with the ambient temperature.

**Condensing temperature:** The temperature inside the condenser at which the gaseous refrigerant gives up latent heat of vaporization and becomes a liquid. This temperature varies with the pressure.

**Condensing unit:** Usually considered as the high side; consists of the compressor, condenser, receiver, fan, and motor, all on a suitable frame, with the necessary accessories.

**Conduction:** The transmission of heat through and by means of matter.

**Conductivity:** The amount of heat in Btu (calories) transmitted in 1 hr through 1 ft$^2$ (m$^2$) of a homogeneous material 1 in. (25.4mm) thick for a difference in temperature of 1°F (.56°C) between the two surfaces of the material.

**Conductor:** A substance or body capable of conducting electrical or heat energy.

**Constrictor:** A device used to restrict the flow of a gas or a liquid.

**Contaminant:** A substance such as dirt, moisture, or some other material which is foreign to the refrigerant or the oil in the system.

**Control:** Any device used for the regulation of a machine in normal operation—manual or automatic. It is usually responsive to temperature or pressure, but not to both at the same time.

**Control system:** All of the components used for the automatic control of a given process.

**Control valve:** A valve that regulates the flow of a medium which affects the controlled process. The valve is controlled by a remote signal from other devices which use as power pneumatic, electric, or electrohydraulic devices.

**Convection:** The transmission of heat by the circulation of a liquid or a gas such as air. When it occurs naturally, it is caused by the difference in weight of hotter and colder fluids.

**Convector:** A surface designed to transfer its heat to the surrounding air by means of convection.

**Cooler:** A heat exchanger that transfers heat from one substance to another.

**Cooling tower:** A device that cools to the wet bulb temperature by means of evaporation of water.

**Copper plating:** An abnormal condition that exists when moisture is present in a refrigeration system. It is a result of the copper being electrolytically deposited on the steel parts of the system.

**Corrosion:** The deterioration of a metal by chemical action.

**Coulomb:** The amount of electrical energy transferred by an electric current at the rate of one ampere per second.

**Counter emf:** The tendency for an electrical current in an induction coil to reverse its flow as the magnetic field changes.

**Counterflow:** The opposing direction of flow of fluids, the coldest portion of one meeting with the warmest portion of the other.

**Coupling:** The mechanical device used to join pipes.

**Crank throw:** The distance between the center line of a main bearing journal and the center of the crankpin or eccentric.

**Crisper:** A drawer or compartment used to store vegetables. It helps to maintain the desired humidity for the storage of vegetables.

**Critical pressure:** The compressed condition of a refrigerant at which the vapor and liquid have the same properties.

**Critical temperature:** The temperature above which a vapor cannot be liquified.

**Cross charged:** The combining of two fluids to create a desired pressure–temperature curve.

**Current:** The flow of electrical energy in a conductor which occurs when the electrons change positions.

**Current relay:** A device used to start an electric motor. It is caused to function because of the change in current flow in the circuit.

**Cut-in:** The temperature or pressure at which a set of contacts completes an electrical circuit.

**Cut-out:** The temperature or pressure at which a set of electrical contacts opens an electrical circuit.

**Cycle:** The complete course of operations of a refrigeration system. This includes the four major functions of compression, condensation, expansion, and evaporation.

**Cylinder:** The chamber in a compressor in which the piston travels to compress the refrigerant gas.

**Cylinder head:** A cap or plate that encloses the open end of a cylinder.

**Dalton's Law:** States that a vapor pressure exerted in a container by a mixture of gases is equal to the sum of the individual vapor pressures of the gases contained in the mixture.

**Damper:** A device used to control the flow of air.

**Deaeration:** The act of separating air from a substance.

**Decibel:** A unit commonly used to express sound or noise intensity.

**Decomposition:** Spoilage.

**Defrost:** To remove accumulated ice from a cooling coil.

**Defrost cycle:** The refrigeration cycle in which the ice accumulation is melted from the coil.

**Defrost timer:** A device connected into the electrical circuit to start the defrost cycle and keep it on until the ice has melted from the evaporator.

**Degree:** The unit of measure on a temperature scale.

**Degree day:** A unit that represents one degree of difference between the inside temperature and the average outdoor temperature for one day. Normally used for estimating the fuel requirements of a building.

**Degree of superheat:** The difference between the boiling point of the refrigerant and the actual temperature above the boiling point.

**Dehumidifier:** A device used to lower the moisture content of the air passing through it.

**Dehumidify:** To remove water or moisture from the atmosphere; to remove water vapor or moisture from any material.

**Dehydrated oil:** A lubricant from which the moisture has been removed to an acceptable level.

**Dehydrator-receiver:** A small tank that serves as a liquid receiver. It also contains a desiccant for removing moisture from the refrigerant and is used on automobile air conditioning systems.

**Deice control:** A device used to control the compressor and allow for the melting of any accumulation of frost on the evaporator.

**Density:** The weight per unit volume of a substance.

**Deodorizer:** A device that absorbs or adsorbs odors. Activated charcoal is usually used for this purpose.

**Desiccant:** A substance used to collect and hold moisture in a refrigeration system. The most commonly used desiccants are activated alumina and silica gel.

**Design pressure:** The highest pressure expected to be reached during normal operation. It is usually the operating pressure plus a safety factor.

**Dew point:** The temperature at which a vapor begins to condense. Usually 100% relative humidity.

**Diaphragm:** A flexible material, usually a thin metal or rubber, used to separate chambers.

**Dichlorodifluoromethane:** A halocarbon refrigerant commonly known as R–12.

**Differential:** The difference between the cut-in and cut-out of a pressure or temperature control.

**Direct expansion evaporator:** An evaporator that uses either an automatic or a thermostatic expansion valve.

**Double duty case:** A commercial refrigerator in which part of the space is used for storage and part is used for display of goods.

**Double pole:** The term used to designate the contact arrangement which includes two separate contact forms, ie., two single-pole contact assemblies.

**Double throw:** A term applied to a contact arrangement to denote that each contact form included is a break–make, ie., one contact opens its connection to another contact and then closes its connection to a third contact.

**Draft gauge:** An instrument used to measure air movement by measuring the air pressure differences.

**Draft indicator:** An instrument used to measure chimney draft or combustion gas movement.

**Draft regulator:** A device that keeps the draft in a vent pipe at the desired volume.

**Drier:** A device used to remove moisture from a refrigerant.

**Drip pan:** A pan or trough used to collect condensate from an evaporator coil.

**Dry bulb:** A thermometer used to measure the ambient air temperature.

**Dry bulb temperature:** The actual temperature of the air, as opposed to wet bulb temperature. The temperature of the air indicated by any thermometer not affected by the moisture content of the air.

**Dry ice:** Compressed carbon dioxide.

**Dry system:** A refrigeration system in which only droplets are present in the evaporator.

**Duct:** A tube or channel through which air is forced in a forced air system.

**Eccentric:** A circle or disc mounted off center on a shaft; used to produce a reciprocating motion.

**Eddy currents:** The induced currents flowing in a laminated core.

**Effective area:** The gross area of the grill minus the area of the vanes or bars. The actual flow area of an air inlet or outlet.

**Effective temperature:** The overall effect on a human being of the air temperature, humidity, and air movement.

**Electric defrosting:** A method of defrosting an evaporator by using electric current to heat the surface.

**Electric heating:** A heating system in which the source of heat is electricity.

**Electrolysis:** The chemical reaction of two substances due to the flow of electricity through them.

**Electrolytic capacitor:** A plate or surface capable of storing small electrical charges.

**Electromagnet:** A coil of wire wound around a soft shaft or core. When electric current flows through the coil an electromagnet is formed.

**Electromotive force (emf):** The voltage or electrical force that causes the free electrons to flow in a conductor.

**Electron:** A basic component of an atom; it carries a negative electrical charge.

**Electronic leak detector:** An electronic instrument that senses refrigerant vapor in the atmosphere. An electronic flow change indicates a leak.

**End bell:** The end structure of an electric motor; it usually holds the bearings, etc.

**End play:** The slight movement of a shaft along its center line.

**Energy:** The ability to do work.

**Enthalpy:** The actual or total heat contained in a substance. It is usually calculated from a base. In refrigeration work the base is accepted as $-40°F$ ($-40°C$).

**Entropy:** The energy in a system. A mathematical factor used in engineering calculations.

**Environment:** The conditions of the surroundings.

**Epoxy:** An adhesive formed by mixing two resins.

**Eutectic:** That certain mixture of two substances which provides the lowest melting temperature of all the various mixes of the two substances.

**Evacuation:** The removal of air and moisture from a refrigeration system.

**Evaporation:** The change of state of a liquid to a gas. The greatest amount of heat is absorbed in this process.

**Evaporative condenser:** A specially designed condenser used to remove heat from refrigerant gas by using the cooling effect of evaporating water.

**Evaporator:** The part of a refrigeration system where the refrigerant boils and picks up the heat.

**Evaporator fan:** The fan that forces the air through the evaporator.

**Exhaust valve:** The outlet port which allows the compressed vapor to escape from the cylinder. *Also called* the discharge valve.

**Expansion valve:** The device in a refrigeration system that reduces a high pressure refrigerant to a low pressure refrigerant.

**External equalizer:** The tube connected to the low pressure side of the diaphragm in the thermostatic expansion valve. It lets the evaporator outlet pressure help control the operation of the valve.

**Fahrenheit scale:** The scale on a standard thermometer. It has the boiling point of water at 212°F and the freezing point at 32°F.

**Fail-safe control:** A device that opens an electric current when the sensing element senses an abnormal condition.

**Fan:** An enclosed propeller that produces motion in air. Commonly designates anything that causes air motion.

**Farad:** The unit of electrical capacity of a capacitor.

**Female thread:** The internal thread on valves, fittings, etc.

**Field pole:** That part of the motor stator which concentrates the magnetic field in the field winding.

**Filter:** A device used to remove the solid particles from a fluid by straining.

**Fin:** The sheet metal extension on evaporator or condenser tubes.

**Finned tube:** A tube built up with an extended surface in the form of fins.

**Flammability:** The ability to burn.

**Flammable liquids:** Liquids whose flash points are below 140°F (60°C), and whose vapor pressures are less than 40 psi (2.81 kg/cm²) at 100°F (37.8°C).

**Flapper valve:** A thin metal valve used as the suction and discharge valves in refrigeration compressors.

**Flare:** An enlargement on the end of a piece of copper tubing used to connect the tubing to a fitting or another piece of tubing. It is usually formed at about a 45° angle.

**Flare fitting:** A type of soft tubing connector that requires that the tube be flared to make a mechanical seal.

**Flare nut:** The fitting placed over the tubing to clamp the flared tubing against another fitting.

**Flash gas:** The gas that is the result of the instantaneous evaporation of refrigerant in a pressure-reducing device. It cools the remaining refrigerant to the evaporation temperature which exists at the reduced pressure.

**Flash point:** The temperature of a combustible material at which there is enough vaporization to ignite the vapor, but not enough to support combustion.

**Float valve:** A valve that is actuated by a float in a liquid chamber.

**Flood:** To allow liquid refrigerant to flow into a part of the system.

**Flooded system:** A system in which the refrigerant enters the evaporator from a pressure-reducing valve and in which the evaporator is partly filled with liquid refrigerant.

**Flow meter:** An instrument used to measure the velocity or volume of fluid movement.

**Flue:** A gas or air passage through which the products of combustion escape to the atmosphere.

**Fluid:** A gas or liquid. Commonly defined as matter in any state which flows and takes the shape of its container.

**Flush:** To remove foreign materials or fluids from a refrigeration system through the use of refrigerant or other fluids.

**Flux:** A substance applied to the surfaces to be joined by brazing or soldering to keep oxides from forming during the joining process.

**Foaming:** The formation of foam in the crankcase as a result of liquid refrigerant being absorbed in the oil and rapidly evaporating. This is most likely to occur on compressor start-up.

**Foam leak detector:** Soap bubbles or a special

foaming liquid brushed over suspected areas to locate leaks.

**Force:** Accumulated pressure, expressed in pounds (grams).

**Forced convection:** The movement of a fluid by a force such as a fan or pump.

**Forced-feed oiling:** A lubrication system that uses a pump to force the oil to the surfaces of the moving parts.

**Freezer:** Any device for freezing perishable foods.

**Freezer alarm:** A bell or buzzer used on freezers to indicate when the storage temperature rises above a safe temperature.

**Freeze-up:** The formation of ice in the flow control device as a result of the presence of moisture in the system. Also, a frost formation on an evaporator which reduces the flow of air through the fins.

**Freezing:** The change of state from a liquid to a solid.

**Freezing point:** The temperature at which a liquid will change to a solid when heat is removed. The temperature at whch a given substance freezes.

**Freon:** The trade name for a family of fluoro-carbon refrigerants manufactured by E.I. DuPont de Nemours & Co., Inc.

**Frostback:** The flooding of liquid refrigerant from the evaporator into the suction line. This condition may or may not be accompanied by frost.

**Frost-free refrigerator:** A domestic refrigerator that operates with a predetermined defrost period.

**Frozen:** Term applied to water in the solid state. Also, seized due to the lack of lubrication, as frozen bearings, etc.

**Fuel oil:** A hydrocarbon oil as specified by U.S. Department of Commerce Commercial Standard CS12 or ATSM D296 whose flash point is not less than 100°F (37.8°C).

**Furnace:** That part of a heating plant in which the combustion process takes place.

**Fuse:** An electrical safety device consisting of a strip of fusible metal used to prevent over-loading of an electrical circuit.

**Fusible:** Capable of being melted.

**Fusible plug:** A safety plug used in refrig-erant containers that melts at a high tempera-ture to prevent excessive pressure from burst-ing the container.

**Gas:** The vapor state of a substance.

**Gasket:** A resilient or flexible material used between mating surfaces to provide a leakproof seal.

**Gas valve:** A device in a pipeline that con-trols the flow of gas to a burner on demand from the thermostat.

**Gauge:** An instrument used for measuring pressures both above and below atmospheric pressure.

**Gauge manifold:** A manifold that holds both the pressure and compound gauges, the valves which control the flow of fluids through the manifold ports, and charging hose connections.

**Gauge port:** The opening or connection pro-vided for the installation of gauges.

**Gauge pressure:** A measure of pressure taken with a gauge. Pressure measured from atmos-pheric pressure as opposed to absolute pressure.

**Grain:** A unit of weight equal to 1/7000 lb (.06480 gram) which is used to indicate the amount of moisture in the air.

**Grill:** A perforated covering for an air inlet or outlet usually made of wire screen, pro-cessed steel, or cast iron.

**Grommet:** A plastic or rubber doughnut-shaped protector placed in holes where wires or lines pass through panels to prevent damage to the wires or lines.

**Ground coil:** A heat exchanger that is buried in the ground; usually the outdoor coil on a heat pump system.

**Ground wire:** An electrical wire that will safely conduct electricity from a structure or piece of equipment to the ground in case of an electrical short.

**Halide refrigerants:** The family of synthetic

refrigerants that contain halogen chemicals.

**Halide torch:** A device used to detect refrigerant leaks in a system. A burner equipped with a source of fuel, a mixing chamber, a reactor plate, and an exploring tube. The reactor plate surrounds the flame. When the open end of the exploring tube is held near a refrigerant leak, some of the refrigerant is drawn into the mixing chamber where its presence changes the color of the flame.

**Halogens:** Substances that contain fluorine, chlorine, bromine, and iodine.

**Hanger:** A device used to support lines by attachment to a wall.

**Head:** Unit of pressure usually expressed in feet of water.

**Header:** A length of pipe large enough to carry the total volume to which several pipes are connected; used to carry fluid to the various points of use.

**Head pressure:** The pressure against which the compressor must deliver the gas.

**Head pressure control:** A pressure-operated control that opens the electrical circuit when the head pressure exceeds preset limits.

**Head velocity:** The height of a flowing fluid that is equivalent to its velocity pressure.

**Heat:** A form of energy produced through the expenditure of another form of energy.

**Heat of compression:** The heat developed within a compressor when a gas is compressed as in a refrigeration system.

**Heat content:** The amount of heat, usually stated in Btu (calories) per pound, absorbed by a refrigerant in raising its temperature from a predetermined level to a final condition and temperature. Where change of state is encountered, the latent heat necessary for the change is included.

**Heat exchanger:** Any device that removes heat from one substance and adds it to another.

**Heat of fusion:** The amount of heat required to change a solid to a liquid or a liquid to a solid with no change in temperature. *Also called* latent heat of fusion.

**Heat intensity:** The heat concentration in a substance; indicated by the temperature of the substance through the use of a thermometer.

**Heat lag:** The amount of time required for heat to travel through a substance which is heated on one side and not on the other.

**Heat leakage:** The flow of heat through a substance when a difference in temperature exists.

**Heat of the liquid:** The heat content of a liquid or the heat necessary to raise the temperature of a liquid from a predetermined level to a final temperature.

**Heat load:** The amount of heat, measured in Btu (calories) or watts, that must be added or removed by a piece of equipment during a 24-hour period.

**Heat pump:** A compression cycle system used to supply heat to a space by reversing the flow of refrigerant from the cooling cycle.

**Heat sink:** A relatively cold surface that is capable of absorbing heat usually used as control points.

**Heat transfer:** The movement of heat from one body to another. The heat may be transferred by radiation, conduction, or convection.

**Heat unit:** Usually refers to a Btu (calorie).

**Heat of the vapor:** The heat content of a gas or the heat necessary to raise the temperature of a liquid from a predetermined level to the boiling point plus the latent heat of vaporization necessary to convert a liquid to a gas.

**Heating coil:** A coil of piping used to transfer heat from a liquid.

**Heating control:** Any device that controls the transfer of heat.

**Heating medium:** A substance used to convey heat from the boiler, furnace, or other source of heat to the heating unit where the heat is dissipated.

**Heating surface:** The exterior surface of a heating unit.

**Heating value:** The amount of heat that may be released by the expenditure of energy. It is usually expressed in Btu/lb, Btu/gal, or cal/g.

**Hermetic compressor:** A unit in which the compressor and motor are sealed inside a housing. The motor operates in an atmosphere of refrigerant.

**Hertz (Hz):** The correct terminology for designating electrical cycles per second.

**Hg (mercury):** A silver-white heavy liquid metal. It is the only metal that is a liquid at room temperature.

**High pressure cut-out:** An electrical control switch operated by the pressure in the high pressure side of the system which automatically opens an electrical circuit when a predetermined pressure is reached.

**High side:** The part of the refrigeration system that contains the high pressure refrigerant. Also refers to the condensing unit which consists of the motor, compressor, condenser, and receiver mounted on one base.

**High side charging:** The process of introducing liquid refrigerant into the high side of a refrigeration system. The acceptable manner for placing the refrigerant into the system.

**High-vacuum pump:** A vacuum pump capable of creating a vacuum in the range of 1000 to 1 microns.

**Holding charge:** A partial charge of refrigerant placed in a piece of refrigeration equipment after dehydration and evacuation either for shipping or testing purposes.

**Holding coil:** That part of a magnetic starter or relay that causes the device to operate when energized.

**Horsepower:** A unit of power. The effort necessary to raise 33,000 lbs (14969 kg), a distance of 1 foot (.304 m) in one minute.

**Hot gas:** The refrigerant gas leaving the compressor.

**Hot gas bypass:** A connection from the compressor discharge directly into the suction side of a compressor. Sometimes used as a capacity control.

**Hot gas defrost:** A method of evaporator defrosting that uses the hot discharge gas to remove frost from the evaporator.

**Hot gas line:** The line that carries the hot compressed vapor from the compressor to the condenser.

**Hot junction:** The part of a thermocouple that releases heat.

**Hot water heating system:** A heating system in which water is the medium by which heat is carried through pipes from the boiler to the heating units.

**Hot wire:** A resistance wire used in starting relay that expands when heated and contracts when cooled. Also, an electrical lead whose voltage differs from that of the ground.

**Humidifier:** A device used to add vapor to the air or to any material.

**Humidistat:** An automatic control that is sensitive to humidity and is used for the automatic control of relative humidity.

**Humidity:** Moisture in the air.

**Humidity, absolute:** The weight of water vapor per unit volume of space occupied, expressed in grains of moisture per cubic foot of dry air.

**Humidity, relative:** The amount of moisture in the air expressed in terms of percentage of total saturation of the existing dry bulb temperature.

**Hunting:** The fluctuation caused by the controls attempting to establish an equilibrium against difficult conditions.

**Hydraulics:** A branch of physics dealing with the mechanical properties of water and other liquids in motion.

**Hydrocarbon:** An organic compound containing only hydrogen and carbon atoms in various combinations.

**Hydrometer:** An instrument used to measure the specific gravity of a liquid by use of floating elements.

**Hydronic:** Term applied to the process of heating and cooling using water as the heat transfer medium.

**Hygrometer:** An instrument used to measure relative humidity.

**Hygroscopic:** Term describing the ability of a substance to absorb and release moisture and to change its physical dimensions as its moisture content changes.

**ICC (Interstate Commerce Commission):** A governmental body that controls the design, construction, and shipping of pressure containers.

**Ice cream cabinet:** A commercial refrigerator that maintains approximately 0°F (−17.8°C) inside temperature and is used for the storage of ice cream.

**Ice water thickness:** A commercial classification applied to molded cork covering and similar insulation for refrigerant and chilled water lines. Somewhat thinner than "brine thickness."

**Idler:** A pulley used on some belt drives to allow belt adjustment and to eliminate belt vibration.

**Ignition transformer:** A transformer designed to provide a high voltage. Normally used to ignite the gas at a pilot burner or to ignite fuel oil at the burner by providing a high intensity spark.

**IME (Ice melting equivalent):** The amount of heat absorbed in melting ice at 32°F (0°C) per pound of ice. It is equal to 144 Btus (36.29 kcal) per pound or 288,000 Btus (72,576 kcal) per day.

**Impedance:** The opposition in an electrical circuit to the flow of an alternating current; similar to electrical resistance in a direct current circuit.

**Impeller:** The rotating part of a pump that causes the water to flow.

**Induced draft cooling tower:** A cooling tower in which the flow of air is created by one or more fans drawing the saturated air out of the tower.

**Induced magnetism:** Magnetism caused in a metal by magnetic induction.

**Inductive reactance:** The electromagnetic induction in a circuit that creates a counter emf as the applied current changes. It is in opposition to the flow of alternating current.

**Infiltration:** The leakage of air into a building or space.

**Inhibitor:** A substance that prevents a chemical reaction such as corrosion or oxidation to metals.

**Instrument:** A term broadly used to designate a device that is used for measuring, recording, indicating, and controlling equipment.

**Insulation, electrical:** A substance that has almost no free electrons.

**Insulation thermal:** A material that has a high resistance to heat flow.

**Interlock:** A device that prevents certain parts of an air conditioning or refrigeration system from operating when other parts of that system are not operating.

**Ion:** A group of atoms or an atom that is either positively or negatively charged electrically.

**IR drop:** An electrical term indicating the voltage drop in a circuit. It is expressed as amperes times resistance ($I$ x $R$).

**Isothermal:** Term describing a change in volume or pressure under constant temperature conditions.

**Isothermal expansion and contraction:** Expansion and contraction that takes place without a change in temperature.

**Isotron:** The refrigerants made by the Pennsalt Chemical Corp.

**Jacket:** To surround with a bath for temperature control or heat absorption, as in water jacketing.

**Joint:** The connecting point between two surfaces, as between two pipes.

**Journal, crankshaft:** That part of a shaft on which the bearing is in contact with the shaft.

**Junction box:** A box or container that houses a group of electrical terminals or connections.

**Kelvin scale:** A thermometer scale that is equal to the Celsius scale, and according to which absolute zero is 0 degrees, the equivalent of −273.16°C. Water freezes at 273.16°K. and boils at 373.16 K on this scale.

**Kilometer (km):** A metric unit of linear measurement equal to 1,000 meters.

**Kilowatt (kW):** A unit of electrical power indicating 1,000 watts.

**Kilowatt hour (kWh):** A unit of electrical energy equal to 1,000 watt hours.

**King valve:** A service valve on the liquid receiver outlet.

**Lag:** A delay in the response to some demand.

**Lamp, sterile:** A lamp with a high intensity ultraviolet ray which is used in food storage cabinets and air ducts to kill bacteria.

**Lantern gland (packing):** A packing ring inside a stuffing box which has perforations for the introduction or removal of oil.

**Lap:** To smooth a metal surface to a high polish or accuracy using a fine abrasive. Usually done on compressor valves and shaft seals.

**Latent heat:** The heat added to or removed from a substance to change its state but which cannot be measured by a change in temperature.

**Latent heat of condensation:** The heat removed from a vapor to change it to a liquid with no change in temperature.

**Latent heat of vaporization:** The quantity of heat required to change a liquid to a gas with no change in temperature.

**Leak detector:** A device used to detect leaks; usually a halide torch, soap bubbles, or an electronic leak detector.

**Limit control:** A safety device used in heating equipment to interrupt an electrical circuit to the main gas valve when an overtemperature condition occurs.

**Liquid:** A substance in which the molecules move freely among themselves, but do not tend to separate as in a vapor.

**Liquid absorbent:** A liquid chemical that has the ability to take on or absorb other fluids.

**Liquid charge:** Usually refers to the power element of temperature controls and thermo-static expansion valves. The power element and remote bulb are sometimes charged with a liquid rather than a gas.

**Liquid filter:** A very fine strainer used to remove foreign matter from the refrigerant.

**Liquid indicator:** A device located in the liquid line with a glass window through which the flow of liquid may be observed.

**Liquid line:** The line carrying the liquid refrigerant from the receiver or condenser-receiver to the evaporator.

**Liquid receiver:** A cylinder connected to the condenser outlet which is used to store liquid refrigerant.

**Liquid receiver service valve:** A two- or three-way manually operated valve located at the receiver outlet which is used for installation and service operations. *Also called* a king valve.

**Liquid sight glass:** A glass bullseye installed in the liquid line to permit visual inspection of the liquid refrigerant. Used primarily to detect bubbles in the liquid, indicating a shortage of refrigerant in the system. Also, a liquid indicator.

**Liquid stop valve:** A magnetically operated valve generally used to control the flow of liquid to an evaporator, but may also be used wherever on–off control is permissible. A solenoid electrical winding controls the action of the valve.

**Liquid strainer:** *See* Liquid filter.

**Liquid-vapor valve:** A dual hand valve used on refrigerant cylinders to release either gas or liquid from the cylinder.

**Liter:** A metric unit of volume measurement equal to 61.03 cubic inches.

**Load:** The required rate of heat removal. Heat per unit of time that is imposed on the system by a particular job.

**Locked rotor amps (LRA):** The amperage flowing in the circuit to a motor-driven apparatus when the rotor of the motor is locked to prevent its movement. This amperage is typically as much as 6 times the full load current of the motor and 4 to 5 times the full load current in hermetic compressors.

**Low pressure control:** A pressure-operated switch in the suction side of a refrigeration system that opens its contacts to stop the compressor at a given cut-out setting.

**Low side:** Composed of the parts of a refrigeration system in which the refrigerant pressure corresponds to the evaporator pressure.

**Low side charging:** The process of introducing refrigerants into the low side of the system. Usually reserved for the addition of a small amount of refrigerant after repairs.

**Low side pressure:** The pressure in the low side of the system.

**LP Fuel:** Liquified petroleum used as a fuel gas for heating purposes.

**Machine:** A piece of equipment, sometimes the total unit.

**Machine room:** An area where the refrigeration equipment, with the exception of the evaporator, is installed.

**Magnetic across-the-line starter:** A motor starter or switch that allows full line voltage to the motor windings when engaging.

**Magnetic clutch:** A pulley operated by electromagnetism to connect or disconnect a source of driving power to a compressor.

**Magnetic field:** An area in which magnetic lines of force exist.

**Magnetic gasket:** A rubber type material with small magnets inside the gasket that keeps the door tightly closed.

**Magnetism:** The attraction between a magnet and materials which are made of iron or are nickel-coated.

**Male thread:** The external threat on a pipe, fitting, or valve that makes screw connections.

**Manifold:** The portion of the refrigerant main in which several branch lines are joined together. Also, a single piece in which there are several fluid paths.

**Manifold, discharge:** A device used to collect compressed refrigerant from the various cylinders of a compressor.

**Manifold, service:** A chamber equipped with gauges, manual valves, and charging hoses which is used in servicing refrigeration units.

**Manometer:** An instrument for measuring small pressure. Also, a "U" shaped tube partly filled with liquid.

**Manual shut-off valve:** A hand-operated device that stops the flow of fluids in a piping system.

**Manual starter:** A hand-operated motor switch equipped with an overload trip mechanism.

**Mass:** A quantity of matter held together to form one body.

**Master switch:** The main switch that controls the starting and stopping of the entire system.

**Mean effective pressure (MEP):** The average pressure on a surface exposed to a varying pressure.

**Mechanical efficiency:** The ratio of work done by a machine to the energy used to do it.

**Mechanical refrigeration:** A term usually used to distinguish a compression system from an absorption system.

**Megohm:** A measure of electrical resistance. One megohm equals one million ohms.

**Megohmmeter:** An instrument used to measure extremely high resistances.

**Melt:** To change state from a solid to a liquid.

**Melting point:** The temperature, measured at atmospheric pressure, at which a substance will melt.

**Mercury bulb:** A glass tube that uses a small amount of mercury to make or break an electric circuit.

**Met:** A measure of the heat released from a human at rest. It is equal to 18.4 $Btu/ft^2/hr$ (50 $kcal/m^2/hr$).

**Meter:** An instrument used for measuring. Also, a unit of length in the metric system.

**Metric system:** The decimal system of measurement.

**Micro:** One-millionth of a specified unit.

**Microfarad:** A unit of electrical capacitance equal to 1/1,000,000 of a farad.

**Micrometer:** A precision instrument for measuring with an accuracy of .001 to .0001 in. (.0254 to .00254 mm).

**Micron:** A unit of length in the metric system which is equal to 1/1000 mm.

**Micron gauge:** An accurate instrument used for measuring vacuum which is very close to a perfect vacuum.

**Milli-:** One-thousandth of a specified unit. For example, millivolt means 1/1000 volt.

**Miscibility:** The ability of several substances to be mixed together.

**Modulating:** Refers to a device that tends to adjust by small increments rather than being full-on or full-off.

**Modulating control:** A type of control system characterized by a control valve or motor that regulates the flow of air, steam, or water in response to a change in conditions at the controller. It operates upon partial degree variations in the medium to which the controller is exposed.

**Mol:** The unit of weight for mass.

**Molecular:** Pertaining to or consisting of molecules.

**Molecular weight:** The average weight of a molecule of a substance; measured against an oxygen atom.

**Molecule:** The smallest particle of a substance that can exist by itself.

**Molliers diagram:** A graph indicating refrigerant pressure, heat, and temperature properties.

**Monochlorodifluoromethane:** A refrigerant that is better known as Freon–22 or R–22. Its chemical formula is $CHCLF_2$; the cylinder color code is green.

**Motor:** A device that converts electrical energy to mechanical energy.

**Motor burnout:** A shorted condition in which the insulation of an electric motor has deteriorated by overheating.

**Motor, capacitor:** A single-phase induction type motor that utilizes an auxiliary starting winding (a phase winding) connected in series with a capacitor to provide better starting or running characteristics.

**Motor, capacitor start:** An induction motor having separate starting windings. Similar to the split-phase motor except that the capacitor start motor has an electrical capacitor connected into the starting winding for added starting torque.

**Motor, capacitor start and run:** A motor similar to the capacitor start motor except that the capacitor and start winding are designed to remain in the circuit at all times, thus eliminating the switch used to disconnect the start winding.

**Motor control:** A device used to start and/or stop an electric motor or a hermetic motor compressor at certain temperature and/or pressure conditions.

**Motor, shaded pole:** A small induction motor with a shading pole used for starting; has a very low starting torque.

**Motor, split phase:** An induction motor with a separate winding for starting.

**Motor starter:** A series of electrical switches which are normally operated by electromagnetism.

**Movable contact:** The member of a contact pair that is moved directly by the actuating system.

**Muffler:** A device used in the hot gas line to silence the compressor discharge surges.

**Mullion:** A stationary frame member located between two doors.

**Mullion heater:** An electrical heating element mounted in the mullion to keep it from sweating.

**Multiple system:** A refrigeration system that has several evaporators connected to one condensing unit.

**Natural convection:** The movement of a fluid which is caused by temperature differences.

**Natural draft cooling tower:** A cooling tower in which the flow of air depends upon natural air currents or a breeze. Generally applied where the spray water is relatively hot and will cause some convection currents.

**Neoprene:** A synthetic rubber that is resistant to refrigerants and oils.

**Neutralizer:** A substance used to counteract the action of acids.

**Neutron:** The core of an atom; it has no electrical potential (electrically neutral).

**No-frost refrigerator:** A low temperature refrigerator cabinet in which no frost or ice collects on the evaporator surfaces or the materials stored in the cabinet.

**Nominal size tubing:** Tubing whose inside diameter is the same as iron pipe of the same size.

**Noncondensable gas:** Any gas, usually in a refrigeration system, that cannot be condensed at the temperature and pressure at which the refrigerant will condense, and therefore requires a higher head pressure.

**Nonferrous:** Metals and metal alloys that contain no iron.

**Nonfrosting evaporator:** An evaporator that never collects frost or ice on its surface.

**Normal charge:** A charge that is part liquid and part gas under all operating conditions.

**Normally closed contacts:** A contact pair that is closed when the device is in the de-energized condition.

**Normally open contacts:** A contact pair that is open when the device is in the de-energized condition.

**North pole, magnetic:** The end of a magnet from which the magnetic lines of force flow.

**Odor:** The contaminants in the air that affect the sense of smell.

**Off cycle:** The period when equipment, specifically a refrigeration system, is not in operation.

**Ohm (R):** A unit of measurement of electrical resistance. One ohm exists when one volt causes one ampere to flow through a circuit.

**Ohmmeter:** An instrument used to measure electrical resistance in ohms.

**Ohms' Law:** A mathematical relationship between voltage, current, and resistance in an electric circuit. It is stated as follows: Voltage $(E)$ = Amperes $(I) \times$ Ohms $(R)$.

**Oil binding:** A condition in which a layer of oil on top of liquid refrigerant may prevent it from evaporating at its normal pressure and temperature.

**Oil check valve:** A check valve installed between the suction manifold and the crankcase of a compressor to permit oil to return to the crankcase but to prevent the exit of the oil from the crankcase upon starting.

**Oil, compressor lubricating:** A highly refined lubricant made especially for refrigeration compressors.

**Oil, entrained:** Oil droplets carried by high velocity refrigerant gas.

**Oil equalizer:** A pipe connection between two or more pieces of equipment made in such a way that the pressure, or fluid level, in each piece is maintained equally.

**Oil filter:** A device in the compressor used to remove foreign material from the crankcase oil before it reaches the bearing surfaces.

**Oil level:** The level in a compressor crankcase at which oil must be carried for proper lubrication.

**Oil loop:** A loop placed at the bottom of a riser to force oil to travel up the riser.

**Oil pressure failure control:** A device that acts to shut off a compressor whenever the oil pressure falls below a predetermined point.

**Oil pressure gauge:** A device used to show the pressure of oil developed by the pump within a refrigeration compressor.

**Oil pump:** A device that provides the source of power for force-feed lubrication systems in refrigeration compressors.

**Oil return line:** The line that carries the oil

collected by the oil separator back to the compressor crankcase.

**Oil separator:** A device that separates oil from the refrigerant and returns it to the compressor crankcase.

**Oil sight glass:** A glass "bullseye" in the compressor crankcase that permits visual inspection of the compressor oil level.

**Oil sludge:** Usually a thick, slushy substance formed by contaminated oils.

**Oil trap:** A low spot, sag in the refrigerant lines, or space where oil will collect. Also, a mechanical device for removing entrained oil.

**On cycle:** The period when the equipment, specifically refrigeration equipment, is in operation.

**Open circuit:** An electrical circuit that has been interrupted to stop the flow of electricity.

**Open compressor:** A compressor that uses an external drive.

**Open display case:** A commercial refrigerator cabinet designed to maintain its contents at the desired temperatures even though the cabinet is not closed.

**Open type system:** A refrigeration system that uses a belt-driven or direct-coupling-driven compressor.

**Operating cycle:** A sequence of operations under automatic control intended to maintain the desired conditions at all times.

**Operating pressure:** The actual pressure at which the system normally operates.

**Orifice:** A precision opening used to control fluid flow.

**Output:** The amount of energy that a machine is able to produce in a given period of time.

**Outside air, Fresh air:** Air from outside of the conditioned space.

**Overload:** A load greater than that for which the system or machine was designed.

**Overload protector:** A device designed to stop the motor should a dangerous overload condition occur.

**Overload relay:** A thermal device that opens its contacts when the current through a heater coil exceeds the specified value for a specified time.

**Oxidize:** To burn, corrode, or rust.

**Oxygen:** An element in the air that is essential to animal life.

**Package units:** A complete set of refrigeration or air conditioning components located in the refrigerated space.

**Packing:** A resilient impervious material placed around the stems of certain types of valves to prevent leakage. Also, the slats or surfaces in cooling towers designed to increase the water to air contact.

**Packless valve:** A valve that does not use a packing to seal around the valve stem.

**Partial pressure:** A condition occurring when two or more gases occupy a given space and each gas creates a part of the total pressure.

**Pascal's law:** States that a pressure imposed upon a fluid is transmitted equally in all directions.

**Performance:** A term frequently used to mean output or capacity, as performance data.

**Performance factor:** The ratio of heat removed by a refrigeration system to the heat equivalent of the energy required to do the job.

**Permanent magnet:** A material whose molecules are aligned to enable it to produce its own magnetic field. Also, a piece of metal which has been permanently magnetized.

**Phase:** A definite part of an operation during a cycle.

**Photoelectricity:** A physical condition by which an electrical flow is generated by light waves.

**Pilot control:** A valve arrangement used in an evaporator pressure regulator to sense the pressure in the suction line and to regulate the action of the main valve.

**Pilot control, external:** A method by which the internal connection of the pilot is plugged and an external connection is provided to make it possible to use an evaporator pressure regulator as a suction stop valve as well.

**Pilot tube:** A device that measures air velocities.

**Piston:** A disc that slides in a cylinder and is connected with a rod to exert pressure upon a fluid inside the cylinder.

**Piston displacement:** The volume displaced in a cylinder by the piston as it travels the full length of its stroke.

**Pitch:** The slope of a pipe line used to enhance drainage.

**Plenum chamber:** An air compartment to which one or more pressurized distributing ducts are connected.

**Polystyrene:** A plastic used as insulation in some refrigeration cabinets.

**Potential, electric:** An electrical force that causes electrons to move along a conductor or through a resistance.

**Potential relay:** A starting relay that opens on high voltage on its coil and closes on low voltage.

**Potentiometer:** A wire-wound coil used for measuring or controlling by sensing small changes in electrical resistances.

**Power:** The time rate of doing work.

**Power element:** The sensitive element of a temperature-operated control.

**Power factor:** The correction factor for the changing of current and voltage values of alternating current electricity.

**Ppm (Parts per million):** A unit of concentration in solutions.

**Precooler:** A cooler used to remove sensible heat before shipping, storing, or processing.

**Pressure:** The force exerted per unit of area.

**Pressure, absolute:** The pressure measured above an absolute vacuum.

**Pressure, atmospheric:** The pressure exerted by the earth's atmosphere. Under standard conditions at sea level, atmospheric pressure is 14.7 psia or 0 psig (1.03 kg/cm$^2$).

**Pressure, condensing:** *See* Pressure, discharge.

**Pressure, crankcase:** The pressure in the crankcase of a reciprocating compressor.

**Pressure, discharge:** The pressure against which the compressor must deliver the refrigerant vapor.

**Pressure drop:** The loss of pressure due to friction of lift.

**Pressure gauge:** An instrument that measures the pressure exerted by the contents of a container.

**Pressure, gauge:** The pressure existing above atmospheric pressure. Gauge pressure is, therefore, 14.7 psi (1.03 kg/cm$^2$) less than the corresponding absolute pressure.

**Pressure, heat diagram:** A graph representing a refrigerant's pressure, heat, and temperature properties. Also, a Mollier's diagram.

**Pressure, limiter:** A device that remains closed until a predetermined pressure is reached and then opens to release fluid to another part of the system or opens an electrical circuit.

**Pressure motor control:** A control that opens and closes an electrical circuit as the system pressures change.

**Pressure operated altitude valve (POA):** A device used to maintain a constant low side pressure regardless of the altitude of operation.

**Pressure regulating valve:** A valve that maintains a constant pressure on its outlet side regardless of how much the pressure varies on the supply side of the valve.

**Pressure regulator, evaporator:** An automatic pressure-regulating valve installed in the suction line to maintain a predetermined pressure and temperature in the evaporator.

**Pressure, saturation:** The pressure at which gas at any specific temperature is saturated.

**Pressure, suction:** The pressure forcing the gas to enter the suction inlet of a compressor.

**Pressure switch:** A switch operated by a rise or fall in pressure.

**Pressure tube:** A small line carrying pressure

to the sensitive element of a pressure controller.

**Pressure water valve:** A device that controls water flow through a water-cooled condenser in response to the head pressure.

**Primary control:** A device that directly controls the operation of a fuel oil burner.

**Prime surface:** A heating surface whose heating medium is on one side and whose air or extended surface is on the other side.

**Process tube:** A length of tubing fastened to the dome of a hermetic compressor; used when servicing the unit.

**Propane:** A volatile hydrocarbon that may be used as a refrigerant, but is used primarily as a heating fuel.

**Protector, circuit:** An electrical device that opens a circuit when an excessive electrical load occurs.

**Proton:** The particle of an atom that has a positive charge.

**Psychrometer or Wet bulb hygrometer:** An instrument used to measure the relative humidity of air.

**Psychrometric chart:** A graphic representation of the properties of mixtures and water vapor.

**Psychrometric measurement:** The measurement of temperature, pressure, and humidity by means of a psychrometric chart.

**Psychrometry:** The study of water vapor and air mixtures.

**Pump:** Any one of various machines used to force fluids through pipes from one place to another.

**Pump, centrifugal:** A pump that produces fluid velocity by centrifugal force.

**Pumpdown:** The reduction of pressure within a refrigeration system.

**Purge:** The discharge of impurities and non-condensable gases to the atmosphere.

**Pyrometer:** An instrument used to measure high temperatures.

**Quench:** To submerge a hot, solid object in a cooling fluid such as oil.

**Quick connect coupling:** A device that permits fast and easy connection of two refrigerant lines by use of compression fittings.

**Radiant heating:** A heating system in which the heating surfaces radiate heat directly into the space to be conditioned.

**Radiation:** The transmission of heat through space by wave motion.

**Ram air:** The air forced through the radiator and condenser by the movement of an automobile along the road.

**Range:** A change within limits of the settings of pressure or temperature control.

**Rankine scale:** The name given to the absolute Fahrenheit scale. The zero point on this scale is −460°F.

**Rating:** The assignment of capacity.

**Reactance:** The part of impedance of an alternating current due to capacitance or inductance or both.

**Receiver, auxiliary:** An extra vessel used to supplement the capacity of the receiver when additional storage volume is necessary.

**Receiver-drier:** A cylinder in an automotive air conditioning system used for storing liquid refrigerant; it also contains a quantity of desiccant and acts as a drier.

**Reciprocating:** Term describing back-and-forth motion in a straight line.

**Recirculated air:** Return air that is passed through the conditioning unit before being returned to the conditioned space.

**Recording ammeter:** An instrument that uses a pen to record on a piece of moving paper the amount of current flowing to a unit.

**Recording thermometer:** A temperature-sensing instrument that uses a pen to record on a piece of moving paper the temperature of a refrigerated space.

**Rectifier, electric:** An electrical device that converts alternating current to direct current.

**Reed valve:** A piece of thin, flat, tempered steel plate fastened to the valve plate.

**Refrigerant:** A substance that absorbs heat as it expands or evaporates. In general; any substance used as a medium to extract heat from another body.

**Refrigerant control:** A control that meters the flow of refrigerant from the high side to the low side of the refrigeration system. It also maintains a pressure difference between the high pressure and low pressure sides of the system while the unit is running.

**Refrigerant tables:** Tables that show the properties of saturated refrigerants at various temperatures.

**Refrigerant velocity:** The movement of gaseous refrigerant required to entrain oil mist and carry it back to the compressor.

**Refrigerating capacity:** The rate at which a system can remove heat. Usually stated in tons or Btu per hour.

**Refrigerating effect:** The amount of heat a given quantity of refrigerant will absorb in changing from a liquid to a gas at a given evaporating pressure.

**Refrigeration:** In general, the process of removing heat from an enclosed space and maintaining that space at a temperature lower than its surroundings.

**Refrigeration cycle:** The complete operation involved in providing refrigeration.

**Register:** A combination grill and damper assembly placed over the opening at the end of an air duct to direct the air.

**Reheat:** To heat air after dehumidification if the temperature is too low.

**Relay:** A device that is made operative by a variation in the condition of one electric circuit to effect the operation of other devices in the same or another circuit.

**Relay, control:** An electromagnetic device that opens or closes contacts when its coil is energized.

**Relay, thermal overload:** A thermal device

that opens its contacts when the current through a heater coil exceeds the specified value for a given time.

**Relief valve:** A valve designed to open at excessively high pressures to allow the refrigerant to escape safely.

**Reluctance:** A force that works as resistance to the passage of magnetic lines of force.

**Remote bulb:** A part of the expansion valve. The remote bulb assumes the temperature of the suction gas at the point where the bulb is secured to the suction line. Any change in the suction gas superheat at the point of bulb application tends to operate the valve in a compensating direction to restore the superheat to a predetermined valve setting.

**Remote bulb thermostat:** A controller that is sensitive to changes in temperature. A thermostat.

**Remote power element control:** A control whose sensing element is located separate from the mechanism it controls.

**Remote system:** A refrigeration system whose condensing unit is located away from the conditioned space.

**Resistance:** Opposition to the flow of electric current. Measured in Ohms.

**Resistor:** A device offering electrical resistance in an electrial circuit for protection or control.

**Restrictor:** A reduced cross-sectional area of pipe that produces resistance or a pressure drop in a refrigeration system.

**Return air:** Air taken from the conditioned space and brought back to the conditioning equipment.

**Reverse cycle defrost:** A method of reversing the flow of refrigerant through an evaporator for defrosting purposes.

**Reversing valve:** A valve that reverses the direction of refrigerant flow depending on whether heating or cooling is needed.

**Riser:** A vertical tube or pipe that carries refrigerant in any form from a lower to a higher level.

**Riser valve:** A valve used for the manual control of the flow of refrigerant in a vertical pipe.

**Rotary blade compressor:** A rotary compressor that uses moving blades as cylinders.

**Rotary compressor:** A mechanism that pumps fluids by means of a rotating motion.

**Rotor:** The rotating or turning part of an electric motor or generator.

**Running time:** The amount of time that a condensing unit operates per hour.

**Running winding:** The electrical winding in a motor through which current flows during normal operation of the motor.

**Saddle valve:** A valve equipped with a body that can be connected to a refrigerant line. Also, a tap-a-line valve.

**Safety control:** Any device that allows the refrigeration or air conditioning unit to stop when unsafe pressures or temperatures exist.

**Safety factor:** The ratio of extra strength or capacity to the calculated requirements to ensure freedom from breakdown and ample capacity.

**Safety plug:** A device that releases the contents of a container to prevent rupturing when unsafe pressures or temperatures exist.

**Safety valve:** A quick opening safety valve used for the fast relief of excessive pressure in a container.

**Saturation:** A condition existing when a substance contains the maximum amount of another substance that it can hold at that particular pressure and temperature.

**Schrader valve:** A spring-loaded valve that permits fluid to flow in only one direction when the center pin is depressed.

**Scotch yoke:** A mechanism that changes rotary motion into reciprocating motion; it is used to connect a crankshaft to a piston in some types of refrigeration compressors.

**Screw pump:** A compressor that is constructed of two mating revolving screws.

**Seat:** The portion of a valve against which the valve button presses to effect a shutoff.

**Seat, front:** The part of refrigeration valve that forms a seal with the valve button when the valve is in the closed position.

**Second Law of Thermodynamics:** States that heat will flow from an object at a higher temperature to an object at a lower temperature.

**Self-contained air conditioning unit:** An air conditioner containing a condensing unit, evaporator, fan assembly, and complete set of operating controls within its casing.

**Self-inductance:** A magnetic field induced in a conductor carrying a current.

**Semiconductor:** A class of solids that have the ability to conduct electricity. This class lies between that of conductors and that of insulators.

**Semihermetic compressor:** A hermetic compressor on which minor field service operations can be performed.

**Sensible heat:** Heat that causes a change in temperature of a substance but not a change of state.

**Sensible heat ratio:** The percentage of total heat removed which is sensible heat. It is usually expressed as a decimal and is the quotient of sensible heat removed divided by total heat removed.

**Sensor:** An electronic device that undergoes a physical change or a characteristic change as surrounding conditions change.

**Sequence controls:** A group of controls that act in a series or in a timed order.

**Serpentining:** The arrangement of tubes in a coil to provide circuits of the desired length. Intended to keep pressure drop and velocity of the substance passing through the tubes within the desired limits.

**Service valve:** A shut-off valve intended for use only during shipment, installation, or service procedures.

**Servo:** A servomechanism is a low-power device, either electrical, hydraulic, or pneumatic, that puts into operation and controls a more complex or more powerful system. A servo motor supplies power to a servomechanism.

**Shell and coil:** A designation for heat exchangers, condensers, and chillers consisting of a tube coil within a shell or housing.

**Shell and tube:** A designation for heat exchangers, condensers, and chillers consisting of a tube bundle within a shell of casing.

**Short circuit:** An electrical condition occurring when part of the circuit is in contact with another part and causes all or part of the current to take a wrong path.

**Short cycle:** Starting and stopping that occurs too frequently. A short on-cycle and a short off-cycle.

**Shroud:** A housing over a condenser, evaporator, or fan to increase air flow.

**Silica gel:** An absorbent chemical compound used as a drying agent. When heat is applied, the moisture is released and the compound may be reused.

**Silver brazing:** A brazing process in which the brazing alloy contains some silver.

**Sine wave:** A wave form of a single frequency alternating current. The wave that has this displacement is the sine of an angle which is proportional to time or distance.

**Single-phase motor:** An electric motor that operates on a single phase alternating current.

**Single-pole, double-throw switch:** An electric switch with one blade and two sets of contact points (SPDT).

**Single-pole, single-throw switch:** An electric switch with one blade and one set of contact points (SPST).

**Skin condenser:** A condenser that uses the outer shell of a cabinet as the heat radiating medium. Usually used in domestic refrigeration.

**Sling psychrometer:** A device with a dry bulb and a wet bulb thermometer that is moved rapidly through the air to measure humidity.

**Slugging:** A condition in which a quantity of liquid enters the compressor cylinder, causing a hammering noise.

**Smoke test:** A test to determine how complete the combustion process is.

**Solar heat:** Heat created by energy waves, or rays, from the sun.

**Solder:** To join two metals by the adhesion process using a melting temperature less than 800°F (427°C).

**Solenoid valve:** An electromagnetic coil with a moving core which operates a valve.

**South pole, magnetic:** The part of a magnet that receives the magnetic lines of force flow.

**Specific gravity:** The weight of a liquid as compared to water, which is assigned the value of 1.

**Specific heat:** The heat required to raise the temperature of one pound of a substance one degree F (.56°C).

**Specific volume:** The volume per unit of mass. Usually expressed as cubic feet per pound ($m^3$/gram).

**Splash system, lubrication:** The method of lubricating moving parts by agitating or splashing the oil around in the crankcase.

**Split-phase motor:** A motor with two windings. Both windings are used in the starting of the motor. One is disconnected by a centrifugal switch after a given speed is reached by the motor. The motor then operates on only one winding.

**Split system:** A refrigeration or air conditioning system whose condensing unit is located apart from the evaporator.

**Squirrel cage:** A centrifugal fan.

**Standard atmosphere:** A condition existing when the air is at 14.7 psia (1.03 kg/cm²) pressure, 68°F (20°C) temperature and 36% relative humidity.

**Standard conditions:** Conditions used as a basis for air conditioning calculations. They are: temperature of 68°F (20°C), pressure of 29.92 inches of mercury (Hg) (1.03 kg/cm²), and a relative humidity of 30%.

**Starting relay:** An electrically operated switch used to connect or disconnect the starting winding of an electric motor.

**Starting winding:** A winding in an electric motor used only briefly during the starting period to provide the extra torque required during this period.

**Static head:** The pressure due to the weight of a fluid in a vertical column or, more generally, the resistance due to lift.

**Steam:** Water heated to the vapor state.

**Steam heating:** A heating system that uses steam as the heating medium.

**Steam trap:** A device that allows the passage of condensed water vapor but prevents the passage of steam.

**Strainer:** A device, such as a screen or filter, used to remove foreign particles and dirt from the refrigerant.

**Stratification of air:** A condition existing when there is little or no air movement in the room.

**Subcooling:** The cooling of a liquid refrigerant below its condensing temperature.

**Subcooling coil:** A supplementary coil in an evaporative condenser, usually a coil or loop immersed in the spray water tank, that reduces the temperature of the liquid leaving the condenser.

**Sublimation:** The change of state from a solid to a vapor without the intermediate liquid state.

**Suction line:** The pipe that carries the refrigerant vapor from the evaporator to the compressor.

**Suction pressure:** Low side pressure. Same as evaporator pressure.

**Suction riser:** A vertical tube or pipe that carries suction gas from an evaporator on a lower level to a compressor on a higher level.

**Suction service valve:** A two-way manually operated valve installed on the compressor inlet and used during service operations.

**Suction side:** The low pressure side of the system, extending from the flow control device through the evaporator and to the compressor suction valve.

**Suction temperature:** The boiling temperature of the refrigerant corresponding to the suction pressure of the refrigerant.

**Sump:** The reservoir of an evaporative condenser in which spray water is collected before recirculation.

**Superheat:** A temperature increase above the saturation temperature or above the boiling point.

**Superheated gas:** A gas whose temperature is higher than the evaporation temperature at the existing pressure.

**Swage:** To enlarge one end of a tube so that the end of another tube of the same size will fit inside.

**Switch, auxiliary:** An accessory switch available for most damper motors and control operators that can be arranged to open or close an electric circuit whenever the control motor reaches a certain position.

**Switch, disconnect:** A switch usually provided for a motor that completely disconnects the motor from the source of electric power.

**Tail pipe:** The outlet tube from an evaporator.

**Temperature:** The measure of heat intensity.

**Temperature, ambient:** The temperature of the air around the object under consideration.

**Temperature, condensing:** The temperature of the fluid in the condenser at the time of condensation.

**Temperature, discharge:** The temperature of the gas leaving the compressor.

**Temperature, dry bulb:** The temperature of the air measured with an ordinary thermometer. It indicates only sensible heat changes.

**Temperature, entering:** The temperature of a substance as it enters a piece of apparatus.

**Temperature, evaporating:** The temperature at which a fluid boils under the existing pressure.

**Temperature, final:** The temperature of a substance as it leaves a piece of apparatus.

**Temperature, saturation:** The boiling point of a refrigerant at a given pressure. It is considered to be the evaporator temperature in refrigeration.

**Temperature, suction:** The temperature of the gas as it enters the compressor.

**Temperature, wet bulb:** The temperature of the air measured with a thermometer having a bulb covered with a moistened wick.

**Test charge:** An amount of refrigerant vapor forced into a refrigeration system to test for leaks.

**Test light:** A light provided with test leads that is used to test electrical circuits to determine if electricity is present.

**Therm:** A symbol used in the gas industry; represents 100,000 Btus (25,200 kcal).

**Thermal overload element:** The alloy piece used for holding an overload relay closed; it melts when the current draw is too great.

**Thermal relay:** A heat-operated relay that opens or closes the starting circuit to an electric motor. *Also called* hot wire relay.

**Thermistor:** Basically, a semiconductor whose electrical resistance varies with the temperature.

**Thermocouple:** A device that generates electricity from heat using two pieces of dissimilar metals that are welded together. When the welded junction is heated, a voltage is produced.

**Thermodisc defrost control:** An electrical bimetal disc switch that is controlled by changes in temperature.

**Thermometer:** An instrument used for measuring sensible temperature.

**Thermometer well:** A small pocket or recess in a pipe or tube designed to provide good thermal contact with a test thermometer.

**Thermopile:** A number of thermocouples connected in series to produce a higher voltage than can be produced by a single thermocouple.

**Thermostat:** A device used to control equipment in response to temperature change. A temperature-sensitive controller.

**Thermostat, modulating:** A temperature controller that employs a potentiometer winding instead of switch contacts.

**Thermostat, on-and-off:** A thermostat designed to open or close an electrical circuit in response to a temperature change.

**Thermostatic expansion valve:** A flow control device operated by both the temperature and pressure in an evaporator.

**Three-phase:** Term designating operation by means of a combination of three alternating current circuits which differ in phase by ⅓ cycle.

**Three-way valve:** A multi-orifice flow control valve with a common and two alternate paths.

**Time delay relay:** A relay actuated after a predetermined time from the point of impulse.

**Timer:** A clock-operated mechanism used to open and close an electric circuit on a predetermined schedule.

**Ton of refrigeration:** A unit of refrigeration capacity measuring 200 Btus (50.4 kcal) per minute, 12,000 Btus (3,024 kcal) per hour, or 288,000 Btus (72,576 kcal) per day. It is so named because it is equivalent in cooling effect to melting one ton of ice in 24 hours.

**Torque:** A turning or twisting force.

**Torque, starting:** The amount of torque available to start and accelerate a loaded motor.

**Total heat:** The sum of sensible and latent heat contained in a substance.

**Transducer:** A device activated by a change in power from one source for the purpose of supplying power in another form to a second system.

**Transformer:** An electrical device that transforms electrical energy from one circuit to another by electrical induction.

**Triple point:** A pressure–temperature condition at which a substance is in equilibrium. Occurs in solid, liquid, and vapor states.

**Tube-within-a-tube:** Heat exchange surfaces or condensers constructed of two concentric tubes.

**Tubing:** A pipe with a thin wall.

**Two-temperature valve:** A pressure-operated valve used in the suction line on multiple refrigeration installations to maintain the evaporators at different temperatures.

**Ultraviolet:** Invisible radiation waves whose frequencies are shorter than visible wavelengths and longer than X-ray wavelengths.

**Unitary System:** A combination heating and cooling system that is factory assembled in one package and is usually designed for conditioning one room or space.

**Universal motor:** An electric motor designed to operate on either ac or dc current.

**Urethane foam:** A type of insulation that is foamed between the shell and the liner of cabinets.

**Useful oil pressure:** The difference in pressure between the discharge and suction sides of the compressor oil pump.

**Vacuum:** A reduction in pressure below atmospheric pressure. Usually stated in inches of mercury or microns.

**Vacuum pump:** A pump used to exhaust a system. Also, a pump designed to produce a vacuum in a closed system or vessel.

**Valve:** A device that controls the flow of fluid.

**Valve, cap seal:** A manual valve whose stem is protected by a tightly fitting cap.

**Valve, charging:** A valve located on the liquid line, usually near the receiver, through which refrigerant may be charged into the system.

**Valve, check:** A valve that permits flow in one direction only; it is designed to close against backward flow.

**Valve, condenser shut-off:** A valve located in the hot gas or discharge line at the inlet of the condenser.

**Valve, cylinder discharge:** The valve in a compressor through which the gas leaves the cylinder.

**Valve, cylinder suction:** The valve in a compressor through which the gas enters the cylinder.

**Valve, expansion:** A type of refrigerant flow control device that maintains a constant pressure or temperature in the low side of the system.

**Valve plate:** The part of the compressor that contains the compressor valves and ports and is located between the compressor body and cylinder head.

**Valve port:** The passage in a valve that opens and closes to control the flow of a fluid in accordance with the relative position of the valve button to the valve seat.

**Valve, purge:** A valve through which noncondensable gases may be purged from the condenser or receiver.

**Vapor:** A fluid in the gaseous state formed by evaporation of the liquid.

**Vapor barrier:** A sheet of thin plastic or metal foil used in an air conditioned structure to prevent water vapor from penetrating the insulation.

**Vaporization:** The changing of a liquid to a gas.

**Vapor pressure:** The pressure imposed by a vapor.

**Vapor, saturated:** A vapor in a condition that will result in the condensation into droplets of liquid if the vapor temperature is reduced.

**Variable pitch pulley:** A pulley that can be adjusted to provide different drive pulley ratios.

**V-belt:** A type of belt commonly used in refrigeration work to transfer power from the motor to the driven apparatus.

**Velocimeter:** An instrument used to measure air speeds on a direct reading scale.

**Velocity:** The speed or rapidity of motion.

**Vent:** A port or opening through which pressure is relieved.

**Voltmeter:** An instrument used to measure electrical voltage.

**Volumetric efficiency:** The relationship between the actual performance of a compressor and the calculated performance of the compressor based on its displacement.

**Walk-in cooler:**  A large commercial refrigerated space often found in supermarkets or places for wholesale meat distribution.

**Water-cooled condenser:**  A condenser cooled by water rather than air.

**Water treatment:**  The treatment of water with chemicals to reduce its scale-forming properties or to change other undesirable characteristics.

**Watt:**  A unit of electrical power.

**Window unit:**  A self-contained air conditioner that is installed in a window to cool a single space.

# Index